C0-CED-852

Machine Tools for High Performance Machining

L.N. López de Lacalle • A. Lamikiz
Editors

Machine Tools for High Performance Machining

L.N. López de Lacalle, PhD
A. Lamikiz, PhD
Departamento de Ingeniería Mecánica
Escuela Técnica Superior de Ingenieros
Industriales
Universidad del País Vasco
Calle Alameda de Urquijo s/n
48013 Bilbao
Spain

ISBN 978-1-84800-379-8 e-ISBN 978-1-84800-380-4

DOI 10.1007/978-1-84800-380-4

British Library Cataloguing in Publication Data
Machine tools for high performance machining
1. Machine-tools
I. Lopez de Lacalle, L. N. II. Lamikiz, A.
621.9'02
ISBN-13: 9781848003798

Library of Congress Control Number: 2008932947

Cover design: eStudio Calamar S.L., Girona, Spain

Printed on acid-free paper

9 8 7 6 5 4 3 2 1

springer.com

"Eman ta zabal zazu".
(in the Basque language)

"Bear fruit and make it known".
(translated)

– from the motto of the University
of the Basque Country

Preface

The machine tool has been, is and no doubt will be, a key factor in industrial and equipment evolution, and as such, improving man's quality of life. Both its evolution and perfectioning have come about due to the sectors where used and have likewise improved their products due to machinery improvements.

Machines have changed greatly in the last 30 years, particularly with the incorporation of numerical control. They have gone from being mechanical machines to real mechatronic systems, where control, drives and sensorisation are key elements. This trend is unstoppable since, thanks to the combination of improvements in materials and mechanical design with control and algorithms executed by the same, better precision, greater speed and worker-friendly inter-relations have been achieved.

However, in the last 15 years an even greater change has occurred, traditional machines, i.e., lathes, milling machines, grinders, *etc.*, have evolved into multiprocess/multitask machines, some of which are capable of milling, drilling, turning, boring, hobbing, measuring and even tempering with laser in the same machine. Every year new concepts appear in this line from the classic machining centre to the turning centre, multitask machines, lathe-milling machines, turning-grinding machines, *etc.* In some cases one hears about the "factory in a machine", which means all operations are performed in the same machine. Designers have done away with many machinery stereotypes, creating designs to solve the user problems. Perhaps we could say "milling machines" are no longer manufactured but rather "machines which mill", or lathes are no longer manufactured but "machines with turning capabilities". However, we may be exaggerating since the production sector tradition and custom tends to see them as classical machine types, depending on the predominance of their functions or machine architectures, *etc.* The tradition is the tradition and metal production is a "conservative" sector.

The wide variety of machinery and options available has complicated the update of classical sources, like books. A single author is unlikely to know all the aspects, technologies or even the production conditions of each sector to write a single book. Thus, the aim of this work has been to seek the contribution of authors specialised in different technological fields and industrial sectors. Each masters one of the technologies comprising machine tools, from control bases to structure, spindles and drives, *etc.* Furthermore, the authors of each chapter have not only had a fluid relation in the past but continue to do so today, thus enabling text coherence and a common view.

In Europe, the USA, and Japan machine tooling is a sector which has undergone a great technological evolution in recent years. In this context, important research and development projects are underway, e.g., the Integrated Project NEXT (Next generation production systems), currently in progression in Europe, or the CENIT "eee-Machine" project in Spain. Asian countries like China and Korea have joined these poles and in recent years India and Turkey too. Competition is high, not only at a technological level but also a monetary level. Two key aspects are: a) cost reduction; this might result from greater production, and b) the need to adapt machines to each customer's needs. Both aspects are contradictory, and are settled using modular design ideas, greater bindings with supplier chains, and the offer of multiple accessories on the same basic machine models.

Nevertheless, we must not lose sight of the importance of environmental impact and machines *life cycle analysis*. Consuming little electricity, reducing coolant use and eliminating electromagnetic radiations are important requirements today. The machine must be "eco-efficient", i.e., with minimum impact and maximum productivity and/or precision.

Machinery precision has also grown. In a hundred years we have gone from tenths of millimetres to below hundredths, and in some cases machines border the micron frontier.

This text is the final result of that work, which attempts to update knowledge on machine tool machine design, construction and use. It is based on the premise that the reader is already familiar with machinery in general and as such familiar with the basic books. Furthermore, it is directed at the reader seeking a source containing the advances of recent years, on display at the main sector fairs, such as the Hanover EMO, Chicago IMTS and JIMTOF. Researchers commencing their work on the machine tool and production sector may find this book useful.

Finally, the authors would like to point out they have gathered information from classical sources and directly from machines existing on the market. The machine tool is a living element with an important industry. It is impossible to generalise without mentioning the companies which invent, improve and re-design these machines. We should also like to express our gratitude to the companies willing to lend their images and ideas. Indeed one of the virtues of this book is its reference to real technology and not solely academic technology.

Bilbao, Spain, April 2008

L. N. López de Lacalle
A. Lamikiz

Acknowledgements

Thanks are given to all the companies cited below for the pictures and information. List of companies with pictures included in this book, using the commercial names in alphabetical order:

ABB
Agie
Airbus
Air Products
Alzmetall
American Axle and Manufacturing
Automation Tooling Systems
Boehringer (MAG Boehringer)
Busak Shamban
Chiron
CMZ Machinery
CMW (Hexapode)
Danobat
Delphi
Dixi
DMG (Deckel Maho Gildemeister)
Doimak
Droop & Rein
DS Technologie (Dörries Scharmann)
Ecoroll
Edel (Die Edel Maschine)
Emag
Etxe-Tar
Fagor Automation
Fanuc
Fatronik-Tecnalia
Fidia
Forest Liné
Fraunhofer IPT
GF AgieCharmilles
GMN
Gnutti
Goratu
Haas Automation
Handtmann
Heidenhain
Heller
Henri Liné
Hermle
Hiwin
Hyprostatik
Ibag
Ibarmia
INA-FAG
Index Werke
ISW-Stuttgart
IWF-Zürich
Jobs
Kaufmann
Kern
KMT Lidköping

Kondia
Kugler
Lagun
Laip
Lealde
M.A. Ford
Magna Powertrain
Makino
Maritool
Mazak
Micromega
Mikron
Mori Seiki
MTorres
NCG, NC Gesellschaft
Nemak
Neos Robotics
Nicolas Correa
Nomoco
Ona Electroerosión
Overbeck Danobat
Precitech
Pinacho
Pietro Carnaghi
Redex Andantex
Röders
Parallel Robotic Systems
SAE International
Sankyo Seiki
Schaudt (Studer Schaudt)
Schneeberger
Shuton
Siemens
Sisamex
SKC
SLF
Spring Technologies
Sumitomo
Starragheckert
System 3R
Tekniker-IK4
THK
Toyoda
Tschudin
Ultra Tech Machinery
Weisser
WFL Millturn Technologies
Zayer

Content

Contributors

Dr. Norberto López de Lacalle
(Chaps. 1 and 6)
Department
of Mechanical Engineering,
University of the Basque Country,
Escuela Técnica Superior
de Ingeniería,
c/Alameda de Urquijo s/n,
48013, Bilbao, Spain
norberto.lzlacalle@ehu.es
www.ehu.es/manufacturing

Dr. Aitzol Lamikiz
(Chaps. 1 and 6)
Department
of Mechanical Engineering,
University of the Basque Country,
Escuela Técnica Superior
de Ingeniería,
c/Alameda de Urquijo s/n,
48013, Bilbao, Spain
implamea@bi.ehu.es
www.ehu.es/manufacturing

Eng. Juanjo Zulaika
(Chap. 2)
Foundation Fatronik-Tecnalia,
Paseo Mikeletegi 7,
20009, Donostia-San Sebastián, Spain
jzulaika@fatronik.com

Dr. Francisco Javier Campa
(Chaps. 2 and 3)
Department
of Mechanical Engineering,
University of the Basque Country,
Escuela Técnica Superior
de Ingeniería,
c/Alameda de Urquijo s/n,
48013, Bilbao, Spain
fran.campa@ehu.es
www.ehu.es/manufacturing

Dr. Joaquim de Ciurana
(Chap. 3)
Department of Mechanical
Engineering and Civil Construction,
University of Girona,
Escola Politècnica Superior,
Av/ Lluís Santaló s/n,
17003, Girona, Spain
quim.ciurana@udg.edu

Eng. Guillem Quintana
(Chap. 3)
Department of Mechanical
Engineering and Civil Construction,
University of Girona,
Escola Politècnica Superior,
Av/ Lluís Santaló s/n,
17003, Girona, Spain
guillem.quintana@udg.edu

Dr. Luis Uriarte
(Chaps. 4 and 11)
Department of Mechatronics
and Precision Engineering,
Foundation Tekniker-IK4,
Fundación Tekniker-IK4,
Avda. Otaloa 20,
20600, Eibar, Spain
luriarte@tekniker.es

Eng. Aitor Olarra
(Chap. 4)
Department of Mechatronics
and Precision Engineering,
Foundation Tekniker-IK4,
Fundación Tekniker-IK4,
Avda. Otaloa 20,
20600, Eibar, Spain
aolarra@tekniker.es

Eng. Ismael Ruiz de Argandoña
(Chap. 4)
Department of Mechatronics
and Precision Engineering,
Foundation Tekniker-IK4,
Fundación Tekniker-IK4,
Avda. Otaloa 20,
20600, Eibar, Spain
iruiz@tekniker.es

Dr. José Ramón Alique
(Chap. 5)
Industrial Computer
Science Department,
Industrial Automation Institute,
Spanish National Research
Council (CSIC),
Arganda del Rey,
28500, Madrid, Spain
jralique@iai.csic.es

Dr. Rodolfo Haber
(Chap. 5)
Industrial Computer
Science Department,
Industrial Automation Institute,
Spanish National Research
Council (CSIC),
Arganda del Rey,
28500, Madrid, Spain
rhaber@iai.csic.es

Eng. Ainhoa Celaya
(Chap. 6)
Department
of Mechanical Engineering,
University of the Basque Country,
Escuela Técnica Superior
de Ingeniería,
c/Alameda de Urquijo s/n,
48013, Bilbao, Spain
ainhoa.celaya@ehu.es
www.ehu.es/manufacturing

Eng. Rafael Lizarralde
(Chaps. 7 and 8)
Ideko-IK4,
C/ Arriaga, 2,
20560, Elgoibar, Spain
rlizarralde@ideko.es

Eng. Oier Zelaieta
(Chap. 7)
Ideko-IK4,
C/ Arriaga, 2,
20560, Elgoibar, Spain
ozelaieta@ideko.es

Eng. Ander Azcarate
(Chap. 7)
Ideko-IK4,
C/ Arriaga, 2,
20560, Elgoibar, Spain
aazkarate@ideko.es

Eng. Jesús Ángel Marañón
(Chap. 8)
Ideko-IK4,
C/ Arriaga, 2,
20560, Elgoibar, Spain
jmaranon@ideko.es

Eng. Alberto Mendikute
(Chap. 8)
Ideko-IK4,
C/ Arriaga, 2,
20560, Elgoibar, Spain
amendikute@ideko.es

Eng. Harkaitz Urreta
(Chap. 8)
Ideko-IK4,
C/ Arriaga, 2,
20560, Elgoibar, Spain
hurreta@ideko.es

Dr. José Antonio Sánchez
(Chap. 9)
Department
of Mechanical Engineering,
University of the Basque Country,
Escuela Técnica Superior
de Ingeniería,
c/Alameda de Urquijo s/n,
48013 Bilbao, Spain
impsagaj@bi.ehu.es
www.ehu.es/manufacturing

Dr. Naiara Ortega
(Chap. 9)
Department
of Mechanical Engineering,
University of the Basque Country,
Escuela Técnica Superior
de Ingeniería,
c/Alameda de Urquijo s/n,
48013 Bilbao, Spain
naiara.ortega@ehu.es
www.ehu.es/manufacturing

Dr. Oscar Altuzarra
(Chap. 10)
Department
of Mechanical Engineering,
University of the Basque Country,
Escuela Técnica Superior
de Ingeniería,
c/Alameda de Urquijo s/n,
48013 Bilbao, Spain
oscar.altuzarra@ehu.es

Dr. Alfonso Hernández
(Chap. 10)
Department
of Mechanical Engineering,
University of the Basque Country,
Escuela Técnica Superior
de Ingeniería,
c/Alameda de Urquijo s/n,
48013 Bilbao, Spain
alfonso.hernandez@ehu.es

Eng. Yon San Martín
(Chap.10)
Foundation Fatronik-Tecnalia,
Paseo Mikeletegi 7,
20009, Donostia-San Sebastián, Spain
ysanmartin@fatronik.com

Eng. Josu Larrañaga
(Chap.10)
Foundation Fatronik-Tecnalia,
Paseo Mikeletegi 7,
20009, Donostia-San Sebastián, Spain
jlarranaga@fatronik.com

Eng. Josu Eguía
(Chap.11)
Department of Mechatronics
and Precision Engineering,
Foundation Tekniker-IK4,
Fundación Tekniker-IK4,
Avda. Otaloa 20,
20600, Eibar, Spain
jeguia@tekniker.es

Eng. Fernando Egaña
(Chap.11)
Department of Mechatronics
and Precision Engineering,
Foundation Tekniker-IK4,
Fundación Tekniker-IK4,
Avda. Otaloa 20,
20600, Eibar, Spain
fegana@tekniker.es

Dr. Justino Fernández Díaz
(Chap. 12)
Department
of Mechanical Engineering,
University of Navarra,
TECNUN-School of Engineering,
Paseo Manuel de Lardizábal 13,
20018, Donostia-San Sebastián, Spain
jfdiaz@tecnun.es

Dr. Mikel Arizmendi
(Chap. 12)
Department
of Mechanical Engineering,
University of Navarra,
TECNUN-School of Engineering,
Paseo Manuel de Lardizábal 13,
20018, Donostia-San Sebastián, Spain
marizmendi@tecnun.es

Dr. Ciro Rodríguez
(Chap. 13)
Centro de Innovación
en Diseño y Tecnología,
ITESM – Campus Monterrey,
Ave. Eugenio Garza Sada #2501 Sur,
Monterrey, NL 64849, México
ciro.rodriguez@itesm.mx

Dr. Horacio Ahuett
(Chap. 13)
Departamento
de Ingeniería Mecánica,
Centro de Innovación
en Diseño y Tecnología,
ITESM – Campus Monterrey,
Ave. Eugenio Garza Sada #2501 Sur,
Monterrey, NL 64849, México
horacio.ahuett@itesm.mx

Chapter 1
Machine Tools for Removal Processes: A General View

L. Norberto López de Lacalle and A. Lamikiz

Abstract In this chapter machine tool basic design principles, technology history and current state-of-the-art technology are described. History refers to the last two centuries; nevertheless, dramatic changes have taken place over the last ten years. One by one the main aspects involved in machine design and construction will be explained in-depth over the following chapters, completing a general view of the machine tool world, making for easy comprehension of the whole book. A new classification of machines for removal processes is proposed, including the new concepts shown in recent industrial fairs, such as multi-task and hybrid machines. At the end, some typical machines for today's important sectors are described.

1.1 Basic Definitions and History

In the Encyclopaedia Britannica the following description for machine tools is given: any stationary power-driven machine that is used to shape or form parts made of metal or other materials. The shaping is accomplished in four general ways: 1. by cutting excess material in the form of chips from the part; 2. by shearing the material; 3. by squeezing metallic parts to the desired shape; and 4. by applying electricity, ultrasound, or corrosive chemicals to the material.

This definition has remained unchanged for the last two hundred years although new advances in the last fifteen years may make the following definition more appropriate (except for forming machine tools): a servo-controlled spatial mechanism that guides and drives a cutting tool along a complex trajectory creating a new

---

L. Norberto López de Lacalle and A. Lamikiz
Department of Mechanical Engineering, University of the Basque Country
Faculty of Engineering of Bilbao, c/Alameda de Urquijo s/n, 48013 Bilbao, Spain
{norberto.lzlacalle, aitzol.lamikiz}@ehu.es

L. N. López de Lacalle, A. Lamikiz, *Machine Tools for High Performance Machining*,

shape in the raw material. Both definitions are complementary; however, the latter matches the new machine tool concepts to be developed in the forthcoming chapters better. Another traditional and somewhat philosophical definition is "the machine tool is the only one capable of building other machines similar to itself".

On the other hand, manufacturing processes can be divided into five main groups with respect to their physical action on raw materials. Some of them require a special machine to be applied. Thus:

1. Material deposition technologies, such as casting and sintering, or the new "rapid manufacturing" techniques developed rapidly in the last ten years.
2. Joining techniques, in which riveting, friction stir welding (FSW), welding and assembly are classified.
3. Forming, where hot working techniques like forging and cold working techniques such as forming, spinning and involute splines production are placed.
4. Material separation, including shearing, nibbling, punching, cutting by laser and cutting by high pressure water or abrasive waterjet (AWJ).
5. And finally material removal processes, also known as machining processes. Here the use of cutting tools with a "defined cutting edge" or a "undefined cutting edge" leads to two main technique groups, i.e., the cutting and abrasive processes respectively. A third group is the "non-conventional machining" processes, which can be also defined as erosion processes. This book covers those machines designed to apply these three process types.

Before presenting a general classification of machine tools for material removal processes, some historical remarks about machine evolution should be mentioned.

1.1.1 Historical Remarks

Making forms on hard materials has been a constant throughout the history of mankind. Perhaps five periods can be distinguished.

1.1.1.1 From the Middle Ages to the Industrial Revolution

The "violin arch" as the basic mechanism to achieve a rapid rotation motion in a wooden axis has been used for several thousand years [1]. Circa 1250 A.D. a foot-driven variation meant a great advance, since the user's hands were free to perform other operations on the wooden workpiece.

The great Leonardo da Vinci designed several machine tools, but most of them were not built for technical hindrance. Father Plumier in 1701 made reference to the difficulties in turning iron workpieces in his book "L'Art de Tourner" when in those years the use of the foot-operated rocket-crank mechanism was generalized. Ramsden (c. 1777) invented the screw lathe, where a rotation and a longitudinal movement were synchronized.

Hand boring had been used for internal cannon finishing since 1372, but was quickly implemented in rotary shafts being driven by hydraulic power.

1.1.1.2 The Industrial Revolution

The 18th century saw the birth of the steam engine birth due to Papin, Newcomen and finally, the Scotsman James Watt. This power source was essential in moving machine tools throughout the 19th century; however, it was also important because it required the construction of a horizontal boring mill by John Wilkinson (1775) to achieve good precision on the engine pistons.

Later, Henry Maudslay in 1797 developed a screw turning lathe that included four main machine tool concepts: an all-metal frame and structure, flat linear guides for the tool carriage movement, interchangeable gears and a screw-thread feed mechanism. Other inventions by this British forerunner were the micrometer and the vertical slotting machine (1803).

In France some new concepts were also devised, like Rehe's grinder (1783). Some time later in 1789 in Spain M. Gutiérrez designed an indexer teeth cutting device for clock gears.

1.1.1.3 The 19th Century

In the 19th century due to the need for greater precision, productivity and repeatability in the construction of other machines, several advances were made by three disciples of Maudslay, i.e., Roberts developed the metal planing machine, Witworth the gear cutting machine with involutes cutters, and Nasmyth the shaping, steel-arm filing, drilling and disc-grinding machines.

Regarding the first turret lathe construction, some contradictory dates are found: Fich before 1854, Robbins and Laurence also on that date and Stone in 1854 in America share the fatherhood of this invention.

In the USA the milling machine was invented by E. Whitney (1818) and later perfected in 1830. In 1862 J. R. Brown introduced the first universal milling machine, with an indexer, a bed with vertical displacement, and transversal table movement via a cardan joint. The engineer of the Cincinnati Screw and Tap company F. Holz included an axial ram in this design in 1884. Frenchman P. Huré introduced an original headstock to work in all spatial angles, which could be adjusted in several positions by means of a 45° rotary movement of the head in 1894. In the 1860s new horizontal boring machines with workpiece movements defined the classical structure upheld till now.

The first universal grinding machine was launched by Brown and Sharpe in 1870. The centreless concept was developed several years later, in 1915, by L. R. Heim in the USA. Automation using kinematic chains came into play with automatic lathes first developed by Miner in 1870, then two years later by J. Schwizer. At the end of the 19th century, the Acme Company manufactured a four multi-spindle lathe.

Fig. 1.1 **a** View of the belt driven machines in use today at the Elgoibar (Spain) machine tool museum. **b** Detail of a drum turret lathe

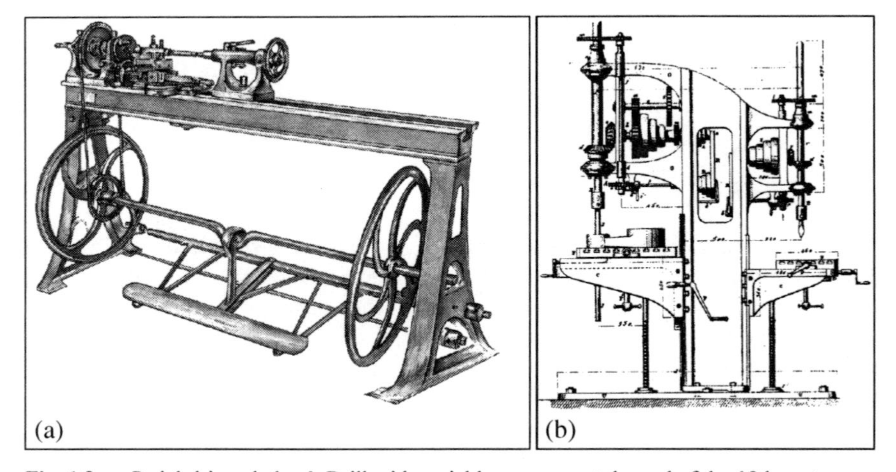

Fig. 1.2 **a** Pedal-driven lathe. **b** Drill with variable pressure, at the end of the 19th century

The manufacturing process for gear manufacturing "by generation" was first applied by Bilgran in 1884 and immediately afterwards by Fellow. Ten years later the German firm Reickner introduced the hobbing principle using a worm-type cutting tool. Figures 1.1 and 1.2 show some 19th century machines.

1.1.1.4 1900–1990s

The 20th century saw the great development of the automotive industry, since in 1908 H. Ford manufactured the first mass produced car. The new mass production, along with tight dimensional and form requirements, led to improvements in the same types of machines known since the end of the previous century. However electrical engines (fully introduced in 1920) were used instead of steam power.

A new concept was to transfer the engine blocks components through autonomous units, combining the tool rotation and feed movement, giving rise to the use of transfer machines.

In 1933 L. Wilkie, working for Do-All, developed another basic machine tool, the metal cutting contour band sawing machine.

In 1948 J. Parson, an engineer at the Bendix Corporation, developed an automatism for controlling a 3D machining operation, improved by MIT over the following three years. In those years, the programming binary code was supported by punched cards and later by perforated tapes. But the real spread of numerical control (NC) was in the 1970s and 1980s, when the microprocessor became the brain of the control mechanism and the CNC (Computer numerical control) concept was fully developed. In the 1990s, the open architecture of controls based on PC buses and cards enabled the integration of machines in intelligent manufacturing systems.

1.1.1.5 The Last Fifteen Years

With the introduction of CNC, machine tools could be fully automated including "automated tool change" (ATC), the optional "automated part change" (APC) and other auxiliary features such as measurement probes, network capabilities and other advanced functions. The programmable controller is currently integrated in the same architecture of the CNC to make all machine operations automated and therefore programmable.

The "machining centre" is currently the most common machine, combining a CNC milling machine with an automatic tool change, ready for drilling, milling, boring and threading. Likewise, a CNC lathe with C axis control and rotary tools placed in a motorized drum turret is called a "turning centre".

The high requirements in time and precision for complex parts on the one hand, and the power of controls on the other, along with the designers' creative minds, has brought in this present decade the concept of "multi-tasking machines", with milling, turning and drilling capabilities, and recently even grinding wheels. In this group, the early types were lathes where an additional milling head was included (now they are called turning centres), but recent developments are directly designed as complex multi-axis machines different of lathes and milling machines.

Figure 1.3 is a good example by Mori Seiki®. The frame is a three-axis box-in-box structure, moving a rotary axis where the spindle head swivels ±120°. Placed here, a turning, milling or drilling tool can be moved in a big workspace. Another main motion is provided by a power lathe headstock placed on a horizontal flat bed; if a turning operation is performing this headstock turns the workpiece, whereas if a milling operation is performing this headstock slowly moves the workpiece controlling its angle position. At the same time a drum turret moves along the horizontal guides. In this design a motor is built in the turret (called a "built-in motor turret").

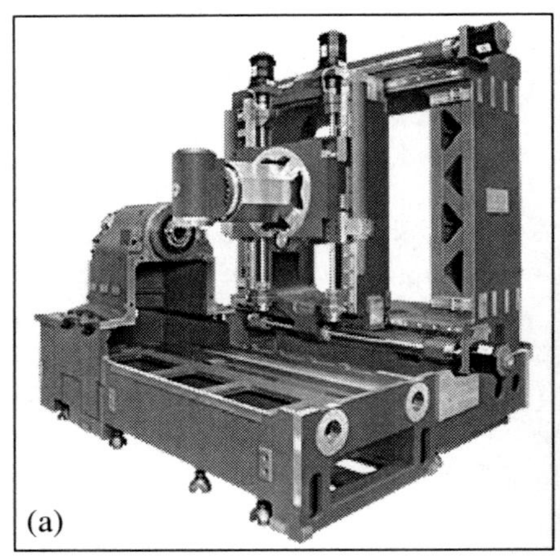

Fig. 1.3 The multi-task machine Mori Seiki® Series NT. **a** Box-in-box structure with an octagonal section ram. **b** Flat bed with a lathe headstock and the possibility of a bottom motorized turret

Recently cutting and grinding operations are included in the same machine, as in machines shown in Fig. 1.4, by Schaudt®.

Finally, evidence of how the designer's mind is open to exploring new concepts is in the use of parallel kinematics structures. The first application was the Variax prototype by Gidding & Lewis®, presented in 1994 at the IMTS of Chicago. Parallel kinematics is not a global solution for all machine tools; however, it can be applied to big ones.

At present, the US industry classification (class. n. 333512) collects 55 types of material removal machine tools [17] related to the final operations to be performed. Some types are disappearing, such as the planing or shaping machines, but others are more or less in use. It is difficult to be strictly academic when defining the types

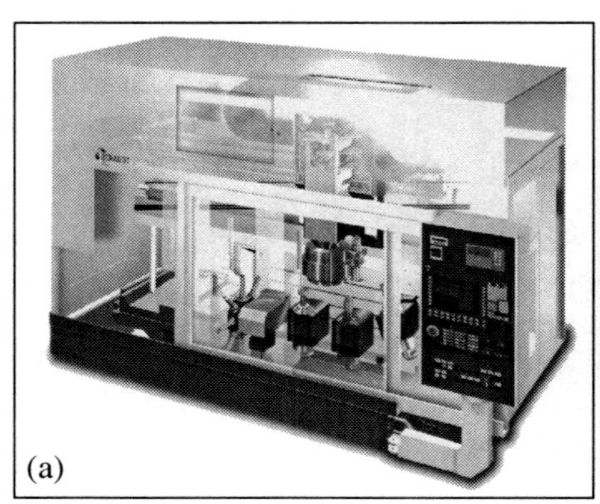

Fig. 1.4 Hybrid machines. **a** Combining turning and internal grinding, model ComboGrind v, by Schaudt®. **b** Combining turning and cylindrical grinding, CombiGrind h of Schaudt®

Table 1.1 Up-to-date classification of current machine tools

Defined cutting edge (cutting)

Main motion: translation

- Broaching machine
- Band saw and Hacksaw
- Planer and Shaper
- Slotting machine

Main motion: rotation

- Turning:
  - Engine universal lathe
  - Vertical lathe (vertical boring mill)
  - Drum turret lathe
  - Multi-spindle lathe
- Milling:
  - Universal knee milling machine
  - Vertical milling machine
- Boring:
  - Horizontal boring machine
- Drilling:
  - Bench drill
  - Drill press (upright drill press)
  - Radial drill press
  - Multi-spindle drill
  - Drum turret drill
  - Deep drilling machine
- Sawing:
  - Circular or disk sawing machines (coldsaws)

Machining centre: machine designed to use rotating tools, with capability of milling, drilling, boring and tapping:

- Vertical
- Horizontal

Turning centre: Machine derived from a lathe with capability of turning and milling, including:

- Motorised tools in a drum turret or/and
- A milling headstock

Transfer machines and systems

Gear manufacturing machines

Undefined cutting edge (abrasive)

- Grinding
  - Cylindrical grinder:
    - ¬ External
    - ¬ Internal
  - Surface grinder:
    - ¬ Rotating
    - ¬ Reciprocating
    - ¬ Creep grinding
  - Point grinder
  - Centreless grinder
  - Tool grinder
- Honing
  - Short stroke
  - Long stroke
- Lapping
  - Single side
  - Double side
- True friction sawing machines (disk and band)
- Abrasive disk sawing machines

Non-conventional (erosion)

- Electrodischarge machining:
  - Wire (WEDM)
  - Sinking (SEDM)
- Electrochemical machining (ECM)
- Electronbeam machining (EBM)
- Ultrasonic machining (USM)

Laser: This new tool can be used for cutting metal sheets, welding, material deposition and material ablation

Multi-task machine: Machine that combines two machining processes:

- Milling and turning[1]
- Turning and grinding
- Milling and grinding

Hybrid machine: machine combining a machining process and other manufacturing processes

[1] When it is impossible to define if the machine structure comes from a milling or a turning machine, and therefore is impossible to be defined as machining or turning centres

of machine tools, because in this new century new hybrid and complex machines are presented at each of the main industrial fairs, especially the EMO in Europe, the JIMTOF in Japan, the IMTS in the USA and those nationals held in the industrialised countries. It would be easier to classify machining processes instead of those machines which apply them. A very up-to-date classification, where the classical machine types are included along with new ones, is shown in Table 1.1.

This book is focused on the general types, i.e., machining centres, lathes, grinding machines and electrodischarge machines. However the basic design and construction principles are common for all, even those only barely mentioned in this book.

1.2 The Functions and Requirements of a Machine Tool

The basic function of a machine tool for removal processes is to move a cutting tool along a more or less complex trajectory with sufficient precision, withstanding the forces from the removal material process. This must be done reaching the required precision and/or material removal rate.

The global scheme for designing, manufacturing and verification of a machine tool is summed up in Table 1.2 and detailed in the following subsections; however all start from user requirements.

Table 1.2 The main steps in the design and construction of a material removal machine tool. Inputs and factors are related

Definition of requirements	***Features of the machine tool and systems***
Workpiece size	*Machine size*
Workpiece geometry	*Milling or lathe, other type*
Removal rate	*Roughing or finishing machine*
Precision	*Assembly, thermal aspects*
Kinematic behaviour	*Element masses, drives*
Batch size	*Automation systems, ATC and APC*
Price	*Life cycle cost analysis*
Selection of the basic mechanism	*Machine degrees of freedom*
Definition of the main motion	*Main motor, workpiece or tool rotation*
Definition of the structure	
Configuration	*Knee, gantry, fixed or travelling column, etc.*
Bed	*Cast iron, polymer concrete, others*
Structural elements	*Cast iron, cast steel, polymer concrete, welded steel*
Guideways selection	*Friction, roller bearings, hydrostatic*
Definition of drive trains	*Drive motors, ball screws or linear motors, couplings*

Table 1.2 (continued)

Selection and implementation of the CNC control	
Requirements and selection	*Number of axes and interpolation complexity*
Adjustment of CNC to machine	*Machine parameters, axes strokes*
Users interface	*Customization and application oriented interfaces*
Definition of loop controls	
Displacement measurement devices	*Measurement rules and encoders*
Control of each axis	*Kv factor and other axis control parameters*
Basic automation by auxiliary functions	
Sensors and inputs/outputs	*Automated tool change or part change*
PLC programming	*Safety systems*
Machine elements manufacturing and assembly	
Testing and verification	*Roundness tests, ISO tests and test parts machining*

1.2.1 User and Technological Requirements

As in the case of other machines, the input for machine tool designers are user requirements, which lead to the definition of the machines main features, always described in machine commercial catalogues. Users might be non-specific customers but included in typical sectors with common requirements, or an individual customer with highly defined specifications. Small and medium machines are usually produced considering hypothetical customers, offering some options in catalogues, on the contrary to large machines which are usually designed to order. Anyway, requirements are related to the following aspects:

- The maximum part size to be machined. Machining would be needed at each point of the part, so the machine workspace must be larger than the workpiece size. In some machines modifications are done to be able to accommodate workpieces larger than the workspace, for example the gap-frame lathes, where the maximum swing diameter is larger than the maximum working diameter.
- Workpiece main geometry. The global shape of the part is the key point to select one type of machine. If the part is cylindrical, the lathe is the first machine to be considered. If it is prismatic, a milling centre may be the most adequate. Only in some cases may there be a doubt, in the light of the capabilities of the new five-axis milling centres provided with high-torque rotary axes tables, which are able to make turning operations as well as milling.

- The second geometry aspect to be taken into account is the number and complexity of the details. If the features are few and simple they do not contribute to the complexity increase (i.e., the *degrees of freedom*, or DOF) of the machine to be used. On the contrary, if they are numerous and/or very complex, the required machine-tool must be structurally much more complex. One example is again the five-axis milling centres, where two orientation axes are added to the basic three-axis Cartesian machine, making it possible to machine all faces of complex spatial forms in one setup. Other examples are the so-called "turning centres". In this case, milling tools are integrated in a lathe to perform details like key-slots, polygonal faces or inclined planes in the same setup, at the same time adding additional DOF to the lathe such as the controlled rotation of the C-axis and the Y-axis linear movement out from the plain XZ.
- Material removal rate. In some applications the design driver is precision, but in others it is mainly to achieve a high productivity. The former lead to defining "finishing machines", in which the main motion does not require high torque and power because the chip section will be small to keep cutting forces low. However precision driveways, machine structures with both a high stiffness and a damping and tight toolpath control are highly recommended. The latter lead to the design of a strong "roughing machine" ready for heavy-duty work, with a high torque and powerful main motor and with a robust spindle. Solutions for precision and productivity are conflicting. Usually users require a machine suitable for several applications, requiring a "sufficiently accurate" and "highly productive" machine, at the lowest price. This the basic pattern for most of the current milling and turning machines.
- Precision. This is a commonly used word which really involves two different concepts: *accuracy* and *repeatability*. Accuracy is the capability of being on target with a specification, quantified by the bias or difference between the obtained and desired result. However, repeatability is the ability to reach the same value over and over again. Therefore a machine may be repeatedly inaccurate in one extreme case, or imprecise yet very accurate at the other extreme. A somewhat related concept is *resolution*, the smallest difference between two following values that can be distinguished by measuring devices.

  Accuracy and precision are the main objectives of machine tool constructors. The guidelines and methodologies for machine design and assembly, the machine elements manufacturing, the testing procedures and the use of auxiliary systems are inspired by this requirement [21, 24]. High precision is very expensive; it requires identifying, controlling and reducing all error sources. Hence, to achieve a determined grade or accuracy requires a careful study of the following:

  - The effects of assembly errors of the machine mobile components on tool position. Good assembly practices, the tight adjustment of carriages to guides and the measurement of errors after assembly are the basic techniques to reduce them.

- The structure deformations under the action of cutting and inertial forces. High stiffness is always desired.
- The dynamic behaviour of the system under the excitation of cutting forces, because most of the machining operations produce highly variable forces. High damping ratios of machine frame elements and joints enable vibration reduction.
- Friction and backlash effects in the guideways and drive trains.
- Non-deformability with respect to heating from thermal sources. In machine tools there are five heat sources: the main motor-spindle, the drive motors, the process, material removed as chips and finally the temperature changes of the workshop.
- The tool trajectory control. Errors in part shape and dimensions may occur if sharp or very rapid direction changes in tool movement are programmed. Smoothing functions are offered to reduce these errors by moderns CNCs.

A very illustrative picture about precision in machining was given by McKeown [16] after Taniguchi [22], lightly modified in [6]. As shown in Fig. 1.5, in the last sixty years, a precision of 1 μm has been achieved for conventional machining, while in ultraprecision a hundredth of a micron is achieved in some cases. Evidently, these good numbers are located in the segment of high quality and expensive machines; however they are a good indicator of the current technology level. A comparison with the boring machine by J. Wilkinson in 1775 "which bores with a thickness error of one shilling in a fifty-seven inch diameter" (i.e., 2,000 μm), previously cited in Sect. 1.1.1, is illustrative.

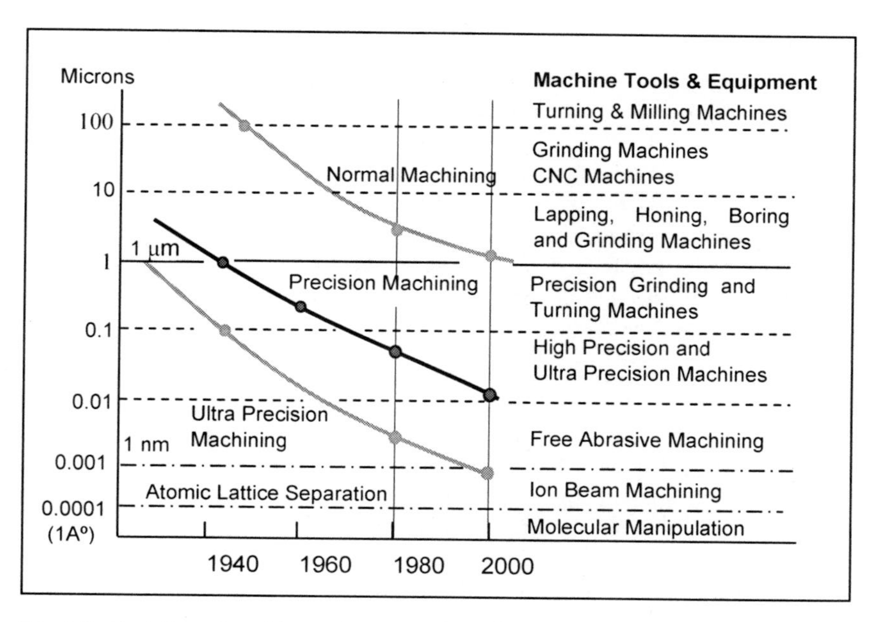

Fig. 1.5 Evolution of machine tool accuracy, from [6]

- Kinematic behaviour (i.e., speed and acceleration). This requirement regards the need for fast machine "idle movements" between successive machining operations and for tool change. Some machines usually make a few but long operations on unique parts while others makes a lot of short operations on each workpiece. An example of the first case is the big milling centre for stamping die finishing. Examples of the second case are milling centres for automotive iron cast parts. The latter cases are where rapid idle movements are much recommended.

  Acceleration is not a direct user requirement; nevertheless it should always be considered by the mechanical engineer when designing a spatial machine, since all inertial forces are directly related to acceleration by Newton's second law. This is an important aspect in sculptured surface milling [13, 14] because the tool is constantly changing trajectories on the complex surfaces; therefore machine agility greatly depends on axes accelerations. The same can be said of machines for automotive components, where the tool must move quickly between the multiple machining points.
- Batch size has an influence on the automation level of machine tools and the use of auxiliary devices. Generally, a universal CNC machine is able to process a wide range of different workpieces, only downloading another CNC program and with few changes in part workholding devices (see example 1.10.6, later). On the opposite, a transfer line (see example 1.10.9) offers a high production rate for a specific part but needs a long time to adapt stations and transfer systems to new parts. The term "hard automation" makes a reference to install special devices for the manipulation and machining of specific parts, whereas "soft automation" refers to the use of CNC machines with universal fixtures and automated tool change (in machine catalogues this is abbreviated to ATC) and part change (APC).
- Price. This is an important factor that depends on machine size in linear proportion and precision in exponential relation. Today the "life cost" concept includes all machine life stages, i.e., the initial investment, maintenance costs, fixed costs and costs of the machine retrieval. Several software utilities for a correct *life cycle cost* analysis (LCC) are available.

  At present, the environmental impact of machine tools themselves is not a key factor; however that of their machining process is taken into account in several innovation projects.

In most cases, the aforementioned requirements lead to a definition of traditional common-sense solutions, because all applications look similar to the previous. This similitude makes designing of a new machine for a "traditional" application easy. However, as with other mechanism and machine problems, new solutions are launched radically rethinking all the design steps from the starting point, i.e., the consideration of user requirements. This fact has become evident in the development of multi-task machines, like those shown in Figs. 1.3 and 1.4, or that of the example in Sect. 1.10.10. See also Sect. 7.2.2, and Sect. 12.5.2.

The first machines of this type (in the 1990s) were modified lathes, with an additional spindle head and a new Y-axis controlled movement placed on the slant bed. Nowadays, the structure of these machines is absolutely different from the classic lathe. In some ways it can be said that a second generation of multi-task machines has been born in the last four years, starting their design from the user requirements, which must be collected from the machine company sales departments.

1.3 The Basic Mechanism

Machine tools are spatial mechanisms with several degrees of freedom (called axes) with sufficient workspace to accommodate or move the workpiece. The function of this mechanism is moving either the tool or the workpiece, or both simultaneously [19].

Three references systems can be defined: 1. an observer placed in the workpiece (WRS, or workpiece reference system), 2. placed on the machine bed (MRS, or machine reference system) and finally 3. placed at the tool tip (TRS, or tool reference system). The three are very useful. Thus, the first is where all machining operations are defined and programmed. The second is that used by an external observer. Finally, the last is used by machine control to move the tool tip. A conversion from the former into the latter must be always performed. In manual machines it is the user who does that, interpreting the workpiece drawings, but in CNC machines it is this control device which performs this function, known as the *interpolation* function.

The machine reference (i.e., the reference with respect to the fixed element, the ground) is useful for the machine tool manufacturer, since it defines which degrees of freedom are related to tool movements and which to workpiece movements. Figure 1.6 shows a DMG machine called Duoblock® due to the disposal of machine axes and consequently the machine two main substructures, two axes affecting the workpiece and three acting on the tool.

The names of machine degrees of freedom are defined with respect to the main motion that provides torque and the power to remove material. The nomenclature and positive and negative directions of axes are defined in the ISO 84:2001 standard [10], being the Z-axis which matches the main motion axis.

The required movements lead to typical kinematics solutions and machine configurations [4, 24]. Thus, for cylindrical parts the basis is 2 DOF mechanisms, resulting in the lathe as the machine to be used. For a milling machine mainly three-axis mechanisms are used. The three-axis movement is solved with a Cartesian configuration, with at least two of the axes mounted in serial. In some cases all the movements are applied to the tool, others are divided among the tool and workpiece, and on the odd occasion all are applied to the workpiece.

The five axes, like the machine shown in Fig. 1.6, is an optimal mechanical solution for milling, since it allows the rotary tool, a cylinder with five degrees of

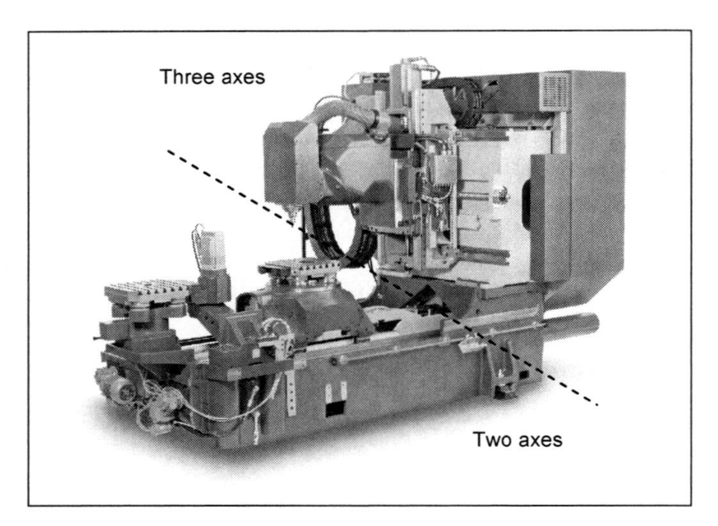

Fig. 1.6 The DMG 60 U Duoblock®, a *RLLLR* machine

freedom, to be moved in any orientation and position in the workspace. In this way the milling tool can be placed in a specific orientation with respect to each workpiece surface. This is really the best solution for milling, yet at the same time the complexity of the mechanism and its control has increased.

Three configurations are common in five axes milling centres; the kinematic chain is going to be defined starting from the workpiece towards the tool tip, where *L* means a linear axis and *R* a rotation axis. Thus, three types are defined in Fig. 1.7.

- *LLLRR*: A two rotary axis head is fitted onto the end of a Cartesian mechanism. One axis rotates (twists) the head while the other tilts it. On the other hand, Cartesian motions may be produced either at the tool or machine table (see diagram of Fig. 1.7a, where motion is of the tool). This configuration is used in large gantry machine tools, usually for machining large moulds and dies.
- *RRLLL*: The workpiece is supported by a double rotation table. One rotation is a cradle-like movement whereas the other is around an axis perpendicular to the plate. This configuration is commonly used in small compact machines, or in three-axis machines provided with accessory rotary tables. The three mean Cartesian axes can be solved by a travelling column configuration, but in other cases the cradle base is provided with one linear degree of freedom.
- *RLLLR*: The workpiece is supported by a rotary turning table and at the same time a swivelling head provides another rotational degree of freedom. These five-axis machines are very suitable for tall workpieces or for cylindrical parts with faced plates and holes around their perimeter.

Current machines tools are really simpler in design than fifty years ago. Then, only mechanical devices were available for the automation of machine movements

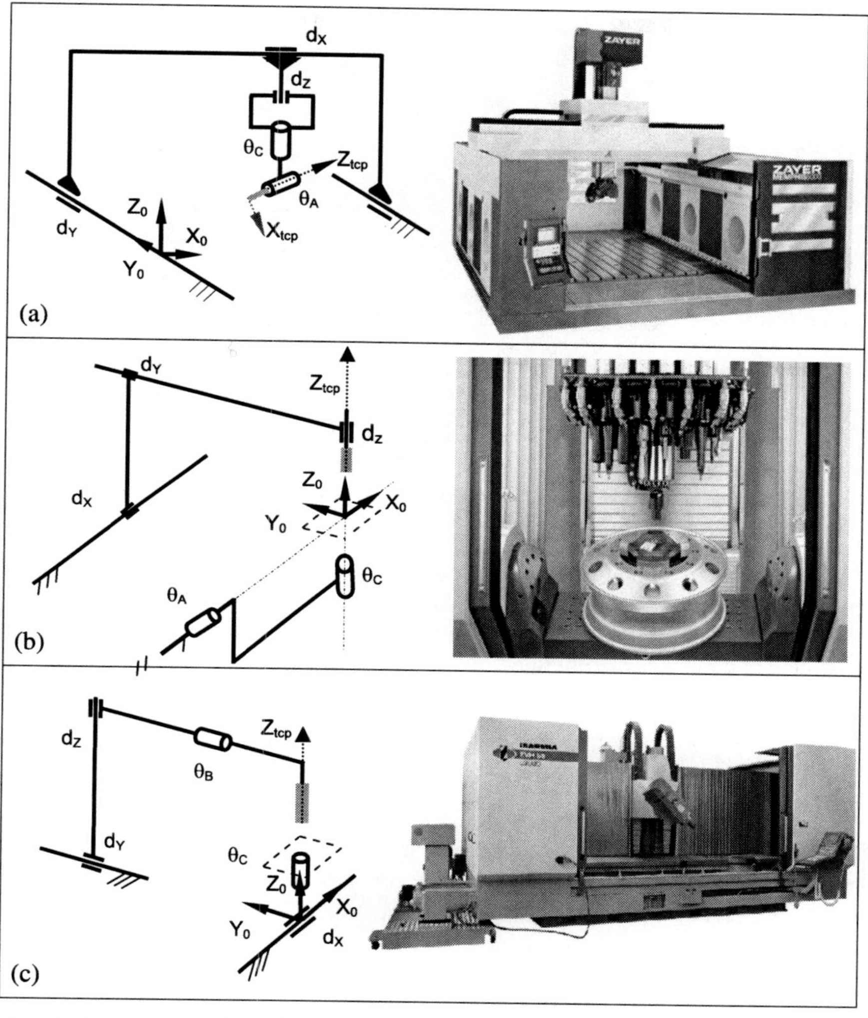

Fig. 1.7 Three mechanisms for a five-axis milling machine. **a** *LLLRR*, model Memphis of Zayer®. **b** *RRLLL*, model Chiron® Wheel. **c** *RLLLR*, the Ibarmia® ZVH55

while now the servocontrol of machines axes is easy to implement and not too expensive. For example, the mechanism of the gear shaper shown in Fig. 1.8a enabled all the process movements to be achieved from a single electrical motor:

- The shape-like movement of the gear cutter by means of a slider-crank mechanism.
- The gear cutter backward movement at each vertical stroke by a reversing dog.
- The feed movement along the gear generation.
- The cycle control by using a flat cam.

Fig. 1.8 Two machines where kinematics solved all the movements. **a** Gear shaper. **b** Hacksaw

In this case, to synchronise all movements a complete set of interchangeable gears and a belt placed on interchangeable pulleys was used. Another example of a full mechanical solution is the hacksaw of Fig. 1.8b, where the main sawing movement and the feed are obtained from a single motor using a planar mechanism.

1.4 The Machine Structure

The structure of machine tools must hold all machine components and at the same time withstand the forces coming from the process, maintaining enough stiffness to keep the required precision. On the other hand, a high damping ratio and low thermal distortion must be achieved. Two main types of elements are included in the structure:

- The frame and bed. The main body of the structure constitutes the machine frame. It can be built in one block or assembling several individual sub-frames (see Fig. 1.9). One important component is the *bed*, where all others components rest. It is the solid base of machine after construction, placed on the ground of the workshop using some kind of isolated supports.
- The structural components. They are part of the mechanism, being linked with relative movement between them. The interface of those elements with relative movement must be very stiff and damped along the perpendicular direction to slide while allowing a smooth motion along it.

Two structure design concepts are used, the *open-loop* and *closed-loop* configurations. In the first case the process forces are conducted to the ground through just one structural way, whereas in the closed case forces are derived by

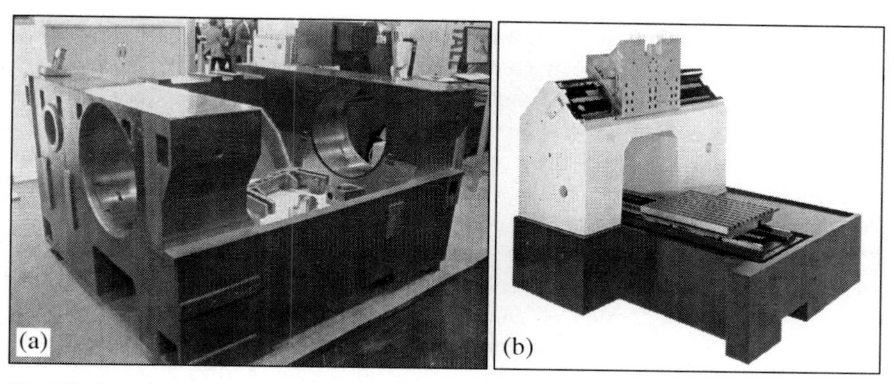

Fig. 1.9 Machine structures. **a** Bed in ductile iron by Alzmetall®. **b** Machine structure by Röders®

several ways to the ground. Obviously, in the first case the structure is weaker, therefore a higher error measured at the tool tip position is produced by machine deformation due to the cutting forces; in the second case, stiffness measured at the tool tip increases. On the other hand, and for the same machine size, the first type usually offers more workspace and workzone accessibility for part handling than the second.

The characteristic open-loop for milling machines is the C or G (knee) frames, very common in small machines. The access to the workzone is easy, but this structure is sensitive to thermal and mechanical charges (torsion and flexion) with an asymmetrical response. The frame overhang produces Abbe errors on the workpiece (see Sect. 6.2.3). In Fig. 1.10 two designs are presented. Thus, case (a) is the "fixed column", stiffer but affected by heat coming from the machining process, and moving different masses depending on the workpiece weight as well. The second is the "travelling column", less stiff but moving the same mass all the

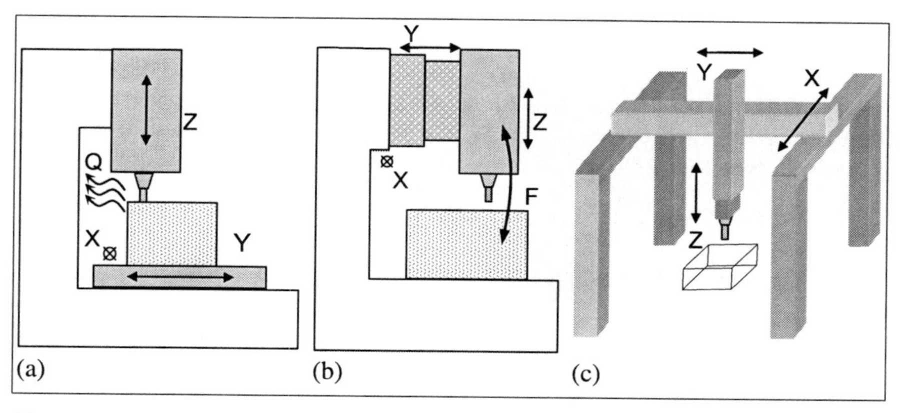

Fig. 1.10 Three machine frames. **a** Fixed column C-frame. **b** Transverse column C-frame. **c** Gantry with a travelling beam, with all movements in the tool

time; another advantage is that it allows the part to be set up in one zone while milling runs in the other if the machine table is sufficiently long.

With respect to closed-loop frames, the bridge or gantry structure is used for medium and large machines, which usually perform heavy-duty work or finishing on big parts. In some cases the bridge is fixed and table moves, in others the workpiece is fixed and all movements are by the bridge or a travelling beam placed on the bridge (case c in Fig. 1.10).

Nowadays, there are also some new architectures using parallel kinematics, where stiffness, kinematic and dynamic principles are somewhat different. In these machines the use of isostatic structures prevents spatial distortion of machine bodies. Chapter 10 is dedicated to this type of machine.

As for lathes, structures are open-loop for horizontal models and closed-loop for the huge vertical ones (also called vertical boring mills). In lathes cutting forces are translated into torsion to the bed through the carriage guideways. For the last fifteen years horizontal CNC lathes have had slant beds in which the turret moves along; this fact makes the part handling and chip evacuation easy; however, at recent fairs some developments with a horizontal bed, traditional for engine lathes, and turrets placed under the workpieces have been shown (e.g., the Mori Seiki NT series; see Fig. 1.4b).

1.4.1 Machine Foundations

As for machine tool foundations, some of them are support-critical whereas others are not. For small and medium lathes and machining centres, isolation pads or blocks are usually enough to reduce vibration transmission to and from the machine tool. These supports have some simple height adjustment to make the alignment.

When vibration isolation is desired in a support-critical installation, an inertial block foundation system is often the best option. The machine is jointed to a concrete basement by anchor bolts, levelling screws or levelling wedges to adjust and align the machine bodies. This is the case with big milling or boring machines where the workpiece table rests on the ground and the column structure is a separate group resting on the ground as well. Another case is the long table milling machines with travelling column, where good alignment is required.

1.4.2 Structural Components Materials

Both the bed and structural components must be stiff, lightweight and easy to manufacture. Taking this into account, four materials groups are mainly used in machine structures:

- Grey cast iron is the most common material due to its stability, easy casting, high damping ratio, self lubricity and economy in machining. Ductile cast iron can be an option to increase the stiffness of some components. Cast steel is used in headstocks.
- Welded frames are used in large machines due to the usual casting problems of large components or when a very short production time is required. The main disadvantage is the lack of damping. Some solutions use fillings, like sand or polymers, to improve damping and attenuate vibrations. Other problems are derived from the residual stresses and distortions typical of welding and the non-homogeneous behaviour of the weld seams.
- Polymer concrete, also known as "mineral casting", has been the subject of several research projects in recent years. Now it is used in some lathe or milling machine beds. The positive feature is its high damping, but its main drawback is the low thermal conductivity. In some cases this material can reinforce the cast iron, for example in the lathe series Quest by Hardinge® Inc. (Harcrete™ material). Another brand is Polycrete™ by Cleveland Polymer®, a material formulated by combining quartz aggregate, a high bonding-strength epoxy resin system and selective additives.
- Granite is used by some manufacturers of special lathes, grinders and other high accurate finishing machines, for example for glass turning and polishing.

1.4.3 Structural Analysis

The analysis present three stages, resistance analysis of the structure withstanding static forces, analysis of natural frequencies and modes, and finally dynamic analysis of the machine with respect to the cutting process.

Structural behaviour under static or inertial loads is currently carried out with the Finite element method (see Sect. 2.3.2). Although cutting forces are variable, both in modulus and direction, the maximum values can be considered inputs for the 3D model. The structure equivalent tensions and deformations are mapped as a result of the analysis (Fig. 1.11), which can be used to redesign the structural components. Currently, even the simplest software packages are able to perform a good calculus. In FEM, the most difficult aspect to define is related to contacts between structural components along the DOF, where stiffness, damping, backlash and other construction details are difficult to estimate.

Typical machine stiffness values, measured like the displacement of the spindle nose with regard to the machine bed due to force action, are as follows. For a vertical machining centre, stiffness values of approx. 62 N/μm in X, 33 N/μm in Y and 67 N/μm in Z are calculated in [7]. In another case, a travelling column milling machine [20], the experimental stiffness is 16 N/μm in the X-axis, 40 N/μm in the Y-axis and 93.7 N/μm in the Z axis. Other values for 3-axis machines range between 15–25 N/μm in the axes X and Y and 70–100 N/μm in the Z axis, Z always being the stiffest in milling machines. With respect to grinders, external cylindrical ones are in

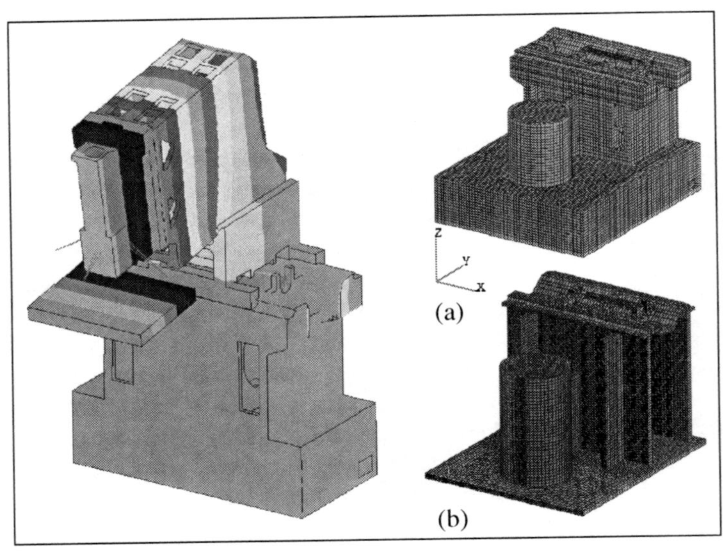

Fig. 1.11 Deformation of an electro-discharge machine, calculated by FEM. Maximum deformation is 12 μm when it holds a heavy electrode. **a** Former design. **b** Optimised case

the stiffness range of 50–60 μm, the flexibility mainly coming from the setup between points (placed onto the headstock and tailstock) where the part is supported.

With some variations depending on the machine tool type, the sources of flexibility are: the tool, ram and column, the spindle and toolholder interface, the axis carriages and rails, ball screws and finally the bed.

The analysis of natural modes and frequencies is easy to run (see Fig. 1.12), but difficult to adjust to reality [15]. In finite element models damping is always an

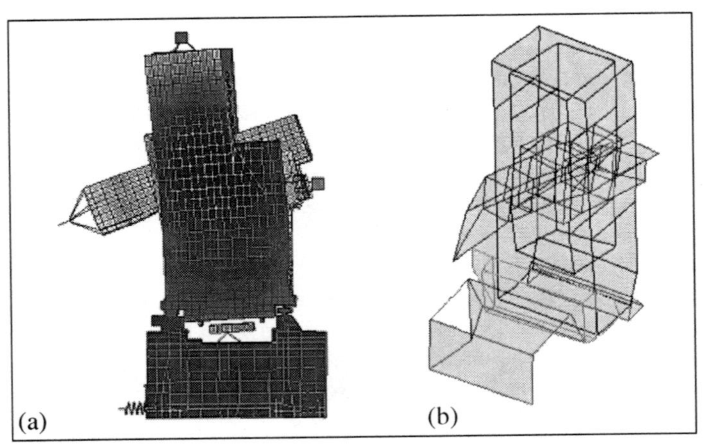

Fig. 1.12 Modal analysis of a column-horizontal ram milling machine. **a** Finite element analysis (mode of 80 Hz). **b** Experimental analysis (mode of 33 Hz)

input, so it must be included as an approximate value. However, in real machines damping comes mainly from the contact in the guideways, where rolling bearings in one case or friction sliding ways in the others include uncertainty in the model. On the other hand, a careful experimental modal analysis can measure the natural frequencies and modes of a just constructed real machine with sufficient accuracy, and at the same time could be used for updating the FEM model used in design; however this latter is currently only used by research centres [9].

Thermal analysis is a classic but still "hot topic" in research. The usual procedure is to use the thermal capabilities of FEM packages, including some assumptions about the maximum heat sources as inputs. Some changes in design could be derived from this analysis.

After machine assembly, the final analysis step is to check the real behaviour of the system formed by the machine and tool, in a test working under aggressive conditions. Then, two vibration problems can happen. The first is the forced structural vibration under the action of the periodical cutting forces. This is a problem common to all mechanical systems, being studied using the FRF (function response function). To prevent resonance, natural frequencies of the system must be far from force frequencies, which can be done varying the stiffness or mass of machine elements.

Moreover, under particular conditions of chip section and spindle rotation speed, the phenomenon known as "regenerative vibration" or "chatter" may appear. Chapter 3 explains the basis of chatter in depth, due to its relation with spindle vibrations and damage. With respect to the machine, excitation of the machine structural modes may occur when machining conditions are aggressive enough. Then, low frequency

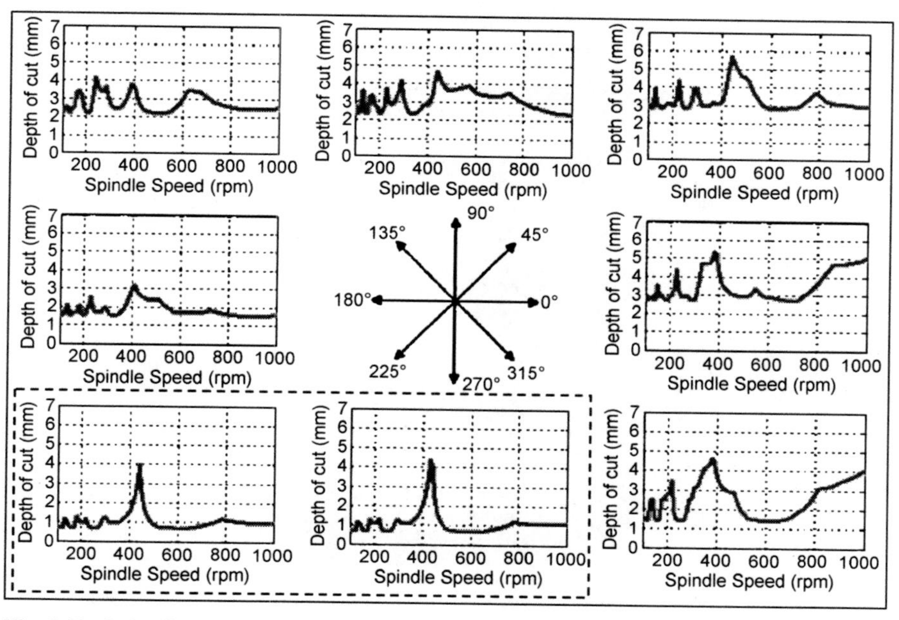

Fig. 1.13 Lobe diagrams for a position of a ram milling machine, along several feed directions

vibrations affecting machine life and workpiece roughness appear. In order to study this damaging circumstance, after solving a model based on a second order differential equation, the so-called "lobe diagrams" are obtained [2, 23]. In them, spindle speed is referred in the abscises and the depth of cut in the Y-axis, showing the borderline between stable and non-stable (i.e., where regenerative vibrations appear) cases.

An example is described in Fig. 1.13, for a milling machine making a rough downmilling on steel along several directions. In the dashed square in this figure are the two most restrictive directions. Along them, an axial depth of cut less than 1 mm must be applied to avoid chatter.

1.4.4 Modularity

Machines must be customised for success in the highly competitive global market, but with short production times and reduced costs. These factors have lead to a policy design based on different machine modules placed on common beds. Three main advantages are achieved with this:

1. Beds for several machine models are produced using the same polystyrene model and making optimal use of each iron casting. The direct advantage, a better price, is obtained from the foundries.
2. Some subcontractor companies reach a high specialisation producing elements and units that can be installed on different machines and for different machine manufacturers, such as rotary heads, tool changers, indexers, tables, rams, *etc*.
3. The elaboration of offers for customers in different areas of the world can be rationalised, with good customisation to user requirements yet at the same time avoiding special engineering for each case.

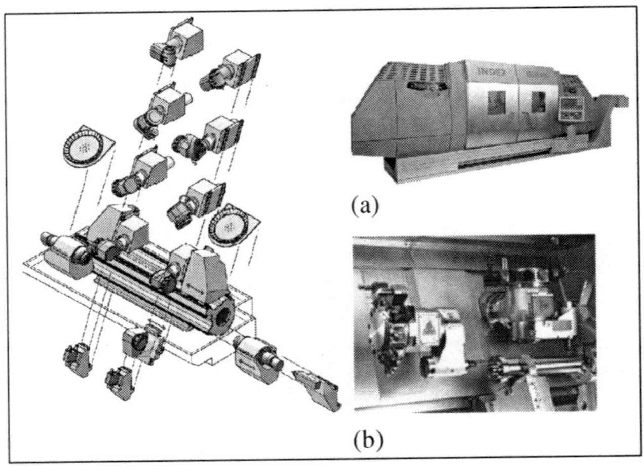

Fig. 1.14 Modularity of the lathe G400, by Index®. **a** Machine. **b** Workzone detail

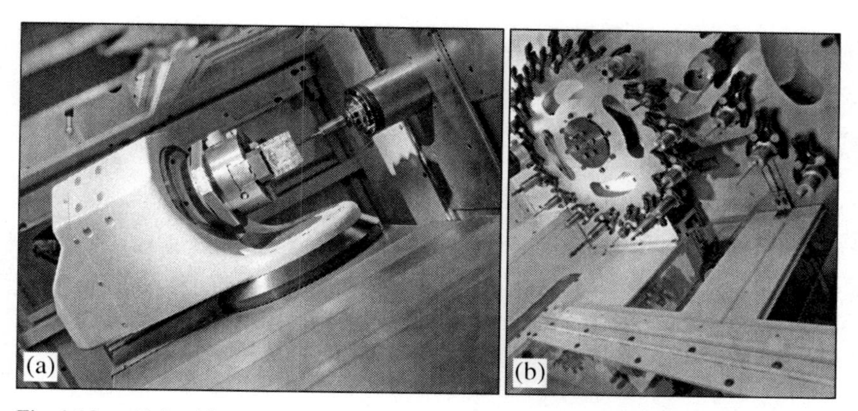

Fig. 1.15 **a** Unit with two rotary axes B and C of the transfer machine Multistep XT-200, of the Mikron Company®. **b** Multi-spindle option

The two former points are very interesting for small and medium companies, whose main competitive factor is the closeness and machine adaptation to a specific customer, whereas the latter is especially interesting for large producers of standard-type machines.

In Fig. 1.14 the modularity of the lathe family Index® G400 is shown. Several options can be implemented on the same bed, like two turrets, milling heads, steady rests and different types of tailstock.

Another example is the two rotary-axis machining unit shown in Fig. 1.15a, by Mikron®. This can be placed in transfer machines, where 5 sides of $200 \times 200 \times 200$ mm parts can be milled, drilled, threaded, recessed, knurled or engraved without reclamping. Moreover, the concept used in this machine, the Multi-step XT-200, is based on individual-spindle modules. They can be used either in "standalone" mode or expanded up to four modules.

1.5 Guideways

The guides are directly responsible for the precision and smoothness of the machine axis movements. Guideways are sliding systems where two surfaces are in contact; the fixed is known as the *guide* whereas that placed on the sliding component (carriage) is known as the *counterguide*. The required functional parameters for the longitudinal guides are the following:

- Geometric perfection, since every defect on a guideway will be translated into part inaccuracy.
- Stiffness, since guides must withstand the cutting forces and inertial loads without deformations. But some deformation is inevitable and if it happens a constant deformation value or a symmetrical behaviour with respect to the guide length is highly recommended.

- Wear resistant and good frictional feature with respect to the counterguide surface or rolling elements. The gripping of slides on guides due to either a low lubrication or a geometrical distortion coming from frictional heating must be taking into account when guides are designed.
- Enough toughness to withstand small impacts coming from the machining process, especially in case of highly interrupted cutting.

And for the counterguides, additional requirements are as follows:

- Adjustment to guides with enough geometrical compliance. To achieve that, usually the guide is hard but the counterguide is softer or more flexible.
- A preload system must be considered, to ensure precise contact with the guide. This is a key element to refit the guide along machine life, by compensating wear induced clearances.
- The couple guide-counterguide must provide as high a damping as possible to reduce the vibration transmission.
- The most important is the smoothness of its movement along the guide. The opposition to sliding must be as low as possible.

Machine movable components are 6 DOF spatial bodies; therefore the mechanical joints must restrict five of them while allowing the desired one, a rotation in the case of bearings or a longitudinal movement.

Usually guides between structural elements are placed in parallel pairs, sufficiently separated to provide the correct support of one element on the other. The use of the kinematic coupling concept, explained in Chap. 6, is the basic design aim. For precision machines the couple V-shape and flat guide (see the example in Fig. 1.16) is often used, but for conventional machines two parallel flat guides between bed and carriage are mainly used (see Fig. 1.9b), remembering to prevent hyperstatic constrains when installing. Usually one guide is the master in the assembling of elements whereas the other is adjusted to the final position.

In the case of rotational joints, ball or roller bearings are the standard solution. Hydrostatic bearings are only used for specific applications in grinders or rotary tables. Hydrodynamic bearings are used only in the wheel axis of grinders, because starting is problematic. For longitudinal movements the three main types of guideways now being used in machine tools for material removal processes are explained below (see also Sect. 4.6).

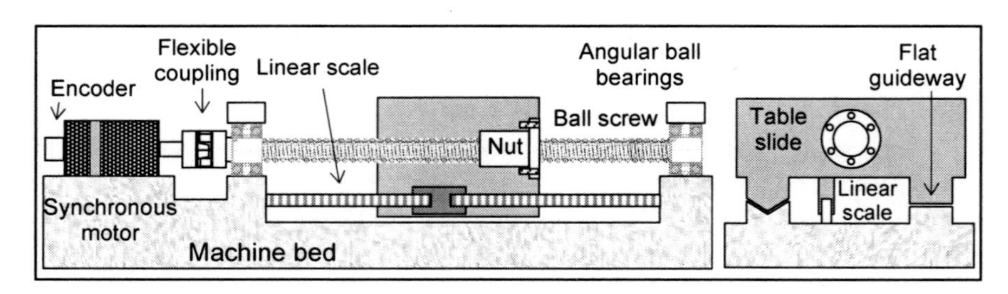

Fig. 1.16 The basic drive train, with lateral and front views

1.5.1 Guides with Limit Lubrication

In this guideway type, an oil film a few hundredths of a millimetre thick between guide and counterguide hugely reduces contact between surfaces. Although contact between roughness peaks cannot be totally avoided, reduction of the frictional work and friction coefficient is achieved. Oil must be periodically injected onto the counterguide to ensure this functional regime is reached. At the same time the counterguide has some small channels on its internal contact surface, called *spider arms*, to provide a uniform oil supply on the entire contact surface.

The main advantage of this classic guide type is the high damping ratio. As a main drawback, friction is too high at high speeds and friction heat can affect precision and even the guide life. The *stick and slip* phenomenon also occurs at low speeds.

Guides can be machined and grinded directly on the structural material, or built on hardened steel and bolt on the machine structure, where they are finally ground to achieve the final straightness. Currently, a few millimetres thick polymer sheet (type PTFE, polytetrafluoroethylene) is bonded to the counterguide contact surface. These materials greatly reduce friction with steel or cast iron. Some frequently used brands are Turcite®, Moglice®, Glacier DP4™ and Glacier DX®.

The final adjustment of the carriage on the guide is an important operation, being manually performed by skilled operators using chisel shape scrapers and tincture (Prussian blue or vermillion). The amount of material trimmed off by each movement of the scraper is around 1–3 μm, making it possible to create any desired shape or form. The surface pattern resulting from scraping is also beneficial for retaining oil along the carriage movement.

This type of guide is used in lathes, likewise in high precision machines, because scraping is the way to achieve very good straightness and flatness. In high speed machines the continuous inversions of movement with the consequent reversal and stick and slip problems do not recommend this solution.

1.5.2 Rolling Guides

A ball or roller linear guide is the "linearization" of the rolling bearing concept, where limit lubrication is substituted by the rolling of balls or rollers on the guide; rolling elements are separated by a flexible cage or retainers, recirculating inside a channel included in the carriage. Balls are adequate for light loads and high velocities, and rollers for high loads but lower velocities, just the same as in the case of bearings.

This sliding system comprises a rail guideway, with one carriage with integrated elastic wipers on the end faces, sealing strips on the upper and lower faces of the carriage and closing plugs to close off the fixing holes in the guideway.

The carriage and guideway of a linear recirculating ball or roller are matched and fitted to each other as a standard system due to their close tolerance preload. Rolling guides are stiffer than friction slides, with a lower opposition to displacement. The installation is easy, as well as the maintenance by the rapid substitution in case of damage to the guides or carriages. Usually machine tool manufacturers buy them just assembled with the required guide length, with the carriage mounted on the guide and preloaded to a specific value depending on the estimated load to move.

The main drawback of this sliding system is its low damping due to the direct contact of metal-to-metal of the bed-to-guide-to-roller-to-carriage. Some additional carriages specially designed for a high damping, using polymers or inducing an oil film, can be inserted among rolling sliders.

The life of this type of guides is determined by the fatigue of the rolling elements or races, in the same way as the rolling bearings. Therefore its selection follows a variation of Palgrem and Miner's approach which is usual for rotary rolling bearings, being standardised in the ISO 14728-1:2004.

1.5.3 Hydrostatic Guides

In hydrostatic lubrication an oil film always separates the sliding elements with sufficient thickness to totally avoid contact. To keep the film, an external pump continuously injecting the oil onto the bearing is required.

When this technique is used on rotary joints, the result is an extremely stiff design with almost no radial error motion; the pressure exerted by the film automatically centres the spindle in the bearings. On the other hand, the stiffness of the bearing is proportional to supply pressure. However the main advantage of oil hydrostatic bearings is their extremely high damping ratio, very important when hard and/or brittle materials are machined. Therefore hydrostatic supported spindles are used in high-cost grinders.

On the other hand, new linear guides designs based on the hydrostatic principle are currently under development. The problem is to inject and collect oil along the sliding of the carriage on the guide. Hydroguide™ and Hydrorail™ are recent

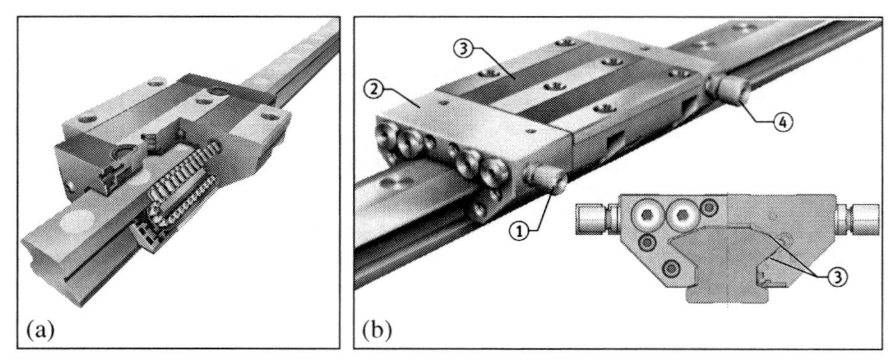

Fig. 1.17 Linear guides by INA®. **a** Roller guidance model RUE. **b** Special hydrostatic carriage model HLE45: points 1 and 4 are the oil input and output

products being installed in high precision milling machines and grinders, providing them with high damping, zero static friction, zero reversal error, a high reduction of friction heat and achieving a high straightness. In these designs, the carriage contains all the hydrostatic compensation and pockets on its internal surface shape. For conventional machines, carriages based on hydrostatics have also been developed, such as the model HLE45 of INA® shown in Fig. 1.17b.

For large diameter rotary plates, hydrostatic support of the flange disk enables very heavy parts to be supported without friction on the external radius (friction torque depends on the radius of the force application ring), whereas roller bearings are used on the plate central axis.

1.6 The Definition of the Main Motion

The headstock makes the essential machine function possible, to remove material evacuating it in the form of chips or dust. To do that a relative motion between the cutting edge (or edges) and the workpiece must be applied. A required high torque at a desired rotation speed to provide the recommended cutting speed (Vc) are the two inputs leading to select a motor + spindle. Motor power derives from both figures, since this power is the product of torque and speed.

In a few cases the tool is fixed on some sort of carriage and workpiece moves, as in lathes. Otherwise, in drilling, milling and most operations, a rotary tool is moved directly by an electrical motor. Asynchronous induction motors are the most widely used, consisting of two components, an outside stationary stator having coils supplied with AC current to produce a rotating magnetic field and an internal rotor attached to the output shaft that is given a torque by the rotating field.

In machine design, a good selection of the motor torque and power is very important. For this purpose a simple calculation can solve the basic motor specifications. The following example shows how to define the torque and power of a lathe for large diameter workpieces. For example, a hypothetical customer desires to turn a ∅ 290 mm part made of Inconel 625 (250 HBN). The tool manufacturer Sandvik Coromant® recommends the P25 carbide insert RCMX 120400 GC235, applying f 0.3 mm/rev and Vc 40 m/min. Rotation speed must be 275 rpm for this cutting speed on this diameter. Two depths of cut are studied, 2 and 4 mm in diameter. Some simple calculi are made:

*Torque = cutting force * (diameter/2)*

*Cutting force = specific cutting force * undeformed chip section*

*Undeformed chip section = feed * (depth of cut/2)*

*Power = torque * rotational speed*

Table 1.3 Example of the required torque and power for a main drive

Depth of cut	2	4
Undeformed chip section (mm^2)	0.3	0.6
Specific cutting force (N/mm^2)	5,073 (±3%)	4,775 (±4%)
Cutting force (N)	1,522	2,850
Torque (Nm)	221	415
Power (W)	6.364	11.988

Using values for the "specific cutting force" given by the tool manufacturer Sandvik, reflected in its catalogue, the results are shown in Table 1.3. There is a high specific cutting force because this is a difficult-to-cut alloy. Bearing these results in mind, a motor can be selected for this lathe applying a safety factor, for example a power of 17 kW and maximum torque of 500 Nm are feasible values.

Basically there are three types of main motion setup, as shown in Fig. 1.18. The first is the conventional and more widely used, where the motor is connected to the spindle by a timing belt. This rubber belt is a good vibration insulator; besides, no special care must be taken in the assembly of the group. This solution is possible up to a 6,000 rpm rotational speed.

The second type is the direct drive using a flexible coupling between motor and spindle. Here, a more reliable and better torque transmission is achieved, 12,000 rpm and even 16,000 rpm in special cases being possible. At the same time the flexible coupling is an effective heat insulator between motor and spindle. Furthermore, with the correct spindle assembly to the machine structure, all its thermal growth is derived towards the coupling and the motor, with no influence on the tool position. For this reason this is the most widely used configuration for precision machines.

Finally, high speed machines with rotational speed higher than 18,000 rpm require compact electrospindles, where the electrical motor is embedded in the spindle. This mechanical solution provides a very good concentricity of the group, needed for high speeds. The angular bearings are hybrid, with steel races

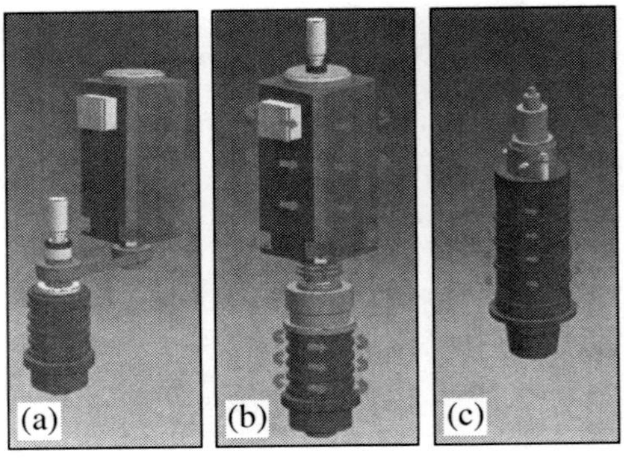

(a) (b) (c)

Fig. 1.18 Three solutions for the machine tool main spindle. **a** Motor and belt, **b** direct coupling, **c** Electrospindle (courtesy of Ibarmia®)

and ceramic balls to reduce the high friction at these speeds, which makes this solution very fragile when faced with an eventual tool collision against the workpiece. On the other hand, the heat from the motor itself and friction in bearings is great, requiring an internal water circuit with external cooling. More details are in Chap. 3.

1.7 The Definition of the Drive Trains

Classically, ball screws (see Fig. 1.19) have been one of the symbols of modern machine tools, very widely used because they are a reliable mechanical solution to convert rotational movement (from the electrical motor) to displacement movement required by the machine axes. Today linear motors make longitudinal displacements possible without other transmissions, but ball screws are still a less expensive and at the same time efficient solution.

In Fig. 1.16 the kinematic chain (drive train) of one machine axis is shown. There, the feed motor, usually a synchronous brushless motor, is coupled to the screw (shaft) by a flexible coupling, which at the same time is a good torque transmitter, a thermal insulator and eliminates the necessity of a perfect alignment of the motor with respect to the screw. In other cases the connection of the motor to the screw is by means of a timing belt, with the same advantages and drawbacks that the case of use belts for the main spindle indicated in the preceding subsection.

The screw is supported at both ends by angular ball or conical rollers bearings, in configuration double back (DB) in the former case to increase the rigidity of the axis. For this same reason the screw is usually pre-stressed. This fact also reduces the impact of screw thermal dilation.

The sliding component is bolt to the nut; nuts can be square, round or with flanges for a stiffer adjustment to the machine component.

The screw diameter ranges from 6 to 100 mm, whereas the screw lead does it between 1 and 60 mm, keeping a certain proportionality between these two features. If the lead is high, fast movements can be obtained but with a low mechanical advantage. Small leads imply slow movements but with a high mechanical

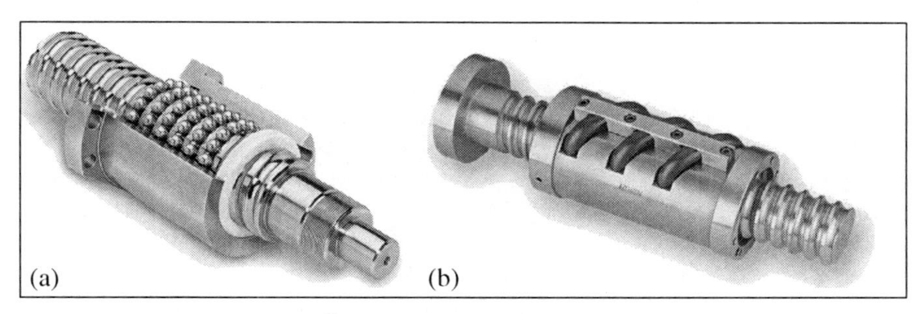

Fig. 1.19 Ball screws, by Hiwin® **a** Image of the nut. **b** A design for high loads

advantage; at the same time small ones are more highly recommended to ensure precision and controllability of the drive train.

The calculation of screws is based on preventing two mechanical problems. First, buckling due to shaft working under compression when the nut displaces along the screw. Second, critical whirling speeds must be avoided, only in the case when the shaft rotates and the nut moves along the screw. The other case is when the screw is fixed and the motor rotates the nut; evidently the motor must be placed onto the carriage. See Chap. 4 for further details about ball screw design and use.

The thermal growth of the screw is a key aspect to be considered and reduced. During the normal function of ball screws, heat coming from the friction of balls movement on the thread causes a significant longitudinal dilation of the screw. Approximately 7–10% of the power is converted into heat. If the shaft has been placed with pre-stress, some dilation is absorbed by this good practice. However some microns always appear. To reduce dilation to a minimum, one solution in machines for high precision is the forced fluid cooling of the screw, internal to the hollowed out shaft, or even the forced cooling of the nut. This increases the price, but maintains thermal growth within machine requirements.

Nevertheless, if the axis position is measured by a direct linear scale, thermal growth becomes considerably less important, because CNC works to reach the desired axis position until it matches with that measured by the linear scale. Therefore, the thermal insulation of external measurement scales from machine heat sources is essential.

1.8 The CNC Implementation

Usually the CNC and drive supplier can be selected by the user, who chooses the brand and model of the CNC for a machine. This is the case of manufacturers of small and medium machines, who try to be flexible regarding user wishes. This is an important competitive factor for their businesses. However in the case of large machine tool companies, the use of their own controls and the need to simplify the production chain means this option is not suitable. Big machine manufacturers usually sign an agreement with a CNC provider to develop specific CNCs for their machines, being commercialised with the machine tool trademark.

Basic CNC functions are shown in Fig. 1.20, particularised for a three-axis machine. In more detail:

- The basic operating system: All current controls are based in microprocessor architecture similar to a PC; therefore they need a basic operating system.
- The program translator: The part is programmed in ISO code (the so-called *G-code*) or a special NC language. The former is useful to learn how to use machines; unfortunately it is too simple for the new machine capabilities. That is the way CNC manufacturers have developed their own languages. The translator always reads some blocks ahead to give time for performing all trajectory calcula-

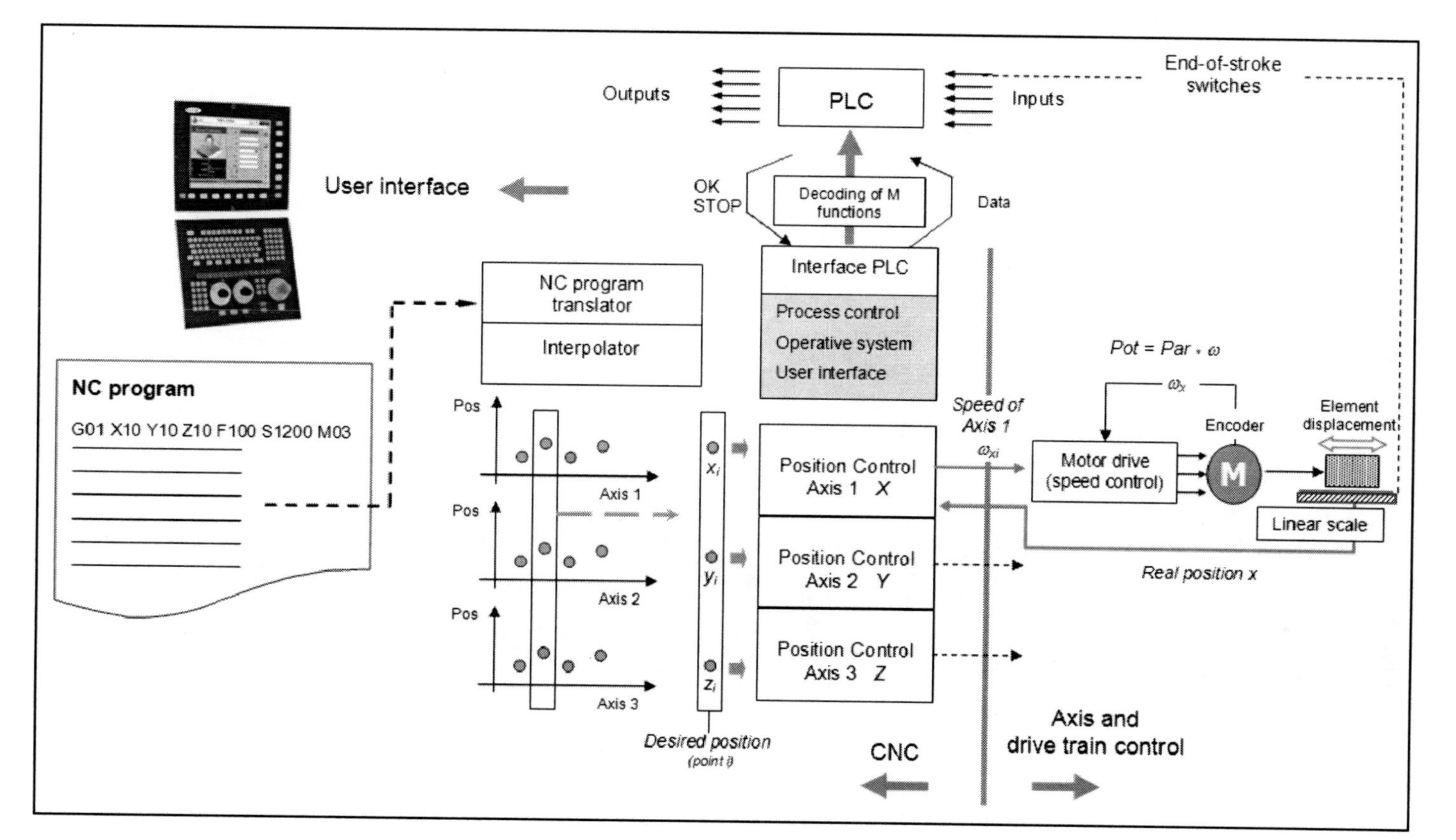

Fig. 1.20 The CNC integration

tions before the tool reaches the next positions. The higher the machine feed is, the further ahead the blocks must be processed.

- The interpolator. This is the most intelligent CNC function, when 3D, 5D (usually three translations and two tool orientations) or even multi-axis trajectories must be converted into axis positions. The high precision required leads interpolator to make calculations at every few microns along toolpaths. Calculation methods are spatial transformations using homogeneous matrices, tool radius compensation and other geometrical corrections (see Chap. 5).
- Axes control. When the position to be reached is obtained by the interpolator the speed of each axis motor is also obtained. The speed value is the input for the motor drive. Through control techniques, PDI being the simplest, motion in each axis is kept up to the real position reaches the target position. Therefore there must be a position measurement device, an encoder or a linear scale shown in Fig. 1.19, to know the real axis position at any given time.
- The user interface. This is today a decisive selling factor due to the high complexity of machines and operations to be done. Users need on-line assistance utilities to make programs free of errors for the new developed machines. Machines for special applications, like tool grinders or gear shapers or the new multi-task machines (see Figs. 1.3 and 1.4), require programming utilities which enable users to generate machining programs for complex parts.
- The Programmable Logic Controller (PLC) which is included in the same hardware architecture today with the rest of the control cards, for enabling a high speed communication with the basic NC modules. PLC controls the auxiliary machine functions (generally coded as *M functions* in ISO), running a program implemented by the machine builder and taking into account the value of digital inputs collected from simple sensors placed in the machine. The PLC digital outputs are connected to switches or electrovalves to command the tool magazines, the coolant circuit, the pallets transfer system and other auxiliary devices. Another important surveillance function is related to the end-of-stroke switches, placed at the ends of all guides to prevent moveable components from leaving the slideways: when a microswitch detects a sliding component is on an end-of-stroke the PLC immediately stops the axis feed motor.

Nowadays, control manufacturers produce two or three basic CNC platforms which can include several options for different kinds of machines. Therefore the machine tool manufacturer must adapt them to each specific machine model, following several steps:

1. The definition of machine axes in relation to CNC ones, giving also the positive and negative senses and strokes of each axis.
2. Control parameters of each axis must be adjusted to the machine dynamics.
3. PLC programming for each machine model with its user options.
4. Customization of the user interface. General CNC control models are prepared for the milling machines, lathes or universal grinders most commonly sold. However, for specific applications special help utilities to prepare NC programs are absolutely necessary.

5. Finally, tuning tests using idle movements and some machining tests are performed. These tests allow the fine adjustment of CNC parameters to machine to achieve the maximum precision and speed, as is explained in the next section.

To improve machining process reliability and facilitate the maintenance of machines, several sensors and monitoring techniques can be managed by the CNC. Spindle consumption, collision detection, thermal growth of spindle and the balance of the main spindle are monitored by the CNC in high end machines.

1.9 Machine Verification

After machine construction, a thorough verification of machine precision and performance must be made. For this purpose some tests have been defined. An important result is the determination of the CNC compensation parameters for machine axes. Nonetheless, tests can also be used for several purposes, thus as acceptance tests by machine buyers, for comparisons between similar machines, for periodical machine checking throughout machine life, *etc*. Three kinds of tests can be mentioned.

First, geometric tests for measuring straightness, parallelism or concentricity errors. ISO 230:2006, ISO 10791:1998, ISO 13041:2004 and other standards defined by the ISO Technical Committee 39/SC 2 ("Test conditions for metal cutting machine tools") specify methods for assessing the accuracy and positioning repeatability of numerically controlled machine tool axes by the direct measurement of individual axis. In the standards, methods for linear and rotary axes are suggested.

The second group includes procedures for studying the interpolation control, based on laser interferometers, the "Ball-Bar" device for the relative movement of two Cartesian axes, or position sensors for measurement of machine repeatability. When several rotation axes are simultaneously considered there are no standard tests, excepting those for bi-rotary heads in machines with a horizontal or vertical Z axis. However, standards do not include tests to check five-axis machines with tilting tables (*RRLLL* configuration in Fig. 1.7) or other structures.

Another important datum to achieve the best machine-CNC adaptation is the lag-error measurement using the internal utilities implemented in modern CNCs for monitoring the behaviour of the axis drives with respect to the axis positions.

The last group includes "testparts" (see Chap. 6) to check the behaviour of lathes and machining centres during process. Currently, milling tests are only focused on three-axis milling centres. The most well-known is the NAS workpiece (defined in ISO 10791-7:1998). But this test does not include complex surfaces, which is why different testparts have been designed in the last years, such as the so-called Mercedes or NCG parts (available at NC-Gesellschaft association [18]). However, none of them are yet included in the ISO standard normative. Furthermore, there is no specific part for testing five axes machining centres, so customers have to define their own tests to check the machine they are interested in buying, which is highly time consuming.

1.10 Typical Machines for Several Applications and Sectors

At the design stage, a machine is considered for a particular industrial application, but using rather non-specific requirements. As was explained in Sect. 1.2.1, the workpiece type and size, operations to be performed, productivity and the precision to achieve will define the machine specifications. In the following section some machines typical of important sectors are described; likewise other examples will be described in depth over the next chapters.

1.10.1 A Machine for Big Structural Turbine Parts

The vertical lathe is a typical machine for this application. The workpiece comes from forging, being disks or cases. Machining operations are turning, boring and milling. The maximum turning diameter can be up to 3 m and part weights of 30 tons are not uncommon. The plate rotation speed is in the range of 200 rpm with a power of 50–100 kW. The ram stroke is long, more than 1,500 mm. In the ram a turning tool or a milling headstock can be placed, in the latter case with a power ranging from 20 to 50 kW. Two machines of this type by P. Carnaghi® are shown in Fig. 1.21.

Another example of a big machine with two rams mounted on the same screw is shown in Fig. 1.22, by Ibarmia®. This is a drilling and milling machine focused on the drilling of circular rings for wind turbine support towers. Both rams are independent although they can work in parallel.

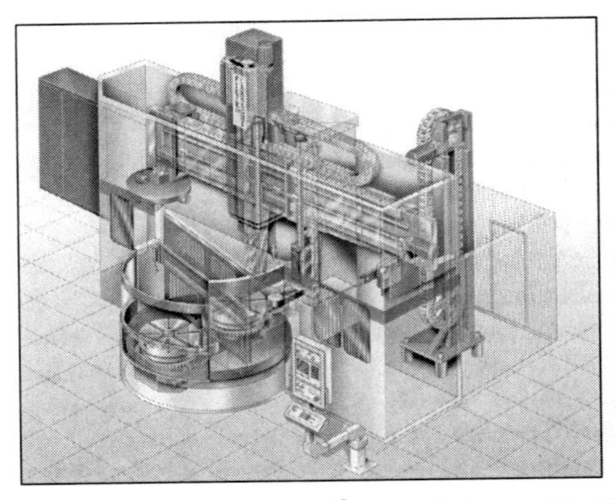

Fig. 1.21 The Pietro Carnaghi® vertical lathes, models ATF14 and AP100

Fig. 1.22 The Ibarmia® drilling machine

1.10.2 A Horizontal Milling Centre for Automotive Components

In the highly competitive automotive sector, companies are encouraged to reduce times and prices while maintaining sufficient quality. Therefore, machines must provide high metal removal rates, fast feedrates, greater accuracy and stability, increased overall productivity and finally, the flexibility to meet changing production volumes.

Commonly produced parts are blocks and cylinder heads in iron or high silicon casting, component cases, gears, crankshafts, rods, or small pieces for the transmission and suspension systems. In Chap. 13 more examples are described.

Automotive components are usually machined in horizontal three-axis machining centres, provided with a B axis for part rotation or in some cases with a simple rotary indexing table (Fig. 1.23). Horizontal centres bring benefits to an application which vertical ones are usually unable to, including faster power spindles and more power spindles, larger tool magazines, a smaller footprint and chip disposal. Parts are held and fixed on universal pallets, which can be transferred by automated systems. In some cases the inclusion of a swivel head or rotation table enables the production of more complex shapes.

In these machines the tool change time "chip to chip" must be as low as possible; currently times of 2–4 seconds are usual. Very rapid idle movements between operations or tool positioning are also required.

Spindles are in the range of 6–12,000 rpm powered with 60–90 kW motors to be able to move large diameter multi-insert rotary tools. Versions with a geared drive or a direct connection of main motor to spindle are possible, the torque being higher in the geared versions (in the range of 600–1,100 Nm).

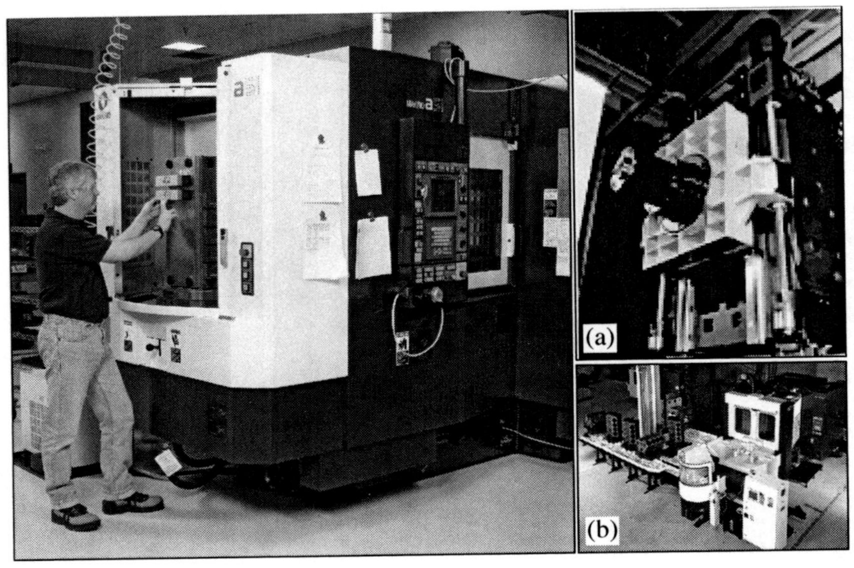

Fig. 1.23 Horizontal machining centre, model Makino® a51. **a** Detail of the spindle and column of Makino's A77. **b** Automated system for production of engine blocks

Tool magazines are the largest in machine tools, with an optional storage of up to more than 400 tools.

Some machines include as an option two twin spindles and tables or drums to fix several parts in one setup; see Fig. 1.24. When this is done, the borderline between individual machines and transfer machines (see Sect. 1.2.9) becomes diffuse.

On the other hand, vertical centres are applied to wheel, steering, brakes or suspension components, for small and medium production.

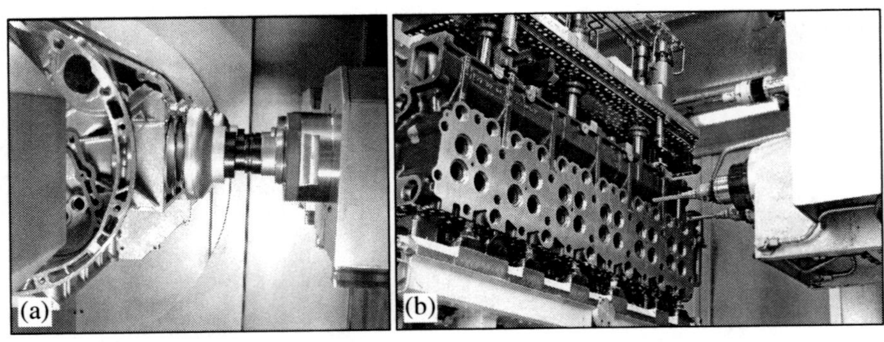

Fig. 1.24 **a** Twin spindle and drum part change system. **b** Double spindle for engine head machining, by Heller Maschinenfabrik®

1.10.3 A Milling Centre for Moulds

In this case, a CNC milling machine with a $1 \times 1 \times 1$ metres working zone is common, with three axes and a vertical spindle as the main solution, and five axes as a widely used option. For small moulds the C-frame is typical, such as the model shown in Fig. 1.25. For medium size moulds the gantry frame becomes more interesting due to its improved stiffness, with a sliding table where the mould is placed with bolts or fixtures.

For big moulds the gantry type with a travelling beam and with a two rotary head allows machining complex shapes. However, for medium and small moulds the two-axis rotary bed is stiffer and therefore more widely used

As to the main motion, an electro spindle with 20,000–25,000 rpm and 17–25 kW is a typical picture but a direct drive by a flexible coupling is also very widely used. This latter solution is cheaper, and more robust against collision and allows 12,000 rpm and even more. Currently, electrospindles are able to give enough torque even at a low rotational speed allowing roughing to be performed in the high speed milling centre starting from the hardened steel material.

Ball screws are the economical and high performance solution for drive trains, allowing an acceleration of 0.8–1 g. In high end machines linear motors are used.

Machines are not usually provided with pallet transferring, because machining on moulds is a long operation not requiring too many tools and, therefore, the setup time is not critical.

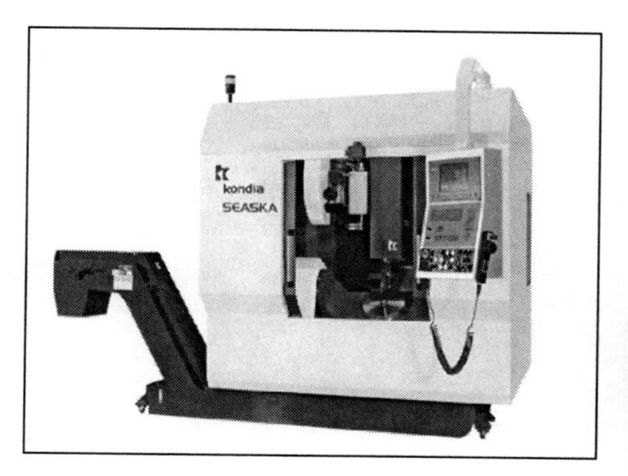

Fig. 1.25 The five-axis milling machine Seaska, by Kondia®

1.10.4 A Milling Machine for Big Dies and Moulds

As mentioned above, these machines are gantry-type, with a transversal travelling beam and a long ram for the Z movement. The heavy die is fixed on the table,

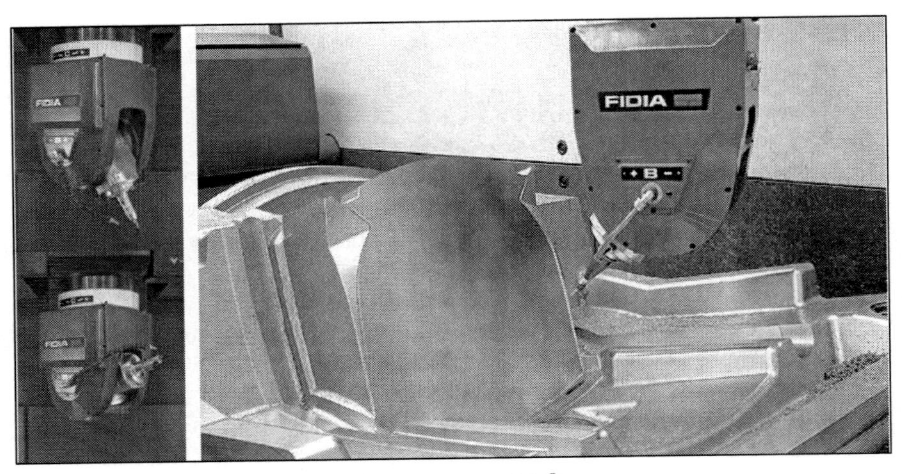

Fig. 1.26 Rotary axis head in a milling machine, by Fidia®

whereas all movements are done by the tool. Usually a two rotary head is used for sculptured surface milling, as shown in Fig. 1.26. There, to obtain the maximum possible accuracy, the B and C axes are both driven by brushless motors and the transducers are mounted directly on the axes. In addition, the axis kinematics chains have a crown gear-worm screw system with a variable pitch for the compensation of the backlash.

New models of rotary heads based on torque motors for the orientation axes have been launched in the last four years. This is going to be the main design solution in the near future, at least for medium and high level machines. Torque motors are ring-shaped and can be placed directly on the rotary joints, eliminating gear transmissions.

Some machines for large dies can change the spindle head automatically, selecting a head with a geared spindle for roughing and another with two axes and a high speed spindle for finishing. Rapid-change connectors must be mounted at both the ram end and base of the head for the electricity, oil and coolant supplies.

1.10.5 Conventional Machines for Auxiliary Operations

Today, manual machines are mainly used as auxiliary machines, for special operations. This group includes the universal milling machine, the press drill and the engine lathe which are always present in workshops, with only slight modifications in relation to the inventions of the second half of the 19th century. An easy and inexpensive option is the inclusion of measurement rules in each axis for the digital display of axis positions, as in the case shown in Fig. 1.27.

Fig. 1.27 Milling machine FU152 of Lagun®

Basically all metal technicians have learnt their basic skills and the "art of working metals" on these machines, since machining operations are more similar in these machines than in the recently launched CNC machines. In Fig. 1.27 a current universal milling machine is shown, with automatic feedrates on 3 axes (X, Y, Z) and continuous variation speeds for main spindle and feedrates. Options for this model are a universal milling head, a universal milling head with quill feed, a universal dividing head, a precision circular table and attachments for slotting, rack milling and gear hobbing. In Fig. 1.28 an engine lathe and its feed gearbox and turret is also shown. This model reaches 2,200 rpm with a 5.5 kW headstock motor.

Fig. 1.28 Engine lathe, model SP250 by Pinacho®. **a** Detail of the feed gearbox. **b** Universal turret

1.10.6 CNC Milling Machines for General Production

Some machine manufacturers have recognized the need for a dependable, low-cost production machining centre. Here the compromise of technology-price-reliability and functionality must be balanced, with parts for multiple applications in mind. A lot of small and medium subcontractors do not know what kinds of parts will be produced in the medium term, and therefore they prefer versatile machines.

An example is the Haas® Toolroom Mill series; the model TM-3P is shown in Fig. 1.29. The TM-3P's reliable 5.6 kW vector drive spindle spins to 6,000 rpm and uses standard 40-taper tools. Rapids and cutting feedrates are 10 m/min. The machine *1015 × 510 × 405* mm travels and a *1465 × 370* mm T-slot table provide plenty of room for workholding. A 10-pocket tool changer is standard. The heavy cast-iron bed and column damp vibration and provide rigidity.

On the other hand, in some areas of the world, machine size and footprint are important due to the expensive industrial ground. For this reason compact machines are produced by several manufacturers. Usually, these ultra-compact machines are small enough to fit through a doorway, can be easily moved using a pallet jack or equipment dolly, can be provided with a caster kit to be rolled from one location to another and even fit into most freight elevators. An example is shown in Fig. 1.29, the Haas® OM-2A Office Mill. In these machines particular attention must be made to the work envelope, typically a cube with sides ranging from 150–250 mm.

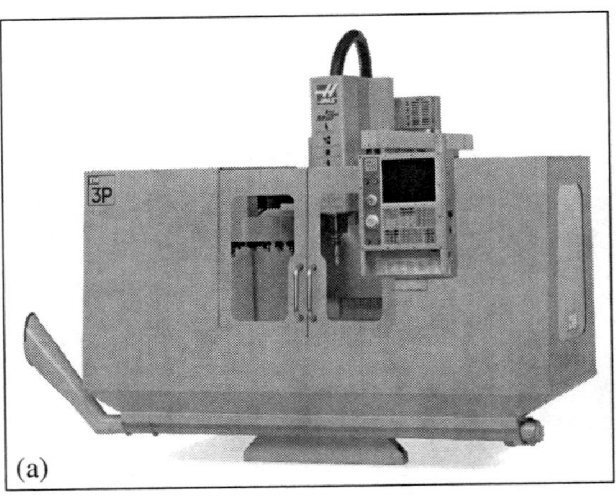

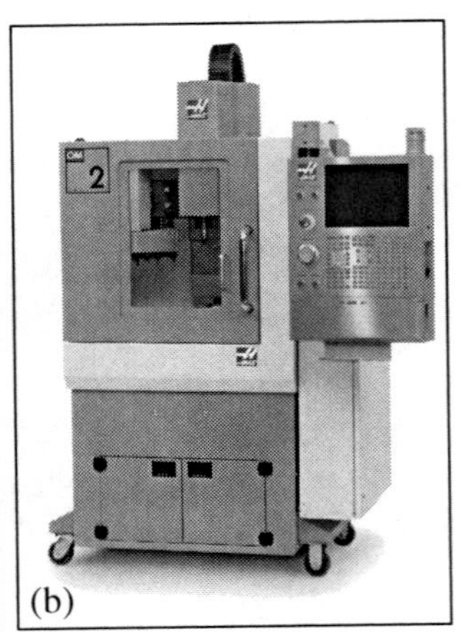

Fig. 1.29 **a** The Haas® TM-3p. **b** The Haas® OM-2A Office Mill

1.10.7 A Heavy-duty Lathe

Lathes for long big parts, such as lamination rods or crankshafts for ship propulsion, are horizontal with flat beds, some models with even a 25 m distance be-

Fig. 1.30 Heavy-duty lathe of Goratu®, model GHT11

tween centres and a maximum swing over bed of 2 metres. Workpieces weighing over 45 tons can also be machined on. Of course, the size and weight of parts requires the use of big steady rests.

The main motion is a powerful motor in the range of 100–150 kW. The carriage moves at moderate feeds, usually no more than 6 m/min. An operator can travel on it, as shown in Fig. 1.30. The size and weight of components requires the motorised movement of the tailstock and its automatic locking to the bed.

1.10.8 A Mitre Band Saw

Sawing is an important operation during the daily work of metal companies. Moreover, it is an essential operation for steel structure constructors. A typical band saw

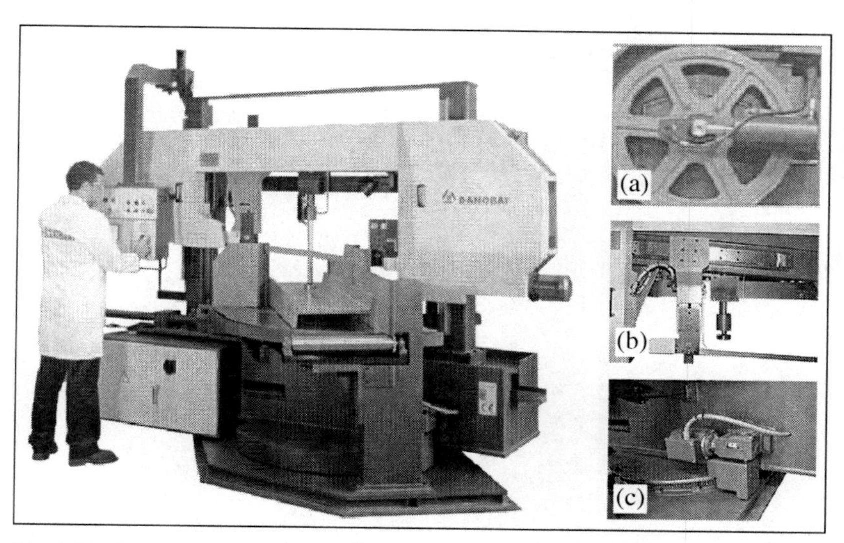

Fig. 1.31 Bandsaw CPI of Danobat®. **a** Detail of the tensioning blade device. **b** Guide of the bandsaw. **c** Manual or NC controlled mitre positioning

is the twin column with mitre cutting between (+45°–60°), focused on the cutting of any type of section, both single and bundles (Fig. 1.31). Machine mitre positioning is carried out manually or NC controlled, depending on customer needs. Main motor power ranges in 5–8 kW and blade speed in 15–100 m/min. The tensioning device of the saw blade is hydraulic, manual or automatic. A coolant pump provides a constant flow of liquid coolant over the blade when metals are cut.

1.10.9 Transfer Machines

Transfer machines are usually designed as per the view of the specific part drawings to be produced. Companies specialised in transfer machines work only as per customer order. The main requirement is high productivity, so hard automation is recommended here. Starting from this, the machine manufacturer's engineering department tries to design a solution adapting several machining units and auxiliary feed devices on a common bed.

Machine modularity makes it possible to obtain different solutions without a lot of changes in the design and production. Regarding the machine frame, a ring bed is a classic solution for rotary indexing table models, made in cast iron or weld steel. These are called dial-index machines. Other machines are in-line systems, with a U-shape layout.

Drum or crown type tool turrets, as shown in Fig. 1.32, can be placed to perform several machining operations like milling, drilling and tapping. All tools are driven by a common motor, using a gear transmission. These modules can be used in different machines with little changes.

On the other hand, the use of parallel spindles when performing the same operation leads to high productivity.

Fig. 1.32 Transfer machines by Etxe-tar®. **a** Machining unit for cylinder blocks. **b** Parallel spindles

1.10.10 A Milling and Boring Centre

The last machine of this chapter, shown in Fig. 1.33, is a turning and milling centre with an additional ram to increase the application scope of the machine, for those part zones with complicated access to deep internal zones.

The 7.5 kW ram spindle with a stroke of 900 mm is mounted on the side of the milling spindle housing, which enables the turning and milling using an angle head. This spindle also has a dedicated 40 position tool magazine. This ram spindle is in addition to the standard B-axis spindle with its 10,000 rpm and 37 kW motor and the turning table with 500 rpm and 37 kW power.

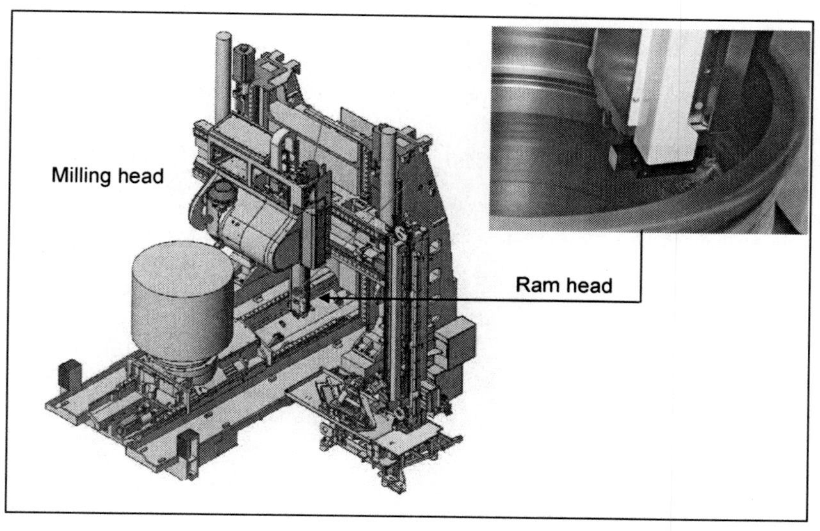

Fig. 1.33 Five-axis milling centre with an additional ram spindle, Mazak® model Integrex Ramtec V/8 (courtesy of Intermaher®)

1.11 The Book Organisation

After the main aspects involved in machine tool manufacturing are introduced, they are be presented in depth over the next chapters. Chapter 2 is about new structural concepts currently being developed in some research projects. Chapter 3 is focused on the basics of spindles.

Chapter 4 explains the classic and recent advances on guides and drive trains. The current CNCs are described in Chap. 5, concluding the main descriptions of machine elements. Chapter 6 describes the most important aspects in design and assemble precise machines.

Chapters 7, 8 and 9 aim to present either some particular machine types or those machines specialised in the most important industrial sectors. Thus the grinding machines, lathes and electrodischarge machines are updated. Chapters 10 and 11 describe two small machine groups but at high technological level: parallel kinematics and micromilling machines. Here special concepts are used, complementing all others presented in previous chapters.

Finally, a chapter devoted to aeronautical machines followed by one more on automotive machines complete this book, hoping the reader has encountered updated knowledge on machine tool construction over these pages.

Important references about machine tools are the books by Weck [24], Boothroyd and Knight [5], Arnone [3], King [12] and Slocum´s book about precision machine design [21]. The basic machine tool theoretical bases are explained, but unfortunately recent advances are not included in them. More recently, books by Altintas [2], Tlusty [23], Kibbe *et al.* [11], Marinescu *et al.* [15], Erdel [8] and finally by this chapter's authors [14] update the state of the art about machine tools.

Acknowledgements Our thanks to Mr. Pedro Ortuondo of the Elgoibar Machine Tool Museum for the information provided; thanks to all the companies cited for their pictures and information. Special thanks to Prof. José Luis Rodil and Xabier Ibarmia for their time dedicated to discussing several aspects of this chapter. Thanks are also addressed for the financial support of the Basque Government (Etortek 2007/09). Also thanks to the financial support of the CENIT project eee-Machine.

References

[1] Adabaldetrucu P, Ortuondo P (2000) Máquinas y Hombres (Machines and men, in Spanish, Basque and annotations in English), Ed. Machine Tool Museum of Elgoibar

[2] Altintas Y (2000) Manufacturing Automation: Metal Cutting Mechanics. Machine Tool Vibrations and CNC Design, Cambridge University Press

[3] Arnone M (1998) High Performance Machining, Hanser Gardner Publications

[4] Artobolevsky II (1979) Mechanisms in modern engineering design, Mir publishers, Moscow

[5] Boothroyd G, Knight WA (1989) Fundamentals of Machining and Machine Tools. Marcel Dekker, Inc.

[6] Byrne G, Dornfeld D, Denkena B (2003) Advancing Cutting Technology. Annals of the CIRP 52/2:483–507

[7] Huang DT, Lee JJ (2001) On obtaining machine tool stiffness by CAE techniques. Int J of Mach Tool Manufacture 41:1149–1163

[8] Erdel BP (2003) High-speed machining. Society of Manufacturing Engineering, Dearborn, Michigan

[9] Garitaonandia I, Fernandes MH, Albizuri J (2008) Dynamic model of a centreless grinding machine based on an updated FE model. Int J of Mach Tool Manufacture doi:10.1016/j.ijmachtools.2007.12.001

[10] International Standard ISO 841: Industrial automation systems and integration – Numerical control of machines – Coordinate system and motion nomenclature (2001) International Organization for Standardization

[11] Kibbe RR et al. (2006) Machine Tool practices (8th edition). Prentice Hall

[12] King R (Ed.) (1985) Handbook of high speed machining technology. Chapman and Hall
[13] López de Lacalle LN, Lamikiz, A, Sánchez. JA, Salgado MA (2007) Toolpath selection based on the minimum deflection cutting forces in the programming of complex surfaces milling. Int J of Mach Tool Manufacture 47:388–400
[14] Lopez de Lacalle LN, Sanchez JA, Lamikiz A (2004) High performance machining (in Spanish). Ed. Tec. Izaro, Bilbao, Spain
[15] Marinescu ID, Ispas C, Boboc, D (2002) Marcel Dekker
[16] McKeown PA et al. (1987) The Role of precision engineering in manufacturing of the future. Annals of the CIRP 36/2:495–501
[17] NAICS, North American Industry Classification System (2002) 333512 Machine Tool (Metal Cutting Types) Manufacturing
[18] NC-Gesellschaft, www.ncg.de
[19] Reuleaux F (2007) Theoretische Kinematik (edition classic in German, of original of 1875). Verlag Dr. Müller
[20] Salgado M, López de Lacalle LN, Lamikiz A, Muñoa M, Sánchez JA (2005) Evaluation of the stiffness chain on the deflection of end-mills under cutting forces. Int J Mach Tool Manufacture 45:727–739
[21] Slocum A (1992) Precision Machine Design. Society of Manufacturing Engineers
[22] Taniguchi N (1974) On the Basic Concept of Nanotechnology. Proc. ICPE Tokyo
[23] Tlusty G (2000) Manufacturing processes and equipment. Prentice Hall
[24] Weck M (1984) Handbook of Machine Tools. John Wiley & Sons Ltd

Chapter 2
New Concepts for Structural Components

J. Zulaika and F. J. Campa

Abstract This chapter is focused on analysing new concepts and trends related to the structure of machine tools. In fact, the structure of the machine has a decisive influence on the three main parameters that define the capabilities of a machine, which are: motion accuracy, the productivity of the machine and the quality of machining. In this respect, this study on structural components will add a new basic parameter, eco-efficiency, because the structure of the machine also has a decisive influence on the whole life cycle of the machine and especially on the materials and energy resources consumed: an issue of increasing concern among machine tool builders.

2.1 Introduction and Definitions

As seen in Chap. 1, the structure of a machine tool has two main functions, to hold the components and peripherals involved in the machine and to withstand the forces which are produced by the process and from the machine motions.

Within this view, the basic challenge for machine tool builders is to conceive machine tool structures that are capable of withstanding with minimum possible deflections the effects of the foreseen forces and of heating foci and at the same time consume the minimum possible in terms of materials and energy resources.

J. Zulaika
Fundación Fatronik, Paseo Mikeletegi 7, 20009 Donostia-San Sebastián, Spain
jzulaika@fatronik.com

F. J. Campa
Department of Mechanical Engineering, University of the Basque Country
Faculty of Engineering of Bilbao, c/Alameda de Urquijo s/n, 48013 Bilbao, Spain
fran.campa@ehu.es

L. N. López de Lacalle, A. Lamikiz, *Machine Tools for High Performance Machining*,

As these two aims are opposing, the process of designing a machine tool is a trade-off between these two targets, so that the specific characteristics of a machine tool will define that balanced position between these two approaches.

In this respect, the final characteristics of a machine tool are defined by means of the following parameters:

- Productivity: The productivity of a machine is measured in terms of its *metal removal rate* (MRR), which largely depends on its kinematic and dynamic capabilities and especially on its static stiffness, which in fact is the primary reason to define the dimensions and shapes of the structural components. To this end, higher stiffness involves bigger masses, which in combination with faster motions aimed at achieving higher productivities, leads to motors of higher power that generate higher inertial forces. This in turn demands stiffer structures, which again demand a larger amount of material, so that stiff machines become productive and at the same time very energy and resource consuming.
- Accuracy: The accuracy of a machine is defined according to the deviations of the tool with respect to a desired profile while it is being moved and positioned (see Chap. 6). These deviations are largely associated to thermal effects and to the mechanical deflections that the machine components bear when inertial and process forces act on them. Therefore, the basis necessary in order to achieve accurate machines will lie in conceiving both thermally-stable and stiff structures.
- Eco-efficiency: The eco-efficiency of a machine is measured in terms of energy and material resources used and the waste and pollution created in the process. As sustainability is an issue of increasing concern in the manufacturing sector, this book will add a new aim for machine tool builders: to reduce the environmental impact associated to a machine tool throughout its entire life cycle.

The structural components of machines have a twofold impact on the global eco-efficiency of a machine: on the one hand, the largest amount of material resources in a machine is associated with its structural components. On the other hand, the energy that a machine consumes during its use phase is largely associated with the motion of movable structural components both in positioning motions and in machining motions.

Moreover, according to a study of the machine tool builder Nicolas Correa®, the use period of a machine represents more than 90% of the total impact associated with their machines. Figure 2.1 shows both graphically and numerically (by means of a *life cycle analysis,* or LCA), the total impact associated with an average medium/large size machine of Nicolas Correa® throughout its entire life cycle.

Accepting thus that the largest portion of the environmental impact of a machine tool is associated with the motions of its movable components, the reduction of each gram of material in these movable structural components will have a decisive role in the final ecological impact of the machine tool.

As a conclusion, the structure of a machine plays a key role on the final functionalities of the productivity, accuracy and eco-efficiency of that machine. The following sections will explain different strategies that can be employed to con-

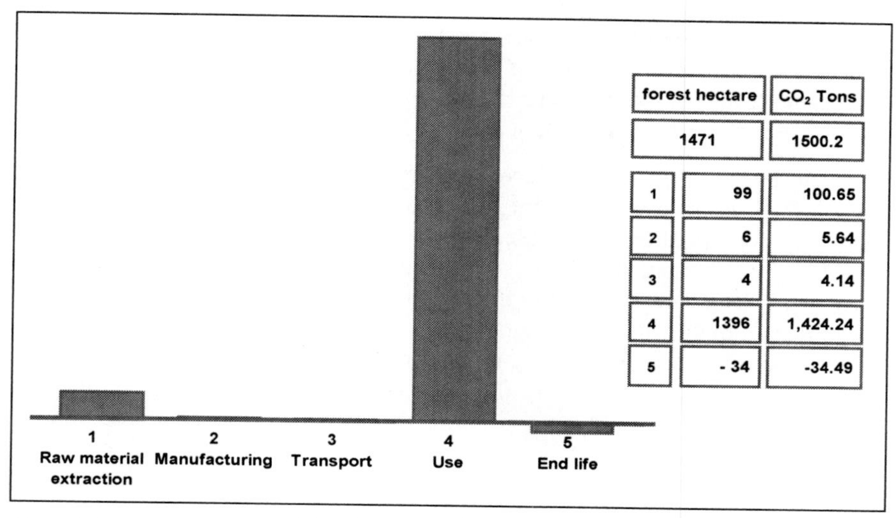

Fig. 2.1 LCA of a machine by Nicolas Correa®

ceive and produce machine structures that achieve appropriate values of productivity and accuracy by means of good thermal and vibration behaviour combined with low amounts of movable masses.

2.2 Optimised Machine Structures

Machine tools provide a relative motion between the tool and the part that is to be machined. With regard to these motions, milling machines have three Cartesian motions, with at least two of the axes mounted in serial. In some cases, the three motions are applied to the tool (really to the spindle onto which the tool is fitted); in other cases the motions are divided into the tool and the part, and only on a few occasions are the three motions applied to the part, because it is quite unusual to vertically move the table on which the part lies. Additionally, it is increasingly common for machine builders to add two additional rotational degrees of freedom to the machine, either to the headstock, or to the table or to both. They thus become machines with five degrees of freedom, i.e., five-axis milling machines.

The structure of a machine covers the components that allow the achievement of the degrees of freedom. These elements are mainly of two types:

- *The frame and bed.* They are the static part of the machine. The frame constitutes the main body of the structure, in which the bed is the solid base of the machine. The frame is usually composed of several bolted or welded components.

- *The structural components*. They are the movable parts of the machine structure, and are linked in different frame configurations. The interfaces among elements with relative motion must be as stiff as possible and highly damped along the perpendicular direction to the sliding one.

For an analysis of the advantages and drawbacks of different machine structures, it is useful to divide the structural design approaches into two groups: the *open-loop* configurations (examples in Chap. 1, Figs. 1.25 and 1.27) and the *closed-loop* configurations (Chap. 1, Figs. 1.9b and 1.22). The main conceptual difference between them is that in the open-loop case, process and inertial forces are conducted to the ground through in just one structural way. In the closed-loop concept, forces are conducted to the ground in several ways.

For the same machine strokes, the main advantage of the open-loop concept with respect to the closed-loop one is the easy access to the workzone and the lower cost. Otherwise, the closed-loop concept presents a higher stiffness at the tool tip and symmetrical behaviour with respect to thermal and mechanical loads.

Finally, the main structural components in an open-loop concept are the bed, table, column and ram, linked with sliding guideways. In the closed-loop concept, the main structural components are the table, bridge and guideways. Any of the involved components in any of the concepts can be either static or movable, i.e., the column will be movable in some machines and static in others, and the same for the rest of the structural components. Indeed, the different relative motions among these structural components will define the different machine architectures and configurations.

2.2.1 A Comparison Among Different Machine Configurations

Apart from the general characteristics associated with the closed-loop and open-loop structures mentioned above, there are few *a priori* rules that state which is the best option to select for a specific architecture and to define the relative motions between components. There are some general rules detailed here that can help the designer to select the optimal machine configuration:

- Symmetrical configurations lead to lower temperature gradients as well as to reduce bending moments.
- The overhang of cantilever-type components must be as short as possible. Indeed, cantilever components are the most critical components from a mechanical point of view, since they generate *Abbe* type errors (see Sect. 6.2.3), in which an angular compliance is amplified as a linear error by leverage effect. Several machines include a ram element, which is always a source of flexibility (machines in Chap. 1, Figs. 1.21 and 1.33). The overhang of this moveable component depends on the point to be machined, so this is a variable stiffness element.

- The path length from the tool to the workpiece through the structure should be minimal with the aim of minimising thermal and elastic structural loops [5]. The less material that separates the parts of the structural machine components that hold the tool and the part the quicker the machine will reach a stable equilibrium.

With these basic rules, and taking into account that there is not a unique optimal architecture, the design task will focus on making the components of the selected configurations as stiff and light as possible, as well as other considerations such as the cost, the ease of assembly, the workzone accessibility, the footprint and the total space occupied by the machine.

Figure 2.2 shows the different architectures that a machine builder considers when designing a hybrid heavy-duty milling and friction stir welding (FSW) machine. The company studied six architectures defining seven indicators to allow an even weighting of the drawbacks and advantages of each architecture. The seven indicators considered were the following: i) stiffness at the tool tip, ii) cost, iii) accessibility, iv) flexibility, v) homogeneous and symmetrical behaviour, vi) occupied space and vii) safety for the end user. Moreover, due to the extraordinarily high forces associated with FSW processes, which are more than the double the forces of heavy-duty milling processes, the machine builder has ranked as the "most important" the characteristics of stiffness, and after that, the characteristics of cost and risk, and has ranked as the "least important" the flexibility of the machine and the space that the machine occupies.

The main characteristics of the six considered architectures were:

1. *C Structure*, with a fixed column, and crossed slides on which a table moves in X and Y, and vertical ram in Z
2. *Fixed bridge* structure, with a movable table in X and crossed slides in Y and Z axes of the tool
3. *Fixed bridge* structure, with a movable table in X, a movable slide in Y and a vertical embedded ram in Z
4. *C Structure* with a fixed column, a movable table in X, and a movable embedded slide in Y and vertical ram in Z
5. *Travelling bridge* structure, with fixed table, travelling bridge in X, embedded movable slide in Y and vertically movable crossbeam in Z
6. *Fixed bridge* structure, with movable table in X, movable embedded slide in Y and vertical ram in Z

The main conclusion of this comparison study was that number 5 produced the stiffest architecture. The most economical solution was produced by number 1, which also provided the most workspace accessible solution. Furthermore, the most homogeneous behaviour was seen in solutions 5 and 6, both of which also provide the safest solution. Finally, there are no notable differences as regards the room that each solution occupies.

Table 2.1 summarises, using a ranking system 1 to 4 (where 1 means a "poor performance" and 4 means an "excellent performance") the main results of this study:

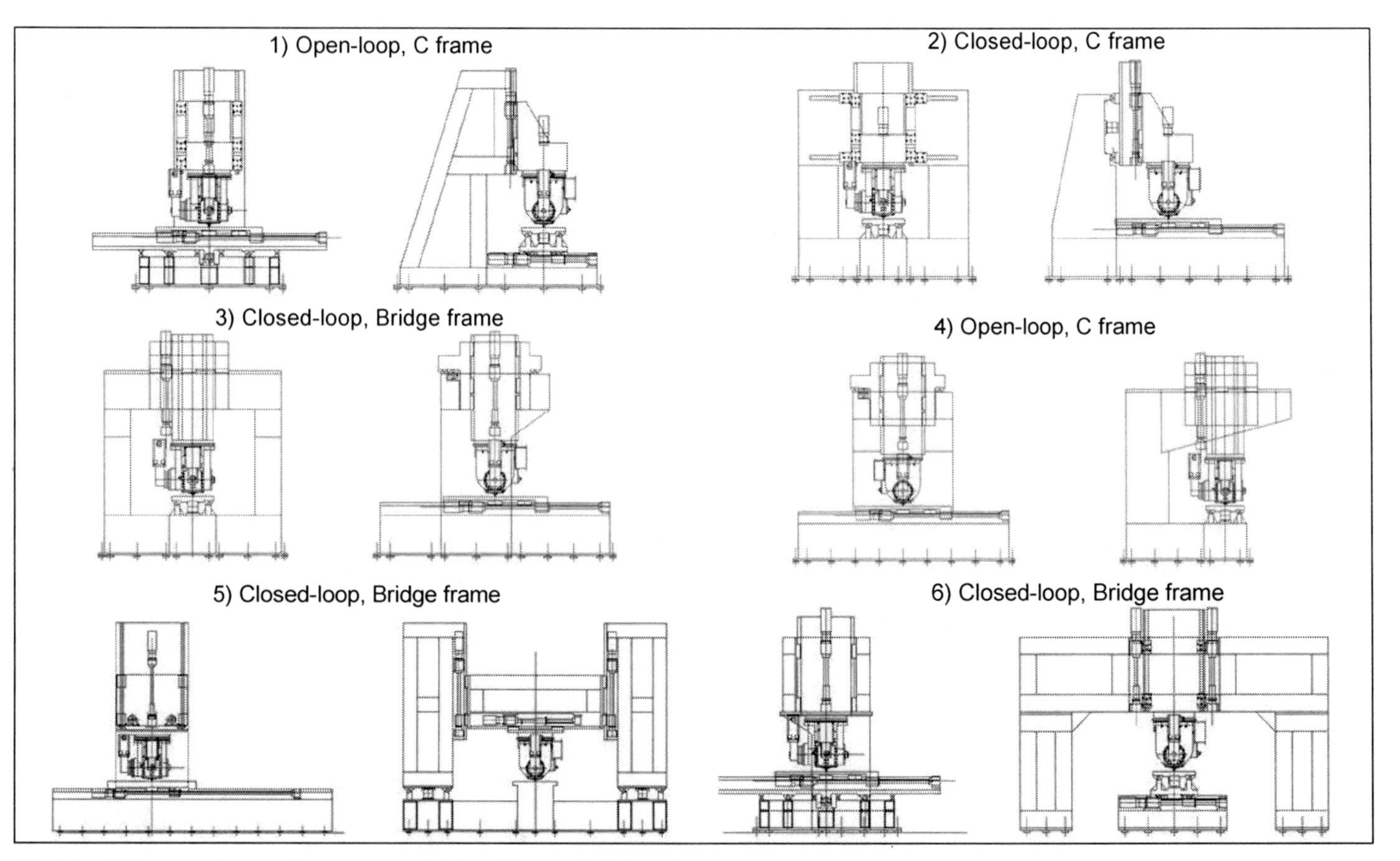

Fig. 2.2 Different machine architectures for a hybrid milling and FSW machine

Table 2.1 Comparisons among different machine architectures

	Machine architecture					
	1	2	3	4	5	6
Stiffness	1.0	2.0	2.5	2.0	4.0	1.0
Cost	4.0	2.0	1.5	3.0	2.0	2.5
Accessibility	4.0	1.0	1.0	2.0	2.0	2.5
Flexibility	3.0	1.0	1.0	1.0	1.0	3.0
Homogenous behaviour	3.0	1.0	2.0	1.0	4.0	4.0
Occupied room	2.0	2.0	2.0	2.0	2.0	2.0
Risk	3.0	1.0	2.0	1.0	4.0	4.0

In the case of selecting adequate machine architecture for a heavy-duty milling and FSW process and taking into account the high importance of stiffness for these processes, the selected architecture was the 5th.

2.2.2 Structural Components in Machine Structures

Figure 2.3 shows two machines built by the machine tool builder Nicolas Correa®. The drawing in Fig. 2.3a depicts a *C type* machine: an asymmetrical, open-loop machine, with a fixed column, a movable table in X, a movable outer slide in Z and a movable ram in Y. The drawing on the right depicts a *Bridge type*, symmetrical, a closed-loop machine, with a fixed bridge, a movable table in X, a movable outer slide in Y and a vertical ram in Z.

Despite their remarkable differences in dimensions and occupied room, their stiffness values at the three axes are somehow similar. Therefore the information of interest consists in knowing the most compliant elements of the machine, because as structural components are mostly in serial, the total compliance of the machine will be larger than the most compliant component. Moreover, when reinforcing the machine, the maximum effect is achieved when the most compliant component is acted upon.

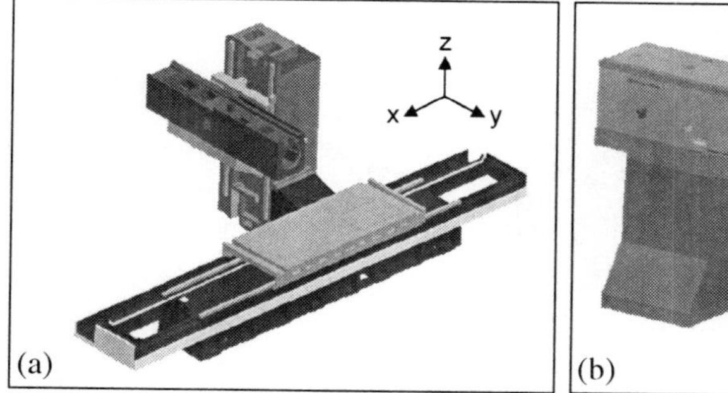

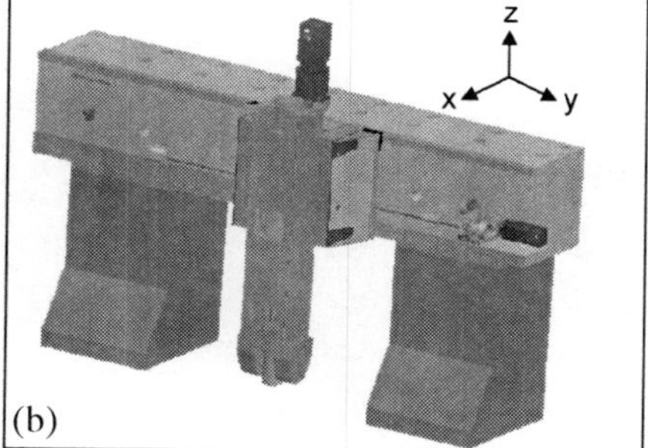

Fig. 2.3 Two machines of Nicolas Correa®: **a** C-type. **b** Bridge type

Table 2.2 Comparisons among different machine architectures

C-type machine				Bridge type machine			
Component	% Static compliance			Component	% Static compliance		
	X	Y	Z		X	Y	Z
Ram	49	27	43	Ram	53	42	26
Slide	16	12	24	Slide	10	30	18
Column	24	31	19	Bridge	35	26	46
Others	11	30	14	Others	2	2	10

Table 2.2 shows the distribution of compliance among the different structural components of the machine.

The data in this table confirms that concerning the static compliance of machines, the ram is the most critical component in almost all directions and architectures. There is also a notable influence from the column with respect to the bending directions of C-type machines. Therefore, designing optimised rams and columns for the case of C-frame machines is an issue of special relevance.

The following section will analyse the main rules for designing robust, stiff and lightweight rams and columns.

2.2.3 Robust Rams and Columns

Rams and columns have in most cases a square section to achieve a symmetrical and balanced behaviour concerning bending and torsion resistance. As the length of these components is defined by the strokes of the machine, there are two important aspects in designing rams and columns: 1) to define the appropriate thickness for their outer walls, and 2) to appropriately place internal ribs as a means of reaching an optimised stiffness-to-mass ratio.

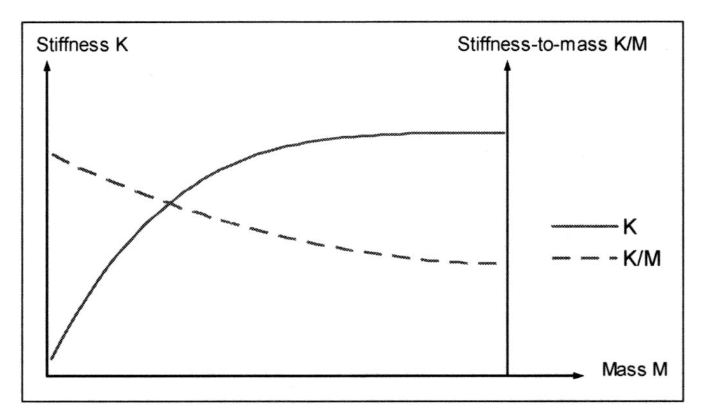

Fig. 2.4 Influence of wall thickness (and mass) on ram stiffness

With respect to the thickness of the walls, maintaining as a constant the outer dimensions of the beam, increasing the thickness of outer walls involves an increase of stiffness that is lower than the increase of mass, so that excessively thick outer walls are not an optimal solution from the eco-efficiency point of view, as Fig. 2.4 shows. Thickness and mass are directly related.

With respect to ribs, they can be either longitudinal or transversal. Longitudinal ribs are not an optimal solution for rams and columns from the mechanical point of view, because for the same amount of mass, an appropriate increase of the thickness of the outer walls allows for achieving a higher stiffness both for bending and torsion loads. Therefore, longitudinal ribs are not an optimised solution according to the stiffness-to-mass ratio.

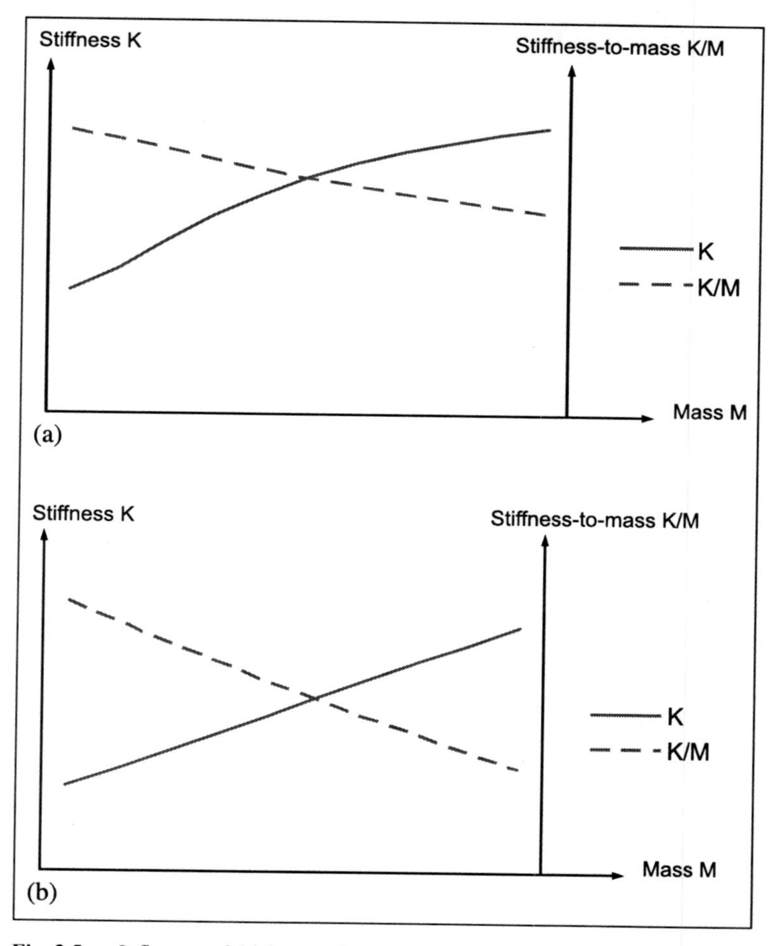

Fig. 2.5 **a** Influence of thickness of transversal ribs on stiffness of a ram. **b** Influence of number of transversal ribs on stiffness of a ram

Unlike longitudinal ribs, transversal ribs do improve optimally the mechanical behaviour of rams and columns concerning their bending and torsion resistance. Therefore, an important issue when designing columns and rams is to select the appropriate thickness of these transversal walls as well as the distance between these walls. As a reference, for an average machine ram with a square section, an optimal thickness for transversal ribs is 10 mm. There is also an optimal space between transversal ribs, which is of the order of 230–240 mm.

Similar to the case of longitudinal ribs, either an excess in the amount of transversal ribs or an excessive thickness of theirs will increase the total stiffness of the ram or column at a lower pace than the involved mass, as shown in the charts of Fig. 2.5.

2.3 Structural Optimisation in Machines

A machine tool is a spatial manipulator that is going to support high loads due to either the cutting forces or the inertial loads. The optimisation of the machine constitutive structures is an important task for designers.

2.3.1 Mechanical Requirements for Eco-efficient Machines

When designing the structural components of the machine, the target is to minimise the amount of mass of moving machine structures while static and dynamic mechanic properties maintain the desired threshold values.

As explained in the chapter introduction, the reduction of moving mass has an extraordinary impact on the environmental impact of a machine. Therefore it is worthwhile passing from a strategy based on "the stiffer the machine, the better" to a strategy based on "the lighter the machine, the better". Since lighter machines mean less stiff machines, the latter strategy requires the definition of minimum "threshold values" for the static and dynamic stiffness, so that an optimal balance between eco-efficiency, productivity and accuracy can be achieved.

With the aim of contributing to the definition of those threshold values for stiffness, Table 2.3 shows the average process forces associated with milling operations when cutting AISI 1045 steel, which have been experimentally measured on

Table 2.3 Average forces and acceptable deformations for different milling operations

Process	Average force in feed direction		Acceptable deformation
Roughing	Conventional tools:	1,500 N	Average: 100 μm
	100–125 mm. diam. tools:	3,000 N	
Semi-finishing	Conventional tools:	1,000 N	Average: 50 μm
Finishing	Conventional tools:	200 N	Average: 10 μm

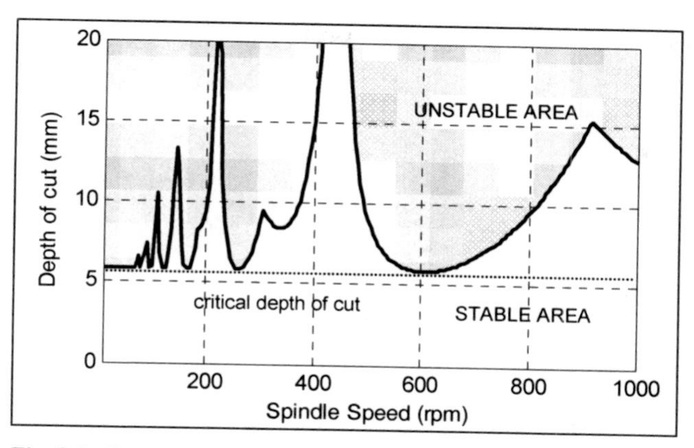

Fig. 2.6 Example of a stability lobe diagram

different machines and workshops. Table 2.3 also collects the accuracy requirements that machine end-users have provided in the survey reported in [7].

Combining the experimental forces and the average requirements coming from the end users, it was stated that a static stiffness of 20 N/μm at the tool tip can be considered as a current typical threshold value for general-purpose milling machines.

In addition to the requirement for static stiffness, dynamic stiffness is also an issue of special concern. Indeed, it is the primary reason to define the dimensions and shapes of the machine tool structural components. For an accurate definition of the threshold values for dynamic stiffness, the best utility available is the stability lobes diagram.

The stability lobes diagram is a plot that separates stable and unstable machining operations for different spindle speeds. Stable cuts occur in the region below the stability boundary, while unstable cuts (*chatter*) occur above the stability boundary [3]. An example of the stable and unstable regions is shown in Fig. 2.6. In Chap. 3 the current methods to obtain these diagrams are explained. Stability lobes are functions of the dynamic stiffness at the *tool center point* (TCP), the tool geometry, the radial immersion of tool into material, as well as of the material to be machined.

Machine manufacturers and users distinguish between two types of chatter, the so-called "structural chatter" or "machine chatter", which is a low frequency chatter, and the so-called "tool chatter", which appears at higher frequencies. In the former case, recognisable for a low-pitched sound, the chatter is associated with the structural modes of the machine, shown in Fig. 2.7. In the latter case, recognisable for a high-pitched sound, the chatter is associated with the modes of the tool or spindle. The type of chatter depends on the cutting frequencies; low cutting frequencies excite structural modes and high cutting frequencies excite spindle-toolholder-tool modes.

Therefore, the aim of an optimised machine is to ensure an acceptable critical depth of cut in every cutting direction, using the minimum movable mass. As a conclusion, the "minimum static stiffness" mentioned at the beginning of this section and the "minimum critical depth of cut" mentioned just above are the two thresh-

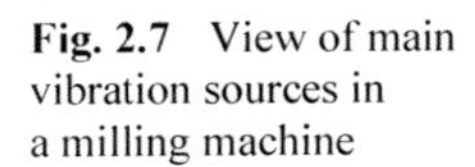

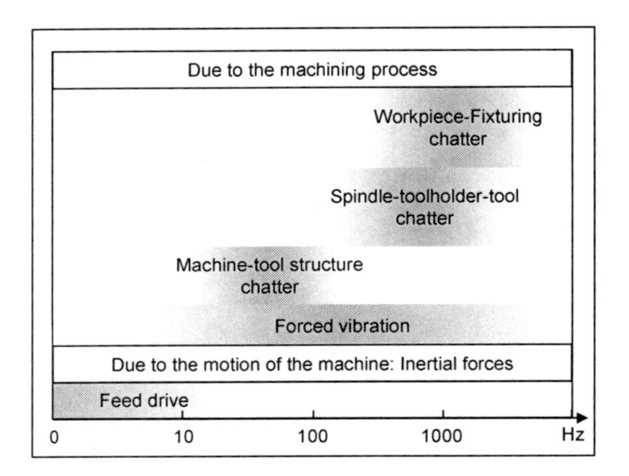

Fig. 2.7 View of main vibration sources in a milling machine

old values to consider in the machine design. That assures that an eco-efficient and lightweight machine is, at the same time, productive and accurate enough.

The machine mechanical modelling is a key to estimate the static and dynamic behaviour of a machine. Finite element method (FEM) based models are widely used, and will be explained over the course of the next section.

2.3.2 FEM Modelling

FEM aims at an approximation of the actual behaviour of a mechanical structure by assembling discrete and simple elements through nodes. The mechanical analyses that are conducted on a machine tool by means of FEM analyses commonly cover three stages:

1. The analysis of the static deformations and strain in a structure that is bearing static forces.
2. The analysis of natural frequencies and modes of the machine structure.
3. The analysis of the dynamic stability of cutting processes by means of analytical stability diagram lobes. The FEM can be used here as a calculation tool for the assessment of the *frequency response function* (FRF), instead of using the experimental modal analysis.

With FEM, the most complex aspect is that related to contacts between structural components along the degrees of freedom, in which stiffness, backlash and friction have a decisive influence on the damping of the machine. More than 90% of damping in a machine tool comes from bolted joints and from contacts in guideways [4]. In this respect, experimental modal analysis allows for the measuring of the natural frequencies and modes of an already-built machine, and above all allows for the measuring of their associated damping coefficients, which is the most difficult mechanical parameter to estimate. The experimental dynamic parameters allow the calibrating of the FEM models that were developed in the design phase.

The updated FEM models can be used to calculate the analytical stability lobes of the machine. Figure 2.8a shows the actual (left) and FEM modelled (right) FRFs of a milling machine at the TCP. Figure 2.8b and c show the stability lobes associated with previous FRFs for the downmilling case 45° and 225° with respect

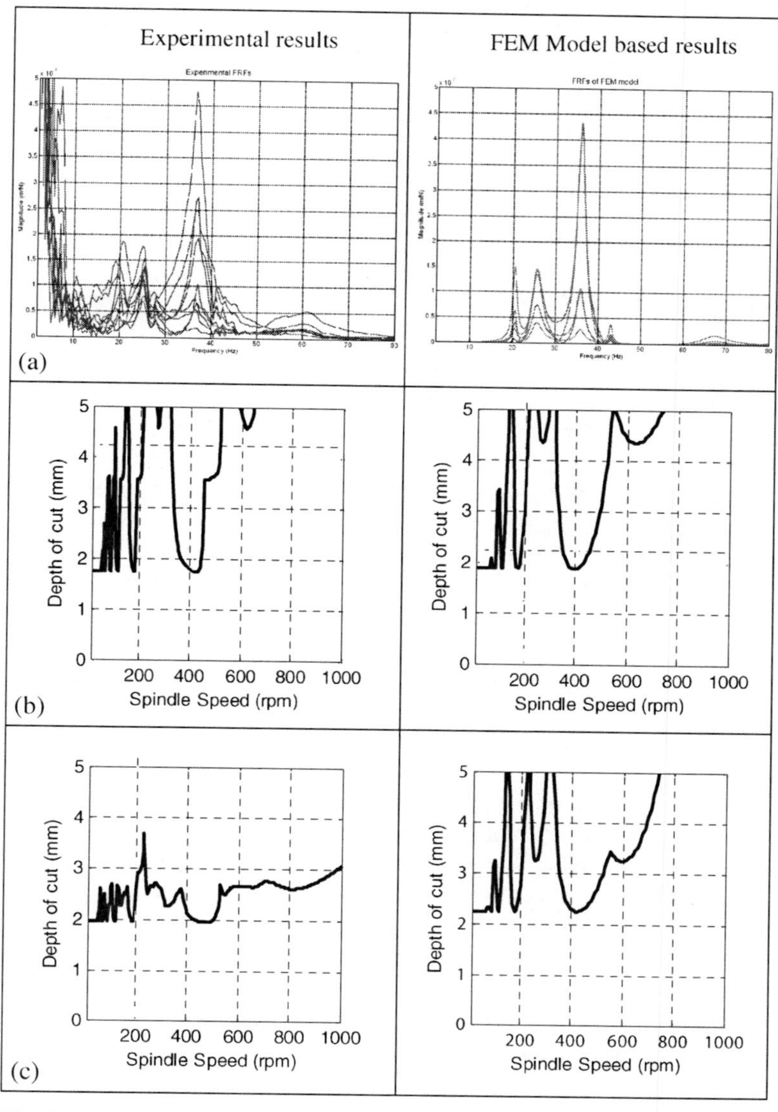

Fig. 2.8 **a** Experimental and FEM FRFs. **b** Corresponding stability lobes in the feed direction 0°, in up-milling. **c** Corresponding stability lobes in the feed direction 270°, in up-milling

to the machine axes. The experimental data shown in the left part of the chart have been achieved for a specific tool and a specific material to be machined. The right part shows the same stability lobe diagrams, with the only difference being that in this case data has been achieved from a FEM-modelled machine, in which the damping associated with each mode has been obtained from an experimental modal analysis on an already built machine.

As is shown, there is a good approximation in the 0° case and a 30% error in the 270° case. Otherwise, both the modelled-FRF and stability lobes diagrams show an acceptable level of coincidence with experimental-FRF and stability lobes. The advantage associated with modelled diagrams is that the machine models enables an evaluation of the effect of mechanical and architecture changes on the stability of cutting processes. Thus, it is possible to test several design approaches to lighten a specific machine and at the same time to validate that the aimed values of productivity are also achieved.

2.3.3 Topological Optimisation

As shown in previous sections, static stiffness and the critical depth of cut within stability lobe diagrams offer quantitative and objective parameters that allow a comparison of the performances among different machine concepts and above all, allow defining minimum thresholds to be fulfilled by eco-efficient machine tools. In fact, a reduction of masses that at the same time allows surpassing the threshold values defined for those parameters will allow an optimisation for that machine tool from the point of view of the eco-efficiency.

Focusing on reducing the movable masses, the topological optimisation of structural components is a critical issue, because this optimised mass-to-stiffness ratio will be the key to tackle effective strategies to reduce the total masses.

With the aim of supporting this objective, there are commercial programs with optimisation algorithms that starting with a given structural component remove material on that component until no further removal is possible without deteriorating the static and dynamic properties, achieving thus a topologically optimised component. What is more, an important additional objective of topological optimisation programs is to assure that the topologically and dynamically optimised structure can be manufactured in an economic way.

As an example, Fig. 2.9 shows the optimisation result for C-frame machine.

This typical structure, with a fixed support on one end and forces on the other end, has been used to analyse the performance of optimisation tools. This example shows that the topologically optimised structure (Fig. 2.9c) maintains almost the same stiffness as the original one, but with much less material [2].

With regard to the manufacturability and the economical feasibility of these optimised structures, the truss-like structures, which are the most frequent results of these automated topological optimisations, present some difficulties. As an example, cavities inside structures are difficult or impossible to manufacture. To allevi-

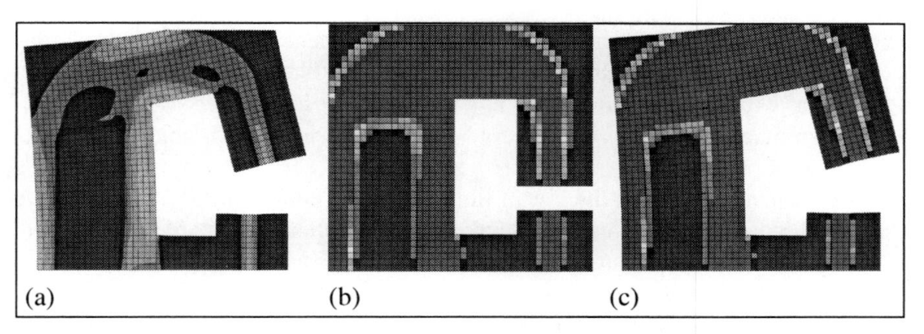

Fig. 2.9 Topological optimisation of a C-frame machine. **a** Stresses distribution map. **b** Results of the mass distribution to achieve the maximum stiffness. **c** Deformation of new distribution

ate this limitation, the automation programs include demoulding functionalities that make manufacturing possible, though the development of these parts will probably lead either to worse static and dynamic properties or to larger masses.

2.4 Structural Materials

When designing structural components to achieve eco-efficient, accurate and productive machines, the materials to be used play a key role in the final properties of components and those machines in which they are assembled. Structural materials have a decisive influence on movable masses, moments of inertia, static and dynamic stiffness and both the modal and thermal properties of the machine.

In the machine tool sector, the most commonly used materials are steel and cast iron, both of which offer an excellent stiffness-to-mass ratio as well as a good quality-to-price ratio. Nevertheless, there are materials whose properties can fit better to the specific needs of a concrete machine, as is explained over the course of the next section.

2.4.1 Involved Parameters

Regarding material properties, the most relevant features are shown, along with the influence on the behaviour of the machine:

a) Young's modulus E: High values of E have a positive influence on the static and dynamic stiffness of the machine.
b) Poisson's ratio ν and shear modulus G: High values for both are a positive influence on the machine torsional stiffness.
c) Density ρ: Low values of density within movable structures have a positive influence on the dynamic properties of the machine as well as on the bandwidth

of control loops, and at the same time high value density have a positive influence on the static elements of the structure, i.e., the base frames and beds.

d) Thermal expansion coefficient α: High values of α have a very negative influence on machine geometrical accuracy, so that the lowest possible value is desired for any case.
e) Specific heat capacity c: the low or high value of c neither positive nor negative per se. Indeed, high values of c make machines thermally stable to changing environmental temperature. At the same time, high values of c mean that machines take a long time to reach a steady state after they have been turned on, so that a trade-off is required between these two opposite effects. In this respect, the machine users usually prefer thermally robust machines when faced with changing environmental conditions, though in order to achieve stable conditions a longer period will be necessary. In such cases, a high value of c is desired.
f) Thermal conductivity k: Similar to the previous case, the low or high value of k is neither positive nor negative per se. Indeed, high values of k make machine temperatures become quickly homogeneous throughout the machine, thus avoiding partial and asymmetrical elongations in the machine. At the same time, high values of k make machines heat up in the presence of non-desired sources of heat such as motors, bearings *etc.* so that a trade-off is required between these two opposite effects. Machine users usually prefer thermally robust machines though it will lead to heat concentrations in the machine, so that in that case, a low value of k will be desired. One possibility to reach a trade-off between these two opposite effects is to have materials with low thermal conductivity k and in parallel to isolate heat sources or to evacuate heat by means of cooling systems.
g) Material and structural damping: High values of damping have a positive influence on the dynamic properties of the machine as well as on productivity, because high values of damping implies that stability lobes rise for a given cutting speed.

These properties are analysed for several materials and classified into two groups, the currently common structural materials and the innovative materials.

2.4.2 Conventional Materials for Structural Components

The most typical materials for machine tool structural components are, without any doubt, steel and above all cast iron. Steel is commonly used in welded structures; whilst for cast iron, the most common solutions are the sand casts obtained from grey cast iron and spheroidal graphite (ductile) cast iron [6]. Some parts such as headstock housings are made of cast steel.

The main advantages of these conventional materials are their low cost in comparison with other materials and their very good machinability, with possibility to

Table 2.4 Properties of materials based on Fe-C

	Steel	Grey cast iron	Ductile cast iron
Young's modulus	$2.1 \cdot 10^5$ MPa	$0.8–1.48 \cdot 10^5$ MPa	$1.6–1.8 \cdot 10^5$ MPa
Density	$7{,}850\ kg \cdot m^{-3}$	$7{,}100–7{,}400\ kg \cdot m^{-3}$	$7{,}100–7{,}400\ kg \cdot m^{-3}$
Damping ratio	0.0001	0.001	0.0002–0.0003
Thermal exp. coeff.	$11 \cdot 10^{-6}\ K^{-1}$	$11–12 \cdot 10^{-6}\ K^{-1}$	$11–12 \cdot 10^{-6}\ K^{-1}$

machine very accurate dimensions and geometric tolerances. Moreover, steel excels in terms of its high value of elasticity modulus and excellent mass-to-stiffness ratio, and the cast iron has a more than acceptable material-damping ratio, especially when compared to steel.

The main disadvantages are their relatively high thermal expansion coefficients, and in the case of steel, its very low material-damping ratio. Table 2.4 shows the main properties of these conventional materials based on Fe-C alloys.

2.4.3 *Innovative Materials for Structural Components*

In the sector of machine tools, materials other than steel and cast iron are not conventional for structural components, with probably some specific exceptions such as aluminium, granite and polymer concrete.

2.4.3.1 Polymer Concrete or Mineral Casting

Polymer concrete is a combination of mineral fillers graded according to their size distribution (flour, sand and different grits). The combination is bonded together using a resin system. Polymer castings are appropriate for precise machines due to their extremely low thermal diffusivity, which makes this material very stable and robust from the thermal point of view. Furthermore, its structural damping is similar to that of cast iron, although the Schneeberger® company claims its mineral casting achieves up to 10 × better values of vibration damping than steel or cast iron. Moreover, mineral-cast elements are resistant against oils, coolants and other aggressive liquids. This material includes different types of compounds and in the near future others will be developed, which will increase its application for machine tools.

2.4.3.2 Granite

Similar to the previous case, granite is appropriate for very accurate machines such as high precision milling machines and measuring machines, as a result of the excellent time stability of its properties and its good material damping (Table 2.5).

Table 2.5 Properties of polymer concrete and granite

	Polymer concrete	Granite
Young's modulus E	$0.4–0.5 \cdot 10^5$ MPa	$0.47 \cdot 10^5$ MPa
Density	$2{,}300–2{,}600\, kg \cdot m^{-3}$	$2{,}850\, kg \cdot m^{-3}$
Damping ratio	D = 0.002–0.03	D = 0.03
Thermal expansion	$11.5–14 \cdot 10^{-6}\, K^{-1}$	$8 \cdot 10^{-6}\, K^{-1}$

2.4.3.3 Fibre Reinforced Composites

Fibre reinforced composites have very high values of a specific modulus of elasticity and specific strength. Nevertheless, wider use of composites in machine tools is complicated due to some remarkable limiting factors such as their high price, their complicated joining and their complicated recycling. Indeed, the technical community is not familiar with common commercial machine tools with applied carbon fibre composites (CFCs), and in fact only some experimental research prototypes have been designed. Their mechanical properties in the fibre direction are collected in Table 2.6.

Table 2.6 Properties of CFC plates made from Prepreg[1]

	Middle modulus	High modulus	Ultra-high modulus
Young's modulus	$1–1.8 \cdot 10^5$ MPa	$1.7–2 \cdot 10^5$ MPa	$2–3.7 \cdot 10^5$ MPa
Density	$1{,}550–1{,}600\, kg \cdot m^{-3}$	$1{,}550–1{,}600\, kg \cdot m^{-3}$	$1{,}550–1{,}600\, kg \cdot m^{-3}$
Damping ratio	0.001–0.05	0.001–0.05	0.001–0.05
Thermal expansion	$12 \cdot 10^{-6} \cdot K^{-1}$	$12 \cdot 10^{-6} \cdot K^{-1}$	$12 \cdot 10^{-6} \cdot K^{-1}$

2.4.3.4 Hybrid Materials and Structures

Structures using hybrid materials are usually developed and designed for specific elements. Therefore, it is important to know first the exact functionalities of the part that is to be developed, and then to find proper topological shapes as well as a macroscopic combination of involved materials [6]. A practical strategy is to use low cost materials such as steel or cast iron for the majority part, and then to use a minimal amount of high-cost materials in order to tune the properties of the part by means of computational analyses.

In the field of machine tool structures, the following hybrid structures are already known:

- Steel weld structure with polymer concrete fill: As a reference, the Spanish machine tool builder Nicolas Correa® has developed a ram made of sandwich of steel and polymer concrete. This ram displays higher damping, that has been

[1] A technique of draping a cloth over a mould with epoxi preimpregnated into the fibres

noticed in the stability lobe diagrams. Moreover, the thermal diffusivity of the machine has decreased considerably, which was one of the aims of Nicolas Correa® with respect to their customers. Furthermore, though the mass-to-stiffness ratio of polymer concrete is worse than of steel, they have achieved a 20% ram mass reduction by means of topological optimisation rules.

- Steel weld structure with foamed aluminium filling: the same company has also developed a ram made of sandwiches of steel and aluminium foam, which has shown higher damping for the whole machine. Nevertheless, this increase of damping has been anyway lower than the increase achieved when using polymer concrete.

2.4.4 Costs of Design Materials and Structures

A global view of the purchasing costs of materials and semi-finished structures, such as welded structures, is shown in Table 2.7. The table summarises the values of minimum and maximum costs found on the European market. Moreover, a value called the "specific cost" has been added, which contains information regarding the material cost and the value of the specific modulus. This is an interesting piece of information significant to the design of eco-efficient machines, since the high life cycle cost associated with the machine production time, highly recommends the use of lightweight materials.

Table 2.7 Cost of materials and structures suitable for machine tool design

	Cost [€/kg]		Specific modulus [MPa/kg·m^{-3}]	Specific cost [€/MPa·m^{3}]	
Design materials	Min.	Max.	Average value	Min.	Max.
Steel-welded structures	3.5	7.0	26.5	0.13	0.27
Grey cast iron	2.0	4.0	16.0	0.10	0.37
Spheroidal graphite cast iron	3.0	6.0	23.5	0.12	0.28
Polymer concrete	2.0	5.0	18.5	0.09	0.33
Granite	3.5	6.0	20.5	0.14	0.36
Technical ceramics	10.0	36.0	107.0	0.07	0.47
CFC plate, middle modulus fibre	110.0	150.0	89.5	0.95	2.40
Hybrid materials & structures	21.0	130.0	62.5	0.21	5.20

2.4.5 The Influence of Innovative Materials on Productivity

The inclusion of materials with high internal damping is always interesting to increase the productivity of the machine from the point of view of the stability

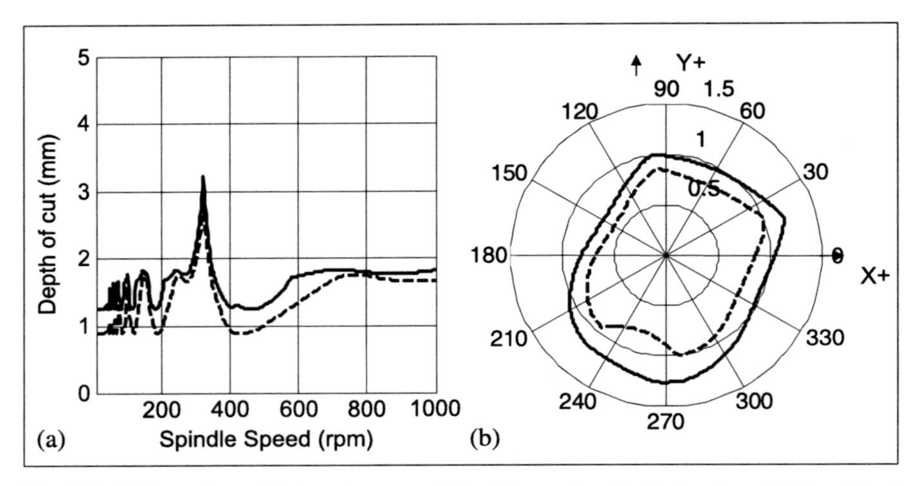

Fig. 2.10 Effect of material damping on stability lobes. **a** Feed in Y-direction. **b** Polar plot of the critical depth of cut at every direction

lobes diagram. On the other hand, the balance of internal energy dissipation is low with respect to the total energy dissipation of the machine.

As a reference, Fig. 2.10 shows the stability lobes of a machine with a structural ram bending mode of 37 Hz. The initial ram, made of welded steel, was substituted by a steel-polymer concrete sandwich. After this material substitution, the machine stiffness as well as the total mass remained almost the same, whilst the damping increased due to the effect of the polymer concrete. The stability lobe diagrams show that for a tool with a 125 mm diameter, the increase of damping associated with the polymer concrete allowed an increase in the critical depth of cut in all the speed range and for all feed directions (see Fig. 2.10b).

2.5 Active Damping Devices

In previous sections, the reduction of mass in movable components of a machine has been demonstrated to be a key point in the achievement of high eco-efficiency. Where static stiffness is concerned, the aim is to remove mass maintaining a threshold value of stiffness, and for dynamic stiffness, the aim is to maintain the productivity, limited by the stability lobe diagrams, in which the minimum values of the stability limits are dependent on the structural stiffness and damping ratio of the machine.

Within this view, an interesting machine tool design approach is to conceive it to be as light and low-stiff as possible, while at the same time increasing the damping of the machine to the equivalent proportion. Therefore, additional damping passive systems such as friction slides, viscoelastic materials and viscous fluids are interesting, as well as the use of *active damping devices* (ADDs). A typical ADD consists of a vibration sensor, an inertial actuator and a controller. ADDs are based on the principle that an acceleration of a suspended mass results in a reaction force towards the supporting structure. In order to tune the acceleration an embedded sensor monitors the supporting structure vibration; the sensor's readings are sent to an external feedback controller that drives the internal electromagnetic actuator of the ADD. As a result, these devices can damp the vibration modes that they observe in an open-loop transfer function [1].

2.5.1 The Implementation of ADDs to Machine Structures

Active devices add damping to the machine independently of its dynamic properties; therefore the correct implementation of ADDs to machine structures lies in defining how to locate the sensor and the actuator. In this respect, the best option for locating an actuator and a sensor is to know the FRFs of the machine at the *tool center point* (TCP).

From the FRFs, the maximum modal amplitudes to be reduced can be observed, which in most cases are associated with the compliance of the most flexible components such as rams and columns. The ideal place to locate the actuators is the *tool center point*, so that ADDs are placed as close as possible to it. Thus, in machines that have a movable ram, a good place to allocate ADDs is at the end of the ram close to the headstock, as illustrated in Fig. 2.11.

In Fig. 2.11, two ADD-45N of Micromega® have been placed on both sides of a horizontal ram to damp the two main ram bending modes well above 20%. More concretely, this additional active damping has allowed an average 2% of modal damping in the bending modes' rams to become an average 4% in terms of modal damping (see Fig. 2.12).

This data is very valuable when integrated in an FEM model of a machine, because if ADDs have enabled damping on modes above 20%, then the ram stiffness is able to be reduced by 20% maintaining machine productivity and therefore achieving a remarkable reduction in the total mass of the machine. In the following sections, this

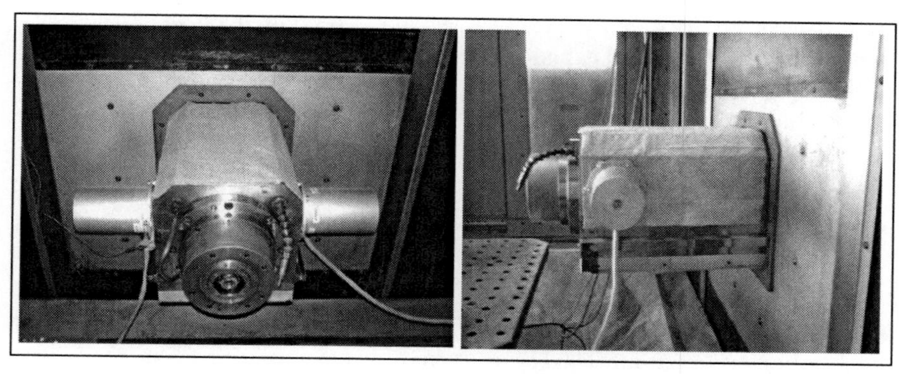

Fig. 2.11 ADDs placed on the ram of a milling machine, by Fatronik®

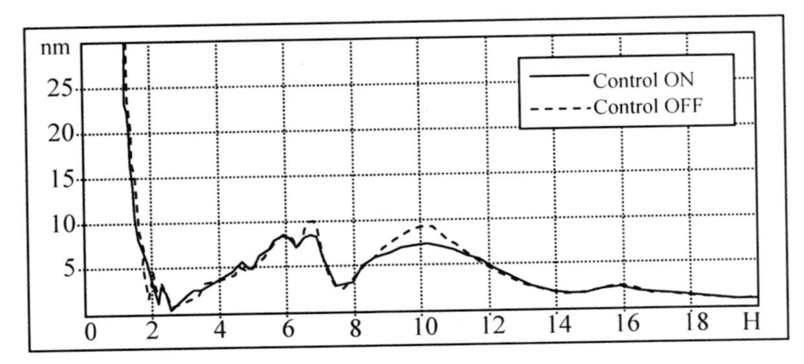

Fig. 2.12 FRF of a horizontal milling machine with and without active control

modelling of machines that incorporate passive and active damping devices are analysed thoroughly.

2.6 The Influence of New Structural Concepts on Productivity

Previous sections have analysed how the dynamic behaviour of the machine structure affects the dynamic stability of the cutting processes, especially in the 0–100 Hz frequency range, at which low-frequency modes of the machine structure can be detected. These structural machine modes can be modelled by means of either FEM or lumped-parameter models, allowing both models the calculation of the compliance and the FRFs at the TCP.

Though FEM models are more accurate, lumped-parameter models are more flexible, in the sense that they can provide qualitative information that can be very useful for designers in the machine conception phase.

Thus, taking as reference the machine whose FRFs were shown in Fig. 2.8, they showed a main mode at 37 Hz that a modal analysis confirmed that was associated to the bending of its horizontal ram. For that machine and that mode, a mechanical system of one degree of freedom has been used, on which modifications of mass, stiffness and damping can be applied. On the other hand, in order to obtain the stability lobe diagrams, a 125 mm diameter cutting tool with 9 edges that machines AISI 1045 steel has been considered.

2.6.1 The Influence of New Design Concepts for Structural Components

Taking as our reference the machine mentioned above, and focusing on the 37 Hz mode, which has been modelled as a system of one degree of freedom, the following design approaches may be conducted:

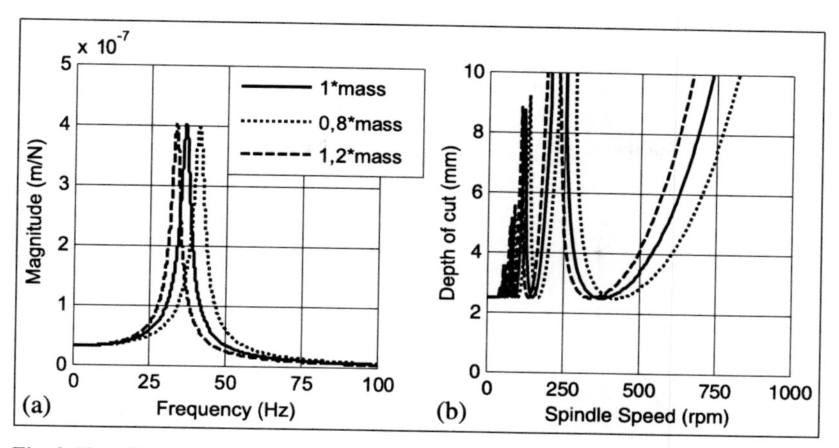

Fig. 2.13 Effect of variable mass and same stiffness on the following. **a** Direct FRF in X. **b** Stability lobes

If this design criterion is applied, the stability areas among lobes increases and the critical depth of the cut remains steady, so that the influence of this design concept is relatively low.

1. *To keep the stiffness associated to the mode, modifying the modal mass.* This makes the frequency response move horizontally, i.e., towards lower frequencies for higher masses and towards higher frequencies for lower masses (see Fig. 2.13). Its effect on the stability lobes is that they also move horizontally along the graph, and in the same direction as in the frequency domain.
2. *To maintain the mass associated with the mode and increase the stiffness associated with the mode.* This design concept increases both the natural frequency

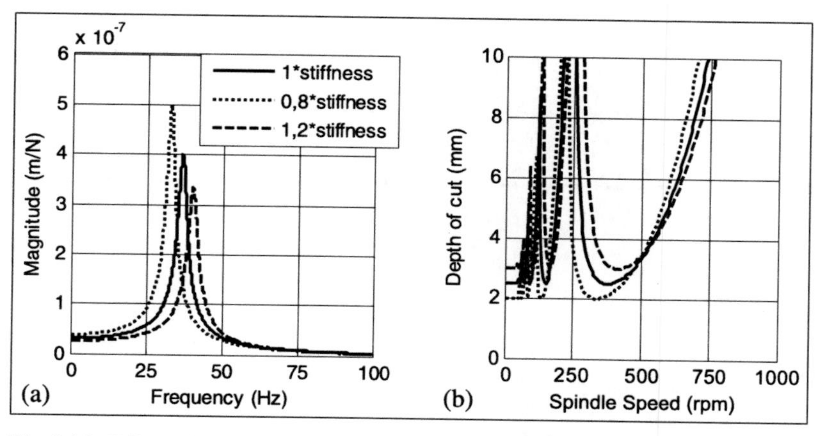

Fig. 2.14 Effect of variable stiffness and same mass on the following. **a** Direct FRF in X. **b** Stability lobes

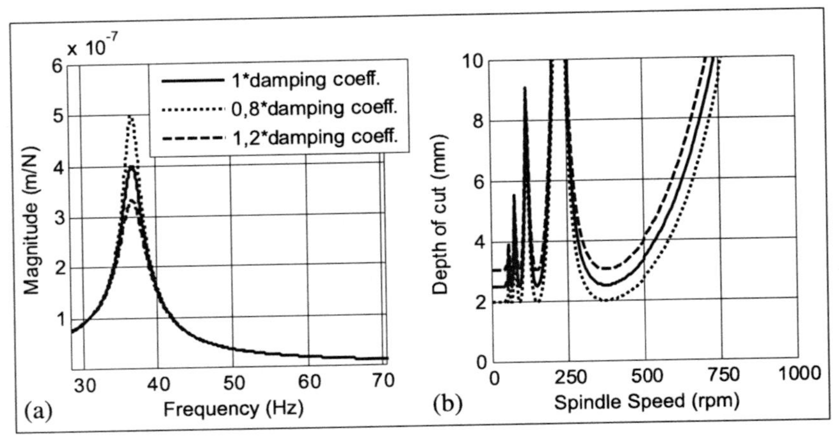

Fig. 2.15 Effect of variable damping on the following. **a** Direct FRF in X. **b** Stability lobes

and dynamic stiffness of the ram, and makes the stability lobes move horizontally towards higher spindle speeds and, above all, makes the lobes move upwards, so that the critical depth of cut increases (see Fig. 2.14).

When this design criterion is applied, the stability areas between lobes increases and simultaneously the critical depth of cut increases, so that this design concept allows for an increase in the productivity of the machine. In fact, the percentage increase in terms of stiffness is directly translated into an increase in the critical depth of cut. As an example, if the stiffness increases by 20% with the same mass, the critical depth of cut also increases by 20%.

3. *To increase damping associated to the modes.* This design concept reduces modal deformation at natural frequencies and causes the stability lobes to move vertically, increasing thus the critical depth of cut (see Fig. 2.15). Static stiffness does not change.

Table 2.8 Overview of the effect of the three design approaches on the dynamics of a 1 degree of freedom modelled machine

Design approach	Frequency, ω	Static deflection, δ	Max. dynamic amplification, D
Mass variation: $m = x \cdot m_0$	$= \left(1/\sqrt{x}\right) \cdot \omega_0$	$= \delta_0$	$\approx D_0$
Stiffness variation: $k = x \cdot k_0$	$= \left(\sqrt{x}\right) \cdot \omega_0$	$= (1/x) \cdot \delta_0$	$\approx (1/x) \cdot D_0$
Damping variation: $\xi = x \cdot \xi_0$	$= \omega_0$	$= \delta_0$	$\approx (1/x) \cdot D_0$
Nomenclature:			Basic formulation:
ω: Natural frequency δ: Static deflection D: Max. dynamic amplification	k: Stiffness m: Mass ξ : Damping ratio x: Variation ratio (per unit)		$\omega = \sqrt{(k/m)}$ $D \approx (1/2\xi) \cdot \delta$

Similar to the previous case, the percentage increase in damping is directly translated to an increase of the same percentage in the critical depth of cut. Therefore, it is possible to reduce the machine stiffness by a specific percentage, with its subsequent associated reduction of mass, maintaining the productivity of the machine.

The three approaches are summarised in Table 2.8, which provides an overview of the influence of the mass, stiffness and damping variation on the natural frequency, the static deflection and the maximum dynamic amplification.

Given the fact that the damping variation approach increases the productivity of the machines without a penalisation of mass, the method of active damping will be analysed thoroughly here.

2.6.2 The Influence of ADDs on Productivity

The influence of material damping is low on the total damping of a machine. Thus, in the case that a higher amount of damping is required to further increase the productivity of the machine, ADDs are an interesting option.

Figure 2.16 shows the influence of having placed the two ADD-45 of Micromega® shown in Fig. 2.11 in the direction X of an actual horizontal milling machine. As can be seen, the additional damping that these two actuators have added to structural modes has allowed a move in the critical depth of cut in all cutting directions, (the effect being especially high in the direction in which the actuators were placed). As a result, it can be seen in the diagram in Fig. 2.16a, that the critical depth of cut in direction X (0° in the polar diagram) has passed from the initial 1.7 mm (dashed line) to almost 8 mm (continuous line). The introduction of damping can be extended to the rest of the directions by placing additional ADDs on the ram of the machine.

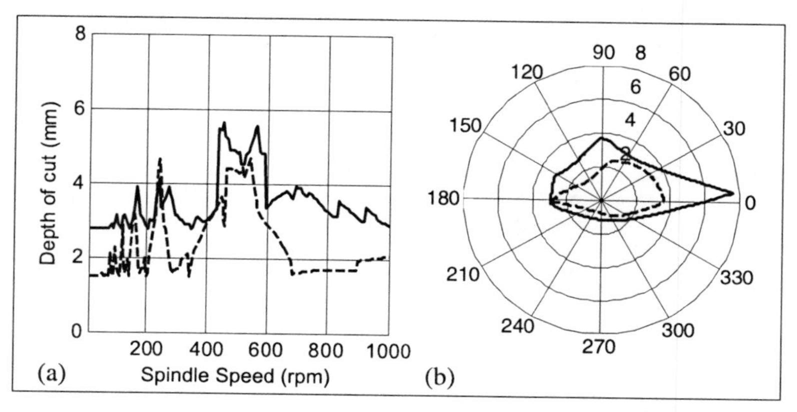

Fig. 2.16 Effect of active damping on stability lobes. **a** Feed in Y+ direction. **b** Polar plot of the critical depth of cut at every direction

2.7 Future Trends in Structural Components for Machines

The main trend in the field of structural components is the *de-materialisation* of their structural components, passing from stiffness-aiming machines to lightweight and robust machines, as seen in Fig. 2.17. This can be tackled mainly by integrating in the structural components active elements such as active piezo-layers, active damping devices and magneto-rheological fluids. In fact, these ubiquitous sensors and actuators will allow the passing from current machine conceptions based on mechanical stiffness to innovative machine conceptions based on mechatronic robustness, similar to axes levitating on magnetic bearings, which show a precise and robust behaviour despite not having any mechanical support.

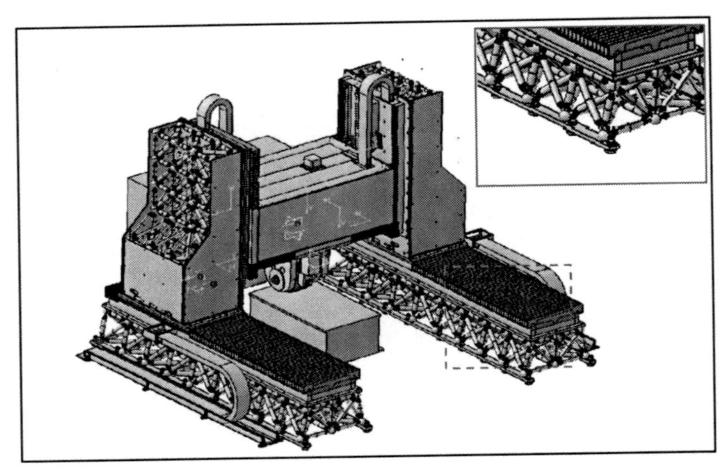

Fig. 2.17 Prototype with structural parts made with ball-bars frame, by Fatronik®

Finally, current machine specifications are defined as a trade-off between productivity, accuracy and surface quality. Within a view of sustainable machine tools and processes, this chapter has introduced a new triangle of specifications based on productivity, accuracy and above all, eco-efficiency.

Acknowledgements Thanks are addressed to Rubén Ansola for his contributions on the topological optimisation of structural components. Thanks are also addressed for the financial support of the Ministry of Education and Research of Spain (DPI2007-60624).

References

[1] ADD Catalogue (www.micromega-dynamics.com)
[2] Ansola R, Veguería E, Canales J, Tárrago JA (2007) A simple evolutionary topology optimization procedure for compliant mechanism design, Fin Elem Anal Design, 44:53–62

[3] López de Lacalle LN, Lamikiz, A, Sánchez. JA, Salgado MA (2007) Toolpath selection based on the minimum deflection cutting forces in the programming of complex surfaces milling, Int J Mach Tool Manufact, 47:388–400
[4] Rivin E. (1999) Stiffness and Damping in Mechanical Design, Marcel Dekker Inc.
[5] Slocum A (1992) Precision Machine Design, Society of Manufacturing Engineers
[6] Smolik J (2007) High Speed Machining 2007: Sixth International Conference, 21–22 March 2007, San Sebastian (Spain)
[7] Zulaika J (2006) Project NMP2-CT-2005-013989, Internal Report D1.1

Chapter 3
Machine Tool Spindles

G. Quintana, J. de Ciurana and F. J. Campa

Abstract This chapter deals with spindle technologies for machine tools. The machine tool spindle provides the relative motion between the cutting tool and the workpiece which is necessary to perform a material removal operation. In turning, it is the physical link between the machine tool structure and the workpiece, while in processes like milling, drilling or grinding, it links the structure and the cutting tool. Therefore, the characteristics of the spindle, such as power, speed, stiffness, bearings, drive methods or thermal properties, amongst others, have a huge impact on machine tool performance and the quality of the end product. Machining requirements differ greatly from one sector to another in terms of materials, cutting tools, processes and parameters. Nowadays, the spindle industry provides a large variety of configurations and options in order to meet the needs of different industries. Therefore, it is crucial that companies correctly identify their machining requirements and make well-informed decisions about which spindle to acquire. In this chapter, some of the main spindle characteristics that are the basis of a well-informed decision regarding spindles are introduced and discussed.

3.1 Introduction

Machining is applied to a wide range of materials to create a great variety of geometries and shapes, practically without restrictions on complexity. Typical work-

G. Quintana and J. de Ciurana
Department of Mechanical Engineering and Civil Construction, University of Girona,
Escola Politècnica Superior, Av/Lluís Santaló s/n, 17003 Girona, Spain,
{guillem.quintana, quim.ciurana}@udg.edu

F. J. Campa
Department of Mechanical Engineering, University of the Basque Country,
Faculty of Engineering of Bilbao, c/Alameda de Urquijo s/n, 48013 Bilbao, Spain
fran.campa@ehu.es

L. N. López de Lacalle, A. Lamikiz, *Machine Tools for High Performance Machining*,

piece materials are: aluminium alloys, cast iron, titanium, austenitic stainless or hardened steels, copper, carbon graphite and also plastics, woods and plastic composites. Machined components can be either simple forms with planes and round shapes or complex shapes. Two types of surfaces are usually defined: ruled surfaces, (e.g., for blades) and sculptured surfaces or free-form surfaces (e.g., for moulds and dies). This enormous number of combinations is able to meet the specific manufacturing requirements of a wide range of manufacturing industries, such as automotive, aerospace or the die and mould industry.

The machine tool spindle plays an important role in machining operations because it provides the cutting speed of the tool and is part of the force chain between the machine tool structure and the tool or the workpiece. The finished product is created by removing material from a blank workpiece with a cutting tool through the relative motion between the tool and the workpiece. The relative motion can be divided into a feed motion, provided by the drives of the machine, and

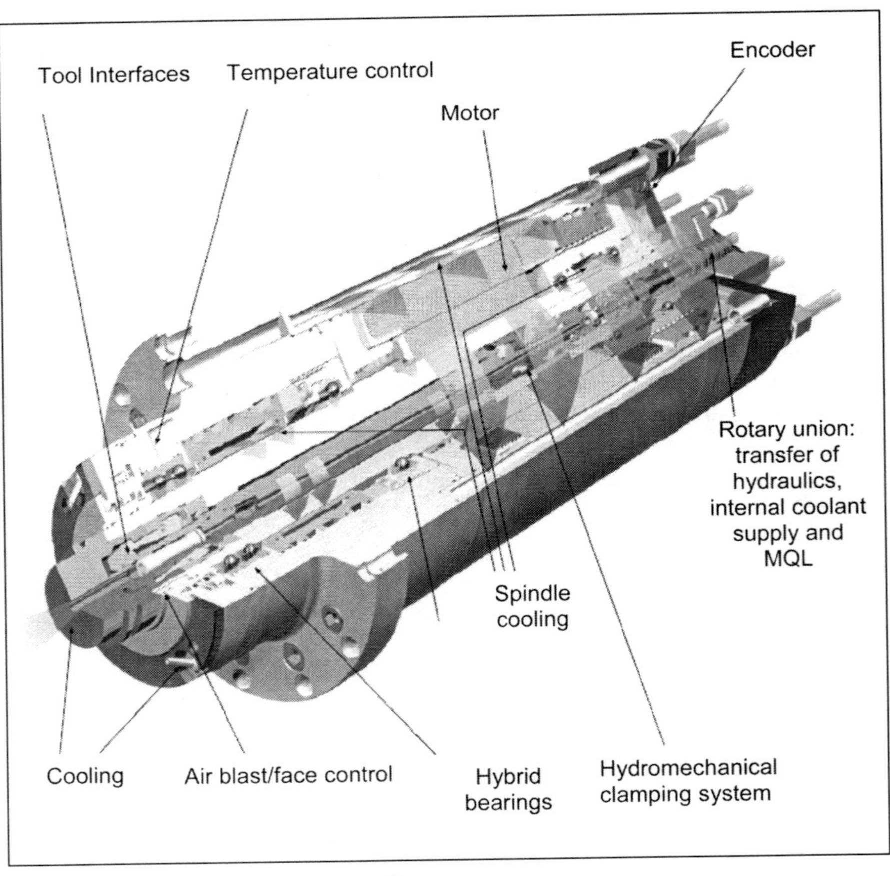

Fig. 3.1 Main parts of a spindle, by Edel®

a rotating motion provided by the spindle, which is responsible for the cutting speed that permits the material to be removed.

Each type of machining process (e.g., drilling, turning, milling, grinding, boring, etc.) has specific characteristics regarding feed rate and cutting speed. In basic turning processes, the spindle rotates the workpiece to provide the cutting speed and the cutting tool is fed by the drives to remove material. In drilling and milling, the spindle rotates a cutting tool with several cutting edges to provide the cutting speed. In drilling the feed motion moves in the direction of the spindle axis, and in milling, it generally moves in a perpendicular direction. In grinding processes, the spindle is also the element that provides the cutting speed to the grinding wheel.

The characteristic elements of a spindle are the tool interface, the drawbar, the shaft, the bearings, the driving system, the cooling system and the housing. There are several types of driving systems, basically with a motor coupled, directly or indirectly, to the spindle or with an integrated motor. Figure 3.1 shows the main elements of an integrated motor spindle.

The spindle industry provides many options and configurations to meet the requirements of each sector. As an example, Fig. 3.2 shows a comparison of the maximum speeds of the direct drive spindles and electrospindles typically used in automotive, mould and die, and aeronautical industries. Hence, the selection of the most suitable spindle is a complex task. In this chapter, spindle features such as motors, bearings, speed, power, stiffness and thermal behaviour are described and the basic criteria for a well-informed decision regarding spindles are presented.

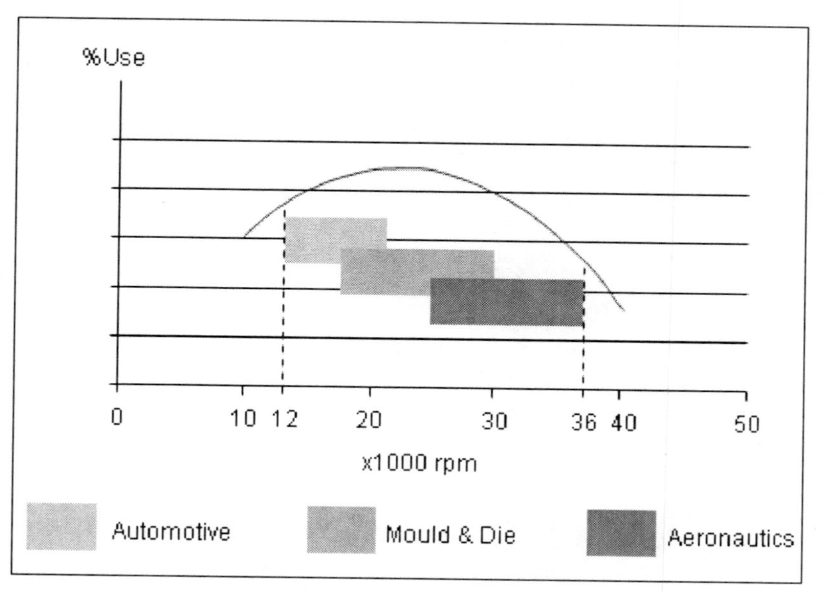

Fig. 3.2 Speed requirements of different sectors for electrospindles

3.2 Types of Spindles

The spindle drive is the mechanism that provides and transmits movement to the spindle. The drive consists of the motor and the coupling. In this way, the speed of rotation, the torque and the power are finally transferred to the cutting tool, via the toolholder.

In general, there are four types of spindle depending on the type of drives used, belt drive, gear drive, direct drive and integrated (built-in) drive. In Chap. 1, Fig. 1.17 shows an example of the belt drive, the direct drive and the integrated drive. Various characteristics have to be taken into account to evaluate the performance of spindles, for example:

- Transmission performance in terms of movement, force, torque, power and speed.
- Heat loss and expansion.
- Vibrations at various speeds.
- Noise.
- Others, such as maintenance and cost.

Below we briefly describe the common features of each type of drive and the most suitable applications for each one.

3.2.1 Belt-driven Spindles

This spindle setup transfers the movement of an external motor to the main spindle by means of a cogged or V-belt. It is widely used in conventional machining due to its low cost and good performance when it comes to transferring the nominal power of the motor into the useful power of the spindle. The efficiency of belt-driven spindles, in terms of transmitting motor power to the spindle is around 95%. This is a little less efficient than direct drive spindles (nearly 100%) but clearly better than gear-driven spindles (less than 90%) [4].

A belt-driven spindle can reach moderate speeds (15,000 rpm) and perform well or with high torque at lower speeds (1,000 rpm), depending on the belts and the transmission ratio. In contrast, at low speeds, gear drives transmit the torque better and at high speeds, direct drives are better (above all in situations where dimensional precision and surface quality requirements are high) because they produce less vibration and noise. However, as belt drives are very versatile, they are used for a wide variety of jobs whose requirements range between high torque/low rotation speeds and low torque/high speeds [10]. The main disadvantages of the belt-driven system are:

- They undergo significant thermal expansion, compared to other drive systems due to the constant contact of the belt.
- They are noisier due to the movement of the belt.
- The tensioning of the belt makes a radial force on the shaft that takes some of the available loading capacity of the bearings.

As the motor and spindle are separate, the housing and maintenance of this kind of drive is simple, although on the other hand, more space is required.

3.2.2 Gear-driven Spindles

Gear-driven spindles (see Fig. 3.3) can reach high torque at low revolutions and they characteristically have multiple speed ranges.

The gears, however, may cause vibrations which have a negative effect on the finished surface of the workpiece. In addition, as we mentioned above, they are less efficient when it comes to converting the nominal power of the motor into the cutting power of the tool, due to its constructive nature. Such power is lost as heat, with all the negative effects that this creates, such as reduced precision due to thermal expansion. For all these reasons, gear-driven spindles are unsuitable for very high velocity machining operations although very adequate for heavy-duty work.

Fig. 3.3 Haas EC-630 gear-driven spindle with two-speed gearbox: 610 ft-lb of torque for heavy machining or 6,000 rpm for finish cuts, by Haas Automation®

3.2.3 Direct Drive Spindles

Direct drive spindles have almost 100% efficiency in terms of transmitting power from the motor to the cutting tool. They can work at high rotation speeds but at lower torques.

Since there is no transmission chain, it is not possible to increase torque mechanically in response to reductions in motor speed. This drive system behaves well in terms of vibrations, which means high speeds can be reached and still achieve good surface finishing.

3.2.4 Integrated (Built-in) Drive Spindles

In these spindles, also called electrospindles, the motor can be a synchronous or an asynchronous electric motor that is integrated into the spindle structure between the front and rear bearings (see Fig. 3.4). In this way, vibrations and noise are reduced and work can be carried out at higher rotation speeds, from 15,000 rpm. This is why integrated spindles are very common in high speed machine tools.

Controlling heat transference within the spindle and the subsequent thermal expansion is a key factor for getting good performance from this kind of drive. With the motor inside the housing, the auxiliary system for removing heat is the first priority (see Sect. 3.5.4).

The great precision required for the assembly of these spindles and the necessity of auxiliary systems for cooling and monitoring, makes them very expensive, despite their excellent mechanical performance.

Fig. 3.4 Integrated spindle with the motor between the front and rear bearings

3.3 Spindle Configurations

The arrangement of the spindle in the machine tool is an important feature that is, in fact, related to the versatility and the final applications of the machine tool. We describe here the usual configurations of the spindle and also some alternatives.

3.3.1 Common Configurations: Vertical and Horizontal Spindles

The arrangement of the spindle depends on the purpose of the machine tool but, as a general rule, horizontal spindle configurations are used in machines with great power and palletised workpieces for a greater flexibility (see Fig. 1.23 in Chap. 1). One of the advantages of this configuration is that the removed chip falls away from the cutting zone. Otherwise, vertical spindles are generally used in machines with less power which need more accessibility, typically for machining dies and moulds; see Fig. 1.25 in Chap. 1. Both configurations can gain accessibility through the use of indexed angular headstocks, as the Huré-type for horizontal spindles, or the swivel-type for vertical spindles. These solutions are still widely used and they allow the tool to be orientated in fixed lockable steps to cut inclined surfaces, which allows the (five faces) of a cube, i.e., the workpiece, to be machined.

3.3.2 Machines with Rotary Headstocks

Many milling machines incorporate rotating axes in the headstock to provide more operation possibilities, so that the orientation of the tool can be changed continuously as required by the surface being machined; see Fig. 1.7 in Chap. 1. These machines, with 5 or more axes, can produce more complex shapes although they require more sophisticated computer numerical control (CNC) systems in order to control the direction of the tool according to the trajectories required. Figure 3.5 shows two classic examples of bi-rotary headstocks: the twist head and the swivel head. The twist head allows higher accessibility and positioning possibilities than the traditional 3 axis machines. However, these heads lack a high stiffness, so they are not suitable for high power applications due to their limitations in terms of accuracy and dynamics. In comparison, the swivel head with its closed design is a much stiffer solution and more indicated for high power machining.

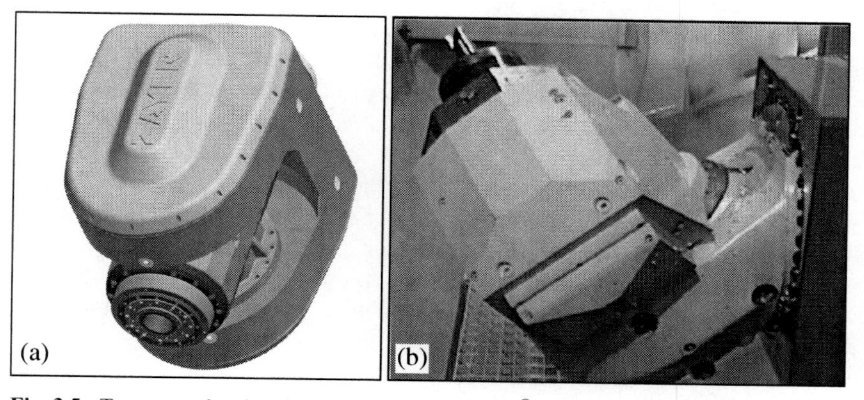

Fig. 3.5 Two rotary headstocks. **a** Twist head by Zayer®. **b** Swivel head by Droop + Rein®

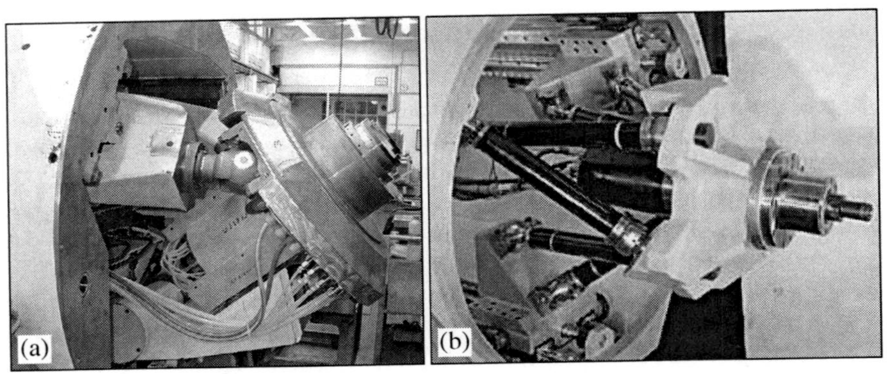

Fig. 3.6 **a** Detail of the DS Technologie® Sprint Z3 head and spindle. **b** Hermes head by Fatronik®

In the last two years the traditional use of gear transmission for the two rotary axes is being substituted for the use of the torque motor directly inserted in the rotary joints. Eliminating gears increases the stiffness and avoids backlash, with more precise rotation; see Sect. 4.5.

Over the last years, several alternative solutions based on the parallel kinematics mechanisms have been developed, such as the Hermes head by Fatronik® or the Sprint Z3™ head by DS Technologie®, as shown in Fig. 3.6. These solutions allow fast orientation of the tool up to ±40° in every position with minimal motion of the headstock.

3.3.3 A Main Spindle with an Auxiliary Spindle

In addition to the spindles described in the previous section, there are other configurations that combine a main spindle with an auxiliary spindle mounted alongside it in parallel, with different transmission systems to take advantage of the possibilities of each spindle according to the specifications of the job. For example, there are machines which combine the different qualities of belt drive and direct drive with a pair of spindles mounted together – one is the main spindle, the other the auxiliary – so it can be used in a combination of high speed/high power applications.

The main spindle is belt-driven since, as we have explained, this drive provides high power at low speed. This spindle would be suitable for roughing and semi-finishing with roughing mill tools of considerable diameters.

A direct drive spindle acting as an auxiliary spindle, completely integrated into the machine tool, can be coupled to the main spindle. The direct drive spindle will carry out low power/high speed operations. In this way, the auxiliary spindle will be ideal for high speed operations providing high quality surface finishing and dimensional precision, using smaller-diameter ball end tools.

3.3.4 Twin Spindles and Multi-spindles

Some machine tools can be equipped with two or more spindles also known as "twin spindles" or multi-spindles; see Fig. 3.7, and Figs. 1.15b and 1.32b in Chap. 1. The main advantage here is that two or more identical components can be machined at the same time in the same machine, thereby increasing productivity considerably. On the other hand, these machines usually lose dimensional precision in comparison with conventional machine tools. There are several reasons:

- The distances between the axes of the spindles are not exactly the same, due to the design and build of the machine.
- The diameter, length and wear of the tools in the twin and multi-spindles used are also variable.
- The temperature and therefore the effects of thermal expansion at specific points throughout the machine are not the same.

Fig. 3.7 Two images of the Nicolas Correa® Neptuno, an example of multi-spindle configuration

3.3.5 Automatic Head Exchange

In certain situations, automatic head exchange may be the best option; for example, when there are fluctuations in the production system requirements in terms of

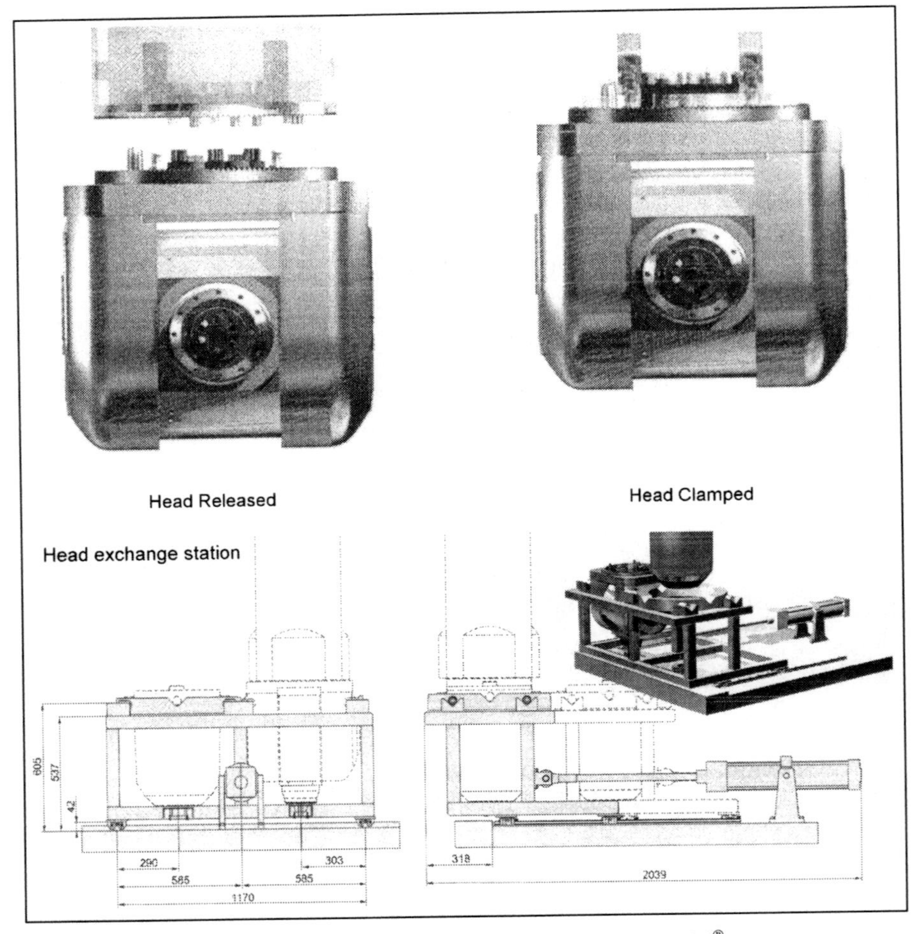

Fig. 3.8 The Cyport CP machine automatic head exchange system by Edel®

flexibility or productivity, demand for the machined parts, bottlenecks in the production plants or stoppages for maintenance. This system allows the heads of the machine tool to be changed automatically using hydraulic toolholder systems (i.e., hydraulic chucks). Figure 3.8 shows the automatic head exchange system of the Cyport CP machine by Edel®.

3.4 Basic Elements of the Spindle

The main components of a modern spindle are described in this section: a) the motors, which propel the spindle, b) the bearings, which allow the relative motion between shaft and machine, c) the shaft, d) the drawbar and e) the tooling system, f) the sensors, nowadays essential to monitor the process and diagnose spindle problems, and g) the housing which not only accommodates all the elements but also contributes to the cooling and refrigeration of the spindle itself.

3.4.1 Motors

The motors provide the mechanical motion to the spindle-tool system. They rotate the shaft to create relative motion between tool and workpiece and provide power and torque to the cutting tool in order to perform operations such as drilling, milling, and grinding.

Machine tool drives use electric motors that may be either direct current (DC), brushed and brushless motors, or alternating current (AC), and either synchronous or asynchronous motors. We will now describe the main characteristics of these different motors and their utility in the field of machine tools.

3.4.1.1 DC Motors

DC motors, whether brushed or brushless, are common in machine tools and although they are more frequently used as feed drives, they have also applications as main drives. Brushed motors, also known as conventional DC motors, use internal commutation, while brushless motors use external commutation to create the AC current from the direct current. Brushless motors have many applications in the field of machine tools, despite the fact that they are more expensive than conventional DC motors, since their design features provide greater performance and

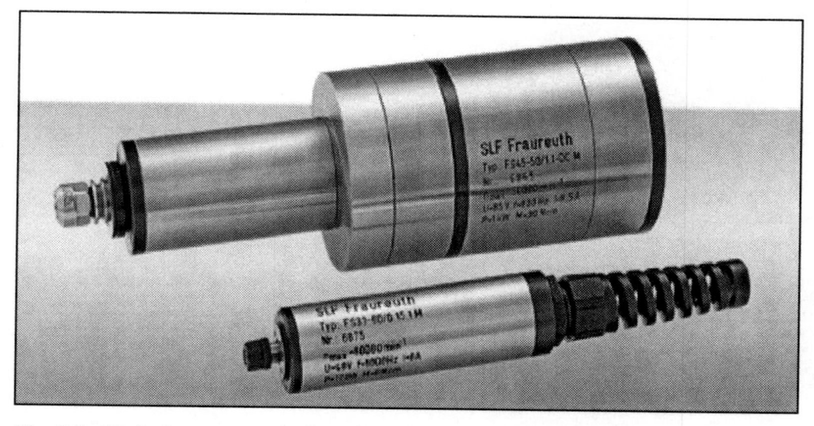

Fig. 3.9 High frequency spindle units FS45-50/1 Brushless DC and FS33-60/0.15 by SLF® (Spindle-und Lagerungstechnik Fraureuth GmbH)

efficiency. The main characteristic is that the change in polarity of the rotor can be carried out without brushes. In this way, efficiency is enhanced by reducing friction. Figure 3.9 shows two high-frequency spindle units used in applications for small CNC milling and engraving machines, the FS45-50/1 Brushless DC and the FS33-60/0.15 from SLF®.

3.4.1.2 AC Motors

AC motors convert alternating current into rotating torque. The AC motor consists of a stator, which is the fixed part that produces a rotating magnetic field by means of an alternating current, and the rotor, which rotates due to the rotating torque generated by the magnetic field. In machine tools, synchronous and asynchronous AC electric motors are currently used.

Synchronous AC motors are distinguished by having a rotor that spins synchronously with the oscillating field of the current that drives it. They have a constant rotation speed determined by the frequency of the grid to which they are connected and by the number of pole pairs of the rotor. The rotational speed N, also known as the synchronism speed, is easily calculated in rpm using Eq. 3.1:

$$N = \frac{f}{p} \cdot 60 \tag{3.1}$$

where f is the grid frequency (e.g., 50 Hz in Europe) and p is the number of poles. The main advantages of synchronous AC motors are that their position and speed can be accurately controlled with open loop controls and that speed is independent of the load, so it can be accurately maintained.

Asynchronous AC electric motors, also known as induction motors, differ from synchronous motors in that the latter have current supplied to the rotor which

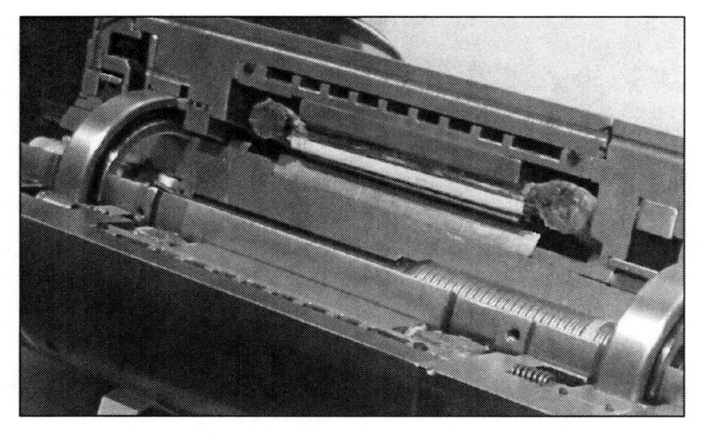

Fig. 3.10 Section of an AC motor of a machine tool spindle. The cavities for cooling can also be seen around the motor

creates a magnetic field. The type of rotor winding, slip ring motor or squirrel cage motor, conditions the motor behaviour. Figure 3.10 shows an AC asynchronous motor.

The rotating speed of these motors is managed by frequency converters, which are the electronic power stages that control the voltage and frequency provided to the motor. The "vector control" techniques have improved the performance of these motors, due to the improved speed control and reduced acceleration and deceleration times.

3.4.1.3 Torque Motors

Torque motors are direct drives based on permanently excited multi-pole brushless synchronous AC motors with a hollow-shaft rotor. As eddy-current losses increase with the number of pole pairs, the maximum number of pole pairs is limited, although they still have a relatively large number of magnetic pole-pairs. Consequently, torque motors are mainly designed for low speed applications, usually below 1,000 rpm.

They have relatively large diameter-to-length ratios and large outer and inner diameters, so it can be said that they are large hollow shafts with very low inertia. In terms of performance, they combine high torque levels, high dynamic responses due to small electrical time constants, high efficiency due to permanent magnets and high angular and dynamic stiffness. Finally, the large mechanical air gaps (0.5 to 1.5 mm) make mounting and aligning them easy. More information on torque motors is given in Sect. 4.5.2.

3.4.2 Bearings

Bearings are mechanical elements designed to reduce friction between an axle and its support. For this reason, the efficiency of the bearings is especially important for the spindle performance. Thus, although it is not a new mechanical element, bearings technology has been and is being studied comprehensively [13, 21].

The bearings used in a machine tool have to meet the demands of the spindle in terms of rotational speed, load capacity and life. That is the reason why rolling bearings are the most common solution. Using taper, roller, deep groove bearings or angular contact ball bearings in machine tool spindles depends on the application. Angular contact ball bearings combine precision, load carrying capacity and high rotational speed. On the other hand, taper bearings provide higher load carrying capacity and stiffness, but they do not allow high rotational speeds. The raceways of deep groove bearings are of a similar size to the balls that run in them. They can take heavy loads but may present misalignment problems. Roller bearings allow a certain amount of axial movement of the axis with respect to the support. They are suitable for large radial loads.

The manufacturing tolerances for bearings dimensions and characteristics are defined by the American Bearing Engineers Committee (ABEC). High speed spindles bearings meet ABEC 9 to have rotational precision and minimise runout.

The maximum speed at which bearings can operate is the $d{\cdot}N$ factor, also known as *speedability*, where d is the mean bearing diameter and N is the speed in rpm. A high speed spindle requires a $d{\cdot}N$ up to 1,500,000 and the only rolling bearings that can reach those values are the angular contact ball bearings. These bearings can operate under axial and radial loads. They have a nominal contact angle, typically 15° to 25° defined by the ball-raceways contact line and are perpendicular to the axis. In drilling, the spindle has higher loading in the axial direction, so higher contact angles should be selected. In milling, radial loading is higher so lower angles are required.
In order to operate correctly, bearings must be preloaded by applying a constant thrust load. The preload eliminates axial and radial play, increases the bearing stiffness, reduces runout, prevents ball skidding under high accelerations and reduces contact angle variation at high speeds. However, too much preload leads to excessive heat generation in running conditions and shortens the bearing life. The preload can be applied a) using springs, b) through axial adjustment or c) using duplex bearings. The first method uses a spring that loads the non-rotating raceway. It is the simplest and provides a constant preload. For high speed applications a hydraulic or a pneumatic system can be used to optimise the preload according to the rotating speed. Axial adjustment provides a fixed preload by using spacers or precision lapped shims. This method requires great precision to avoid excessive preloading due to the setup or the thermal growth in running conditions. Duplex bearings combine several angular contact bearings and the preload is given by the mounting configuration. With this method the bearings share the loads and an overall increase of stiffness is achieved. The bearings can be stacked in several ways; see Fig. 3.11, Fig. 3.12 and Fig. 3.13:

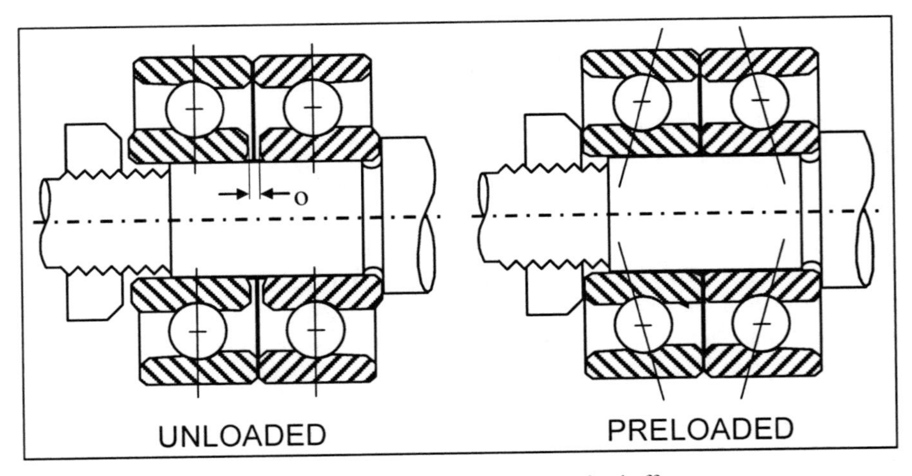

Fig. 3.11 Back-to-back bearing mounting; “o” is the total preload offset

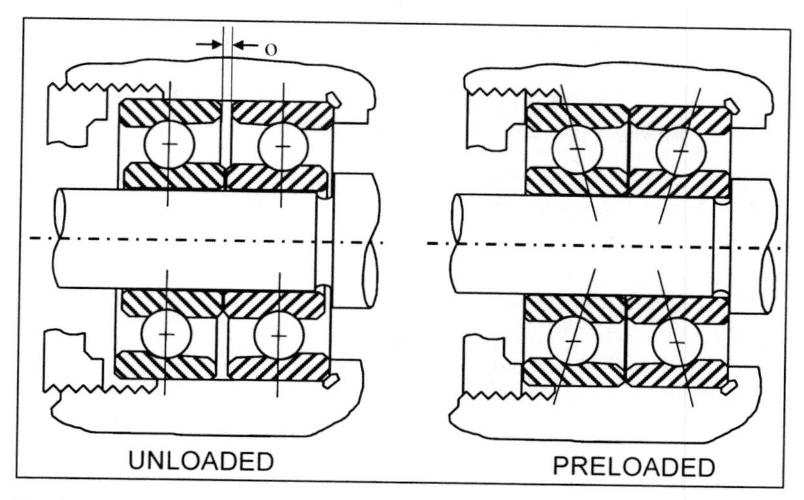

Fig. 3.12 Tandem bearing mountings; "o" is the total preload offset

- "Back-to-back" or DB mounting is the most common technique. The inner races are first relieved and then clamped together eliminating relief clearance. This is also called "O" mounting due to the shape of the contact lines. It is used when there is a good alignment between housing and shaft. With the correct preload, this method provides good accuracy and high rigidity.
- "Face-to-face" or DF mounting: Outer races are relieved and then clamped together. Also called "X" mounting due to the shape of the contact lines. It provides a higher radial and axial stiffness and good accommodation of misalignment. However, the speedability is lower than back-to-back mounting.

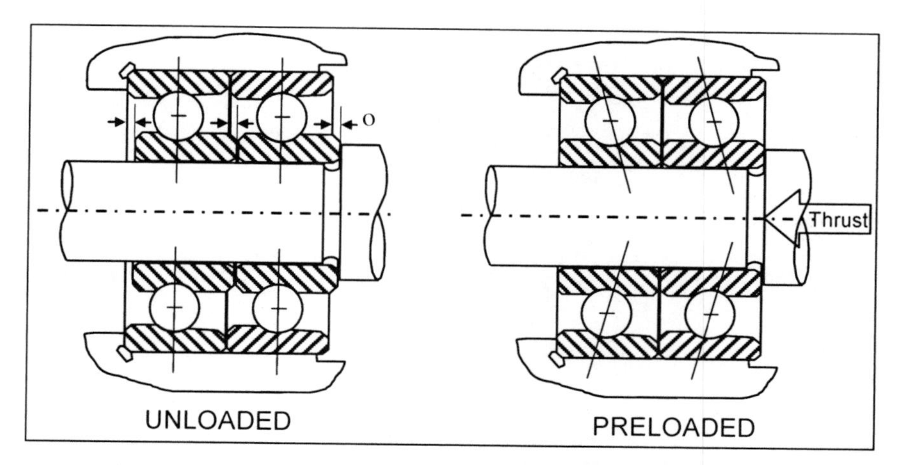

Fig. 3.13 Face-to-face and tandem bearing mounting; "o" is the total preload offset

- "Tandem" or DT mounting: the inner and outer races have proportional offsets, so the contact lines are parallel. This provides higher load-carrying capacity due to load sharing. Heavy trust loads are countered in one direction but reversing loads cannot be taken like DB or DF mountings.

If the spindle is going to operate under high loads, einstead of a set of two bearings, three or more bearings can be assembled following the mentioned arrangements.

Regarding the arrangement of the bearings in the spindle, the spindle housing and the shaft locate the bearings. Generally they are arranged as follows: at the nose of the spindle there are one or more pairs of angular contact bearings and, near the rear of the spindle shaft, there is another pair that may be angular contact bearings, deep groove bearings or roller bearings. In fact, a usual and simple configuration consists on locating a tandem pair at the spindle nose and another at the rear, set to a global DB setup, as shown in Fig. 3.14. However, the selection and

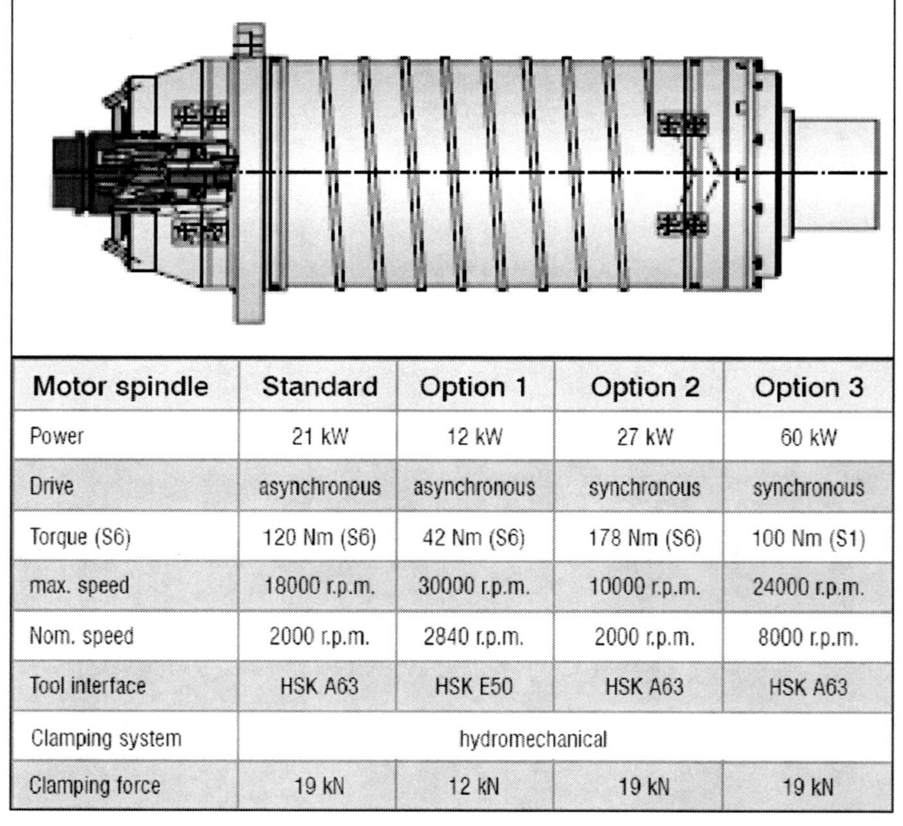

Motor spindle	Standard	Option 1	Option 2	Option 3
Power	21 kW	12 kW	27 kW	60 kW
Drive	asynchronous	asynchronous	synchronous	synchronous
Torque (S6)	120 Nm (S6)	42 Nm (S6)	178 Nm (S6)	100 Nm (S1)
max. speed	18000 r.p.m.	30000 r.p.m.	10000 r.p.m.	24000 r.p.m.
Nom. speed	2000 r.p.m.	2840 r.p.m.	2000 r.p.m.	8000 r.p.m.
Tool interface	HSK A63	HSK E50	HSK A63	HSK A63
Clamping system	hydromechanical			
Clamping force	19 kN	12 kN	19 kN	19 kN

Fig. 3.14 The spindle industry provides different spindle configurations (Courtesy EDEL®). Top, the configuration of angular bearings in the spindle: global DB configuration with DT arrangement in the front bearing, and DT in the rear bearing

location of bearings depends on the requirements of the spindle design and there are a huge variety of arrangements.

3.4.2.1 The Service Life of the Bearings

The common cause of bearing failure is the fatigue. The bearings' life is affected by the radial and axial loads, vibration levels, working speeds, temperature and lubrication. Generally, a spindle bearing reaches 5,000 to 7,000 working hours of good use. Now we will briefly describe the methods used in the DIN/ISO 76 (static load rating) and DIN/ISO 281 (Dynamic load rating, life rating), which are usually provided by the bearings manufacturers. Equation 3.2 allows us to calculate the dynamic load rating C in Newton's for two or more bearings. Generally, C is a factor specified by the manufacturer.

$$C = i^{0.7} \times C_s^{bearing} \tag{3.2}$$

where $C_s^{bearing}$ is the load rating of single bearing in Newton's and i is the number of bearings in the bearing set. The equivalent dynamic load P is calculated as follows in Newton's.

$$P = X \times F_r + Y \times F_a \tag{3.3}$$

where X and Y are the radial factor and the axial factor, respectively. F_r, F_a are the radial load and the axial load, respectively and are also given in Newton's. The nominal life rating L_{10h} can be calculated in hours using the following equation which is based upon a 10% probability of failure:

$$L_{10h} = \frac{10^6}{60 \cdot N} \cdot \left(\frac{C}{P} \right)^3 \tag{3.4}$$

where N is the spindle speed in rpm, C is the dynamic load rating and P is the equivalent dynamic, both previously calculated. The adjusted life rating L_{nah} in hours is calculated as:

$$L_{nah} = a_1 \cdot a_{23} \cdot f_t \cdot L_{10h} \tag{3.5}$$

where f_t is the factor for operating temperature and the factors a_1 and a_{23}, are provided by the supplier.

3.4.2.2 Metallic and Hybrid-ceramic Bearings

The bearings used by machine tool manufacturers have traditionally been metallic, but in recent years, hybrid-ceramic bearings have been gaining popularity due to their improved performance and suitability for high-speed machining; see Table 3.1.

Table 3.1 Comparison of properties of bearing steel balls and ceramic balls

Properties (units)	Conventional steel bearing	Hybrid ceramic (Si_3N_4) bearing
Young's modulus (GPa)	208.00	315.00
Hardness (VickersRc)	60.00	78.00
Density (g/cm^3)	7.80	3.20
Max. usage temperature (ºC)	120.00	800.00
Coefficient of expansion (10^{-6}/K)	11.50	3.20
Poisson's ratio	0.30	0.26
Thermal conductivity (W/mK)	45.00	35.00
Chemically Inert	No	Yes
Electrically conductive	Yes	No
Magnetic	Yes	No

Typically, steel raceways and silicon nitride Si_3N_4 balls are used. The ceramic balls may weigh up to 40% less than the metallic ones; therefore they need less energy to keep moving and there is a reduction in centrifugal loading and skidding. What is more, ceramic balls have a higher stiffness so, at high speeds, they suffer less deformation due to centrifugal forces. As a consequence, ceramic balls can roll up to 30% faster than steel balls with the same service life.

The ceramic balls have a lower affinity with the steel raceways so there is very little wear from adhesion. In steel balls the microscopic cold welds between ball and raceways break during the rotation creating surface roughness that increases friction forces and the risk of excessive heat.

They also have a lower friction coefficient, which means that they operate at lower temperatures. This is an interesting feature, as it increases not only the bearings life but also the lubricant life.

Experimental tests have demonstrated that hybrid bearings operate at lower vibration levels. As they have a higher stiffness and less mass, they also have higher natural frequencies and dynamic stiffness.

All these factors mean that with ceramic balls precision is improved and the bearing reaches higher speeds and has a longer service life, often twice that of the conventional metallic bearings. However, using them requires special attention from the machinist to avoid collisions because their main drawback is the high risk of brittle fracture when there is a collision.

3.4.2.3 Lubrication

Lubrication of the bearings plays a key role in how they perform and function. The lubricant prevents direct contact in bearings forming a fluid film between ball, race and cage. In steel bearings, direct metal-to-metal contact is avoided so wear and corrosion are reduced. Lubricating systems may use either grease or oil/air lubricants.

Grease lubrication is the most commonly used for conventional machining and for some HSM cases where the required spindle rotational speeds are not too high. This kind of greasing is cheap and permanent, so it needs minimal maintenance. However, they cannot run continuously at a $d{\cdot}N$ factor above 850,000, since high temperatures degrade the grease and the risk of bearing damage rises.

If higher speeds are required in the bearings, oil-air lubrication is more suitable, as it lubricates, cleans and cools the bearings. However, this is a more expensive method because it needs monitoring, controlling and maintenance, as well as a continuous supply of clean dry air; see Fig. 3.15. It is sensitive to variations in

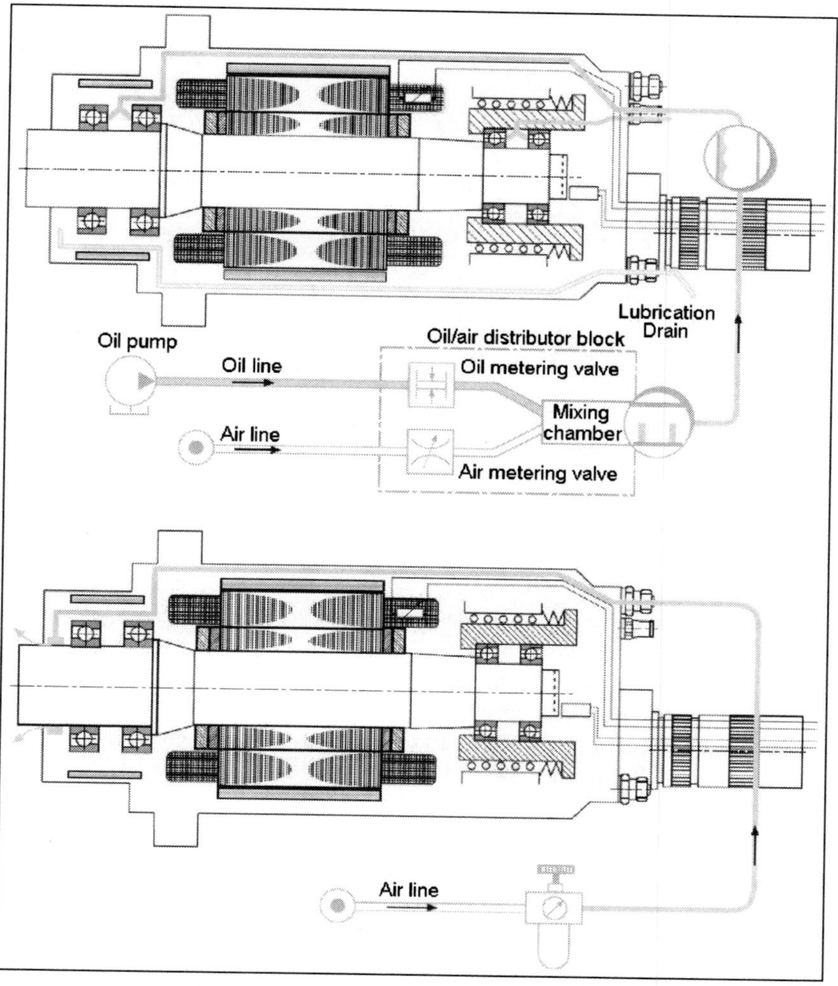

Fig. 3.15 The oil-air lubrication method, above, requires adding a lubrication system to pump, control and mix the lubricant. In comparison, grease lubrication, below, is simpler. Here the circuit of air that creates an overpressure in the spindle nose is presented. Schematics by GMN®

the composition of the air-oil mixture; an excessive amount of oil can heat the bearing while insufficient oil leads to premature wear. There are several methods for mixing the oil and air and transporting it to the bearing: oil-mist, oil-air, oil-jet and pulsed oil-air. The oil mist lubrication is difficult to control, measure and transport, since the mist can degrade into large drops of oil and the uniform distribution of the lubricant becomes difficult. The oil-air method is more controlled and requires a high pressure pump and mixing valves. The advantage is that the mixture is produced close to the bearings and an accurate quantity is provided. The oil-jet method is based on a high pressure pump that provides lubricant directly to the bearing races. It is used in high performance spindles with high loads and speeds. However, it requires a complex pump, an oil tank and a temperature control system. Finally, the pulsed oil-air method injects oil in very small quantities with a frequency of injection that can be periodic or dependent on the spindle working conditions.

Performance evaluation of the lubricant and of the lubrication method must take into account factors such as the length of operation times between maintenance operations, the maximum rotational speed, the heat generated and loss through friction, as well as certain environmental aspects such as recycling, disposal, etc.

As important as the lubrication is a correct sealing of the spindle to avoid impurities from entering the bearings. The most common method is based on creating an over-pressure inside the spindle in such a way that flow comes out in the rear and the spindle nose so that particles cannot enter; see Fig. 3.15. Another method usually combined with the over-pressure is the use of labyrinth seals in the spindle nose.

3.4.2.4 Magnetic Bearings

Active magnetic bearings (AMBs) work on totally different principles than conventional rolling bearing systems, as they support loads thanks to magnetic levitation. AMBs consist of electromagnets which produce a magnetic field using a constant electric power flow. At least front and rear radial magnetic bearings and an axial magnetic bearing are needed. A set of power amplifiers supply current to the electromagnets, and a controller and a set of gap sensors work together in a closed loop, monitoring and controlling the position of the rotor within the levitation gap. Since physical contact is avoided, the effects derived from contact, such as friction, temperature increases and wear are also reduced. Those characteristics make them a good solution for high speed spindles. What is more, the electromagnets can be used as actuators and sensors of the spindle running conditions. That is the reason why the use of hybrid spindles with rolling bearings and AMBs is being explored, using the AMB to absorb the process dynamic loads and the rolling bearings to position and support the shaft [1]. However, the main drawback is that they are expensive and complex.

3.4.2.5 Hydrostatic Bearings

In hydrostatic spindles, a fine layer of oil is injected at high pressure to produce what is known as a *hydrostatic equilibrium* in the interior of the bearing. These bearings have ceramic pads which contain the oil injected at high pressure to support the shaft.

This method avoids some disadvantages of ball bearing systems, such as low damping, a limited service life, and the problems derived from heating and vibration. In comparison, hydrostatic bearings are practically wear-free, operate with little noise, and provide excellent stiffness and damping; hence, they perform better under dynamic loads. The oil is injected at low temperatures so it reduces the thermal expansion of the shaft. They cannot reach the same $d{\cdot}N$ as roller bearings though. What is more, nowadays the cost of the oil supply and sealing system makes then more affordable. Also, at low speeds they perform better than hydrodynamic bearings, due to a high load carrying capacity and friction conditions. Hence, these bearings are nowadays used in applications as grinding, milling and turning machine tools.

3.4.2.6 Hydrodynamic Bearings

The operation mechanism of hydrodynamic bearings is based on the fluid-wedge principle. For machine tool applications, the shaft is supported by tilting-pad elements that can be spring loaded or can move due to the flexibility of the material they are embedded into. The working speed of the spindle must be in the fluid friction range since, at lower speeds or when the spindle is started or stopped, the wear rate increases as it runs under boundary lubrication conditions. These bearings were traditionally used in grinding machines where rotational speeds are high enough and remain approximately constant. However, nowadays, they have generally been substituted by hydrostatic bearings.

3.4.2.7 Aerostatic Bearings

The operation principle is similar to that of hydrostatic bearings. The advantages of these bearings are low friction, low heat generation, accurate running, long life, and simple construction because there is no need for seals. However, they have a small load-carrying capacity, low damping and a tendency to vibrate. Hence, they are suitable for very high speed applications with low loads. In the machine tool industry, they are used for precision grinding machines and ultraprecision applications.

3.4.3 The Toolholder

In milling, the toolholder is the interface between the cutting tool and the spindle and it ensures the transmission of the rotational movement of the spindle to the cutting tool. The selection of the toolholder is defined by the performance and possibilities of a machine tool. It requires a simple system to clamp and unclamp the tool, allows automatic tool changing within the machine tool, ensures a precise and rigid adjustment with the spindle and it must guarantee a proper eccentricity and balancing. A defective toolholder can have negative effects on the dimensional precision and surface quality of the piece, and it can reduce the service life of the cutting tool and the spindle. Nowadays, hollow toolholders are made in such a way that the cutting fluid can flow through the spindle to the tool tip.

A toolholder for milling has three main parts: the taper, the flange and the collet pocket. The taper is the conical part that is introduced into the spindle. The flange is the part that serves to hold the toolholder when it is in the tool changer storage magazine of the machine tool. The collet pocket is the part of the toolholder that actually holds the cutting tool.

3.4.3.1 Tapers

Several technological solutions have been developed from the need to standardise how the cutting tool is held in machine tools. These standards are BT, CAT (also known as V-Flange), ISO (also known as SK) and HSK; see Fig. 3.16. A recent alternative is the Big-Plus® dual contact. We should also mention the R8 collet toolholder technologies and the Morse taper collet which are used in manual machines.

The conical forms such as BT, CAT and ISO, are usually used in conventional milling machines. In contrast, the HSK toolholders are more suitable for high speed milling. The conical toolholders as BT, CAT and ISO are coupled to the spindle via "two cones in contact". The centrifugal force that increases with the rotational speed of the spindle, together with the thermal effects, tends to push the cone into the spindle, thus increasing the retention force. The displacement that the cone experiences within the spindle may provoke imprecision in the machining.

Fig. 3.16 Standard toolholders by Maritool®, from left to right: BT50, CAT50, ISO30, HSK50A

ISO toolholders are more widespread in Europe for traditional machining. They are clamped to the machine by means of hydraulic or pneumatic actuators. The dimensions of the ISO cones are standardised and classed as ISO 30, ISO 40 and ISO 50, according to the external diameter of the cone. In the US, however, CAT toolholders are more widely used in traditional machining.

The CAT toolholder was created by the Caterpillar Corporation in order to standardise the tool holding systems in their machine tools. These holders are designed according to the size of the cones and are classified as CAT-30, CAT-40 and CAT-50. The BT toolholder is similar to the CAT, with slight differences, such as the fact that it is symmetrical with respect to the rotary axis of the spindle, which improves performance.

HSK, now in the ISO 12164-1:2001, stands for *hollow taper shank* in German. This toolholder has certain advantages over the others in terms of HSM where rotational speed is greater than in conventional milling. As rotational speed increases, the centrifugal force increases the retention force on the cutting tool by means of milled drive keys that are located within the shank of the mechanism. The keys tend to expand in the interior of the toolholder because of the centrifugal forces in such a way that the metal-on-metal contact is reinforced and pressured within the spindle axis. Thus, the contact stiffness is increased and more aggressive cutting operations can be undertaken; see Fig. 3.17. The retention force of an HSK toolholder can be up to twice that of a conventional toolholder, such as, for example, a BT. In the BT interface, the clamp pincers are on the outside of the shank, so they tend to lose contact and retention force as centrifugal force increases. The HSK toolholders outperform the ISO toolholders by up to 5 times in terms of radial stiffness. In addition, HSK toolholders offer greater repeatability.

There are various designations for HSK toolholders. There is a number to indicate the external diameter of the flange (25, 32, 40, 50, 63, 80 up to 100 mm) followed by a letter ranging from A to F. The letters most suitable for high speed machining are

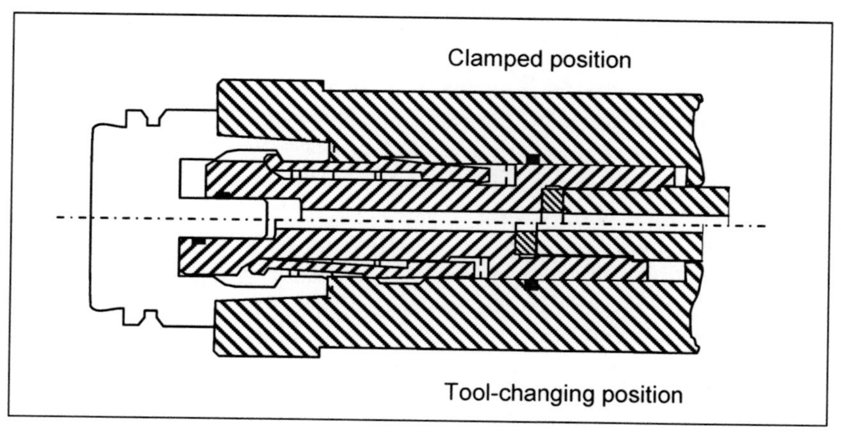

Fig. 3.17 Clamped and tool-changing positions of the HSK toolholder

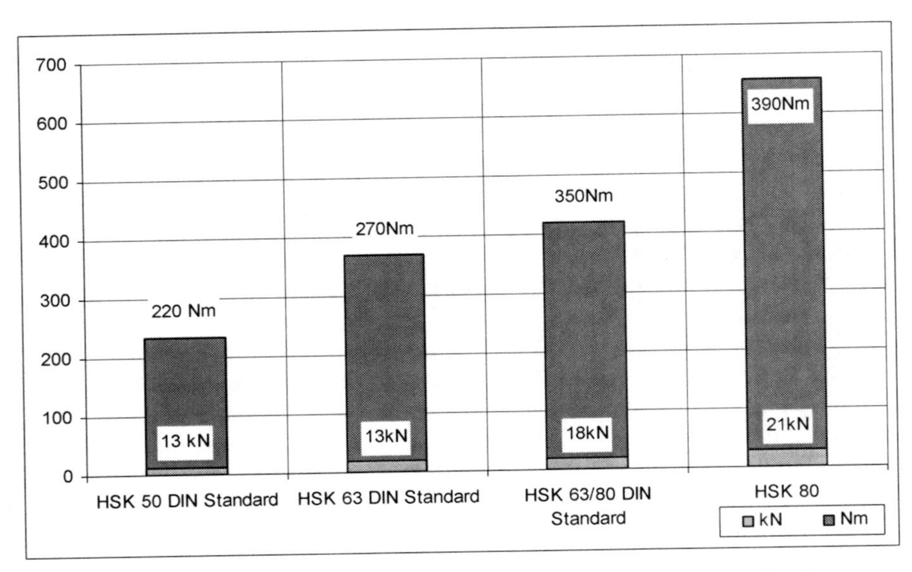

Fig. 3.18 Bending moment on several HSK tapers with the corresponding clamping force

A, B, E and F. The letter A is used for applications requiring moderate torque and high rotational speed with automatic tool change. Letter B is for high torque, high speeds and automatic tool change. Letter C is for moderate torque and high rotational speed, but manual tool change. The letter D is for high torque, high speed and manual tool change. The letters E and F are also for HSM applications, but with low torque and very high speeds. E and F designate HSK toolholders that are designed for very high speed operations that either have no framework or for cutting operations that may negatively affect the equilibrium of the machining process, resulting in the appearance of chatter. The diameter of the toolholder determines the stiffness. In Fig. 3.18, the bending moment of several HSK tapers with different diameters and clamped with the force recommended by the standards are compared.

The recently developed Big-plus™ system consists in simultaneous dual contact to ensure that the toolholder is located accurately in the spindle. The taper and face of the machine tool spindle fit together due to the elastic deformation of the spindle, so that the taper and flange face fit simultaneously. This system permits interchangeability with the standard toolholders and machines and is being offered by many machine-tool manufacturers.

An alternative is the ICTM standard (Interface Committee for Turning-Mills) which is the standard developed by 17 companies in Japan as the interface of turning centres (turning + milling) based on the two-face restraint type ISO 12164-1:2001. It is compatible with HSK-A type.

Finally, there is also a wide range of other toolholders that have been specially designed by companies, like the Sandvik Coromant® Capto™, Kennametal® KM™ or Komet® ABS™. These toolholders feature a polygonal-shaped coupling, and are

Fig. 3.19 Section of a "great power" collet chuck, by Laip®, showing the internal needles to make the radial force

used in lathes and in milling centres, for a quick tool change. Figure 3.19 shows a collet chuck.

When manipulating these toolholders it is important to pay particular attention to cleaning and maintaining the cones (especially high-precision ones, such as those for HSK), since swarf and lubricant can negatively affect their performance. If the contact surfaces are insufficiently clean, the quality of the contact decreases, which will impede the cone settling within the spindle. It is advisable to blow the collet clean before proceeding to clamp it.

3.4.3.2 Clamps

There are various methods for securing the tool in the toolholder: mechanical chucks, hydraulic chucks and shrink-fit chucks. The clamping requirements of the tool in the toolholder are similar to those of the toolholder in the spindle: a) the maximum stiffness of the union; b) the precise alignment of the tool axis with the spindle shaft axis and c) avoiding unbalances.

Mechanical clamping by means of chucks (collet chucks) is the most widespread method. In essence, the tool is loaded into the chuck which holds the tool when tightened. Segmentation of the chuck enables it to deform to a certain extent, allowing the pressure to be distributed uniformly over the tool shank. It is a low-cost system, even for HSM, because the jaws in the chucks are interchangeable, or adjustable, so tools of different diameters can be held.

An improved version of mechanical clamps is the "great power chuck" providing a 10% higher torque and axial rigidity. It is based on needles inside two slightly angular cones, as shown in Fig. 3.20.

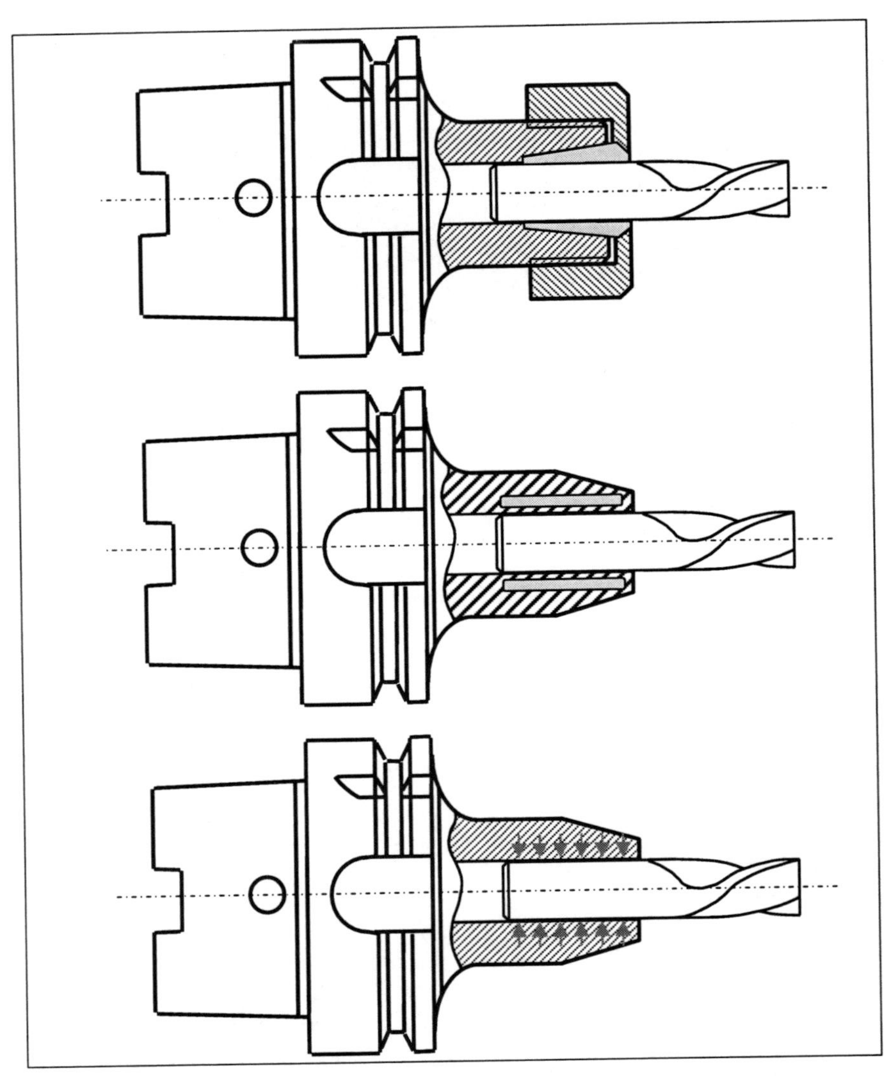

Fig. 3.20 Schematic representations of a collet chuck, hydraulic chuck and shrink-fitted tool shank

Figure 3.20 shows a schematic diagram of the clamp mechanism of a collet chuck, a hydraulic chuck and a shrink-fitted chuck.

Hydraulic chucks employ a metal membrane to envelop the tool shank. Next to the membrane is a perfectly sealed hydraulic reservoir, which is connected to an actuating screw which can increase the pressure of the fluid. The increase in fluid pressure is transmitted to the membrane which secures the cutting tool. Hydraulic chucks are expensive systems. Moreover, it is not possible to secure tools of different diameters in the same chuck, although some manufacturers offer additional

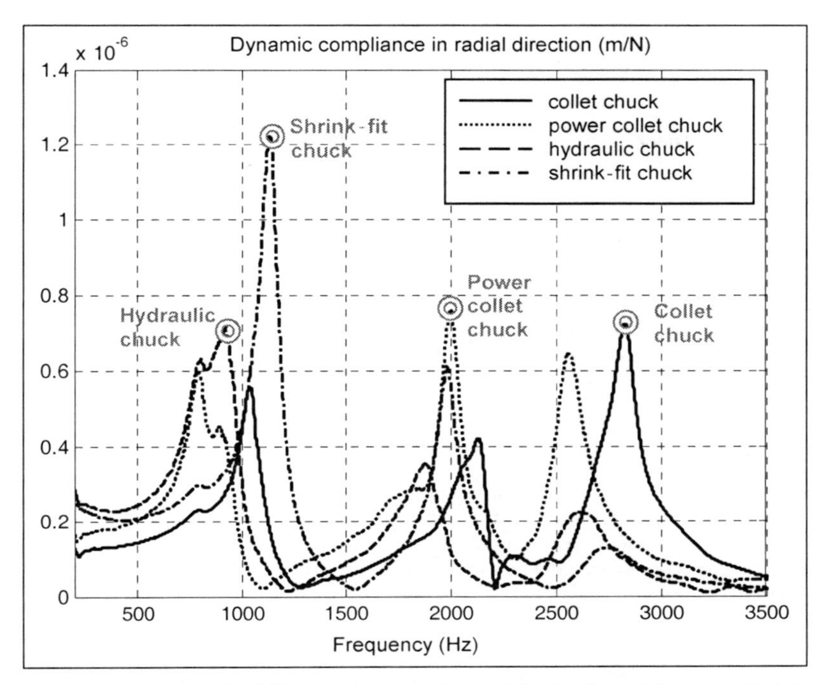

Fig. 3.21 Dynamic flexibility measured at the tool tip for four different toolholders with the same tool in the same spindle

membranes that can be introduced between the tool and the chuck so that the same chuck can accept different diameters of tools. However, this has the disadvantage of an inevitable amount of misalignment.

Shrink-fit chucks are solid blocks with a high-precision opening in which the cutting tool is loaded. At room temperature, the opening is smaller than the tool. The chuck heats up by means of an induction heater, and the thermal expansion allows the tool to be introduced. Then the chuck cools down again (in some systems a cooling mechanism is provided), and the opening shrinks to its original size gripping the tool in place. The stiffness of this holding system is very high. Unlike the hydraulic chuck, which needs an internal structure and other elements, such as actuating screws, the shrink-fit chuck can be much more symmetrical and free from the imperfections that may reduce equilibrium when rotating. Although shrink-fit chucks are simpler than hydraulic chucks, they are also more expensive.

Furthermore, a heating system needs to be acquired and, in some cases, a cooling system as well.

The choice of the toolholder definitely affects the machining dynamic performance and it is conditioned by the spindle characteristics. Comparing the frequency response function (FRF) measured at the tip of a tool of 16 mm diameter and 72 mm overhang in a power collet chuck, a collet chuck, an hydraulic chuck and

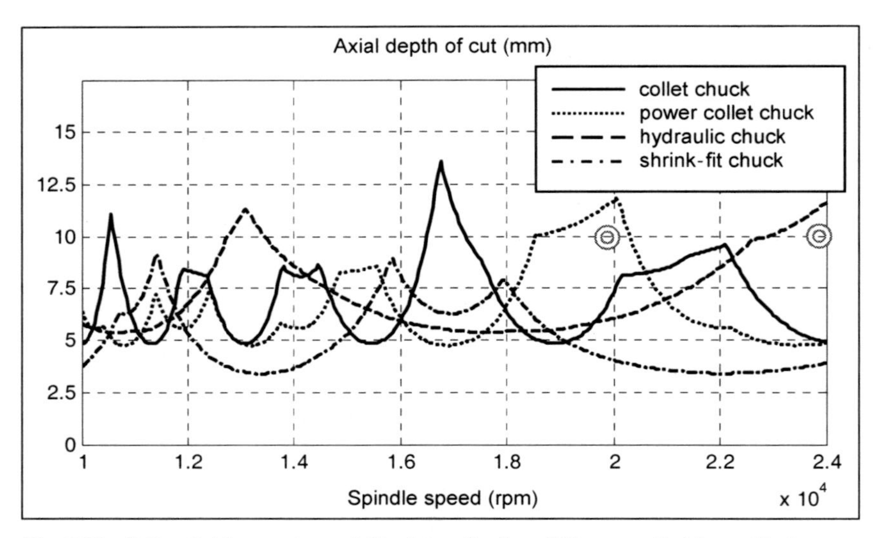

Fig. 3.22 Full radial immersion stability lobes for four different toolholders with the same tool in the same spindle. Circles indicate the best cutting conditions for the hydraulic chuck and the power collet chuck

a shrink-fit chuck for the same spindle, a Step-Tec® HUCS 230-S-40-9/24-4FMMS, the conclusion is that the hydraulic is the most rigid one, followed by the conical, the cylindrical collet and the shrink-fit chuck; see Fig. 3.21.

The stability lobes calculated for aluminium machining reveal that the hydraulic chuck theoretically reaches the most productive conditions with a spindle with a maximum speed of 24,000 rpm; see Fig. 3.22. However, if the user has a spindle that can only reach 20,000 rpm, it is the power collet chuck that is the most productive. Another critical factor to take into account is the balancing. The tests performed with these toolholders revealed that the hydraulic toolholder could not run above 20,000 rpm due to balancing problems, so the power collet chuck was selected as the most productive one for that tool.

3.4.4 The Drawbar

The drawbar mechanism must hold the toolholder within the spindle during machining operations without balance problems. This system basically works with a mechanical chuck and a spring-loaded mechanism. The retaining force that the drawbar exerts at high rotational speeds has to be enough to ensure the stiffness of the coupling. On the other hand, the expulsive force used to eject the tool must not lead to any load on the bearings. Finally, *coolant through spindle* systems are able to convey the cutting fluid through the shaft to the tool; see Fig. 3.23.

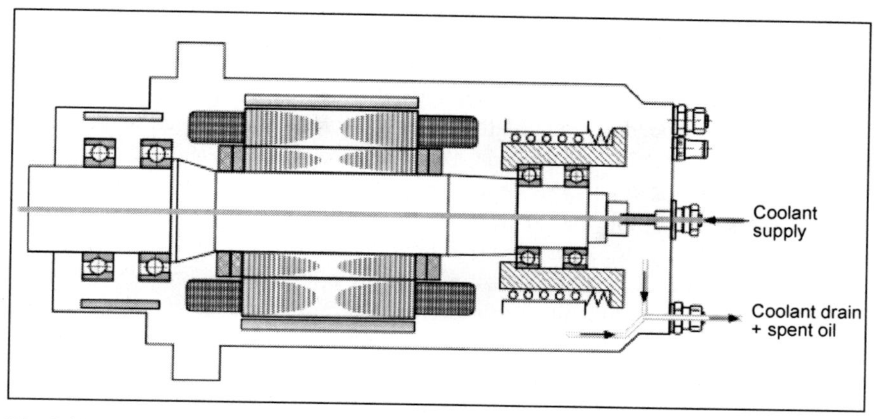

Fig. 3.23 Coolant through the shaft with gap seal, by GMN®

3.4.5 The Shaft

The spindle shaft is responsible for transmitting the power generated in the motor to the tool, and locates the drawbar and tooling system as well as the encoder. The main characteristics that define the behaviour of the shaft are the length and the diameter. Typically, the machining process excites bending modes of the shaft; therefore, in order to avoid dynamic problems, the first natural frequency of the shaft, which depends strongly on the length and the diameter, must be above the maximum speed of the spindle. High speed milling spindles are usually designed to have a maximum speed of 50% of the first natural frequency [28]. Hence, the shorter the shaft and the larger the diameter, the higher the natural frequency will be, although large diameters also mean large bearings that limit the maximum working speed.

Finally, it is important to bear in mind the thermal expansion of the shaft. During the machining, the temperature increases in the interior of the spindle thus provoking thermal expansion. Due to its length (the longer the shaft the bigger the expansion) the shaft may be in large part to blame for any imprecision in the machined end product as a result of excessive heating.

3.4.6 The Sensors

The technology of sensors has advanced greatly in the last ten years (see Table 5.1 in Chap. 5 for a full view). Sensors are now used to register more than the positions of the tool clamp system or the spindle speed. At present, sensors are used to monitor and control spindle behaviour and diagnose possible anomalies during operation that may affect the spindle integrity.

For example, high resolution encoders provide feedback to control the real spindle speed as well as angular position, so the spindle full power capacity can be utilised. The resolution of the encoder determines the precision of the calculation of the position and speed, so the greater the encoder resolution, the more precise is the information received to perform the vector control of the speed. Accurate rotation is achieved in operations that require high torque at low speeds, without fluctuation of the speed. Shafts can be positioned within 0.001 degrees, so an oriented spindle stop to change tools or C-axis operations as thread cutting are possible. Also, quick acceleration and deceleration times are allowed. A more in-depth view of the use of encoders in machine tools is provided in Sect. 6.2.3.

Accelerometers are usually placed near the bearings to test the balancing state, and to measure unbalances and resonances that may damage the spindle. The vibration patterns are registered and visualized for process optimisation and spindle life estimation. Thermocouples and thermistors are used to measure temperatures of the motor, front and rear bearings, the housing, etc. Displacements sensors measure the shaft axial displacement. Sensors for cutting forces and torque measurement are placed for process monitoring.

Monitoring and diagnosis of the condition of the spindle and the operation helps to increase the service life. The trend appears to point to a future in which machine tools will be able to diagnose the spindle condition during the manufacturing process, correct balancing problems, and take decisions that will reduce imperfections and errors to ensure the quality of the workpiece and the good working order of the machine. For example, there are machine tools available that use the temperature sensors information to compensate the thermal expansion of the spindle. AMBs are not only being used as bearings but also can work as sensors to measure vibrations and cutting forces and as actuators for active damping purposes.

3.4.7 The Housing

The housing supports and locates the bearings, the shaft, and in the case of electrospindles, also the motor. It has to provide the following elements:

- A circuit for the cooling of the motor: In electrospindles, the heat generated by the motor, as explained in Sect. 3.5.4, negatively affects the accuracy of the spindle, so a cooling circuit around it becomes essential.
- Pressurised air circuit: Usually, pressurised air is introduced into the spindle in order to seal it from pollution.
- Lubricating circuit for the bearings: Used when the bearings are lubricated using an oil-air solution.
- Cutting fluid circuit: Some spindles provide cutting fluid, not only through the spindle shaft, but also through the housing which has several nozzles in the nose pointing at the tool tip.
- Locations for sensors and corresponding wirings.

3.5 Spindle Properties and Performance

For a machine tool manufacturer and for end users, machine tool spindles are selected according to their features. Here we will explain the main features of spindles and how the mechanical and thermal loads affect their performance.

3.5.1 Spindle Power and Torque versus Spindle Speed Curves

In this section we present the characteristics of spindles in terms of spindle power and torque with respect to spindle speed. Usually, this relationship is expressed using the power/torque curves for the spindles of the machine tools. This information is provided by the machine tool manufacturers in their catalogues as it defines the performance and efficiency of the spindle.

Two spindle power torque curves can be usually found: the constant torque curve and the "knee-type" curve. In constant torque spindles, horsepower increases linearly along the whole speed range. However, despite them providing torque at high speeds, the maximum torque available is lower than those provided by knee-type spindles. In these spindles, torque remains constant up to a point, the knee, where it starts decreasing. Meanwhile, power increases linearly up to the knee and then remains constant. These spindles accommodate well the use of large and small diameter tools. Small-diameter tools require high speed spindles in order to offer acceptable feed rates given that, due to their shape and size, feed per tooth rates are necessarily small. These tools do not need high torque to operate and in any case, compared to larger-diameter tools, they would not be able to support high torque because of their size. On the other hand, large-diameter tools require higher torque but do not need such high spindle speeds as the small-diameter tools. Figure 3.24 shows the power curves (kW) and torque (Nm) with respect to the rotational speed (rpm) for the spindle of a high-speed Deckel Maho® "105 V linear" milling machine.

Factors such as operating times and load must be taken into account in order to quantify the power that the machine can supply. The service class (S1 up to S9) is defined by the set of operation conditions, including the periods of idle running and rest, to which the machine is subjected. The types of service are defined in the standards IEC34-1and VDE 0530 part 1. The most commons ones are S1 and S6:

- *S1, Continuous service*: continuous operation of the machine, i.e., operating at constant load for enough time to reach a thermal balance.
- *S6, Continuous service with intermittent load*: sequence of identical operating cycles, consisting of a period with a constant load and a period of idle running. No rest periods. It is indicated by the relative duration of the load period as a percentage of the whole operating cycle; for example, S6 30% is 45 minutes. The operating cycle is 10 minutes long if no value is given (thus, S6 30% indicates a load period of 3 minutes and a no-load period of 7 minutes).

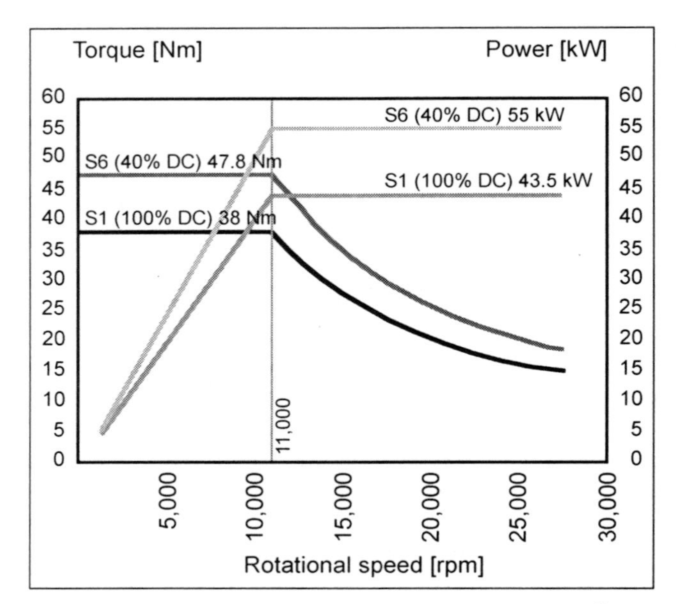

Fig. 3.24 Spindle power and torque curves of a Deckel Maho® "105 V linear"

When evaluating power and torque requirements, it is important to consider a) the range of speeds for which the machine tool is intended, b) the cutting tools that are to be used, c) the shape and precision requirements and d) the material of the pieces that are to be manufactured.

3.5.2 The Stiffness

The static behaviour of the spindle is related to the tool tip deflection under the action of cutting forces. Really the spindle is a hyperstatic beam supported on the front bearings (usually two angular ball bearings in a DF arrangement) and the rear bearing (usually an angular ball or cylindrical roller bearing). Bearings are inserted into the housing, so the stiffness of the spindle is related to the housing, which will be added to that of the headstock itself.

The total deflection is the sum of three effects, a) those derived from the deflection of the beam, b) those derived from the flexibility of bearings, and finally c) those derived from the flexibility of the bearing accommodation into the housing [24]. Assuming the simplification of constant section for the spindle and a short tool, the tool tip displacement due to spindle flexibility is (see also Fig. 3.25):

$$\delta_s = \frac{Fba^2}{3IE}\left(1 + \frac{a}{b}\right) \tag{3.6}$$

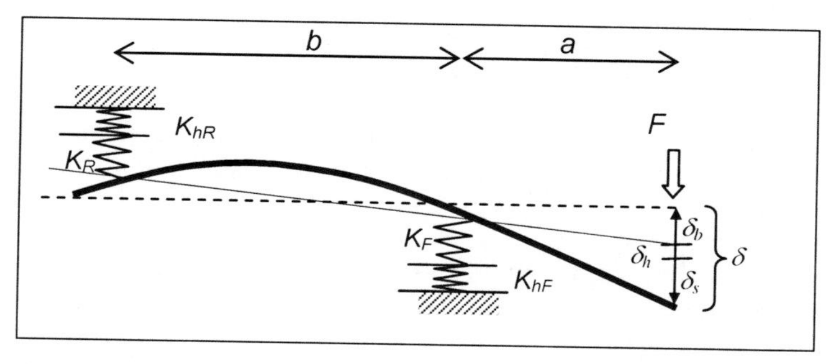

Fig. 3.25 The three components of flexibility, displacements at the tool tip

On the other hand, the displacement due to bearing flexibility is:

$$\delta_b = \frac{F}{b^2}\left(\frac{(a+b)^2}{k_F} + \frac{a^2}{k_R}\right) \tag{3.7}$$

Where k_F is the radial stiffness of the front bearing and k_R is the radial stiffness of the rear bearing, which are also determined by the bearings preload. At the view of this equation, the stiffness of the front bearing is more important than that of the rear bearing. Moreover, the reaction force in this support is higher that in the rear bearing. Therefore, selecting this bearing and the preload is a key point in the design of all types of spindles.

Finally, the stiffness of the housing is difficult to calculate with a simple model, instead with FEM a fine value can be obtained. The displacement at the tool tip δ_{st} due to spindle system flexibility is obtained adding the previously calculated equation:

$$\delta_{st} = \delta_s + \delta_b + \delta_h \tag{3.8}$$

All of them are directly proportional to the acting force F, therefore the final spindle system stiffness k_{st} measured at the tool tip is:

$$\frac{1}{k_{st}} = \frac{1}{k_s} + \frac{1}{k_b} + \frac{1}{k_h} = \frac{\delta_s}{F} + \frac{\delta_b}{F} + \frac{\delta_h}{F} \tag{3.9}$$

Other elements that add flexibility are:

- In milling, the toolholder and tool, and the interfaces of the toolholder fitted onto the spindle and the tool fitted into the toolholder. A useful model for studying tool deflection is that in which the tool is regarded as a cylindrical cantilever beam. Tool tip deflection then conforms to the equation:

$$\delta_{tt} = \frac{64}{3\pi}\frac{F}{E}\frac{L_H^3}{D^4} \tag{3.10}$$

Thus, it can be seen in Eq. 3.10 that tool deflection in the static model is a function of the following three parameters, E is the Young's modulus for the tool material, L_H^3/D^4 is the tool slenderness parameter. D is the equivalent tool diameter (the tool is not a cylindrical beam) and L_H is the overhang length. F is the cutting force perpendicular to the tool axis. In [15] a complete study of the stiffness of machine, tool, and couplings between toolholder and spindle and tool and toolholder is explained.

- In turning or cylindrical grinding, the flexibility of the workpiece. Really the workpiece clamped in the chuck is a prolongation of the spindle with a flexible internal link (the chuck itself). If the part is short with a large diameter, the chuck usually clamps it sufficiently and the tailstock can be used to increase the stiffness. If the part is long or very slender the use of "steady rests"' is recommended, as seen in Chap. 1, Fig. 1.30 in the case of heavy-duty lathes.

3.5.3 Dynamic Behaviour and Vibrations

Vibrations in milling appear due to the lack of dynamic stiffness of some component of the machine tool-tool-workpiece system. They can be divided into two main groups: forced and self-excited vibrations. The former vibration type, supposing that the tool and spindle subsystem is well balanced and supported in a milling machine on an isolating foundation, is due to the variable chip thickness and the interrupted nature of the process. This means that they are always present. Therefore, to prevent damaging the tool and/or spindle, the vibration level must be maintained under a threshold value. Another source of forced vibration is the use of inadequate or misbalanced tools. As previously mentioned, high frequency spindles have the maximum speed limited below the critical speed, but the use of inadequate tools can shift the critical speed near the operating speed range with disastrous consequences.

Regarding the latter type, the most common self-excited vibration in milling is the "regenerative chatter". Huge efforts have been made to understand, predict and avoid this kind of vibration since the beginnings of the 20th century and especially during the last twenty years [2, 5, 7, 9, 17, 18, 27]. Since milling is an intrinsically interrupted cutting operation, vibrations always appear, and therefore the cutting edge leaves waviness in the surface material. The difference of phase between the current cutting edge vibration and the waviness produced in the previous pass, results in a variable chip thickness and a highly variable cutting force; this excites structural modes of the machine, the spindle-toolholder-tool and the workpiece system. If a component of this system lacks dynamic stiffness or damping, displacements and forces become higher at each tooth pass and the process becomes unstable. The relation between forces and displacements was well summarized by Merrit [11], proposing a regenerative loop similar to Fig. 3.26.

Chatter vibration can be completely prevented calculating the "stability lobes" diagrams. These diagrams indicate the limiting chatter-free axial depth of cut at each

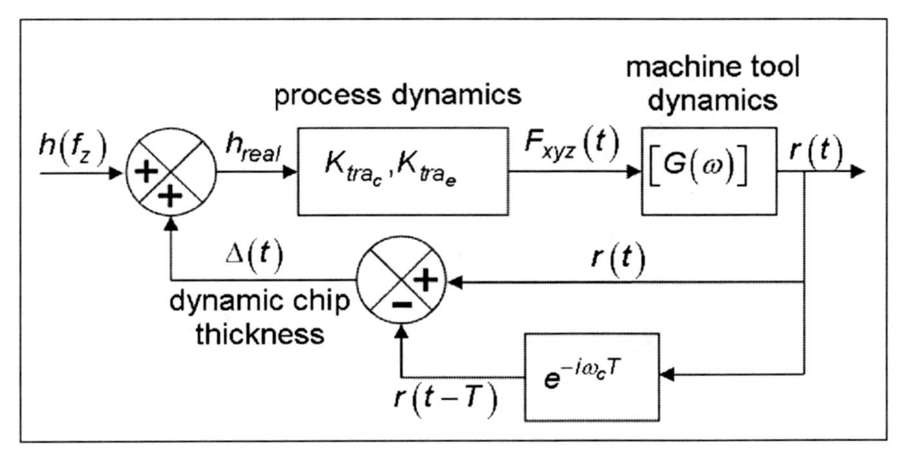

Fig. 3.26 Block diagram of the chatter loop

spindle speed for a given radial depth of cut. Therefore, they graph the borderline between stable and unstable cases, which is composed of several lobe-shape curves.

The traditional approach throughout the 20th century was that chatter in milling was a steel roughing problem, affecting machine tool structural low frequency modes from approximately 15 to 200 Hz. The cutting speeds derived from the machine and tool technology available in those years implied that high frequency modes were not excited.

However, at the end of this century, high speed machining at rotation speeds ranging from 15,000 to 40,000 rpm became possible, and as consequence chatter began to appear at higher frequencies, between 500 and 2,500 Hz (Fig. 3.27). In high speed machining local modes of the spindle-toolholder-tool subsystem are excited because the tooth passing frequency is high above the modal frequencies of the machine structure. The consequences of the high-frequency chatter are bad surface roughness and tool life reduction, but particularly problematic is the spindle bearings damage. A broken tool can be easily substituted, and a mark in the workpiece can be corrected, but the cost of repairing a broken spindle is high and means delays in production. Hence, some companies buy an additional spindle for replacement while damaged ones are repaired.

To prevent the high-frequency chatter, the stability lobes diagram (Fig. 3.27) suggests machining at even higher speeds, at which the stability areas between lobes are wider. Hence, not only a higher productivity is achieved but also chatter vibrations are eliminated. This means that spindle designers should not only ensure a static but also a dynamic stiffness at high rotational speeds. However, spindle stiffness is complicated to estimate in the design stage, although recent advances in modelling have made it possible.

Estimating the dynamic stiffness allows the stability lobes to be calculated under working conditions prior to spindle manufacturing. Thus, spindle models make it possible to observe the effect of changes in the design on the spindle

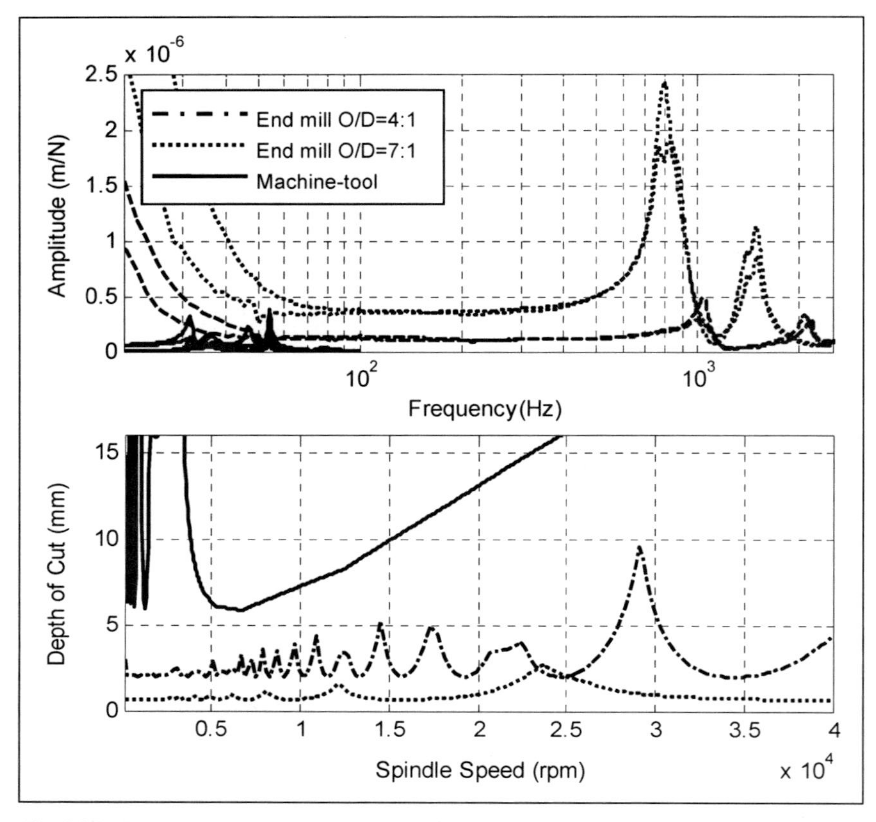

Fig. 3.27 Frequency response functions and corresponding stability lobes of a machine-tool structure and end mills with an overhang/diameter ratio (O/D) of 4:1 and 7:1

performance. In this section, some basic concepts regarding the estimation of the spindle dynamic properties are presented, followed by the current methods to calculate the lobes diagrams.

3.5.3.1 The Spindle Dynamics

This section deals with the identification of the dynamics of spindles with angular contact ball bearings. The dynamic stiffness at the tool tip, where the cutting forces are located, is the sum of the stiffness of all the elements of the system, i.e., the tool, the toolholder, the shaft, the bearings, the housing, and the mechanical joints between them; the machine is not included because of its lower natural frequencies (see the previous subsection). On the other hand, the spindle rotation introduces forces that change the stiffness of both the bearings and the shaft, thus modifying the "0 rpm dynamic response" at the tool tip.

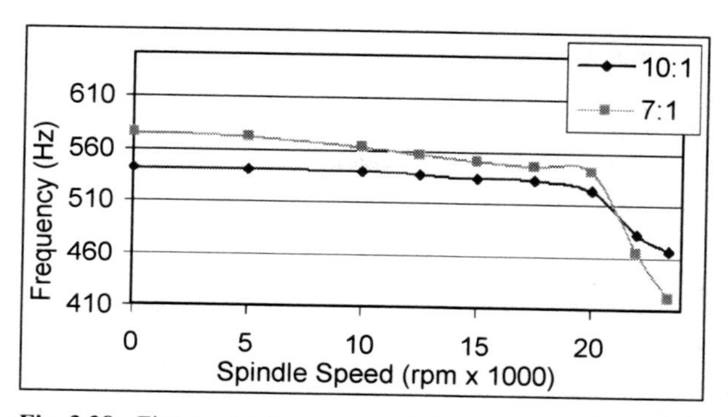

Fig. 3.28 First modal frequency variation on two tools with a different overhang-diameter ratio

The variation of the spindle dynamics is the superposition of the mechanical and thermal phenomena [1, 8, 14, 16, 19]. When the spindle does not rotate, the contact angle of balls with the inner and outer raceways are defined by the geometry and kinematics of the bearing to avoid the balls pivoting in the races, and to ensure the axial and radial preload of the balls with the raceways. When the spindle starts rotating, two additional loads proportional to the rotational speed act on the balls, the centrifugal force and an induced gyroscopic moment; both affect the initial preload. The centrifugal force reduces the initial preload and slightly changes the kinematics of the bearing, varying the contact angles, and thus decreasing the bearing and spindle stiffness.

On the other hand, the contact force in the ball-raceway interface varies due to the cutting forces and the thermal gradient between the inner and outer raceways of the bearing at high speeds. For example, Fig. 3.28 shows the evolution of the measured frequency of the first mode for a tool with an overhang-diameter ratio of 10:1 and 7:1, clamped on a collet chuck toolholder HSK63 fitted onto the Step-Tec® Hucs 230-S-40-9/24-4FMMS spindle.

At the same time, a gyroscopic moment acts on the shaft due to the high rotating speed, which splits the modal frequencies in two, therefore changing the tool tip frequency response.

The experimental measurement of the real FRF at the tool tip when the spindle runs at high speeds becomes very complex. To avoid it, one approach is the modelling of the whole spindle-toolholder-tool system dynamics [8, 14, 16, 19]. Nowadays, there are models developed such as the ISBAP® by Purdue University and commercial solutions available like the Spindle Pro module™ of CutPro® software or the SPA™ by Metalmax® [12, 25, 26].

The substructures coupling techniques allow the dynamics of the spindle and the tool to be studied separately and then combined to obtain the system global response. Consequently, once the spindle dynamics have been studied theoretically or experimentally, it is possible to estimate the response at the tool tip for different tools. Given the importance of both the tool diameter and the overhang on the dynamic behaviour, these techniques are very interesting for machine tool users. If spindle and machine manufacturers integrate the dynamics of each of their products in a software utility, the final user can introduce the specific geometry of his/her tools to obtain the frequency response. This FRF is an input for the stability lobes calculation which leads to the selection of highly productive conditions.

Otherwise, if a model is not provided or it is not accurate enough, the FRF can be obtained with an experimental approach based on the excitation and response measurement. At 0 rpm, it is easily made by tap testing the tool tip with an impact hammer and an accelerometer. However, at higher speeds the tap test may be dangerous and less accurate; therefore, alternative procedures as the use of active magnetic bearings for both non-contact excitation and response measurement are currently being investigated [1, 14].

3.5.3.2 The Estimation of the Stability Lobes

For the evaluation of a spindle in working conditions it is necessary to have a model of the process where the cutting parameters, the workpiece material properties and the tool geometry are involved.

The real milling model is the simulation in the time-domain that integrates the equation of motion [17]. This model reproduces the evolution in time of forces, displacements and chip thickness, making it possible to determine if machining is stable or unstable, and in the stable case if the amplitude of forced vibration is acceptable. The main drawback of time-domain modelling is the high computation time for the stability lobes calculation.

The alternative solution is to calculate the stability lobes directly in the frequency domain or to resolve the periodic delay-differential equation. A schematic description of these models is outlined in Fig. 3.29 and Table 3.2. Models follow the following steps: a) the relation between the chip thickness and the vibration of the system must be formulated, b) a cutting forces model is proposed, c) the system dynamics is introduced, and finally d) the eigenvalue problem that defines the stability of the process is solved.

For spindle behaviour verification purposes in terms of cutting force, a simple semi-mechanistic force model with a linear relation between the force and the chip thickness is accurate enough. Cutting tools for high speed roughing of aluminium alloys have a typical helix angle lower than 30° so, at the usual depths of cut, the effect of this angle on the shape of lobes is minimal and can be neglected [27]. If the purpose is to calculate stability lobes for tools with complex geometries or helical end mills, a more complex model is needed.

Furthermore, it is essential to modify the algorithms to include the dependency of the system dynamics on the spindle speed. In Fig. 3.29 there is a comparison of the stability lobes considering the FRF at 0 rpm and the speed dependent FRF for the tool with an O-D ratio of 7:1 in Fig. 3.30.

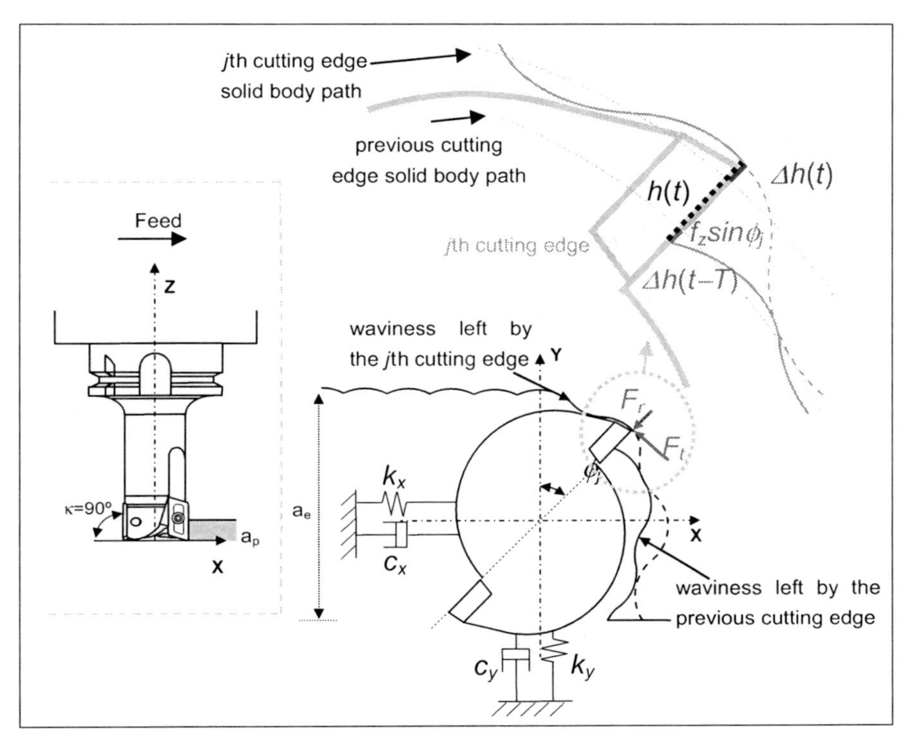

Fig. 3.29 Schematic diagram of a 2D model of milling

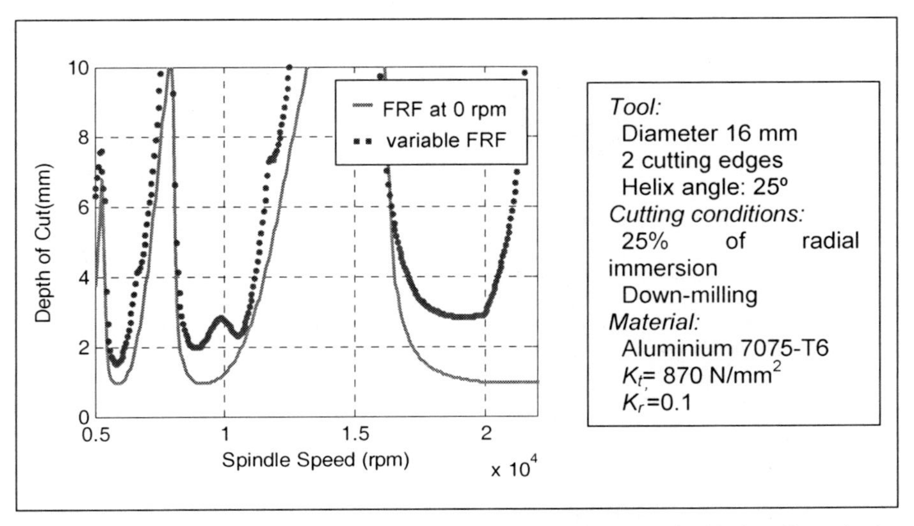

Fig. 3.30 Stability lobes obtained with the FRF measured at 0 rpm and with the effect of spindle speed on the FRF

Table 3.2 Fundamentals of methods for directly obtaining the stability lobes diagram

<table>
<tr><td colspan="5">Approach</td></tr>
<tr><td colspan="2">Resolution of the chatter loop</td><td colspan="3">Resolution of the periodic delay-differential equation</td></tr>
<tr><td colspan="5">Examples</td></tr>
<tr><td>Single-frequency [1]</td><td>Multi-frequency [7]</td><td>Semi discretisation [9]</td><td>Pseudo spectral Chebyshev</td><td>Temporal finite element [5]</td></tr>
<tr><td colspan="5">a) Chip thickness model</td></tr>
<tr><td colspan="5">$h_j\left(\phi_j\right)=\left(\Delta x\cdot\sin\phi_j+\Delta y\cdot\cos\phi_j\right)$; $\Delta x=x(t)-x(t-T)$; $\Delta y=y(t)-y(t-T)$</td></tr>
<tr><td colspan="5">b) Cutting forces model</td></tr>
<tr><td colspan="5">$\begin{Bmatrix}F_t\left(\phi_j\right)\\ F_r\left(\phi_j\right)\end{Bmatrix}=K_t\cdot a_p\cdot\begin{Bmatrix}1\\ K_r^{'}\end{Bmatrix}\cdot h_j\left(\phi_j\right);$
$\begin{Bmatrix}F_x\left(\phi_j\right)\\ F_y\left(\phi_j\right)\end{Bmatrix}=\sum_{j=1}^{Z}g(\phi_j)\begin{bmatrix}-\cos\phi_j & -\sin\phi_j\\ \sin\phi_j & -\cos\phi_j\end{bmatrix}\begin{Bmatrix}F_t\left(\phi_j\right)\\ F_r\left(\phi_j\right)\end{Bmatrix};$
$\begin{Bmatrix}F_x(t)\\ F_y(t)\end{Bmatrix}=a_p\cdot K_t\cdot\left[A(t)\right]\cdot\begin{Bmatrix}\Delta x(t)\\ \Delta y(t)\end{Bmatrix}$</td></tr>
<tr><td colspan="5">c) System dynamics</td></tr>
<tr><td colspan="2">$\begin{Bmatrix}\Delta x\\ \Delta y\end{Bmatrix}=\left(1-e^{-i\omega_c T}\right)\left[G(\omega_c,\Omega)\right]\cdot\begin{Bmatrix}F_x\\ F_y\end{Bmatrix}$
Operates with the FRF at the tool tip</td><td colspan="3">$[M]\{\ddot{x}\}+[C]\{\dot{x}\}+[K]\{x\}=\{F(t)\}$
$\left[M(\Omega)\right]$; $\left[C(\Omega)\right]$; $\left[K(\Omega)\right]$
Modal parameters at the tool tip</td></tr>
<tr><td colspan="5">d) Stability analysis procedure</td></tr>
<tr><td>Sweep the FRF searching the stability limit at each frequency</td><td>Sweep the spindle speed and the frequency of the FRF searching the stability limit at each spindle speed</td><td colspan="3">For a grid of cutting conditions, test if the eigenvalues of the matrix that defines the stability of the process have a modulus lower than 1</td></tr>
<tr><td colspan="5">Nomenclature</td></tr>
<tr><td colspan="5">h_j: Dynamic chip thickness of the jth cutting edge
ϕ_j: Angular position of the jth cutting edge with respect to the Y axis
z: Number of tooth of the tool
T: Tooth passing period, calculated as $T=2\pi 60/\Omega z$
$F_{t,r,x,y}$: Cutting force in tangential, radial, X and Y direction
K_t: Tangential shearing cutting coefficient
$K_r^{'}$: Radial shearing cutting coefficient normalised with respect to K_t
a_p: Axial depth of cut
$g(\phi_j)$: Window function equal to 1 when the jth cutting edge is cutting
$[A(t)]$: Directional coefficients T-periodic matrix
ω_c: Frequency of chatter in rad/s
Ω: Spindle speed in rad/s</td></tr>
</table>

3.5.3.3 Methods for the Reduction and Avoidance of the Chatter Vibration

The estimation of lobes explained above is the off-line approach to prevent the unstable machining. If the lobes are finely calculated, the user can select cutting conditions for stable machining with, at the same time, a very high axial depth of cut, in other words with a high removal rate. However this approach implies a complete analysis of machine dynamics, difficult to be carried out by industrial users, and an in-depth knowledge of the machining process and material. For this reason, several on-line approaches are being developed, monitoring the vibration or noise level and rapidly modifying the process parameters:

- Spindle speed variation: The CNC monitors the vibration or noise level and rapidly modifies the spindle speed. For example, the software Harmonizer™ by Metalmax® searches a stable speed [26]. A variation of the previous one is the continuous spindle speed variation (SSV) that continuously changes the spindle speed during the cutting process [6].
- Increasing damping: There are several methods for this, for example, using an active magnetic bearing to absorb dynamic forces [1], taking advantage of the damping properties of the electro-rheological fluids, or supporting the rolling bearing with a non-rotating hydrostatic bearing and regulating the pressure [20].

Regarding the tool and toolholder, there are several options to increase the dynamic behaviour of the spindle system:

- Using tools with integrated dampening like the Sandvik® Coromant CoroMill 390 or special boring bars.
- Reducing the tool O/D ratio. This is an advantage of five axis machining.
- Using monoblock tools, where the shank and toolholder are in the same body.
- Breaking the periodicity of the tooth impacts using variable pitch and variable helix tools.

3.5.4 The Thermal Behaviour

In machine tools and spindles, the three kinds of heat transference, conduction, convection and radiation, can be present and, furthermore, they may appear simultaneously. Conduction occurs, for example, in the contact area between the cutting tool and the workpiece. Convection occurs when the cooling emulsions, the *minimum quantity of lubricant* (MQL) or air, make contact with the rotating cutting tool, exchange heat and are then expulsed. And finally, radiation, which is the least important as far as machine tools are concerned, occurs when, for example, the temperature of an element rises because of the exposure to the sun or other nearby machines that behave as heat sources. All three forms of heat transference must be taken into account in order to improve machine tool behaviour in response to internal temperature changes from the machine tool elements and machining process, and external temperature changes.

The increase in spindle temperature increases the length of the spindle shaft. The thermal expansion can be represented as follows:

$$L_{T_1} = L_{T_0} \times \left(1 + \alpha \times \left(T_1 - T_0\right)\right) \quad (3.11)$$

where α is the thermal expansion coefficient which varies according to the material and is determined experimentally, T_1 and T_0 are the final and initial temperatures, respectively, and L_{T1} and L_{T0} are the corresponding lengths at each temperature. Equation 3.13 is of limited use since it assumes that the heat distribution is constant throughout the whole body and that it is homogenous and uniform, but it gives a good idea of the main factors involved. The thermal expansion depends on the material, the initial length of the element and the variation in temperature. Table 6.2 in Chap. 6 shows the thermal expansion coefficient for materials commonly found in the design and manufacture of machine tools. Therefore, to minimise the effects of heat, it is necessary to intervene in three areas: the material used should have a low thermal expansion coefficient, the shaft length should be minimised and the increase of temperature should be reduced by a cooling system.

Thermal growth affects the precision of the process. The shaft expansion "pushes" the tool axially toward the workpiece, so the machined surface location differs from the desired. This effect is clearly shown in Fig. 3.31, where a high speed spindle has been heated during 3.5 hours. Infrared pictures were taken in 30 minute increments. Over this time, the spindle temperature was increased from approximately 20°C to 41°C. The error between the initial image and the last for a simple slot machining resulted in a depth variation of 0.1 mm. One solution for reducing this problem is to compensate for the deviation in the spindle, which requires the measurement of the thermal induced deformation. Mazak® has developed and commercialised this solution in the heat displacement control function of their Mazatrol Matrix™ CNC control.

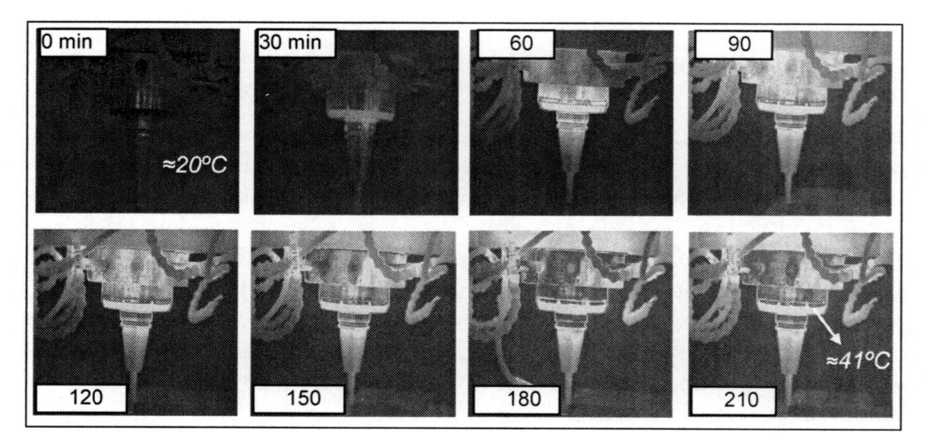

Fig. 3.31 Main spindle temperature during 3.5 working hours. Each image has been taken in 30 minute increments

The bearings are also affected by the thermal growth, changing their mechanical characteristics. The inner and outer rings and the rolling elements suffer a different thermal growth, so the initial bearing preload changes. Also, the shaft thermal expansion adds more preload in such a way that the inner race is forced into the bearing. As a consequence of the excessive preload, friction forces rise, the temperature is higher and the bearing life is reduced. There are several solutions to reduce this effect. For example, one technique used in high speed milling and grinding spindles is to mount the rear bearings on a floating housing preloaded with springs that allow axial movement. Another technique is to use a hydraulic device to compensate the preload on the bearings at high rotational speeds, so it is possible to maintain low friction forces [20].

3.5.4.1 Heat Sources

In order to avoid heat problems and increase the spindle service life, it is necessary to know about the heat sources to reduce their contribution. The main sources of heat that affect the spindle are the motors, the friction in bearings and the cutting process.

The heat generated by electric motors in stator and rotor is a function of torque and speed and it is defined by the motor efficiency. The electrical power that enters the motor is transformed into mechanical power, but also into heat, due to the primary copper losses, iron and stray losses and secondary copper losses. In built-in spindles the heat source is very close to the spindle shaft so the heat has to be properly evacuated. Belt-driven and gear-driven spindles also have heat sources in their transmission elements, pulleys and gearbox, respectively, due to friction. Another heat source is due to the viscous friction between the rotating elements and the air.

The heat generated in the bearings is due to the friction between the rolling elements and the inner and outer raceways and it depends on the speed, the preload and the lubrication. Faster speeds mean higher contact forces due to the centrifugal force and, hence, higher friction. In addition, the larger the diameter of the ball bearings, the faster the linear speed is and, again, friction rises. Although a heavy preload may be desirable to increase the spindle stiffness, it also implies higher contact and friction forces, thus the bearing life is reduced. Finally, poor lubrication drastically reduces the spindle bearings life at high speeds.

The heat produced in the cutting zone due to the chip removal mechanism can be transferred from the tool to the toolholder and the spindle. The magnitude of the problem depends strongly on the workpiece material, the tool and toolholder material, and the cutting fluid. Weck reported a case of thermal growth reduction of up to 50% in finish milling using a toolholder made of low expansion Invar material instead of tool steel [22].

3.5.4.2 The Refrigeration

Cooling the spindle is necessary in order to stabilise temperatures and avoid the negative effects of excessive heat and the temperature variations during spindle operation. That is the reason why coolant fluids are forced to circulate through various zones of the headstock and spindle structure to absorb the heat generated. The objective is to create a cooling "jacket" around the spindle and this is achieved surrounding it with coolant by means of cavities or channels; see Fig. 3.32. There is a variety of coolant fluids, the most typical of which are air, oil and water with additives to prevent oxidation of the conduits.

The most common spindle cooling for a good control of the temperature is based on a closed loop cooling system where the coolant that comes from the spindle is fed into an external refrigeration unit to cool it, and then is pumped back into the spindle. The refrigeration units have heat exchangers where the refrigerant circuit cools the coolant circuit. Closed loop cooling systems have a greater capacity for removing heat and need less time to do so than other heat exchange systems. They are controlled with a thermostat that controls the temperature of the coolant that circulates around the spindle, keeping it stable in such a way that the temperature of the coolant is not affected by room temperature.

Another solution consists on the continuous circulation of cool dry air around the spindle. More economic although more inefficient systems use the cutting fluid to cool the spindle. The cutting fluid is fed into the cavities of the spindle and then is sprayed onto the tool. This solution is a bit outdated as the circulating fluid is heated by the cutting process, so its cooling capability is limited.

3.5.4.3 Spindle Warm-up Cycles and Warm-up Systems

Keeping a stable spindle temperature is more important than simply reducing the temperature. The idea of a warm-up is to achieve a stable operation temperature during the entire machining process in order to reduce the effects of thermal expansion caused by temperature variations. The warm-up cycle should be carried out in conditions similar to the operating conditions, with the same cutting tools, if possible. This is because warming-up at feed rates and rotational speeds that are much higher or lower than subsequent operating conditions means that the warm-up temperature will be much higher or lower than the operating temperature, thus leading to thermal expansion errors. It is advisable to start up the machine in a dry run mode, for example. Nowadays, many machine tools have automatic spindle start-up and warm-up systems that bring the machine to the correct temperature ready for the operator to begin work as required.

There is also another method employed in special machines based on a system of warm-up circuits that carry fluid at a specific temperature to the different elements in the spindle. These systems are permanently engaged and controlled by a thermostat which is responsible for conducting the heated liquid to the spindle jacket and other elements. The fluid recirculates and is reheated to a set temperature considered to be optimum for the spindle operation.

3.5.5 Spindles in Use: Other Problems

Other problems related to the performance of the spindle that the machinist has to face are the runout of both the spindle and tool, which can lead to poor accuracy and collisions with the workpiece or with the elements of the machine, which may damage the spindle.

3.5.5.1 Runout

The term "runout" in machining describes the state of a spindle and tool when they have a rotational axis that differs from their geometrical axis. The *total indicated runout*, TIR, is the total distance measured from the maximum position in radial direction to the minimum position in the same direction; see Fig. 3.32. This factor limits productivity, as it negatively affects the dynamic balance, the chip load distribution, the part finish and the tool life. The runout can refer to the spindle or to the tool.

Eccentricities in the nose of the spindle can arise for various reasons, such as defective elements of the spindle, inaccuracies in the assembly, slack or simply the ageing of the spindle, among other possibilities. Although radial and axial runouts are measured at the nose of spindles, it is the radial runout that is the most significant.

Cutting tool runout is usually measured as a composition of radial and angular runout, and is produced by inaccuracies in the mounting operation. The result is that one tooth supports larger chip loads than the others. This error is due to deficiencies in the tool-collet adjustment, the degree of collet wear, or small impurities

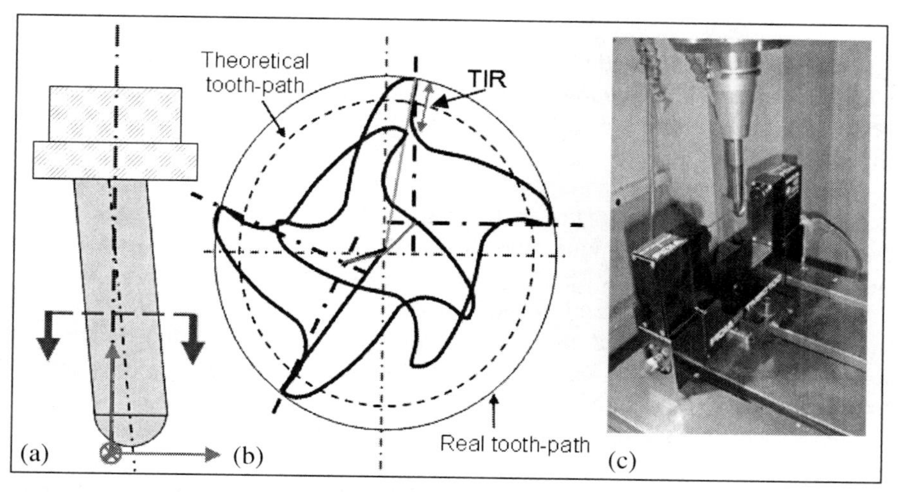

Fig. 3.32 **a** Tool runout schematic. **b** Effect of the tool runout on the real cutting edge path and *TIR*. **c** Laser control of tool runout

of chips or coolant that may have been left on the collet that render the grip of the cutting tool imprecise or less rigid.

Runout can be measured while the spindle is in operation (dynamic measurement) or at rest, using static tests. Dynamic measurement provides more reliable and precise results, although the instruments required are more expensive.

3.5.5.2 Collisions

Accidental collisions produced during material removing operations are one of the biggest worries for machine tool operators. They can have extremely negative, even irreversible, effects on the machine tool structure and especially on the spindle, the cutting tool or the workpiece. Generally, when a significant collision occurs, the machine undergoes an emergency shutdown to reduce damage and allow the operator to fix the problem if possible. The operator can also press an emergency shutdown button to stop the machine when necessary.

Nevertheless, accident prevention and safety involving machine tools has improved in recent years. One way to prevent collisions is to predict them using simulations of the machining process. Computer-aided manufacturing (CAM) programmes provide this kind of evaluation and analysis of the possible problems that may arise, so that they can be resolved before the actual work with the machine begins. Also, there are CNCs that can detect machine interferences online. It is the case of Mazak Mazatrol Matrix™ that uses a synchronised 3D model for checking collisions when an operator manually moves the machine axes.

3.6 Spindle Selection

As mentioned in the introduction, the spindle industry provides a large range of spindle options and configurations to satisfy the particular requirements of each sector. Figure 3.33 shows an example of three added options to the standard spindle, offered by EDEL®.

Selecting the most suitable spindle is an iterative process that must take into account the final characteristics of the end product and the requirements of the production system. Maximum spindle speed, effective power, torque, drive method, type of bearings, stiffness and dynamic behaviour, tool holder technology and thermal properties are all spindle characteristics that should be taken into consideration and evaluated to get the best final product specifications in terms of tolerances, surface finish, etc. and meet the production system requirements in terms of productivity, flexibility, etc.

Hence, choosing the right spindle is an important but far-from-easy task, as the wrong decision can negatively affect the production system and the performance of a company. The decision-maker has to be familiarised with spindle technologies in order to select the most suitable spindle bearing in mind all the factors

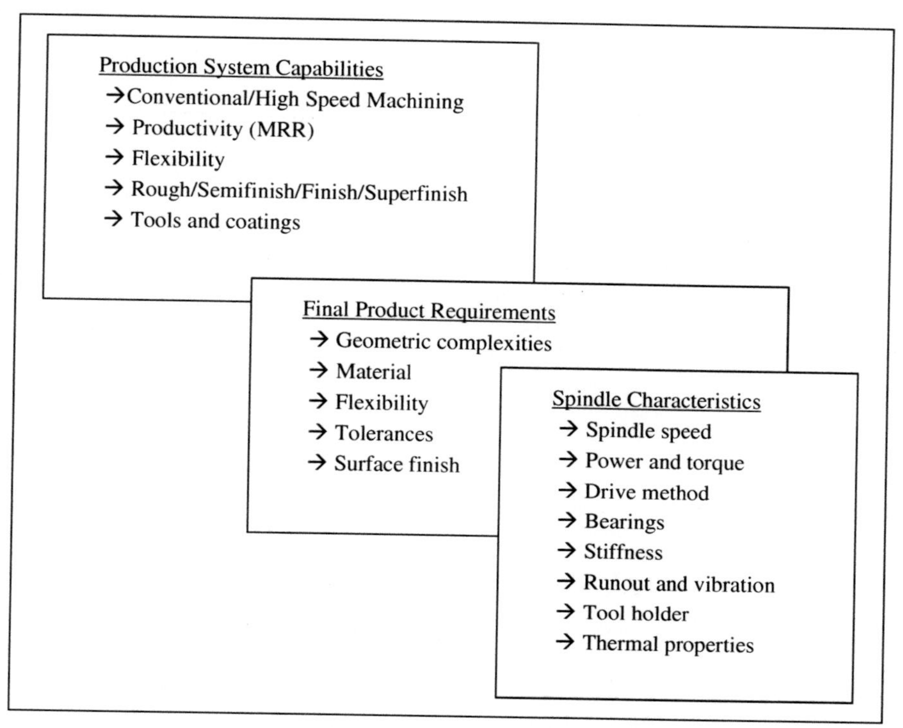

Fig. 3.33 Production system and final product requirements are decisive factors in selecting spindle characteristics

involved; see Fig. 3.33 where the basic steps for the selection of a spindle for a milling process are shown.

3.6.1 Conventional Machining or HSM

Conventional machining processes combine low spindle speeds and feed rates and high depths of cut and tool immersions. These processes require a spindle with high effective power and torque. In comparison, high speed machining means higher spindle speeds and increased feed rates to achieve high material removal rates.

Conventional spindles usually run at speeds lower than 10,000 rpm and their effective power does not exceed 25 KW. There is a relationship, or a compromise, between the effective power of the machine and its maximum rotational speed. As Fig. 3.34 shows, it is unusual to find machines on the market that can offer high spindle speeds coupled with high power. Generally, they are very expensive spindles and are oriented to the aluminium machining for the aeronautical sector, which demands this kind of spindles as the workpiece material and the tooling allow machining at very high speeds.

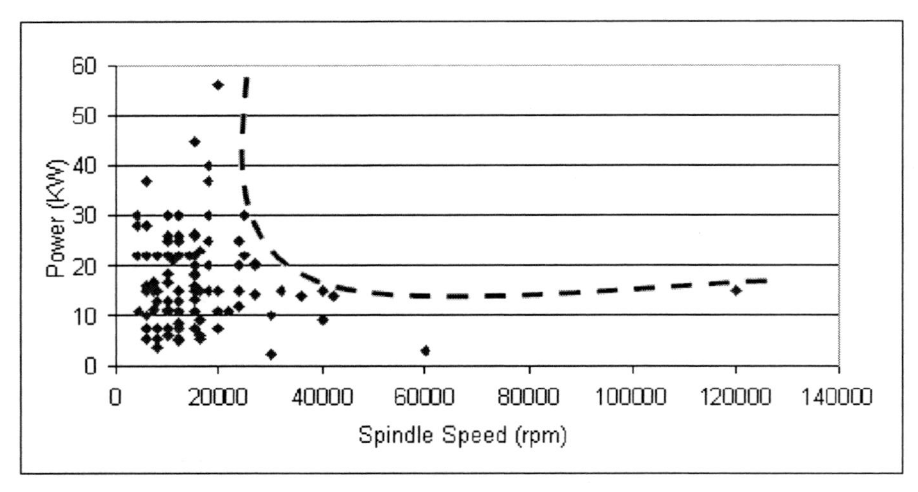

Fig. 3.34 Typical spindle speed vs power in machine tools

The choice of cutting tools, in terms of appropriateness for a workpiece material, geometrical characteristics and cutting parameters, is a factor that must be taken into account in the choice of a suitable spindle.

3.6.2 Tool Selection

The characteristics of the cutting tool used to carry out the machining influence the performance features required of the spindle. There is a huge variety of cutting tools in terms of shape, size, number of teeth, surface coating, performance, usage, manufacturer, etc.

The surface complexity, the accuracy and the surface roughness determine which operation and which cutting tool is needed. Such factors are considered again and again when programming roughing, semi-finishing, finishing and super-finishing operations. For simple geometries that require considerable roughing, large-diameter tools with several cutting edges are used. On the other hand, small-diameter tools are needed for complex geometries to ensure high precision requirements or highly smooth surfaces. For example, the mould and die industry commonly deals with hard materials, above 50HRC, and considerable precision requirements between 0.01 and 0.05 mm and smooth surfaces with a R_a of less than 2 μm.

The cutting parameters can be calculated with the help of the tables usually provided by manufacturers and by using the well-known semimechanistic formulas. The workpiece material conditions the cutting speed Vc which determines the spindle speed N needed in rpm:

$$N = \frac{V_c \cdot 1000}{D \cdot \pi} \tag{3.12}$$

where D is the diameter of the tool. Then, the feed speed V_f, in mm/min is calculated taking into account the recommended feed per tooth in mm and the number of teeth of the tool:

$$V_f = f_z \cdot z \cdot N \tag{3.13}$$

Once the radial depth of cut a_e and axial depth of cut a_p are selected, the volume of material removed Q in cm³/min can be calculated with the feed speed V_f:

$$Q = \frac{a_e \cdot a_p \cdot V_f}{1000} \tag{3.14}$$

3.6.3 The Workpiece Material

The effective power required to remove a certain volume in a specific time depends on the material. That value is easy to estimate from cutting tests for turning, but for milling it becomes more complex to obtain since the number of tooth in cut changes and the chip thickness is also variable. To solve that problem, an average value of the chip thickness h_m is calculated in mm and related to the specific cutting force k_{c1} measured in turning tests.

$$k_c = k_{c1} \cdot h_m^{-m_c} \text{ where } h_m = \frac{\sin\kappa \cdot 180 \cdot a_e \cdot f_z}{\pi \cdot d \cdot \arcsin\left(a_e/d\right)} \tag{3.15}$$

The factor m_c is used also to take into account the "size effect" cutting at low feeds. The second column of Table 3.3 shows the specific cutting force in N/mm² of several materials.

Table 3.3 Specific cutting force coefficients and specific power coefficients of several materials

Work material	Specific cutting force k_{c1} (N/mm²)	Specific powercoefficient K
Structural steels	1,600–1,800	4.0–5.7
Alloy steels	1,950–2,900	5.3–7.4
Cast iron	900–1,100	2.5–3.7
Titanium alloy	1,300–1,400	4.7–5.1
Aluminium alloy	400–700	1.3–2.1

3.6.4 Power and Spindle Speed Requirements

Once the volume of material removed and the specific cutting forces are known, the required power in kW can be calculated using the following equation, where η is the machine efficiency.

$$P = \frac{Q \cdot k_c}{60 \cdot 1{,}000 \cdot \eta} \tag{3.16}$$

However, to simplify the estimation of the specific cutting force in milling, nowadays tool manufacturers provide directly a specific power coefficient K dependent on the feed per tooth and the radial immersion a_e/D; see Table 3.3. Hence, the power required in kW can be simply calculated as:

$$P = \frac{a_p \cdot a_e \cdot V_f \cdot K}{100{,}000} \qquad (3.17)$$

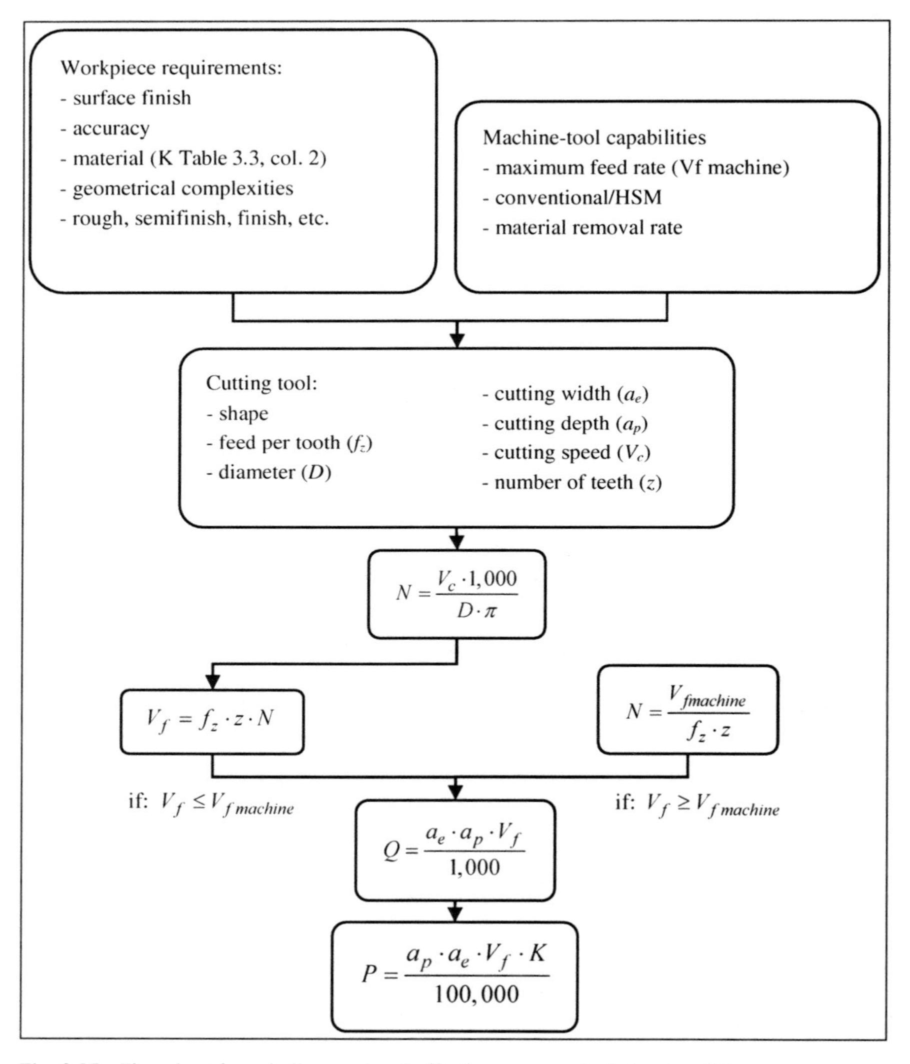

Fig. 3.35 Flowchart for spindle speed and effective power calculation in milling

The power obtained with this expression is then fine adjusted to include the effect of the rake angle of the cutting edges. Thus, the spindle rotational speed N and the effective power P requirements for the machining of a workpiece can be calculated, once we know which cutting tool will be used, which strategies will be employed for roughing and finishing in conventional machining or HSM, the material to be machined and the required precision, smoothness and geometrical complexity. The calculation of these parameters is an iterative process which is shown in the flow chart of Fig. 3.35.

With the spindle speed and the effective power required the spindle can be chosen. It must be remembered that the calculated feed rate cannot exceed the machine maximum feed rate. If a higher feed rate is obtained in the calculations, we will have to recalculate the required spindle speed N using the machine maximum feed rate $V_{f\,machine}$:

In this way, the capabilities of the machine tool and the technological limitations involved in milling a specific material are taken into account. If the milling machine is capable of offering the rotational speed, feed rate and effective power that has been calculated, then machining may proceed.

This process can be carried out in a similar way to calculate the parameters and requirements in other metal removing processes such as those using lathes, drills or grinding machines.

3.7 Brief Conclusions

The machine tool spindle plays an important role in machining operations because it provides the relative motion between the tool and the workpiece and the torque needed to perform remove material. Hence, spindle specifications have an enormous influence on machine tool performance and the production system flexibility. In this chapter, the basic elements of the spindles and the technology behind them have been presented as well as several aspects that characterise their behaviour in cutting conditions and some criteria to make a proper selection of spindles.

The advances in spindle technology have allowed high speed machining technologies to be developed. Higher cutting speeds have been possible due to the advances in bearing technologies, in lubrication systems, in mechatronics, with the vector control of motors with encoder feedback or the recently developed AMBs, in drive systems, in modelling to predict the spindle thermal and dynamic behaviour, etc. Nevertheless, further work is necessary to increase the present limitations of maximum spindle speed, maximum power available and spindle life. Furthermore, spindles have also become a tool for process monitoring and diagnosis. The increasing use of sensors is not only a source of information for the machinist to optimise the cutting process and check the spindle health, but also allows the use of control techniques and actuators to avoid dynamic and thermal problems online.

Acknowledgements Our thanks go to J. Ocaña from DMG®, J. Roca from Intermaher-Mazak®, R. Kortázar from Fidia®, A. Rainer Förster from GMN®, for Correa Anayak®, Zayer® and ASCAMM Technology Centre. Also, special thanks are addressed to J. Pacheco for his contributions on stability lobes identification methods and to Juanjo Zulaika and Jon Ander Altamira from Fatronik® for their valuable contributions on spindles design. Thanks are also addressed to the financial support of the Ministry of Education and Research of Spain (DPI2007-60624) and Etortek Program of the Basque Government. Finally, our gratitude to Professor López de Lacalle for his valuable suggestions.

References

[1] Abele E, Kreis M, Roth M (2006) Electromagnetic actuator for in process non-contact identification of spindle-tool frequency response functions, CIRP 2nd Int Conference on HPC Machining, Vancouver, Canada

[2] Altintas Y, Budak E (1995) Analytical prediction of stability lobes in milling, Annals of the CIRP 44/1:357–362

[3] Altintas Y, Weck M (2004) Chatter stability of metal cutting and grinding, Keynote paper, Annals of the CIRP 53/2:619–652

[4] Arnone M (1998) High Performance Machining. Cincinnati, USA: Hanser Gardner Publications

[5] Bayly PV, Halley JE, Mann BP, Davies MA (2003) Stability of interrupted cutting by temporal finite element analysis, J of Manufact Sci Engineer 125:220–225

[6] Bediaga I, Egaña I, Muñoa J, Zatarain M, Lopez de Lacalle LN (2007) Chatter avoidance method for milling process based on sinusoidal spindle speed variation method: simulation and experimental results, 10th CIRP Int Workshop on Modelling of Machining Operations, August 27–28, Reggia Calabria, Italy

[7] Budak E, Altintas Y (1998) Analytical prediction of the chatter stability in milling – Part I: General formulation, J of Dynam Syst Measure Contr 120:22–30

[8] Cao Y, Altintas Y (2007) Modelling of spindle-bearing and machine tool systems for virtual simulation of milling operations, Int J of Mach Tool Manufact 47:1342–1350

[9] Insperger T, Stépán G (2000) Stability of high-speed milling, Proceedings of Symposium on Nonlinear Dynamics and Stochastic Mechanics, Orlando

[10] López de Lacalle, LN, Sánchez JA, Lamikiz A (2004) Mecanizado de Alto Rendimiento. Procesos de Arranque (in Spanish, High performance machining). 1ª ed. Bilbao: Ediciones Técnicas Izaro, S.A

[11] Merrit H (1965) Theory of self-excited machine tool chatter, J of Engineer Indust 87: 447–454

[12] meweb.ecn.purdue.edu/~simlink/welcome.html

[13] Popoli W (2000) Spindle bearing basics, Manufacturing Engineering, Nov 2000

[14] Rantatalo M, Aidanpää J, Göransson, Norman P (2007) Milling machine spindle analysis using FEM and non-contact spindle excitation and response measurement, Int J of Mach Tool Manufact 47:1034–1045

[15] Salgado M, López de Lacalle LN, Lamikiz A, Muñoa M, Sánchez JA (2005) Evaluation of the stiffness chain on the deflection of end-mills under cutting forces. International Journal of Machine Tools and Manufacture, Vol. 45, pp. 727–739

[16] Shin YC (1992) Bearing nonlinearity and stability analysis in high speed machining, J of Engineer Indust 114:23–30

[17] Tlusty J, Ismail F (1981) Basic non-linearity in machining chatter, Annals of the CIRP 30/1:299–304

[18] Tobias SA, Fishwisck W (1958) A theory of regenerative chatter, The Engineer, London

[19] Wang KW, Shin YC, Chen CH (1991) On the natural frequencies of high-speed spindles with angular contact bearings, Proc Instn Mech Engrs 205:147–154
[20] Weck M, Hennes N, Krell M (1999) Spindle and toolsystems with high damping, Annals of the CIRP 48: 297–302
[21] Weck, M, Koch, A (1993) Spindle-Bearing Systems for High-Speed Applications in Machine Tools, Annals of the CIRP 42/1:445–448
[22] Weck, M, McKeown, P, Bonse, R, Herbst, U (1995) Reduction and Compensation of Thermal Errors in Machine Tools, Keynote paper, Annals of the CIRP 44/2:589–598
[23] Weck M (1984) Handbook of Machine Tools Volume 1: Types of Machines, Forms of Construction and Applications, John Wiley & Sons
[24] Weck M (1984) Handbook of Machine Tools Volume 2: Construction and Mathematical Analysis, John Wiley & Sons
[25] www.malinc.com
[26] www.mfg-labs.com
[27] Zatarain M, Muñoa J, Peigne G, Insperger T (2006) Analysis of the influence of mill helix angle on chatter stability, Annals of the CIRP 55/1:365–368
[28] Zulaika JJ, Azkoitia JM, Rodríguez M, Azpiazu P, Garate A (2002) Diseño de una gama de electromandrinos de alta velocidad (in Spanish, Design of a high speed electrospindles family), Proceedings of the XIV Congreso de Máquinas-Herramienta y Tecnologías de Fabricación, 2:831-850, ISBN: 931828-5-0

Chapter 4
New Developments in Drives and Tables

A. Olarra, I. Ruiz de Argandoña and L. Uriarte

Abstract In this chapter the different alternatives of machine tool drives are described. They show the industrial state-of-the-art in driving and guiding systems. Both linear and rotary drives are analysed from a descriptive perspective and some basic design principles are also explained. Examples are taken from well-known manufacturers. Finally, the latest trends in components for driving systems are shown, giving us a perspective of the near future machine tools.

4.1 Introduction

The present trends in the manufacture of machine tools claim higher performance in regard to the speed and accuracy of the machining. In order to fulfil these requirements, in principle antagonists, important efforts are being made in the development of new driving and guiding components and configurations, for linear as well as rotary drives. This "travel without end" centres its main activities on the following areas:

1. A constant development in servomotors.
2. A tendency toward the modularisation of slides and drives for families of machines. Perhaps it is not optimal from the technical viewpoint, but it is from the economic viewpoint. The advantages of this tendency originate from the advantages intrinsic in all modularisation: the reduction of machine development times, the reduction of design and assembly failures or errors and the

A. Olarra, I. Ruiz de Argandoña and L. Uriarte
Department of Mechatronics and Precision Engineering, Foundation Tekniker-IK4
Fundación Tekniker-IK4, Avda. Otaloa 20, 20600 Eibar, Spain
aolarra@tekniker.es

L. N. López de Lacalle, A. Lamikiz, *Machine Tools for High Performance Machining*,

reduction of part and component references. On the contrary, in an initial stage it requires an in-depth study of the family of machines to be developed, in addition to how to "split" them into modules. The specifications for these modules are even more difficult to obtain than those of the slides for a specific machine.

3. There is also a tendency not to develop individual components but to rather select and integrate; this tendency is above all clear in the case of guiding systems. In addition, for some time the machine tool manufacturers designed and had the means to manufacture the guiding of these, basically through friction systems. Presently, the tendency is to use roller guides developed by a different manufacturer and integrate these in the own designs.
4. The definition of the design specifications of the linear and rotary drives. Still, except for specific-purpose machines, the definition of forces, times, speeds, etc., involves a challenge which totally determines the drive and guiding configuration. As in all design exercises, the solution selected always results in being a compromise between requirements with a certain degree of incompatibility.

4.1.1 Precision and Dynamics

The continuing tendency to reduce machining times and improve accuracy requires optimising the dimensioning of all the machine components, including the driving and guiding systems.

The mechanical design of a drive determines the dynamic characteristics of this and, in turn, these characteristics limit the capacity of the servodrive control to act, that is, to perform appropriately even at the high gain values of the control loops. The basic control loop used to position a machine tool carriage is shown in Fig. 4.1. The main control parameters, *Kv*, *Kp* and *Ti* are shown too. As it can be seen, the position loop controller is a proportional gain (*Kv*) while the velocity loop has proportional (*Kp*) and integral (*Ti*) effects.

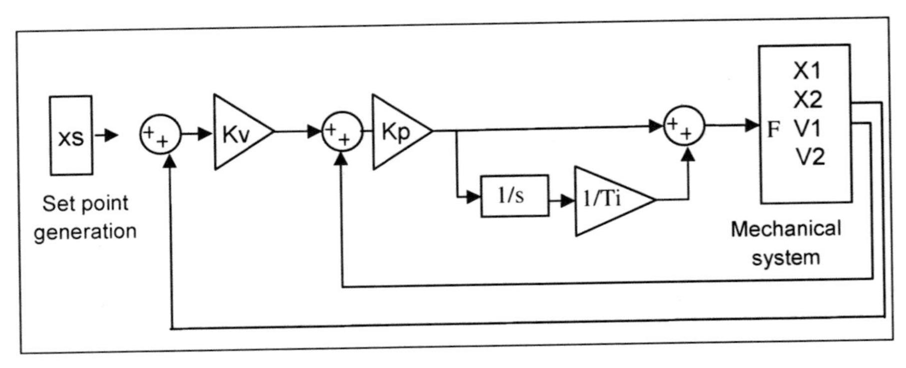

Fig. 4.1 Position and velocity control loops

On the one hand, a high gain value of the control loops results in a drive which responds faster in the case of set points and consequently provides a greater interpolation accuracy, without considering other CNC functions, such as feed-forward.

The control parameters allowable by the drive are related to the natural frequencies of the power transmission chain between the motor and the slide. In order to obtain high dynamics, it is necessary to increase the stiffness of the transmission chain and reduce the masses and inertia of the moving elements.

On the other hand, the high gains provide stiffness against interruptions. When the slide is submitted to a power stage, due to the friction force in a change of direction of movement, for example, the amplitude of the maximum drive excursion is inversely proportional to the product of the proportional and integral actions of the speed loop, while the recovery time is inversely proportional to the position loop gain. This behaviour can be observed in Fig. 4.2.

The drive bandwidth, directly related to the gains of the control loops, in addition to being limited by the natural drive frequencies, should be such that it does not excite the structure modes of the machine tool. Consequently, it is of interest to dimension the drive considering the natural frequencies of the structure.

It can be concluded that in view of the new accuracy and productivity requirements, the design of the drives takes on special importance given the dynamic character of these. For this, mathematical models are used which allow us to determine the natural frequencies of the drive.

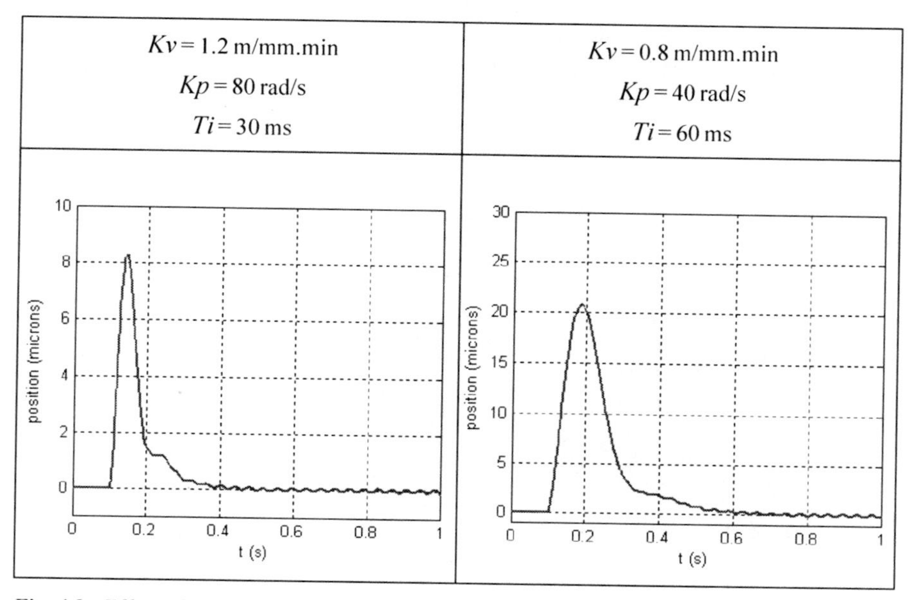

Fig. 4.2 Effect of the gains in the drive response against a force step of 1000 N

4.2 Linear Drives by Ball Screws

The drive by ball screw is the most widespread in the machine tool field for strokes not exceeding 4–5 m. The main characteristics which place it in such a favourable position are the high mechanical reduction provided maintaining a high efficiency and stiffness, as well as sufficient accuracy for existing machine tools.

A ball screw is a mechanical device which transforms the rotary movement into linear movement. Its operation is similar to a leadscrew in which the sliding between the threads of the leadscrew and the nut has been replaced by the rolling of balls which are recirculated in the nut.

4.2.1 Dimensioning

A correct dimensioning and selection of a ball screw must take into account several aspects.

4.2.1.1 The Elimination of the Gap

The accuracy requirements of existing machine tools require the use of drives in which the gap has been reduced to the largest extent possible. The ball screws used for the positioning of the slides of the machine tools tend to be preloaded to eliminate this gap. There are different systems to provide the preload.

The simplest preload system consists in using balls of a slightly larger diameter than the available space. This type of preload is only valid when the necessary preload is small; due to that the balls undergo sliding at the contact points, which generates a high wear and tear.

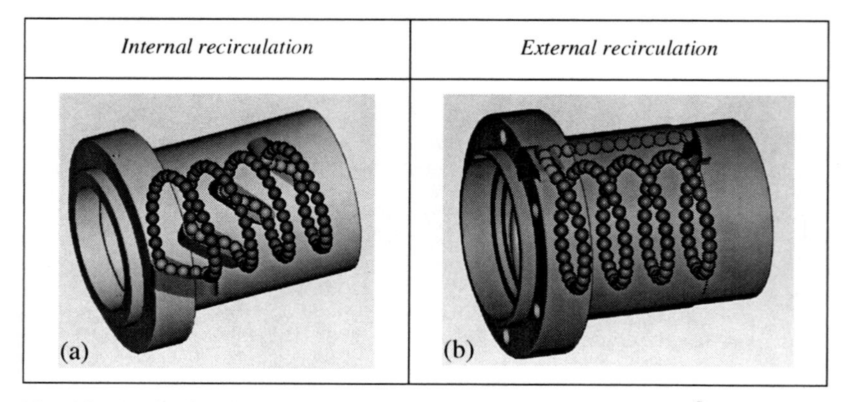

Fig. 4.3 Detail of the ball recirculation. **a** Internal. **b** External (by Shuton®)

When the necessary preload is average, a single nut can be used in which the pitch of the thread has been modified to a single intermediate thread of the nut. Thus, the balls only make contact at two points.

The most effective method for providing an average-high and well-controlled preload consists in using two nuts between which a separator is intercalated, whose thickness determines the preload force. In Fig. 4.3, the two nuts are clamped.

When the screw is subject to forces, the preload on one of the sides of the nut increases and is reduced on the other. It can be demonstrated that the preload is lost when the force applied is 2.83 times the value of this [3]. In general, this situation should be avoided because, in addition to the gap, it generates a premature wear and tear of the balls.

The preload increases the wear and tear and heat generation. For this reason, the preload is limited to approximately 12% of the dynamic load capacity, with a normal value being between 6% and 8%.

4.2.1.2 Types of Recirculation

The recirculation of balls can be carried out either through an internal deflector or, otherwise, through an exterior tube. Whenever possible, it is preferable to use the deflector through an exterior tube, due to the fact that it generates less noise, heat and wear and tear. In spite of this, the diameter of the nut with exterior recirculation is larger, which conditions its use in certain designs. Figure 4.3 shows different types of ball recirculation.

4.2.1.3 Critical Speed and Deflection

When the screw turns, the main limiter of the maximum allowable speed is the "whip" phenomenon. This involves the instability which occurs when the turning speed of the screw is the same as the bending frequency. This frequency depends on the diameter of the screw, the unsupported length and the type of supports. In practice, the turning speed of the screw is maintained under 80% of the calculated limit value. The critical turning speed can be determined by:

$$N_{cr} = \frac{\lambda^2}{2 \cdot \pi \cdot L^2} \sqrt{\frac{E \cdot I}{\rho \cdot S}} \cdot 60 \tag{4.1}$$

where:

N_{cr}: critical speed (rpm)
E:Young modulus
I: Section inertia
S: Section area
ρ: density
L: unsupported length

Table 4.1 Critical speed function of the type of supports

Type of supports	λ	N_{cr}
Fixed-fixed	4.730	$221 \cdot d/L^2$
Fixed-supported	3.927	$152 \cdot d/L^2$
Fixed-free	1.875	$34 \cdot d/L^2$

The factor λ depends on the type of supports used (see Table 4.1). In the following table, the allowable speed (rpm) of a screw is shown, limited to 80% of the critical speed, according to the diameter (mm), the unsupported length (m) and type of supports.

A possibility for increasing the critical speed of a screw is through the use of "rests". In this manner, the distance between the supports is reduced such that the bending frequency of the screw increases substantially. As the slide moves, the rests have to automatically withdraw to prevent colliding with the nut. In practice, the use of rests is quite complicated and not widespread.

Another aspect to consider when long screws of more than 4 or 5 metres in length are used is the deflection they acquire due to their own weight. It is desirable that the deflection does not exceed 1 mm, in order to avoid important radial stress on the nut which give rise to an accelerated wear and tear of the drive. The rests also help to reduce deflection when long screws are used.

4.2.1.4 The Preload

The heat generated during the operation, mainly due to the friction torque between the screw and the nut, results in a temperature rise of both components. A temperature rise between 5°C and 10°C is usual. If the screw is embedded at both ends, the expansion due to the temperature rise may result in buckling which drastically affects the performance of the drive. To avoid this buckling, the screw is usually preloaded to compensate for the estimated expansion.

4.2.1.5 The d·N Value and Lubrication

Another limiting factor related to the speed is the sliding of the balls within the roller races. This condition may occur when due to the high acceleration of these, the necessary forces to guarantee rolling cannot be transmitted. Usually, this is quantified through the $d \cdot N$ product value. This is the product of the diameter (mm) and turning speed (rpm) of the screw. An average value for the leadscrews normally used in machine tools is 125,000 mm·rpm. The allowable value increases when exterior recirculation is used or lubrication is carried out either with oil or oil-air instead of grease. The use of a small-diameter ball or ceramic balls also helps to increase the allowable $d \cdot N$ value. Exceeding the allowed value results in an accelerated wear and tear of the ball screw.

The processes which require positioning with short movements, in the neighbourhood of a few millimetres or less than one millimetre, may give rise to accelerated wear and tear of the screw due to the difficulty in lubricating the work area. In this case, special attention should be given to the type of lubrication used.

4.2.1.6 The Load Capacity and Life

Two load capacities are distinguished. On the one part, the static load capacity (*C0*) is that which generates a permanent deformation of 0.01% of the rotating elements. Operation under this condition should be avoided in order to prevent noise and a premature wear and tear of the drive. For machine tool applications it is necessary to apply a high safety factor to prevent permanent damage in the drive. When the forces to which the drive will be submitted are known, an appropriate safety factor is 2.5. This coefficient should be higher when the forces are not known with certainty, or when the drive will be subject to impact and vibration. In practice, the static load capacity does not tend to be a limitation at the time of selecting a drive.

On the other hand, the dynamic load capacity (*C*) is related to the life of the ball screw. This involves the load under which 90% of the screws reach a period of life of one million revolutions.

The load capacity basically depends on the diameter of the screw, the diameter of the balls and the number of load bearing balls (number of circuits). This is also affected by the surface hardness of the roller races and manufacturing tolerances.

Figure 4.4 shows the dynamic load capacity of ball screws according to the diameter of the screw. In order to show representative data, the diameter of the ball

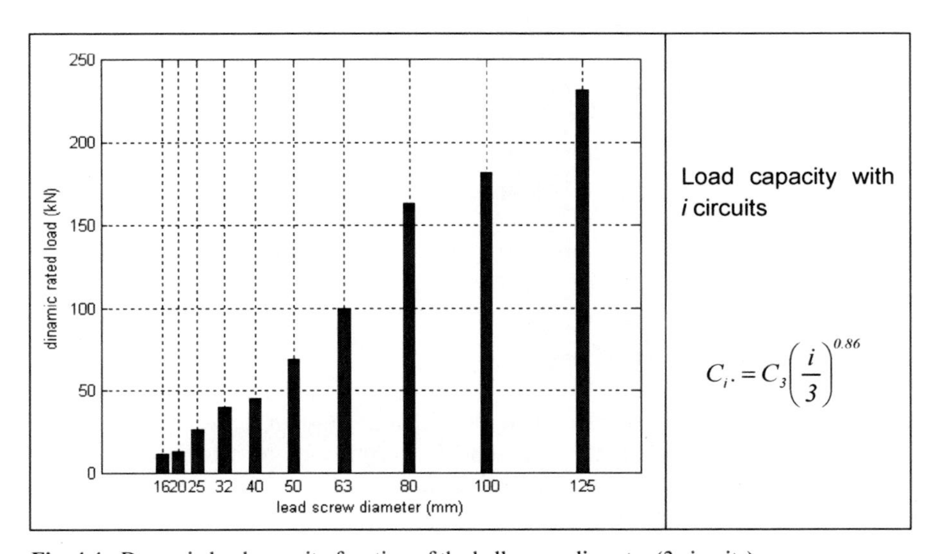

Fig. 4.4 Dynamic load capacity function of the ball screw diameter (3 circuits)

Table 4.2 Life calculation of preloaded ball screws

$F_{ma(1)} = F_{pr}\left(1+\frac{F_m}{3F_{pr}}\right)^{3/2}$	$F_{ma(2)} = F_{ma(1)} - F_m$
$L_{10(1)} = \left(\frac{C_{am}}{F_{ma(1)}}\right)^3 \cdot 10^6$	$L_{10(2)} = \left(\frac{C_{am}}{F_{ma(2)}}\right)^3 \cdot 10^6$
$L_{10} = \left(L_{10(1)}^{\frac{-10}{9}} + L_{10(2)}^{\frac{-10}{9}}\right)^{\frac{-9}{10}}$ (revolutions)	
$L_h = \frac{L_{10}}{n_m}\cdot\frac{1}{60}$ (hours)	$L_{km} = L_{10}\cdot p\cdot\frac{1}{10^6}$ (km)

most often used for each screw diameter has been selected. In addition, the load capacities correspond to preloaded screws with three circuits in each half nut. The ratio indicated in the same figure can be used to estimate the dynamic load capacity for screws with a different number of circuits.

The duration of the ball screw is obtained from the preload force, the average axial force which acts on the ball screw and the load capacity. The calculation according to standard DIN 69051 (ISO 3408-5) is summarised in Table 4.2.

To calculate the life, it is necessary to determine the average axial force which acts on the screw. This average load is determined using the Miner's rule method for fatigue calculations. In this case, the equivalent force is determined according to Eq. 4.2.

$$F_m = \sqrt[3]{F_1^3 q_1 \frac{n_1}{n_m} + F_2^3 q_2 \frac{n_2}{n_m} + \ldots + F_n^3 q_n \frac{n_n}{n_m}} \text{ (N)} \tag{4.2}$$

Similarly, the average speed is obtained as:

$$n_m = q_1 n_1 + q_2 n_2 + \ldots + q_n n_n \text{ (rpm)} \tag{4.3}$$

4.2.1.7 Dynamic Models of the Drives

As was indicated at the beginning of this chapter, the capacity to estimate the dynamic characteristics of the drives in the design phase of the machine is presently of particular importance. Mathematical models, such as that shown in Fig. 4.5, have been developed for this purpose.

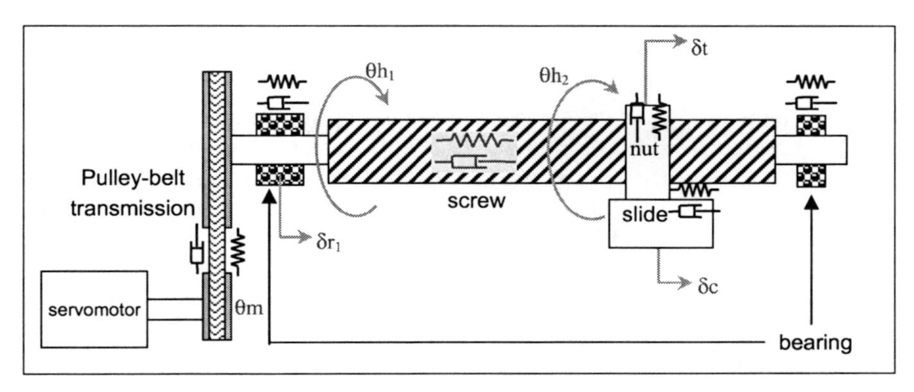

Fig. 4.5 Dynamic model of a ball screw drive with pulley transmission

The models used for this purpose should have sufficient detail to be valid in all cases in which they are used. As a reference, in the model shown previously, the degrees of freedom used can be distinguished in blue.

From the selection of the degrees of freedom, the calculation of the mass and stiffness matrices can be completed, from which it will be possible to obtain the natural frequencies of the drive.

The bandwidth of the control loops, that of speed as well as position, will be limited by the first natural frequency of the system shown in Fig. 4.5. Thus, it can be estimated that the bandwidth of the speed loop can reach up to 80% of the first natural frequency, and the position loop in turn at 25% of the bandwidth of the speed loop.

The dynamic models of the drives [6] also permit us to determine the sensitivity of the allowable gains vs design variables such as either the diameter and pitch of the screw or the rigidity of the coupling. Table 4.3 shows an example of the type of information which can be obtained. The table shows the increase in Hz of the first natural frequency vs changes of 10% of different parameters.

Table 4.3 Sensitivity of the 1st natural frequency vs changes of 10% of different parameters

$\Delta\ \omega_1, \omega_1 = 35.7$	Parameter
1.71 Hz	Coupling torsional stiffness
0.16 Hz	Axial stiffness of the 1st bearing
0.16 Hz	Axial stiffness of the 2nd bearing
0.36 Hz	Ball screw axial stiffness
0.12 Hz	Ball screw torsional stiffness
0.70 Hz	Nut axial stiffness
1.73 Hz	Mass of the slide plus workpiece

4.2.2 The Rotary Screw

The most common way of using a ball screw consists in the rotary screw arrangement (Fig. 4.6). The screw is supported on roller packs in the fixed part of the machine and is operated by a rotary servomotor. The nut is fixed to the mobile slide, such that it is prevented from turning. In this manner, the turning of the screw generates a shift of the slide.

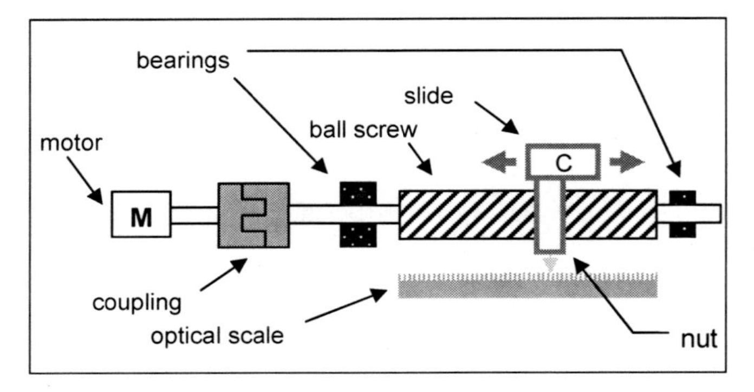

Fig. 4.6 The standard configuration by rotary screw

4.2.3 Other Configurations

In certain cases, the configuration described previously is not the most appropriate. A common variant is a rotary nut (Fig. 4.7). In this configuration, the nut turns instead of the screw. This solution provides two advantages: on the one hand, it eliminates the problem of critical speed of the screw and on the other, it may reduce the total inertia of the drive.

With long rotary screws (in excess of 3 metres), one of the main limitations is the reduced critical speed even with embedded supports. In order to reach the

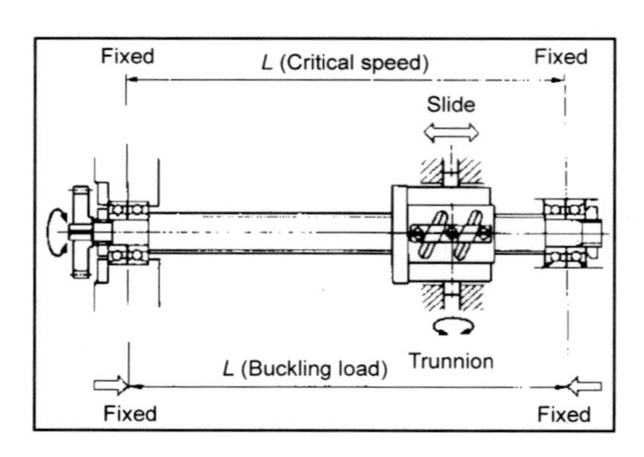

Fig. 4.7 Optional configuration by rotary nut

desired maximum speed without encountering problems with the critical speed it is necessary to use long pitches per turn, which to a great extent conditions the remaining requirements.

On the other hand, when the screws are long or the pitches are very small, the rotary inertia may be large, even more so than that of the slide on moving. In this case, a solution based on a rotary nut provides greater acceleration with a small motor.

When a rotary nut is used, the nut and motor can be shifted, as well as the screw. The second solution is used to drive slides whose length is greater than the required stroke, for example long worktables. Through the rotary nut solution, slides are driven with strokes of up to 10 or 12 metres, although 6 or 8 metres are not usually exceeded.

4.3 Linear Drives by Rack and Pinion

Drives through rack and pinion are preferably used when the stroke of the slide to be used is greater than 4 or 5 metres.

The main advantage of the rack and pinion drive vs ball screw drive is that the characteristics do not depend on the slide stroke. The slide stroke does not limit either the maximum speed or the drive or affect the stiffness, unlike what occurs in case of the ball screw. In addition, it permits us to obtain high feed rates and it is common to reach 120 m/min.

On the other hand, the rack and pinion drive does not provide the high reduction that the ball screw can provide. In order to provide a greater reduction, pinions of as small a diameter as possible should be used, between 15 and 20 teeth. Therefore, a reducer should be used to obtain the same force and acceleration performance as the screw.

4.3.1 The Elimination of the Gap

A basic aspect for any type of machine tool drive is the elimination of gaps. In case of the rack and pinion drive, the solutions to provide an operation without gaps have undergone an important development in the last few years.

In this type of drive, the main gap is between the teeth of the pinion and teeth of the rack. To eliminate this gap, two pinions are used working on the same rack, each supported on a different side of the rack and consequently pushing in a reverse direction. Other systems, such as the split pinion with a single reducer, or any solutions used in the past, did not completely eliminate the intrinsic play which exists on the inside of the reducer and, consequently, does not result in being a valid option for the machine tool.

There are different ways of providing a preload between both gearboxes. This can be provided through mechanical means (Fig. 4.8). The following figure shows two

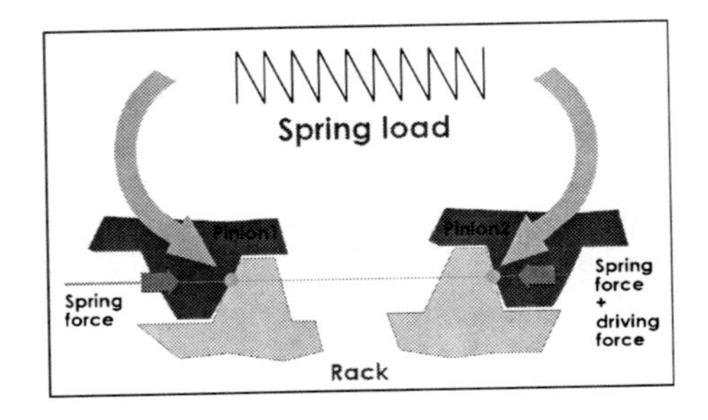

Fig. 4.8 Preload system for backlash suppression by Redex Andantex®

gearboxes, each with its exit pinion meshed in the rack. The entry shafts of both gearboxes are connected by means of an elastic coupling which allows to adjust the desired preload. Both gearboxes are driven by a single motor (Figs. 4.9 and 4.10).

The present tendency to eliminate the gap is through the use of two independent motor-reducer-pinion assemblies. In this case, the preload is managed on an electronic basis, making one motor operate against the other (Fig. 4.11).

This solution permits us to simplify the mechanics of the drive at the cost of greater complexity for control. An advantage of the preload through two independ-

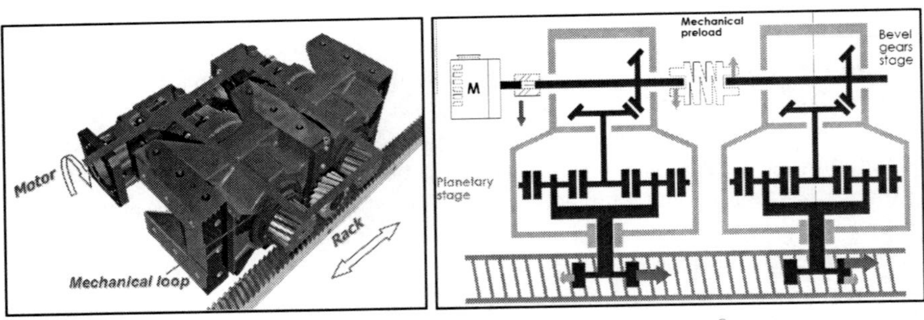

Fig. 4.9 Rack and pinion drive with mechanical preload (Redex Andantex®)

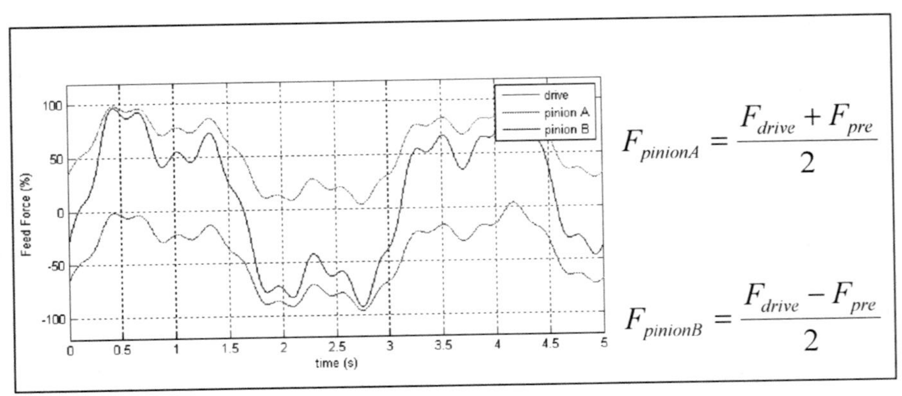

Fig. 4.10 Repartition of the load between pinions with 100% of mechanical preload

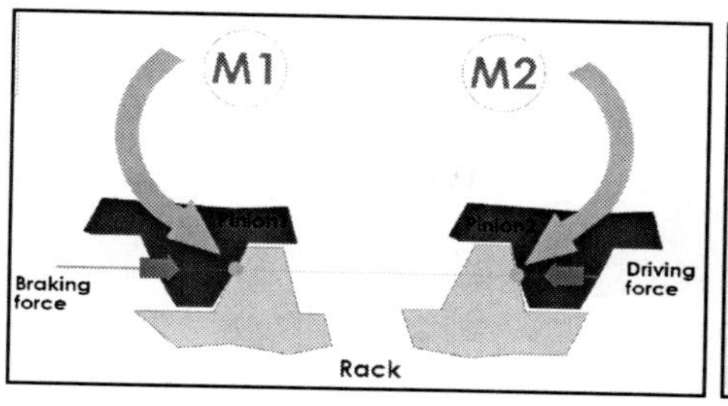

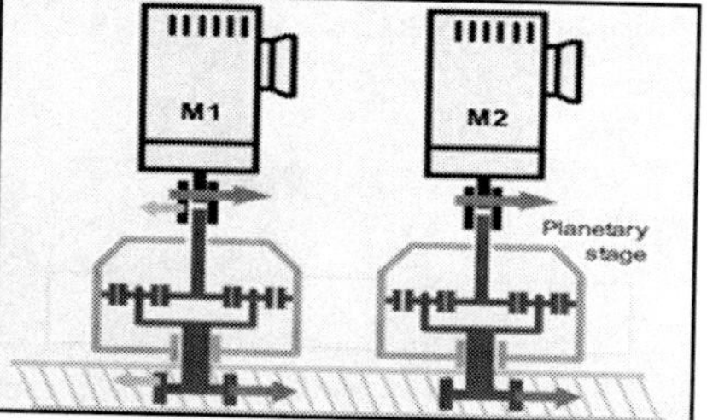

Fig. 4.11 Rack and pinion drive with electrical preload (Redex Andantex®)

ent motors is the possibility of making the two motors operate in the same direction in case it is necessary, for example, to accelerate. It is common to configure the preload between 25% and 50% of the maximum force to be provided by the drive.

4.3.2 Dimensioning

The allowable force to be transmitted by a rack and pinion drive basically depends on the modulus of the teeth. To a lesser extent, it also depends on the number of teeth of the pinion, the pitch angle (straight or helical pinions with an angle of 19° 31' 42'') and the reduction ratio provided by the reducer. Figure 4.12 shows certain illustrative values of the allowable force for each pinion according to the modulus.

The dimensioning of the drive should consider nominal operating conditions as well as intermittent operating conditions. In addition, each pinion should be calculated separately, considering the present forces according to the force required of the drive and preload force.

The nominal force values shown in Fig. 4.12 are estimated to reach a life approximately of 15,000–20,000 hours of operation. These nominal force values

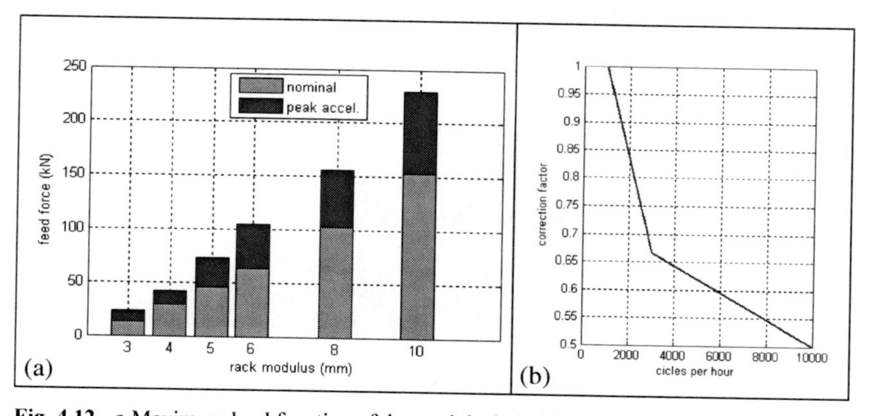

Fig. 4.12 **a** Maximum load function of the module. b Correction factor per cycles

should be compared with the equivalent average force on the pinion, calculated as:

$$F_m = \sqrt[3]{F_1^3 q_1 \frac{n_1}{n_m} + F_2^3 q_2 \frac{n_2}{n_m} + \ldots + F_n^3 q_n \frac{n_n}{n_m}} \text{ (N)} \tag{4.4}$$

On the other hand, the intermittent operating conditions should consider the maximum forces required and the number of cycles per hour. In machine tool applications, it is common to require the maximum drive from its capacities during acceleration and deceleration of the slides. The required force of the drive during acceleration should be less than the peak forces indicated in Fig. 4.12. In addition, if the number of cycles per hour is high, the allowable forces are reduced according to the correction factor shown in the same figure.

In the absence of further information regarding the conditions under which the machine will operate, it can be estimated that the machine will accelerate at the maximum capacity for 20–30% of the total time, with the acceleration time being estimated as per Eq. 4.5.

$$t_a = \frac{v_{\max}}{a_{\max}} \tag{4.5}$$

The number of cycles per hours is:

$$\frac{cicles}{hour} = \frac{20}{100} \cdot \frac{3600}{t_a} \tag{4.6}$$

4.3.3 Dynamic Models of the Drives

As in the case of rotary screw drives, in case of rack and pinion drives, the possibility of estimating the dynamic characteristics of the drive in the design phase is of particular importance. The dynamic models used for this are similar to those described in Sect. 4.2.1.

4.4 Linear Drives by Linear Motors

Today, another important type of drive for machine tool slides is the linear motor (Fig. 4.13). The type of linear motor which has been imposed for machine tool applications is the ironcore synchronous linear motor.

This consists of a coil slide or primary and magnet plates as secondary, both on a linear basis. One of these is fixed on the moving part and the other on the fixed part.

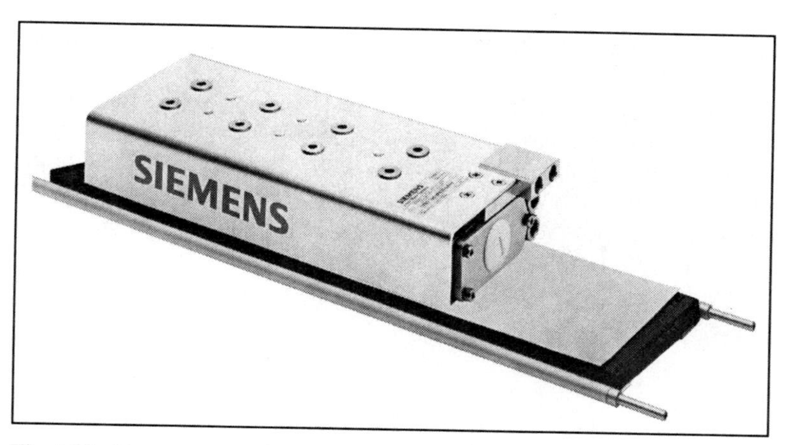

Fig. 4.13 Linear motor series 1FN from Siemens®

The main characteristic of the linear motor is the absence of mechanical elements of transmission of movements and/or forces. On eliminating these mechanical elements, wear and tear, gaps, lubrication requirements and noise are also eliminated. Another important advantage is the absence of natural resonance frequencies of the drive. However, at the same time, the mechanical advantage is lost, such that the force generated by the motor is directly the force exercised on the slide, and under no circumstances can it be higher.

In practice, the linear motor is an interesting option provided that the forces of the machining process are not considerable and high dynamics are required [7]. That is, when a high acceleration substantially increases the productivity of the machine. For example, in the case of a sharpener.

As has been indicated, one of the main drawbacks of the linear motors is the impossibility of multiplying the generated force. Consequently, a large amount of magnetic material and winding is required to provide sufficient force to use these as slide drives of machine tools. On the one hand, this amount of magnetic material noticeably increases the price of the drive and, on the other, the linear motor is less efficient than a conventional motor used with any kind of reducer, particularly due to Joule losses.

On the other hand, as has been mentioned, the linear motors used in machine tools are provided with ironcore coils. Consequently, the magnet plates exercise a force of attraction over the coil on a continuous basis. This force of attraction is more or less constant and depends little on the force that the motor is generating on an instantaneous basis. The force of attraction is approximately three times the maximum force that the motor can provide, that is, for a motor of 6,000 N of maximum force; the force of attraction is in the neighbourhood of 2 Tn. This point should be considered for the appropriate dimensioning of the guidance of the slide driven by the linear motor.

When a linear motor is used, the thermal aspect should be considered. The heat generated by a linear motor is considerable, such that it is necessary to carry out a study to determine the impact that this heat may generate, for example expansion, and assess the need to cool the motor.

The rated force that a linear motor can provide depends to a great extent on the cooling system used. As a reference value, it can be considered that for a motor cooled by means of natural convection, the rated force is approximate 20% of the maximum force. This value may be increased up to 40% if water cooled.

The supply of linear motors for machine tools covers from 300 N of maximum force up to around to 20 kN.

$$F_m = \sqrt{F_1^2 q_1 \frac{n_1}{n_m} + F_2^2 q_2 \frac{n_2}{n_m} + \ldots + F_n^2 q_n \frac{n_n}{n_m}} \text{ (N)} \tag{4.7}$$

The speed that a drive can reach with a linear motor is comparable or higher than the speed provided by rack and pinion drives. It is common to reach 300 m/min in case of medium-sized motors, up to 5,000 N of peak force, and 150–200 m/min in case of larger motors.

4.4.1 Mounting

The mounting tolerances of linear motors are not as narrow as those necessary for the correct alignment of screws or racks. In spite of this, the gap between the coil and the magnets should be maintained within a few tenths of a millimetre. Sufficient stiffness of the slide as well as of the guidance should be guaranteed in order that they do not deform in case of a powerful force of attraction between the magnets and coil.

4.4.2 Configurations

When using a linear motor, the coil may be used as a moving part and the magnets as a stationary part or vice versa (see Fig. 4.14). Each configuration has its advantages.

When the coil is moving, the method of carrying the supply cables and cooling lines up to the coil has to be considered. In some cases, the problem of accessibility to the coil below the mobile slide may be presented.

In case the magnets move, the cables and lines will be on the fixed part, facilitating their installation. In addition, the magnets tend to weigh less than the coil, thus gaining some acceleration. The main drawback of this solution is that the length in which the magnets should be installed is greater. The slide should have this length to install the magnets.

An interesting alternative is to have two linear motors in order that the forces of attraction are mutually cancelled. This point is particularly interesting when

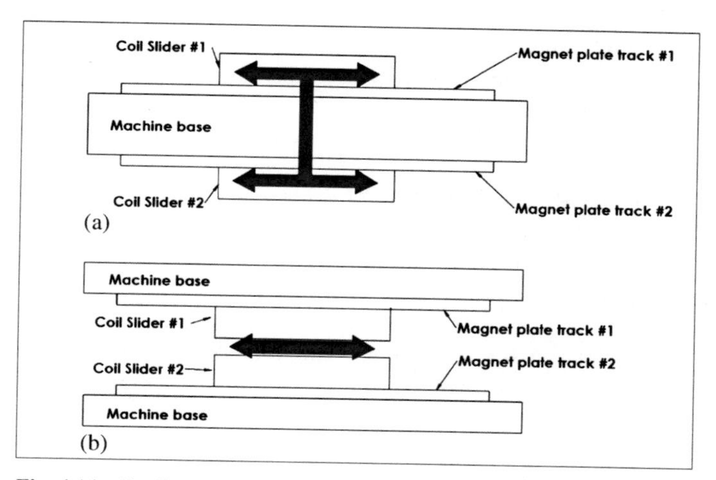

Fig. 4.14 Configurations with linear motors. **a** Movable coil, **b** Movable magnets (by Fanuc®)

a machine which should be lightweight is being designed. Two possible symmetrical arrangements which cancel the forces of attraction are shown below.

4.5 Rotary Drives

The tendency of the requirements stipulated for rotary tables is similar to that required for linear slides. A fast dynamics is desired with a high accuracy of positioning and interpolation.

Depending on the process for which the rotary table is designed, the speed and force specifications can be quite different. In case of tables for lathes, the maximum speed and allowable pairs obtain a greater weight than high dynamics and accuracy. The case is different in case of five-shaft milling tables.

In case of rotary tables, two large drive groups are distinguished. On the one hand are the mechanical transmission drives and on the other direct drives.

4.5.1 Mechanical Transmissions

Two of the most common mechanical transmissions for rotary drives are worm-gear and wheel-gear.

4.5.1.1 Worm-gear Transmission

One possibility is the worm-gear transmission. This type of transmission provides a very large reduction, a high capacity to transmit force and quite reduced wear.

The main difficulty, in addition to the cutting of the teeth, consists in eliminating the gap. There are different possibilities for this. It is possible to cut the worm gear such that the pitch is variable, such that shifting it axially may eliminate the gap up to the desired level. The wear and tear caused by use can be eliminated in the same way. Another option for eliminating the gap consists in using a split worm gear preloaded with a spring against the teeth of the gear.

4.5.1.2 Wheel-gear Transmission

Another option for driving rotary tables consists in using two pinions which act on a gear. The gap is eliminated in a similar manner to that in the rack and pinion drive, preloading the pinions by mechanical means or otherwise using two motors.

4.5.2 Direct Rotary Drives

Nowadays, the drive which provides the best dynamic performance for rotary tables is the direct motor (Fig. 4.15). In addition, there are no gaps or wear and tear.

As occurs in linear motors, in this case special attention should be given to the thermal aspect. Normally, it is necessary to provide water cooling to the torque motors.

Torque motors are supplied in two formats. On the one hand, there are complete motors, in which the exit shaft normally tends to be hollow. On the other hand, in order to be able to provide a more compact solution and better integrated in the product designed by the machine tool manufacturer, the torque motor suppliers also offer the motor in two units, rotor and stator.

The present supply of torque motors provides rated torques of up to approximately 5,000 Nm and a range of speeds from 200 rpm to 1,000 rpm. The peak torque tends to be in the vicinity of 80% higher than the rated.

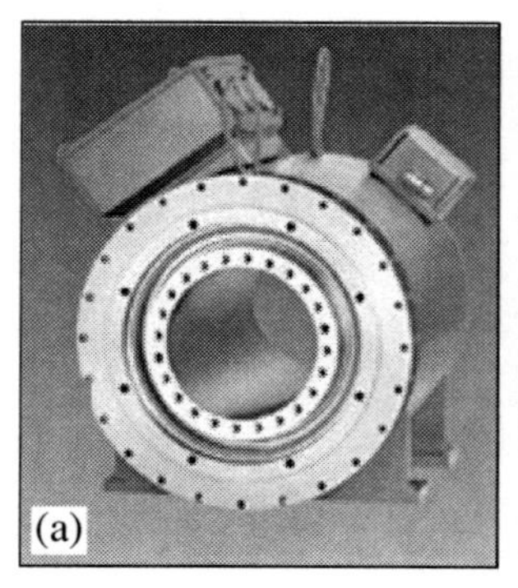

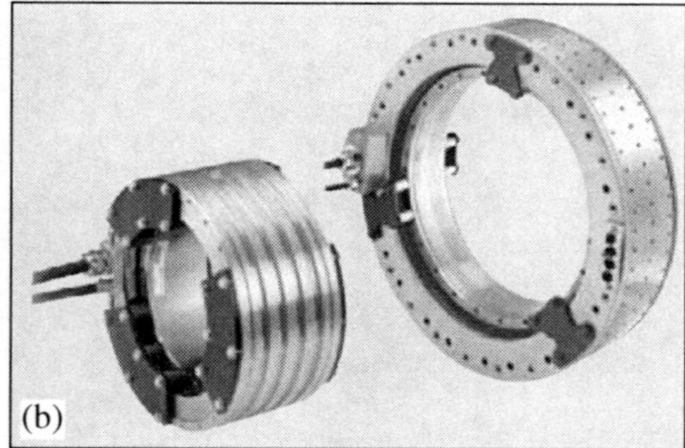

Fig. 4.15 **a** Complete torque motors. **b** Built-in torque motors (Siemens®)

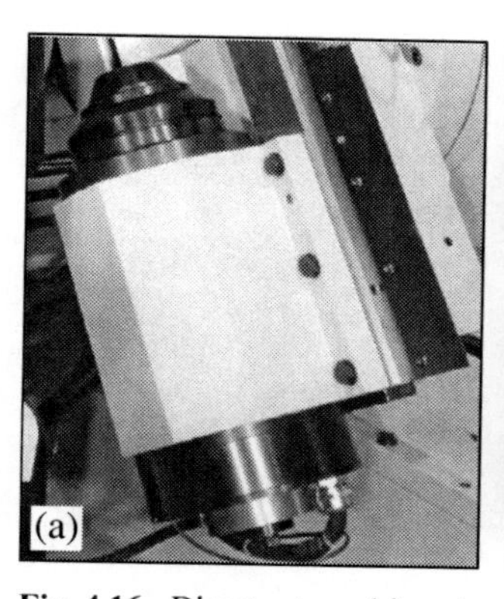

Fig. 4.16 Direct rotary drives in profile grinding machines [5]. **a** Gear grinding. **b** Ball screw grinding (Doimak®)

In Fig. 4.16 two examples of rotary drives with torque motors are shown, in grinding machines.

In Sect. 3.3.2, the application to two-rotary milling headstocks is commented upon, where torque motors are substituting to hypoid gear transmissions. Other example is the machine by Ibarmia in Fig. 1.7 in Chap. 1, where the tilting movement of headstock is provided by a torque motor.

4.6 Guidance Systems

The guidance systems, together with the drive, determine the operation of the machine tool slides. The technologies used for machine tool applications are described below.

Historically, the type of guidance most used has been friction. Normally, the machine manufacturer also fabricates the guidance system. Presently, the roller-based guidance has gained much popularity thanks to the high quality and performance that commercial components provide and the fact that the manufacturer can integrate these into his machine. On the other hand, for very demanding applications in regard to accuracy and stiffness, the hydrostatic guidance has been and continues to be the best option from the technical viewpoint.

4.6.1 Friction Guides

The friction guide (Fig. 4.17) is the most widely used type of guidance in machine tool applications.

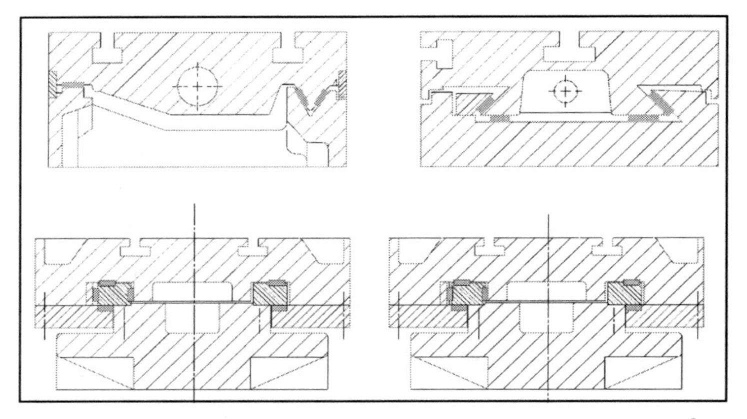

Fig. 4.17 Different configurations of friction guides (by Busak Shamban®)

4.6.1.1 Materials

Different pairs of materials have been used to manufacture friction guides. Casting, steel, bronze and certain polymers has been used as sliding materials. A key factor to ensure the controllability and smooth operation of the guide is to avoid the stick-slip phenomenon which appears when the static friction coefficient is higher than the dynamic friction coefficient (see Fig. 4.18).

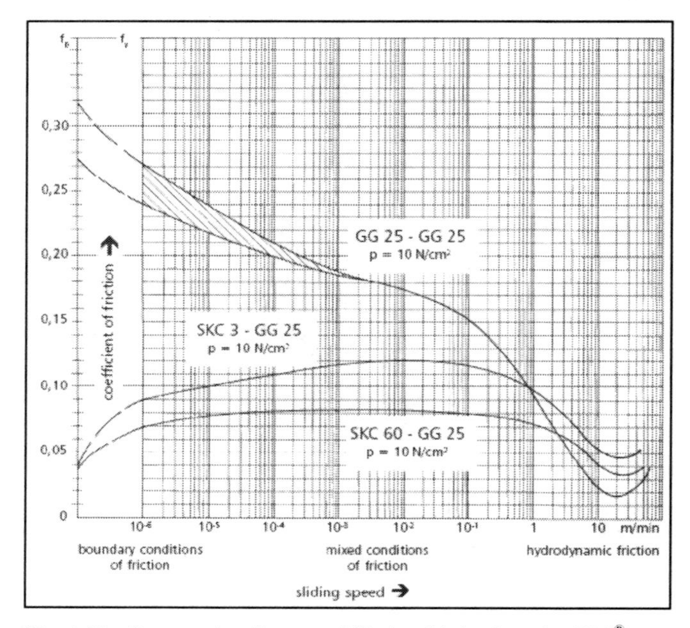

Fig. 4.18 Comparative diagram of frictional behaviour, by SKC®

Fig. 4.19 Coated slideways and worm rack in a table of a gantry milling machine (by SKC®)

The development of polymeric materials with additives which favour lubrication, for example, Turcite™ Slideway or SKC® (see Fig. 4.19) or Moglice®, has allowed to a great extent to reduce the stick-slip phenomenon.

4.6.1.2 The Load and Damping Capacity

Among the favourable characteristics of the friction guides are the high load capacity they provide, up to 140 MPa, its strength against impact and the capacity to provide considerable damping. On the other hand, its use will be limited to slides with relatively low speeds, under 0.5 m/s.

4.6.1.3 The Manufacturing

In order to obtain a good quality friction guide it is necessary to work the slideways to ensure a uniform contact of the bedplate and slide. This finishing operation, the scraping, requires highly experienced labour. In addition to guaranteeing a uniform contact, it is of interest to carry out a series of markings on the surface in order that contact is produced on a multitude of small projections. In this manner, adhesion problems are avoided when specular surfaces are matched, with these being more important when the sliding is at low speeds.

It is appropriate that lubrication be used to favour a smooth movement. For this purpose, slots in the vicinity of one millimetre in depth on one of the surfaces are prepared. In order to force the movement and distribution of the lubricant, it is appropriate that the slots be on the moving part.

The application of polymeric-based material can be carried out in several ways. The most classic consists in bonding a polymer sheet a few millimetres thick over one of the parts, machining this once the bond has set in order to correct any flaws. The surface which slides on the polymer should be slightly warmed, for example, by induction, to minimize wear and tear. Presently, it is more common to inject the polymer in a liquid state in order that once it is cured it adapts to the shape of the guide being manufactured. For this, the surface which will receive the polymer is prepared with a very rough surface finish to favour the adherence and is also provided with injection holes. The bedplate and slide are matched, maintaining the space that the polymeric material will occupy between them. When the injected material has cured, in general after 24 hours, the bedplate and slide are separated and the surfaces are finished.

4.6.2 *Rolling Guides*

Presently, the roller-based guides are the most widely used in machine tool applications. There are a large variety of models and sizes which adapt to most machine tool slide requirements.

Two large families of rolling guides can be distinguished. On the one hand, there are those without recirculation, and on the other, there are guides with recirculation.

4.6.2.1 Linear Bearings without Fecirculation

The rolling guides without recirculation (Figs. 4.20 and 4.21) tend to be used when the stroke of the slide is relatively short. It basically consists of rolling elements, balls, rollers or needles, placed between moving parts. In general, a cage is used to prevent contact between different rolling elements and to handle the assembly with greater facility.

The strong points of this type of guide are the very high stiffness and load capacity, due to the high number of rolling elements which can be used, and high accuracy in the movement.

To guide a slide, it is necessary to use at least two guides, in order to restrict all degrees of freedom except that of movement. The preload is attained by making one guide work against the other; for example, by means of screws supported

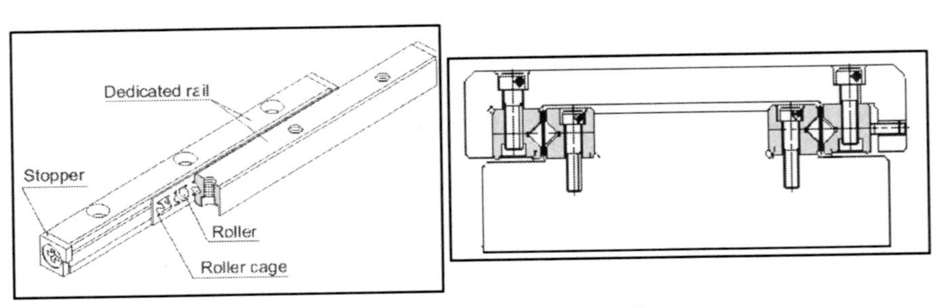

Fig. 4.20 Linear rolling bearings without recirculation, by THK®

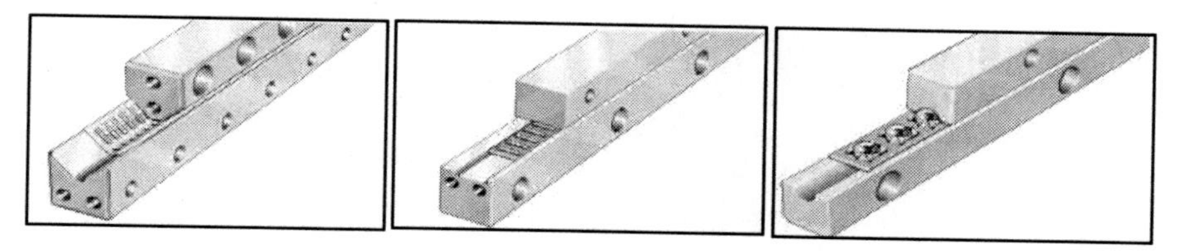

Fig. 4.21 Linear rolling bearings without recirculation, by INA®

directly on the rail or on an intermediate jib which distributes the force on a more uniform basis causing minor deformations of the rail and improving accuracy.

Normally, the supply of this type of guide includes the guide rails. Sometimes, due to lack of space availability, for example, it is necessary to use an element of the machine as a roller track. In this case, it is necessary to temper and harden this area to prevent accelerated wear and tear.

4.6.2.2 Recirculating Rolling Guides

On the other hand, there are the recirculating rolling systems (Fig. 4.22). These consist of a shoe, within which the rolling elements, balls or rollers recirculate, as they move along the length of a guide rail.

The guide rail has at least four roller tracks, such that each shoe restricts all the degrees of freedom except that of movement. In this case, the preload is attained using rolling elements of a slightly oversized diameter. Generally, the manufacturer offers the same product with different degrees of preload.

Presently, the recirculating rolling guide manufacturers have obtained designs with high-load capacities vs forces transverse to the movement as well as vs moments. To refer to the guide size, the tendency is to use the rail width in mm, with there being rails available from size 15 up to size 65 or even larger. The load capacities of a guide family from the manufacturer THK are indicated in Table 4.4.

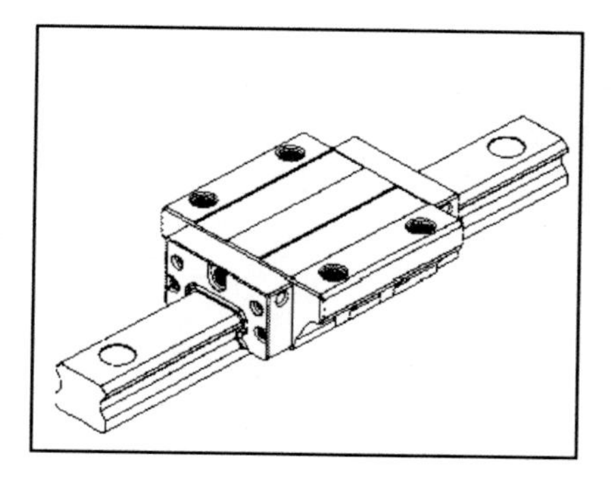

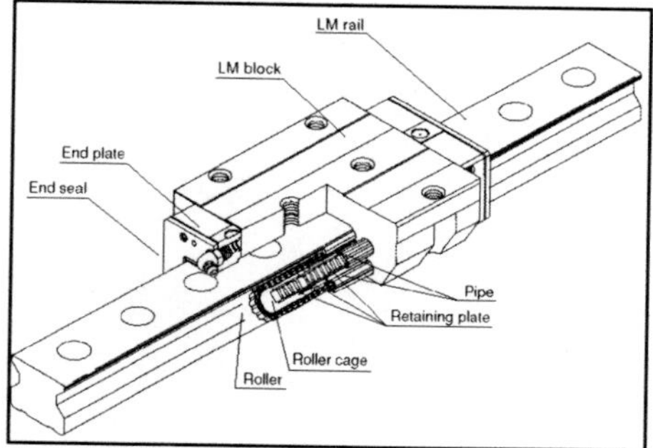

Fig. 4.22 Recirculating rolling guide model, by THK®

Table 4.4 Load capacity of THK® guiding systems

Model No.	Outer dimensions			Basic load rating Radial, reverse radial and side	
	Height M	Width W	Length L	Dynamic rating C	Static rating C0
	mm	mm	mm	kN	kN
SHS 15 V	24.0	34.0	64.4	14.2	24.2
SHS 20 V	30.0	44.0	79.0	22.3	38.4
SHS 25 V	36.0	48.0	92.0	31.7	52.4
SHS 30 V	42.0	60.0	106.0	44.8	66.6
SHS 35 V	48.0	70.0	122.0	62.3	96.6
SHS 45 V	60.0	86.0	140.0	82.8	126.0
SHS 55 V	70.0	100.0	171.0	128.0	197.0
SHS 65 V	90.0	126.0	221.0	205.0	320.0

The preload is necessary to eliminate gaps. In addition, the preload increases the stiffness of the skids. An average preload can increase the shoe stiffness by a factor of three in respect to a shoe which is not preloaded, and makes it less dependent on the load applied.

On the other hand, the drawbacks of a high preload are the increase of friction and reduction of the useful life of the guidance system.

4.6.3 Hydrostatic Guides

In case of hydrostatic bearings (Fig. 4.23 and Fig. 1.16 in Chap. 1), the surfaces of the components which slide are separated by a thin oil lamina. The pressure main-

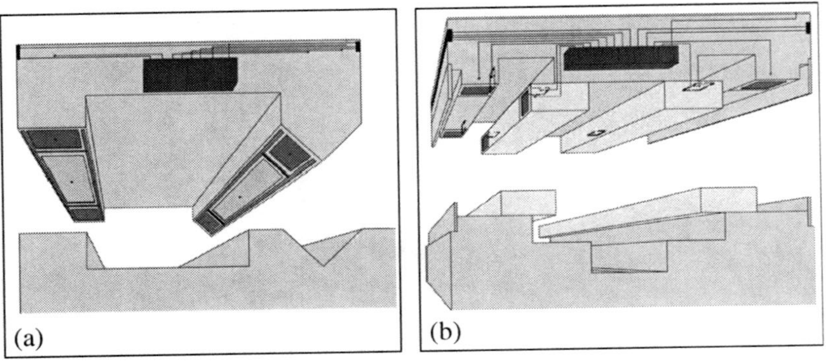

Fig. 4.23 **a** Hydrostatic V-flat guide. **b** Hydrostatic guide with wrap-around (Hyprostatik®)

tained by the oil lamina is provided by an external pump. In this manner, the operation is guaranteed without wear and tear, and without the possibility of the stick-slip phenomenon appearing, due to the fact that the bearing does not present any static friction [2].

One of the surfaces is provided with cavities or cells, known as "oil cells", which are provided with pressurised oil from the outside. Around these cells is the area through which the oil is released as it loses pressure. This area is known as "land". The distance between the land and the surface which slides over it is the "oil gap". Usually, it is 10 to 40 microns and generates a certain resistance to the passage of the fluid. The pressure difference between the cavity and the atmospheric pressure which acts on the outside is known as "cell pressure".

The oil flow resistance along the gap can be estimated as follows:

$$R_c = \frac{\Delta p}{Q} = \frac{12 \cdot \eta \cdot L}{b \cdot h^3} \tag{4.8}$$

The pressure gradient along the length of the land can be assumed to be linear in an initial approximation, such that for purposes of force it can be assumed that the pressure acts on up to one half of the length of the land. The area on which it is assumed that the complete pressure acts is known as the effective area (A_{eff}).

The hydrostatic guides are designed with several cells, such that they can support off-centre forces and moments. Each cell should be supplied at a different pressure, in order that it can withstand a different force according to the operating conditions. For this, usually a single pump is used, with restrictors that allow to supply each cell at the appropriate pressure.

In this manner the pressure in a cell will be:

$$p_c = p_p \cdot \frac{R_c}{R_k + R_c} \tag{4.9}$$

Usually, the restrictors are built in the form of capillaries so that in this manner the resistance depends on the oil viscosity, as occurs in the cells. The resistance to the oil flow of a capillary can be calculated as follows:

$$R_k = \frac{8 \cdot \eta \cdot L_k}{\pi \cdot r_k^4} \tag{4.10}$$

Different types of capillaries are used. The short restrictors should have a very small diameter to provide the necessary resistance. This diameter is limited by the size of suspended particles in the oil which can block the capillary. The use of short capillaries is also limited because the design is very sensitive to the diameter of these; the resistance depends on the fourth power of the diameter. Capillaries with larger diameters and a longer length are also used, which tend to be in spiral form.

The state of equilibrium of a cell vs a specific force can be determined. From this state of equilibrium, the stiffness of the system formed of the pump, restrictor and cell can be determined. In case of a cell supplied through a capillary by means of a pump which operates at a constant pressure, the resulting stiffness is:

$$K = 3 \cdot \frac{P_0}{h_0} \cdot \frac{R_k}{R_k + R_{c0}} \tag{4.11}$$

As can be seen in the previous expression, the stiffness depends on the load supported by the cell. For this reason, it is very interesting to use hydrostatic guides with a "retaining plate" in order that a high preload can be applied. In addition to gaining stiffness, the guide can absorb loads in both directions.

4.6.3.1 The Damping

The hydrostatic guides provide much higher damping values than the roller-based guides against movements perpendicular to the cell; this is due to the friction force that the oil lamina presents on sliding. In case of movements parallel to the cell, the damping is reduced because there is no displacement of fluid.

4.6.3.2 The Energy Consumption

The energy consumed by hydrostatic guide depends, on the one hand, on the work carried out due to friction in the land and, on the other, the work carried out by the pump. This energy is integrally transformed into heat which increases the oil temperature.

The work carried out to overcome the hydrodynamic friction can be determined as:

$$W_r = F_r \cdot v = A_r \cdot \eta \cdot \frac{v^2}{h} \tag{4.12}$$

while the work carried out by the pump is:

$$W_p = \frac{Q \cdot p_p}{\varepsilon} \tag{4.13}$$

The total work results in:

$$W_p = W_p + W_r = \left(\frac{p_p^2 \cdot b}{12 \cdot \varepsilon} \right) \cdot \frac{h^3}{L \cdot \eta} + \left(v^2 \cdot b \right) \cdot \frac{L \cdot \eta}{h} \tag{4.14}$$

The dimensioning of a hydrostatic guide is relatively complex due to the strong dependency shown by the variables which define the status of this. A very important variable is the temperature rise of the oil throughout the hydraulic circuit and the temperature in the restrictors and throughout the hydrostatic guide cells. The oil viscosity is heavily dependent on the temperature such that the calculation of the state of equilibrium requires repeating the calculation several times. The preparation of a computer program which assists the designer in the calculation of this state of equilibrium for different speeds or loads applied to the guidance is of interest.

4.6.3.3 The Hydraulic Circuit

The functions of the hydraulic circuit are to ensure an adequate supply of oil to the hydrostatic guidance, in regard to pressure and flow, to suck from the oil the heat generated by the losses and prevent contact between the sliding elements in case of pump failure. It generally consists of a pressurised oil tank, a pressure control system, a safety valve, a suction pump, dirt and a pump filter. Sometimes, an oil cooling system located after the suction pump is also necessary for the collection of the oil exiting from the guidance. The function of the pressurised tank is to provide oil to the bearing in case of pump failure, as well as to equal the oil flow at the pump discharge. The most common type of pump used is the gear pump.

In general, the cost of the hydraulic system tends to be high compared to other types of guides.

4.6.3.4 The Design Criteria

In general terms, it is of interest to keep the gap as small as possible. However, the minimum value is limited by the accuracy which can be obtained during the manufacture and by the elastic deformation of the components, in order to avoid contact between sliding surfaces at all costs.

In order to determine the stiffness of the guidance, it is appropriate to consider the flexibility of the slides themselves. It is probable that this is comparable with that of the guidance.

On the other hand, special attention should be given to the system for collecting oil released through the hydrostatic guidance and its return without contamination. For this, appropriate seals should be provided in each case. As an example, the characteristics of a rotary table are shown in Table 4.5.

Table 4.5 Design parameters of an aerostatic rotary table by Precitech®

Table outer diameter	380 mm
Axial error motion	0.035 μm
Radial error motion	0.079 μm
Load capacity (safety factor 3)	6,670 N
Axial stiffness (17 bar)	1,315 N/μm
Radial stiffness	700 N/μm
Tilt stiffness	19 Nm/μrad
Effective stiffness 300 mm above the centre of bearing	162 N/μm

4.6.4 Aerostatic Guides

In the case of aerostatic guides, the medium which separates the sliding surfaces is the air. There are no fundamental differences in respect to the hydrostatic guides in that referring to the operating mode.

The following points are highlighted from among the characteristics and advantages of this technology [4]. The aerostatic guides present exceptionally low losses due to friction, including at very high speeds, due to the low viscosity of air. At the ambient temperature, the air viscosity is approximately three orders of magnitude less than those of the oils used in hydrostatic guides. In addition, it is not necessary to consider the means for returning the fluid and seals are not required. An interesting characteristic of the aerostatic guides is their aptitude for applications subject to large differences of temperature, due to that the viscosity of the air presents a good consistency in a wide range of temperatures.

On the other hand, the dimensioning of an aerostatic guide is more complicated than in the hydrostatic case. Air compressibility, for example, contributes to this. This compressibility, combined with the fact that the air barely provides damping, generates different dynamic phenomena which, in general, are problematic. There may be, for example, self-excited excitation, known as "air hammering" which may become audible and on occasions seriously affect the correct operation of the guidance. This vibration can be mitigated through the use of a high number of cells, each with its restrictor. For this reason, nowadays, the guides are manufactured in porous material. This tends to consist of a sintered part in which the porosity emulates the operation of multiple restrictors.

This type of guidance is seldom used in machine tool applications, mainly due to the difficulty in attaining sufficient load capacities at a reasonable cost of the air-supply equipment. For this reason, it is necessary to machine the sliding surfaces to a very high degree of accuracy, such that the gap can be extremely small. As an example, certain characteristics of interest of a workholding spindle of an ultraprecision lathe are presented in Table 4.6.

Table 4.6 Design parameters of a workholding aerostatic spindle by Precitech®

	HSS SP75	HSS SP150
Material	Steel shaft/bronze journal	
Swing capacity	220 mm diameter	
Load capacity	180 N	680 N
Axial stiffness	70 N/μm	228 N/μm
Radial stiffness	22 N/μm	88 N/μm
Motion accuracy	Axial/radial < 50 nm	Axial/radial < 25 nm through dynamic range
Max. speed	15,000 rpm	7,000 rpm

4.7 The Present and the Future

Some new ideas are now developing, with prototypes and even released products with improved features. Some of them are explained below.

4.7.1 Rolling Guides with Integrated Functions

The manufacturers of rolling guides have made an effort to integrate functionalities in their systems (Fig. 4.24). For example, presently, there are rolling guides with position pickup systems. The pickup system offers a similar performance to that of the average level linear encoder and is compatible with the electronics that this uses. Another product are the racks prepared to support a rolling guidance system. These products permits us to reduce the time required for design and assembly and allow the machine tool manufacturer to offer more compact solutions.

Fig. 4.24 Rolling guides with integrated functions, by Schneeberger®

4.7.2 The Hydrostatic Shoe on Guide Rails

Another interesting product which has recently appeared on the market is the shoe designed to operate on guides similar to those used in rolling systems, shown in Fig. 1.16b in Chap. 1.

This product allows the machine tool manufacturer to obtain the performance of the hydrostatic guides without having to design and manufacture the guides themselves.

4.7.3 Guiding and Actuation through Magnetic Levitation

State-of-the-art technology has allowed the development of machine tool slides guided by active magnetic fields [1]. The characteristics of a levitated slide, driven by means of magnetic fields, are shown in Fig. 4.25.

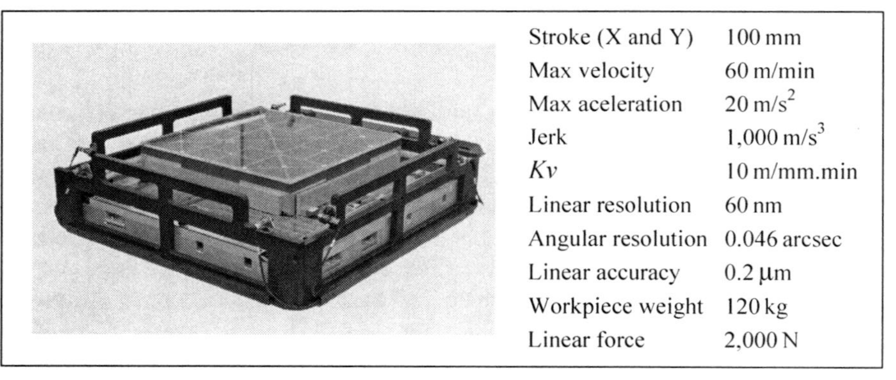

Stroke (X and Y)	100 mm
Max velocity	60 m/min
Max aceleration	20 m/s^2
Jerk	1,000 m/s^3
Kv	10 m/mm.min
Linear resolution	60 nm
Angular resolution	0.046 arcsec
Linear accuracy	0.2 µm
Workpiece weight	120 kg
Linear force	2,000 N

Fig. 4.25 Planar motor magnetically levitated, by Tekniker®

Among the advantages of this type of slide is the fact that there are no forces of friction, the dynamics are very high due to the fact that there are no transmission force elements and the accuracy is only limited by the measurement system used. Today, this type of guidance is still in the research phase.

Acknowledgements Our thanks to the companies which have kindly collaborated, by giving pictures and technical information. Special thanks to Mr. R. Gonzalez from Shuton and P. Rebolledo from Redex-Andantex for their time dedicated in discussing several aspects of this chapter.

References

[1] Etxaniz I et al. (2006) Magnetic levitated 2D fast drive. IEEJ Trans Industr Appl 126(12):1678–1681
[2] Shamoto E et al. (2001) Analysis and Improvement of Motion Accuracy of Hydrostatic Feed Table. Annals of the CIRP, 50/1:285–288
[3] Slocum A (1992) Precision Machine Design. Prentice-Hall
[4] Takeuchi Y et al. (2000) Development of 5-axis control ultraprecision milling machine for micromachining based on non-friction servomechanism. Annals of the CIRP, 49/1:295–298
[5] Uriarte, L et al. (2004) Rectificado de precisión de piezas helicoidales (in Spanish, Grinding of helical parts). IMHE, 305:52–58
[6] Zaeh MF et al. (2004) Finite Element Modelling of Ball Screw Feed Drive Systems. Annals of the CIRP, 53/1:289–292
[7] Zatarain M et al. (1998) Development of a High Speed Milling Machine with Linear Drives. CIRP Int Seminar on Improving Machine Tool Performance 1:85–93

Chapter 5
Advanced Controls for New Machining Processes

J. Ramón Alique and R. Haber

Abstract This chapter describes the basic concepts involved in advanced CNC systems for new machining processes. It begins with a description of some of the classic ideas about numerical control. Particular attention is paid to problems in state-of-the-art numerical control at the machine level, such as trajectory generation and servo control systems. There is a description of new concepts in advanced CNC systems involving multi-level hierarchical control architectures, which include not only the machine level, but also include a process level and a supervisory level. This is followed by a description of the sensory system for machining processes, which is essential for implementing the concept of the "ideal machining unit". The chapter then goes on to offer an introduction to open-architecture CNC systems. It describes communications in industrial environments and an architecture for networked control and supervision via the Internet. Finally, there is a brief summary of the systems available to assist in programming and the architectures of current CNC systems. There is also a description of the most recent developments in manual programming for current CNC systems and possible architectures for these systems, with the different uses of PCs and their various operating systems.

5.1 Introduction and History

A numerical control is any device (usually electronic) that can direct how one or more moveable mechanical elements are positioned, in such a way that the move-

J. Ramón Alique, R. Haber
Industrial Computer Science Department, Industrial Automation Institute
Spanish National Research Council (CSIC) Arganda del Rey, Madrid, Spain
jralique@iai.csic.es

L. N. López de Lacalle, A. Lamikiz, *Machine Tools for High Performance Machining*,

ment orders are put together entirely automatically through the use of numerical and symbolic information defined in a program.

The first attempt to provide a mechanical device with some form of control is attributed to Joseph Marie Jacquard, who in 1801 designed a loom that could be made to produce different kinds of fabrics purely by modifying a program that was entered into the machine by means of punch cards. Subsequent examples included an automatic piano, which used perforated rolls of paper as a way of entering a musical program. While these devices were actually automatic controls, they cannot be regarded as true numerical control systems.

The great leap forward in numerical control evolution came when numerical control was applied to the machining of complex parts. The introduction of automation in general, and numerical control in particular, came about as the result of several different circumstances: 1) the need to manufacture products that could not be obtained in large quantities at high enough levels of quality without resorting to automating the manufacturing process; 2) the need to produce items that were difficult or even impossible to manufacture because they required processes that were too complex to be controlled by human operators; and 3) the need to produce items at sufficiently low prices.

In order to solve these problems, inventors came up with a number of automatic devices using mechanical, electromechanical, pneumatic, hydraulic, electronic and various other kinds of systems. In the beginning, the main factor that drove the whole automation process was the need to increase productivity. Other factors have subsequently emerged that, both individually and as a whole, have been enormously important in the industrial sector, such as the need for precision, speed and flexibility. Viability is not included here, as it has been of little importance from a quantitative point of view, but thanks to these devices it has been possible to manufacture parts with highly complex properties that could otherwise never have been made.

One early attempt to apply numerical control techniques as an aid to part machining was made in 1942, in response to demands by the aeronautics industry. However, the key advance in the automation of machining processes occurred in the 1950s, when Parsons, a company under contract with the US Air Force, asked the Servomechanisms Laboratory at the Massachusetts Institute of Technology (MIT) to develop a three-axis milling machine with numerical control. This controller was the state of the art in the 1960s. Since then the use of computers and particularly the first microprocessors that emerged in the early 1970s has led to spectacular advances, as more powerful, reliable, economical computers have become available.

5.1.1 Computer Numerical Control and Direct Numerical Control

Until the appearance of microprocessors in the early 1970s, numerical control systems were divided into two main groups: 1) systems designed to control costly, sophisticated machine tools in which a minicomputer could be included as a basic control system without excessively increasing the cost of both control and machine; and 2) small and medium-sized control systems designed for simpler machines and invariably hardware implemented. After computers began to be included as a basic element, numerical control became known as CNC (computerized numerical control).

Almost immediately, however, in the mid-1970s, the microprocessor began to be used as a basic unit. This placed very strict conditions on system organisation, the choice of microprocessor and the design of the system with different functional units. In the case of system organisation, depending on the type of numerical control used, this meant the use of parallel processing techniques in the broadest sense of the term. In high-range equipment, parallel processing was achieved using multi-microprocessor architectures, with some specific functions supported by microcomputer LSI peripherals. These peripheral microcomputers considerably enhanced numerical control performance, as they required minimum attention, since they operated in parallel with the central microprocessors.

In the creation of multi-microprocessor systems and particularly monoprocessor systems, there were certain basic conditions that the systems in question had to meet. The most important of these was determined by the size of the numeric values to be handled, always in integer arithmetic. The use of 32-bit microprocessors has offered an essential advantage in performing operations in integer arithmetic.

Furthermore, as shall be observed below, these CNC systems require particular characteristics in the interpolation unit and the position control servomechanisms. The algorithms to be used in the interpolation process must be reference word algorithms, while the servo control systems must be sampling systems, with a sampling period of T_s. The value assigned to the T_s sampling period is of vital importance as regards errors in the contour shape generated by the machine tool.

While the first microcomputer-based CNCs were being used, the first steps were being taken on the road towards optimizing the machining process to some degree. The optimisation of machining processes has traditionally relied on part programs based on unreliable pre-processed data. It was soon found that it would be impossible to optimise such processes, and particularly to implement the concept of the objective function (which shall be discussed at a later point) based solely on the use of existing part programs.

The first serious attempt at optimisation was made in the 1960s during a US Air Force contract with Bendix Corporation. The system that was designed, which was given the name of Adaptive Control Optimization (ACO), included on-line optimisation and adaptive controls that were very advanced for their time. However, this system never went into practical use in the industrial environment, as its operation was based on the existence of sensors that were not yet available, such as tool condition sensors.

This showed that sub-optimum control systems could be more suitable and easier to use in industrial environments, leading to the appearance of adaptive control systems with constraints (ACCs) and geometric adaptive controls (GACs). These systems, which were developed in laboratories back in the 1980s, also failed to win full industrial acceptance due to their highly limited properties.

Another advance in CNC systems came with the creation of distributed numerical control architectures (DNCs), new manufacturing architectures in which several machine tools with CNC controls were run by a central computer via a direct real-time connection. DNC systems connect a group of machines to a central computer that is used to program and record part programs, transmit these programs on demand and, in general, manage the activities of the machines in question.

The general aims sought by DNC systems are: 1) increasing the efficiency of the programmer, the operator and the machine itself; 2) providing a flexible struc-

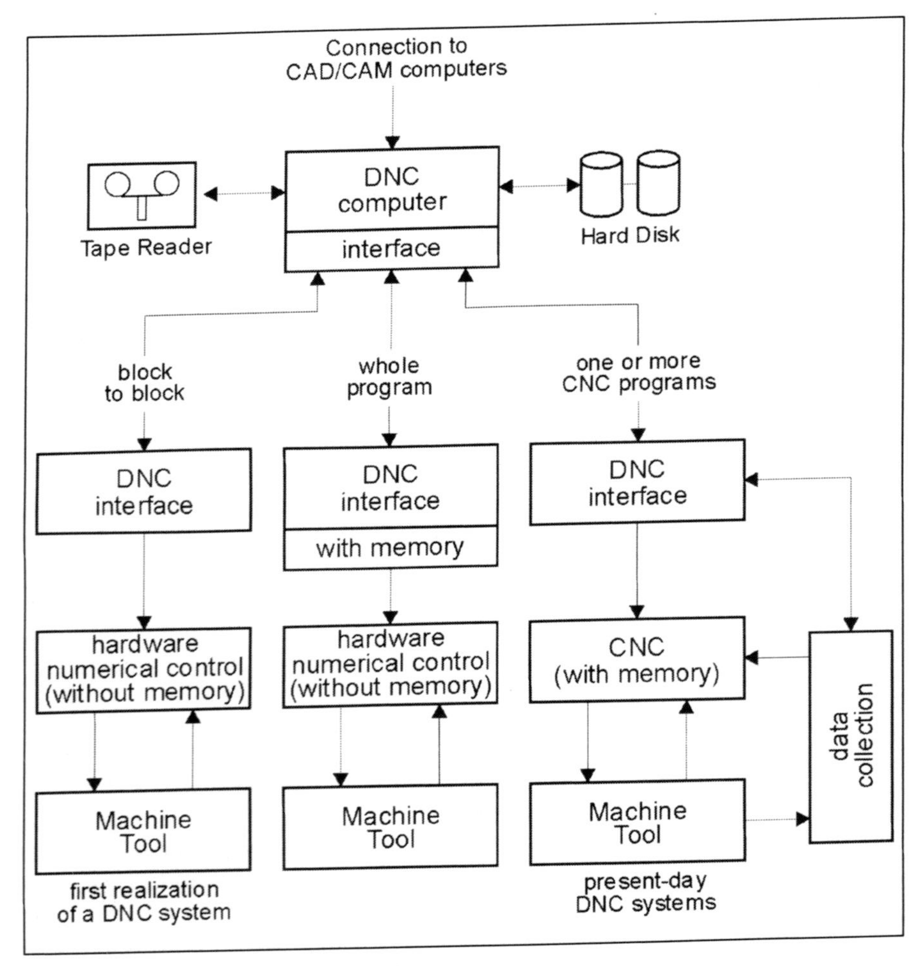

Fig. 5.1 Historical evolution of the DNC concept

ture that can be extended to and integrated with other systems; and 3) allowing for real-time feedback that offers information on everything that is happening in the overall machining process.

This new form of architecture can be dated back to the late 1960s, though it was not clearly consolidated until the early 1980s, with the use of local communications networks using multi-point star, ring and bus topologies. Figure 5.1 illustrates the historical evolution of the DNC concept.

However, the emergence of the local area networks (LANs) changed the world of DNC, and the CNC-DNC idea made way for the networked control and supervision of all machining processes. Since the mid-1990s, a huge amount of work has been done on networked control and supervisory systems, from LANs to wide area networks (WANs) and the Internet itself.

5.1.2 Networked Control and Supervision

CNC and machine tools currently exist and function in a level-based architecture, beginning with the company at the macro level and moving down through to the individual factory, which itself contains business units responsible for supervising the operation of groups of machines. Finally, at the micro level, there is the machine itself, which includes the machine tool and its control systems. Advances in CNCs, PC-based control platforms and new measuring and operational systems have led to an increase in both user time and precision, with the resulting reduction in down time and repairs and an increase in both the quality and quantity of parts produced per measured unit of time.

Is there such a thing as an open, intelligent, distributed CNC? Open control systems have indeed been essential in making the development and implementation of modular and reconfigurable manufacturing systems possible [28]. The development of open platforms has been accompanied by the construction of open development and implementation environments that include standards for interoperability and integration. At the same time, standard-based intermediary systems (*middleware*), which do open up new possibilities such as communications buses for the monitoring and control of complex processes [30], are still a long way from current manufacturing systems, which feature a wide variety of CNC manufacturers and end users [3]. The unification and standardisation of current features through intelligent, distributed decision-making is still in its early stages. Nevertheless, direct numerical control has been evolving towards new networked control and supervision architectures. Using industrial networks (e.g., Profibus), local networks (e.g., Ethernet) and bigger networks like the Internet, it is now possible to carry out networked control and supervision operations. Communications technologies and advanced computational algorithms are increasing in importance.

Until only a few years ago, communications were made purely on a point-to-point basis, but new applications have optimised communications with continuous access to information from all the control devices connected to a network. Despite these advances, there are big drawbacks associated with this new approach that seriously affect the performance and stability of closed-loop systems, such as sampling frequency and the real-time requirements of feedback control systems. The delay generated in shared communications channels entails degradation and instability for a networked control and supervision system. A great many studies are currently under way on the limits of integrated control and communications systems, such as the limitation on bit rate (bandwidth), communication delays, information loss, transmission errors and asynchronisms.

In an ideal case, if the delay is constant and a linear model of the system is available, design and analysis would be simplified considerably. Token ring protocols could be used, which permit a constant delay, but the maximum delay the system can tolerate must be known. Another option might be to use real-time predictions based on predictive controllers and compensate for any delay that occurs in the previous sampling. To do this, a sampling period would have to be established that would be tolerated by the delays occurring in the network (the greater the sampling speed, the more congestion and delays) and would at the same time ensure that the system could reach the desired performance targets.

Studies have been carried out on constraints on communications from the control system viewpoint. Some approaches deal with an analysis of the delays in a switched networked control system with a defined latency time, using the TCP/IP transmission protocol. In general, these delays are modelled using statistical techniques such as negative exponential distribution functions and Poisson and Markov chains.

5.2 New Machining Processes

The difficulty of developing qualitatively superior levels of machining processes has increased even further with the appearance of new paradigms. High speed/high performance machining and micro/nanomachining with the inclusion of objective functions and merit variables have thrown up new computing obstacles.

Today's market not only demands the proper machining of parts, but also imposes very strict requirements as regards productivity, dimensional precision and surface quality, all at the lowest possible cost. It could be said that one of the new machining paradigms, one whose achievement would involve further levels of machining beyond the range of current CNCs, could call for maximising the *material removal rate* (MRR), minimizing tool wear and breakage (TCM) and maintaining quality within given specifications in terms of both the dimensions and the finish of the machined parts, from the very first part produced. The main cause can be found in the new strategies employed in machine tool purchasing. Nobody is buying machines or even machining processes any more. They are now buying production, i.e., parts, so machine tools are evaluated and purchased in terms of production.

5.2.1 High Speed Machining

High performance machining does not involve such high cutting speeds, but as the depth of cut is several times greater, much higher material removal rates can be achieved. The effects of increasing cutting speed are as follows: 1) The material/tool friction coefficient is reduced; 2) the material displaced by cutting disappears, giving a substantial improvement in surface quality; 3) the cutting forces are reduced; 4) almost all the heat is evacuated with the chip; and 5) the most significant wear effect is diffusion, making it necessary to protect tools with special coatings. High speed machining processes offer the following advantages: 1) increased productivity, 2) reduced cutting forces, 3) reduced cutting time, 4) improved surface integrity, 5) a more stable process with less vibration, 6) improved surface texture, 7) the possibility of machining thin walls, and 8) reduced chip thickness and smaller shaving spirals.

Generally speaking, high speed machining is a concept that involves much more than just machining at high cutting speeds. It is in fact a different machining process altogether, with more expensive, more advanced equipment and tools that require the operator to have both greater protection and more training. This is because operators have to use more CAD/CAM systems and spend less actual programming time on the machine itself, though in spite of all the advances some programming time remains essential.

High speed machining processes demand very strict conditions, both from the machine tools themselves and from their control systems, which also include the machine drives. It could generally be said that these new characteristics are leading the way closer to the "ideal machining unit" concept, as shown in Fig. 5.2.

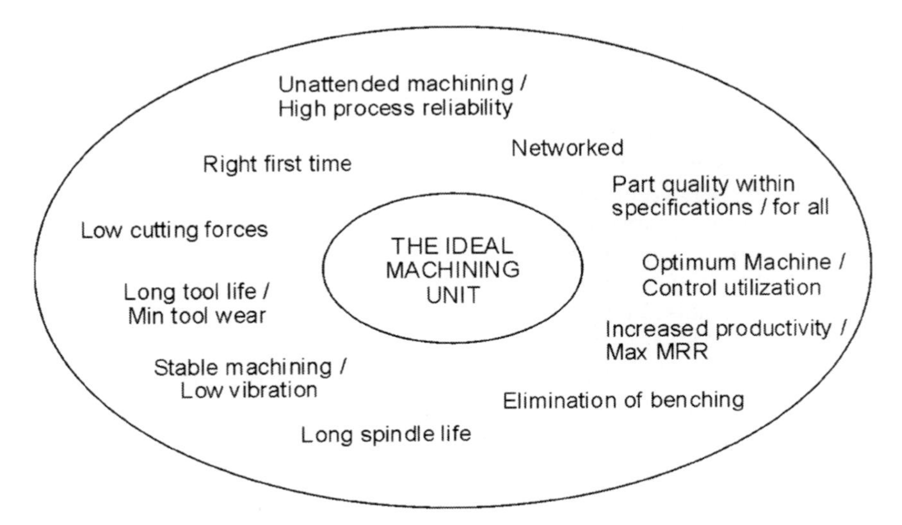

Fig. 5.2 The "ideal machining unit"

Conventional numerical controls cannot be used to attain an ideal machining unit, so multi-level hierarchical control systems must be created in which the normal machine level (current CNCs) will have to be enhanced with at least two higher levels, namely a process level and a supervisory level. If these higher levels are to be introduced, it will be necessary to develop some very powerful machine sensor systems to provide real-time information about what is happening to the machine, the process and the cutting tool. This subject shall be continued in Sect. 5.4.

5.2.2 Micromechanical Machining

Similarly, there is now a marked demand in the marketplace for ultra-precise miniature components, i.e., microcomponent manufacturing. A product is classified as belonging to the category of microcomponents when at least one of its functional characteristics is measured in microns. Generally speaking, the market demands the production of miniature components with micrometric (and even submicrometric) details.

There are fundamentally two types of techniques for manufacturing microcomponents: MST (micro-systems technologies), which covers technologies for the miniaturisation of silicon-based components, and MET (micro-engineering technologies), which covers technologies for the manufacturing of products with high-precision 3D geometries in a wide variety of materials. MEMS (micro-electromechanical systems) technologies are a subdivision of MST and refer to products with both electrical and mechanical components.

As a result, MET technologies tend to be used at present for the production of miniature devices and components with details ranging from tens of microns to a few millimetres in length. Even though these micromechanical machining techniques cannot be used to obtain sub-micrometric details, they nevertheless play an essential role on the mesoscale as a bridge between the macro domain and the nano and micro domains in the production of functional components.

However, the know-how accumulated in the area of macromachining cannot be applied directly to micromanufacturing processes. There are notable differences between high speed machining and micromechanical machining: 1) the depth of cut/tool radius ratio. The minimum thickness required to remove material is not known; 2) tools. Complex geometries are required. Due to the effect of the size of the grain of the material and its coating, the cutting edge is rounded instead of sharp, thus influencing the minimum thickness of the material removed. The tool round-off could be greater than the feed per tooth, causing big differences in the load on the tool teeth; 3) cutting conditions. Spindle speeds are much higher, up to 120,000 rpm, with a depth of cut, both radial and axial, of just a few microns; 4) cutting forces. Cutting forces are on the order of 100 times smaller than in high speed machining – just a few Newtons. Feed per tooth is just a few microns; 5) sudden, unpredictable tool breakage. It is essential to have sensors to detect tool breakage; 6) hard-to-remove burrs. More complex machining processes are needed, with additional operations to remove burrs; 7) material used. In micromachining, the size of the grain of the material to be cut may be on the same order as the radius of the cutting edge and the thickness of the removed chip. Material cannot be regarded as homogenous; 8) precision. In micromachining the tolerance grades required are IT2 and IT3, while on the macro scale they are IT7 and IT8. The thermal distortions or precision errors during machining may be as large as the cutting parameters themselves or the axial or radial depth of cut, thus leading to errors of 100%.

5.2.3 An Introduction to Nanomachining Processes

Advances in both the scientific and technological fields have brought us closer to molecular machining or nanomachining, which can be described as the construction of objects by cutting, profiling or shaping at an atomic level using a sequence of operations (e.g., ultra-fast lasers or chemical reactions) carried out by non-conventional machinery [39, 5, 8].In many cases, the absence of mechanical processes during cutting affects the validity of the definition of nanomachining as worded here.

Figure 5.3 shows some nanomachining cutting operations. However, there is no doubt that, in the short term, the main properties of some machine tools (in many cases non-conventional tools) are moving towards programmable positioning of molecules with a tolerance of 0.1 nm, mechanosynthesis at 10^6 operations/device second, operation at 10^9 Hz, conversion of electromechanical potential at 10^{15} W/m^3 and, in general, production systems that can double production in 10^4 seconds.

From a scientific point of view, the removal of material occurs in the presence of dynamics that are determined by the forces of friction, dragging and spiralling, while non-linear, external attraction and repulsion nanoforces are produced that drag or push a particle with long- or short-range effects. These nanoforces are intermolecular forces such as the "Van der Waals force", the "Casimir force", capillary forces, hydrogen bonding, covalent bonding, the "Brownian movement", the steric effect and hydrophobic forces, forces which are not significant at the macro scale.

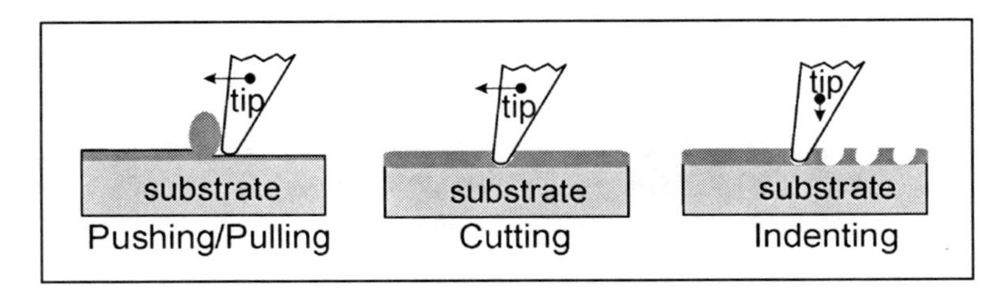

Fig. 5.3 Cutting operations in nanomachining

These properties and characteristics give rise to the following:

- An exponential increase in task specificity at the nanoscale, depending on the predominant intermolecular forces, the physical, geometric and chemical properties of the surface and the atmospheric conditions and disturbances.
- An increase in the functional complexity of manufacturing due to non-linearity.
- An increase in the functional and precision requirements of sensors, activators and computation media.

5.3 Today's CNCs: Machine Level Control

Today's machining processes, particularly high speed machining processes, feature a number of characteristics that have made it necessary to find new concepts and strategies for the mechanical design of some machine components and have forced a drastic redesigning of all the models concerned in process monitoring, control and supervision. These characteristics are:

1. Extremely complex cutting processes. There are almost no models in the process domain. Uncertainty and incomplete information are inherent elements in these machining processes.
2. The great uncertainties of in-process measurements. The need to use empirical predictive models. Noise in measurements.
3. The fact that there are disturbances that affect machines, materials, cutting conditions, *etc*.

At the present time, very few manufacturers and, above all, very few end users of machine tools have accepted monitoring, control and supervision devices, and solutions are few and far between and also very tailor-made. The poor performance of these systems in conventional machining operations combines, at high speeds, with the heavy constraints on processing time and an insufficient knowledge of the actual physical processes taking place.

To tackle all these challenges and to achieve an "ideal machining unit" concept (see Fig. 5.2) it will be necessary to adopt new control and supervision architectures. Multi-level hierarchical control models, supported by powerful machining sensory systems, represent a viable way of resolving the problems raised. A multi-level hierarchical control system would make the process more predictable at the lower control levels and thus reduce the complexity of the control scheme required. It would seem clear that there must be at least two levels of control: the machine level (the level at which CNCs currently operate) and the process/supervisory level. This latter level will be discussed at a later point.

Control at the machine level, the level at which CNCs currently operate, is intended to optimise the dynamic behaviour of the machine tool. It must therefore have two basic, powerful functional units: the trajectory generation unit and the servo control systems, one for each of the machine axes. In current CNCs, these units must possess very special characteristics and be equipped with very powerful interpolation and position control algorithms in order to be able to meet the demands of both high speeds and precision.

5.3.1 The Interpolation Process

Interpolation is a method that allows intermediate points to be found along a particular curve between a starting point and an end point, given in such a way that the trajectory generated offers a satisfactory approximation to the desired programmed curve. These points are usually subsequently joined using linear segments, making it possible to form a more complicated curve merely by using a sufficient number of segments. In general, if a high level of precision is required, hundreds of small linear segments will have to be calculated and programmed. Complex curves would require thousands of separate points connected by straight lines.

There are basically two main families of interpolation methods: 1) *reference pulse interpolation*, also known as *constant displacement interpolation*, and 2) *reference word interpolation*, also known as *constant time interval interpolation*. In reference pulse interpolation, the computer generates a series of pulses, and each pulse causes a basic movement (1 BLU, Basic length unit). These algorithms, which are of a general kind, make it possible to generate highly precise curves. Interpolation speed is, however, quite limited, and this system is therefore not used in today's CNCs.

In reference word interpolation, for each sampling period, the computer gives a new reference position for each of the axes. In general all reference word interpolation methods are based on the resolution of a difference equations system of the following kind:

$$X(i+1) = f[X(i), Y(i)]$$
$$Y(i+1) = g[X(i), Y(i)] \tag{5.1}$$

which the computer solves by iterative procedures. As is to be expected, these are not generalised equations, but are instead different for each kind of curve to be interpolated. For circular interpolation, the following type of difference equation is used:

$$X(i+1) = \cos\alpha\, X(i) - \sin\alpha\, Y(i)$$
$$Y(i+1) = \cos\alpha\, Y(i) + \sin\alpha\, X(i) \tag{5.2}$$

where α is the angle described between two points. There are many approximations for the *sine* and *cosine* values of α: Euler, Taylor, Tustin approximations, *etc.* As a result, there are geometric errors, the radial error *ER*, due to a truncation error, and the chord error *EH*. The radial error is:

$$ER(i) = \sqrt{X^2(i) + Y^2(i)} - R \tag{5.3}$$

where R is the desired radius. It is easy to demonstrate that:

$$ER\ (i) = i\,(\sqrt{\cos^2 \alpha + sin^2\alpha} - 1)R \tag{5.4}$$

and, therefore, the radial error increases with the number of iterations. Chord error EH is not cumulative, and it is easy to demonstrate that:

$$EH\ (i) = R - R\ (i)\ \sqrt{\frac{1+\cos\alpha}{2}} \tag{5.5}$$

It should also always be checked that:

$$\begin{aligned} &(ER)_{max} \leq 1\ BLU \\ &EH\ (i) \leq 1\ BLU \end{aligned} \tag{5.6}$$

In Fig. 5.4, these errors can be observed in the generation of circular arcs.

In modern CNCs, the circular interpolation process does not tend to use difference equations of the type shown, but instead those of the following type:

$$\begin{aligned} X\ (i+2) &= F[X\ (i+1),\ X\ (i),\ Y\ (i+1),\ Y\ (i)] \\ Y\ (i+2) &= G[X\ (i+1),\ X\ (i),\ Y\ (i+1),\ Y\ (i)] \end{aligned} \tag{5.7}$$

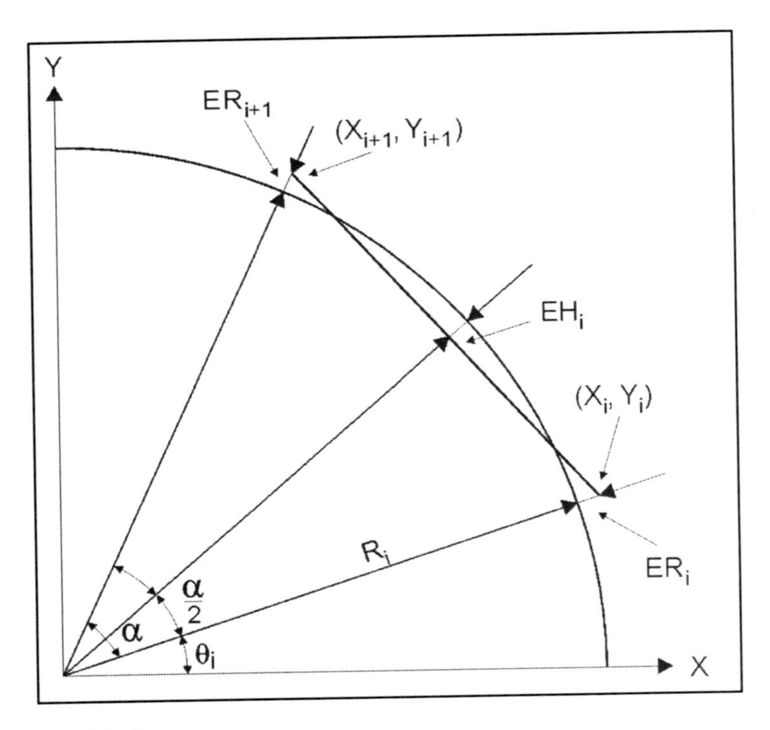

Fig. 5.4 Contour errors

which makes it necessary to calculate a starting point using one of the aforementioned approximations, e.g., a Taylor approximation, though the remaining points are calculated using these new difference equations, which are simpler than the earlier ones.

5.3.1.1 Trajectory Generation

Trajectories in modern CNCs are defined using the CAD/CAM system, which generally breaks the 2D or 3D curve into a series of linear segments that approximate the curve to be described with the desired accuracy. Each segment is subsequently generated by the CNC linear interpolator.

However, there is a conflict in connection with the number of segments into which the CAD system is to divide the curve. It is a good idea to maximise the number of segments as much as possible, both to reduce contour errors and to minimise the effect of segmentation that causes discontinuities in the first derivatives and leads to a reduction in curve smoothness.

The representation of a curve in segments gives rise to a number of problems:

1. The average value of the feed does not achieve the desired velocity, due to the cumulative effect of the reduction in the feed rate in the final iteration of each segment.
2. When machining small segments, the tool never reaches the desired speed, due to the acceleration/deceleration phases automatically applied at the beginning and end of each segment by the position control servomechanisms. As a result, the feed is not constant throughout the curve, leading to a poor surface finish. Furthermore, machining times become longer, because the average speed is lower than desired.
3. When the number of segments is very high, a great deal of points have to be stored in the CNC memory, and CNC memory is quite expensive (though this is now changing). Moreover, the CNC memory could prove to be too small to store the many segments required to describe a part with complex contours.
4. The communications load between the CAD system, where the segments are described, and the CNC system should be reduced. If it increases, transmission errors could easily occur, as, for example, a result of noise perturbation.

For all the foregoing reasons, it would be very useful if the CNC could interpolate general curves in real time, allowing the CAD system to transfer only the curve information to the CNC. The CNC could then generate the trajectory internally, in real time. Furthermore, a very large number of parts currently use free-form (i.e., non-analytical) curves and surfaces. Examples include dies, aircraft models, car models and turbine blades.

Normally there is a set of P_i points to begin with, which are to be interpolated in such a way as to find the mathematical expression of the function that passes through the points and also to meet certain continuity conditions in the slopes and curves (smooth curve). Clearly, one solution would be to generate the required degree polynomial (Cartesian representation). This leads to many problems: it generates many oscillations (various roots), the curve has no vertical tangent, the shape of the curve changes if there is a rotation in the set of points and slopes, and furthermore it is difficult to generate in real time.

The tendency is therefore to extend parametric vector representation. A polynomial function is placed between each two points, and the sections are joined together, with the requirement that certain continuity conditions must be adhered to.

The general expression of this elementary curve is:

$$\overline{P}(u)=\sum_{i=0}^{m}\overline{R}_i\varphi(u) \tag{5.8}$$

where φ(u) is a parametric, normally polynomial function (with uniform parameterisation, $0 \leq u \leq 1$).

Important research is currently being done into parametric interpolators for the generation of curves using Bezier or *non-uniform rational β-spline*, the latter known as NURBS interpolators. In general, the curve to be generated is defined by a sequential series of control points connected by smoothed curves that intersect with the points. In the case of β-splines, they are defined by end points and control points that do not necessarily intersect with the curve but instead act to "push" the curve in the direction of the point. The term "rational" refers to the fact that the weight of the push for each control point can be specified. "Non-uniform" means that the knot vector, which indicates the portion of the curve to be affected by a specific control point, is not necessarily uniform. Thus, a single curve can be used to express a considerably greater number of complex forms. By manipulating the weight values, the knot vectors and the control points, a wide variety of complex forms can be described using NURBS.

NURBS interpolators are generally based on the generation of curves at a constant velocity except in the acceleration and deceleration phases. The first algorithms produced maintained a constant chord length. The tool moves the same distance for each sampling period. As a result, chord errors occur in sections where the curvature is very small, and there could be many such sections in a high speed operation. To solve this problem, algorithms have been developed with an adaptive feed rate that guarantee that chord error remains within maximum limits during the whole interpolation process. The algorithm must at all times identify the ratio between chord error, feed rate and curve radius.

However, the effects of curvature also lead to variations in the rate at which material is removed. The material removal rate is higher in concave regions and lower in convex regions. As a result, some algorithms take account of the curvature at all times and should be able continually to modify the feed rate on this purpose. This is very useful in protecting the cutting tool and the machine from excessively high cutting forces, ensuring part accuracy [35]. As a result, the most recent research in this field has aimed towards the design of hybrid NURBS interpolators, i.e., interpolators that not only control and limit chord error, but also ensure a constant material removal rate, keeping a constant cutting load, depending on the limitations of both tool and machine.

NURBS interpolators offer many advantages over traditional interpolators. The first is the amount of memory required in the CNC. With a linear interpolator, the amount of data to be stored in the CNC is very high (all the tool position data figures), but with a NURBS interpolator all that needs to be stored is the knots, weights and control points, using sentences such as:

NURBS (KNOT ΔXXX, ..., XXX, ΔWEIGHT ΔXXX, ...,XXX

ΔCONTROL POINT ΔXXX, YYY, ZZZ,......,XXX, YYY, ZZZ)

This subject shall be continued in Sect. 5.7. In linear interpolators, control errors are associated with fluctuations in the feed rate. Truncation and roundoff errors can be significant if both high levels of precision and high speeds are sought. However, fluctuations in the feed rate of a parametric interpolator are solely due to truncation errors, which are fairly small when using floating-point calculations. Computation times are clearly much shorter in linear interpolation. However, this used to be more important in the past, when computers enjoyed less processing power.

A controller equipped with a linear interpolator instructs the servo system to track a series of segments. At each joining point, there is a sudden change in the direction of the tracking speed, forcing acceleration at an infinite rate. Given the physical limitations of the motors, there are always significant tracking errors at the joining points between segments. Naturally, acceleration at these joining points does not reach infinite rates because of the limitations of the motors themselves. Nevertheless, they are generally very high, which leads to a high degree of jerk. The jerk problem does not happen with parametric interpolators, as changes in direction occur slowly throughout the arc of the curve. As a result, the acceleration and jerk magnitudes are higher for the linear interpolators.

In spite of the unquestionable advantages, NURBS interpolators are not widely used in the latest generation of CNCs. This is due to the fact that some of the problems described earlier have now been greatly reduced. Furthermore, a tool trajectory represented using NURBS based on a plane intersection with a NURBS surface is not exact and requires the use of a tolerance factor similar to that used in chord deviation, which also leads to a lack of precision. In addition, the large number of blocks that a CNC has to make in a point-to-point representation (a linear interpolation) is no longer a serious problem for the latest generation of CNCs, since block processing speeds have increased considerably (from 20 ms, milliseconds, to less than 1 ms). The reduced memory size required for a NURBS representation is also no longer a key issue, given the large amount of cheap memory now available to CNCs, along with the possibility of using networked connections with high transmission capacities. One important underlying problem with a NURBS representation of curves and surfaces is that it is not easy to interpret, making it practically impossible for the operator to edit a part program directly at the machine itself.

5.3.2 The Position Control Servomechanism

This will be a rough sketch of the current state of servomechanisms for controlling the position of machine axes, particularly in high speed machining. The function of an axis controller is to provide the appropriate drive signals to the actuators (motors) so that the actual position of the axis precisely "tracks" the required axis reference command provided by the interpolation unit, with the aim of eliminating any tracking error (position error) for each driving axis. In general, each axis on the machine is driven by its own controller and is supposed to obey the command signal produced by the respective interpolator. There are basically two types of controllers, the open-loop and the closed-loop. Open-loop controllers use stepping motors, while closed-loop controllers use a position sensor (feedback) to provide information about the actual position of the axis at every point during its motion.

Two basic alternative types of closed-loop position controller can be used. In the first, which is approximately continuous, there is an external up/down counter that works as a position error register. Although, due to quantification of the position references and position feedback all internal signals are discrete, the fact is that the system is a quasi-continuous one.

The second alternative involves the use of sampled-data control systems. Although there are a number of potential alternatives, they all work on a sampling basis with a constant sampling frequency f_s. In these cases, all information processes are carried out every T_s ms, such that position references are updated for each of the axes every T_s ms. In the interim, a constant value is maintained. Here the sampling frequency value f_s plays a significant role, as shall be seen at a later point, in quantifying contour errors, and even in helping to stabilise the system, which could become unstable at high K values in the system open-loop gain.

In a sampled-data CNC system, each axis is controlled independently via computer in a closed system. A typical system is shown in Fig. 5.5, where f_s is the sampling frequency. In a contouring system with two axes, this structure must be duplicated, one for the X-axis and the other for the Y-axis.

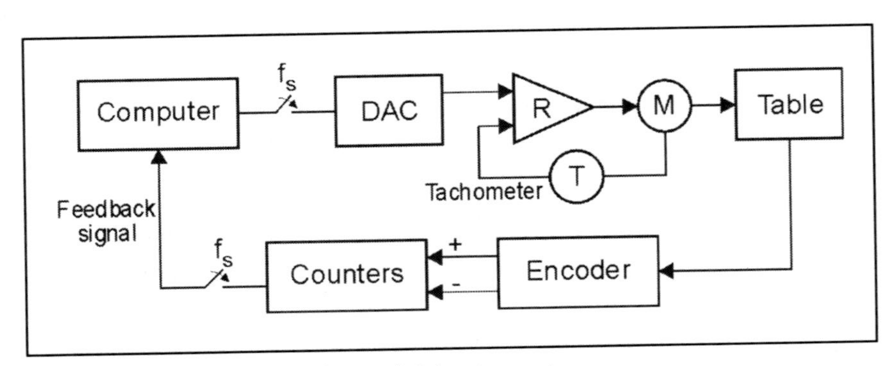

Fig. 5.5 Typical control loop of a sampled-data type system

The most important feature in the performance of contouring systems is the accuracy of the overall system, or the system contour error. Contour error is defined as the distance at any time between the programmed path and the actual path produced by the machine. The problem of contour error can be seen in linear motion, circular motion, in the cutting of angles and, in general, in the cutting of any contour.

To improve the performance of a multi-axis servo control system, the following two approaches can be used:

- tracking-error control
- contouring-error control

5.3.2.1 Tracking-Error Control Systems

With tracking-error control, the system attempts mainly to reduce errors of positioning (tracking error) in each axis, though there will also be an indirect reduction of the contour error. With contouring-error control, the aim is to reduce contour error by estimating it in real time and using this estimate in the feedback control law.

In order to reduce contour error in straight-line movements, it is essential for the open-loop gains for both axes to be equal (perfectly matched) and also allocated the highest possible values. According to [27], the contour error in the generation of straight lines in steady-state mode for a classic control loop is:

$$e_e(t) = \frac{V\sin(2\sigma)}{2}\left[\frac{1}{K_y} - \frac{1}{K_x}\right] \tag{5.9}$$

where V is velocity, σ is the angle and K is the open-loop gains for the axes.

So:

$$e_{emax} = \frac{V}{2K}\frac{\Delta K}{K} \tag{5.10}$$

where:

$$\Delta K = K_x - K_y, K = \frac{K_x + K_y}{2} \tag{5.11}$$

and:

$$K_x K_y \cong K^2 \text{, with } \frac{\Delta K}{K} \text{ small} \tag{5.12}$$

It can be concluded that the contour error depends mainly on the unbalance degree ΔK in the open-loop gains for the axes. The unbalance degree is proportional to the velocity, diminishing in all cases at high K values (average gain value). This is true even where there are individual axis errors with respect to time.

Theoretically, if $K_x = K_y$, the contour error is zero, although only during the steady-state mode. However, in practice there are a number of key issues that are not taken into account here: friction in the guideways, cutting forces, *etc*. It should also be borne in mind that a quasi-continuous system has been assumed for the foregoing formulae, i.e., $T_S \cong 0$.

In the case of quasi-continuous systems, it can be demonstrated that the radial error takes the following form [26]:

$$\frac{e_r(t)}{R} = 1 - \frac{1}{\sqrt{1 + (\frac{\omega}{\omega_n})^4 - 2(\frac{\omega}{\omega_n})^2 + (\frac{2\zeta\omega}{\omega_n})^2}} \tag{5.13}$$

Where ω is angular velocity, ω_n is natural frequency and ξ is the damping factor.

In the case of sampled-data-type systems [16]:

$$\frac{e_r(t)}{R} = 1 - |M| \tag{5.14}$$

where $|M|$ is the system response amplitude module or gain.

To continue, after several approximations, the following expression is obtained:

$$\frac{e_r}{R} = \left(\frac{T_s + 2\tau}{2K} - \frac{1}{2K^2}\right)\omega^2 \tag{5.15}$$

where $\tau_x = \tau_y = \tau$ (time constant for the motors), $K_x = K_y = K$ and T_s, is the sampling period.

As can be seen, radial error in the steady-state mode depends heavily on angular velocity ω. The "sampling period" effect can also be seen. Its value has a negative influence on contour errors. *K* values need to be as high as possible. Nevertheless, due to problems of imbalances in the axis dynamics, contour error is not eliminated, and elliptical profiles are generated. In the operation of cutting a corner, if *K* gain is too low, the system is sluggish, leading to undercutting. If the gain is too high, the axis overshoots the desired profile, causing overcutting. To obtain high contouring accuracy, the system should be critically damped, and the system gains, perfectly matched. Furthermore, reducing the feed rate will reduce the contour error by limiting centripetal acceleration.

However, *K* values cannot be increased too much in order to try and optimise the dynamic behaviour of the servo system. Research has been going on for several decades into other non-conventional structures for servo controllers. The main aim has been to develop control algorithms that seek to improve the tracking accuracy for each individual axis (minimising axial tracking error).

The first solution proposed was feed-forward controllers. As a general rule, feed-forward controllers focus on disturbances that are known to affect the process and can be measured in advance, and they use a model to determine the command signal required to reduce error. A pioneering contribution came from Tomizuka

[36], who developed a *zero phase error tracking controller* (ZPETC) system to reduce tracking errors to zero. To do this, he used a feed-forward controller together with a conventional controller. The controller used the desired future outputs to compensate for the delay in the closed-loop transfer function. The ZPETC utilised $(d+s)$ steps ahead in the desired output $Y_d\,(k+d+s)$, where d was the number of delay steps in the closed-loop transfer function and s was the number of zeros in the closed loop that were unacceptable for a pole/zero cancellation. As a technique for cancelling poles/zeros, this strategy failed to eliminate amplitude error, nor did it take account of the saturation problems typical of feed-forward controllers. In [37], an improved algorithm called "extended bandwidth ZPETC" was introduced. This algorithm computes the feed-forward signal in two stages. For the first it uses the ZPETC algorithm, and then it compensates for the residual error by adding additional feed-forward signals that repeatedly and gradually reduce the tracking error.

To reduce the saturation problem, [40] designed a compensation method for tracking errors using a filter, which reduced the problem but failed to take account of the coordinated movement of the axes.

Feed-forward is an open-loop compensation method and therefore does not take account of contour error when calculating control signals. [29] proposed a new design for a PD controller that includes a feed-forward velocity loop and takes advantage of information on current and voltage limits, the process model and future path information. They demonstrated that the largest source of path error in perfectly tuned systems is control signals that exceed the velocity or acceleration capabilities of one or more of the axes. These sources of error can be minimised by calculating the capabilities of the axes and modifying the feed speed of the tool so that these limits are never exceeded. The same authors also reported that limits on velocity and acceleration should never be fixed, but should instead be adapted to all the system's capabilities as a whole when planning trajectories. Figure 5.6 shows a diagram of this feed-forward controller for an X-axis.

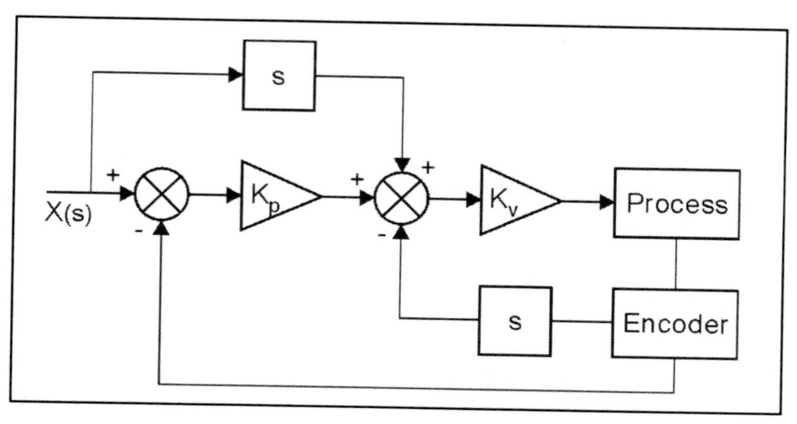

Fig. 5.6 Feed-forward controller

In order to implement these strategies, it is essential for CNCs to use the look-ahead technique, evaluating any change in movement in the axes in advance. When changes in trajectory are confirmed, the control adjusts the feed speed value, giving the servos sufficient time to accelerate or decelerate in order to maintain the trajectory. The machine can then make straight line motions, track corners or make broad curves at the highest feed speed value requested, and the feed speed will be automatically reduced when the curvature of the trajectory requires it. Without the look-ahead function, the feed speed value would have to be programmed for the worst possible case, such as an abrupt change of direction. The number of blocks to be analysed in advance is a dynamic figure, as it will vary depending on the contours of the part, the precision requirements and the characteristics of the machine itself. The slower the machine and the smaller its acceleration, the greater the number of look-ahead blocks required.

Another way of improving the accuracy of the servo control system is by estimating the real position of the tool (tool centre point). Feeding back information on the tool centre point will prevent errors due to the flexibility of mechanical elements between the machine structure and the tool. However, the measurement elements that could be used to measure the exact position of the tool point are very complex and expensive, not to mention difficult to implement, making their practical use unfeasible just yet.

In order to overcome this restriction, work is currently underway [2] to design strategies for estimating the position occupied by the tool centre point instead of measuring it using the ruler attached to the machine structure. To this end, mathematical models are being developed of the dynamics between the two points, and the tool centre point position is being estimated using a status observer based on a previously identified model. The aim of this predictor will be to estimate inertial deformations and other externally caused distortions machines are subject to during the machining process. The aforementioned work includes an example of this predictor's design that factors in the effects of disturbances such as cutting forces and friction.

5.3.2.2 Contouring-Error Control Systems

The methods seen so far offer significant improvements in each individual axis' performance, though they do not ensure good overall control of all the machine's axes, and that is what will ultimately afford real precision to the machine as a whole. For real overall precision, as mentioned, contour error has got to be reduced to a minimum. The first solution was presented by Koren [15], who proposed a new architecture known as cross-coupling controller (CCC) architecture, which was constructed between and parallel to the axial control loops. The CCC calcu lates the contour error in real time and activates a control command to eliminate the error. The CCC system estimates the point in the desired trajectory that is nearest the plant position and uses this to determine the error for each axis. This error signal is subsequently used, in combination with others, to control the coordinated

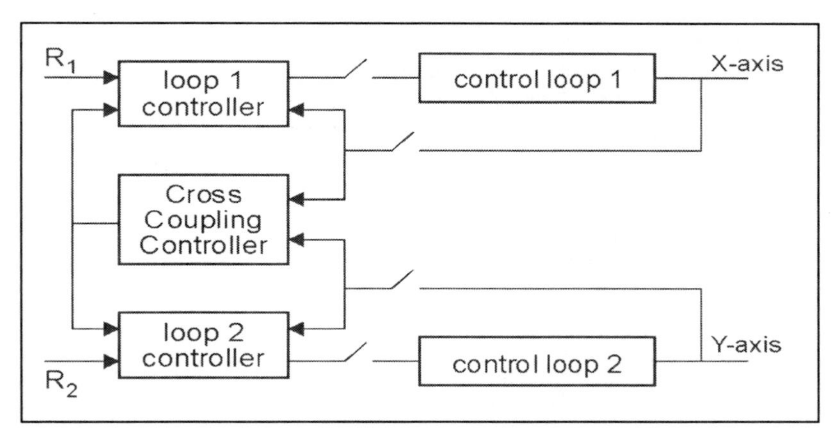

Fig. 5.7 Basic structure of a CCC controller

position. The operation of a CCC is based on providing corrections proportional not only to the individual axial errors but also to the contour error. One problem with CCCs is that they do not work to compensate future trajectory changes. They do not provide for actions to reduce speeds, for example, at corners or any other obstacle that could require more deceleration than one of the axes is capable of providing. The recommendation, therefore, is to use CCC strategies in combination with some kind of feed rate planning. Figure 5.7 shows the basic structure of a CCC.

In addition, other control algorithms, such as repetitive control, predictive control, adaptive control and optimum control, have been developed with the aim of improving contouring performance in new machine tools, particularly in new high-speed machining processes. The use of PC-based CNC systems has furnished the necessary computing power. The ultimate aim will be to machine parts with maximum levels of both surface accuracy (e.g., surface roughness) and dimensional accuracy.

5.4 Advanced CNCs: Multi-level Hierarchical Control

When designing systems that could lead to the development of an ideal machining unit, it would seem clear that the kind of performance sought cannot be achieved using the machine-level control structure described above. As already mentioned, a hierarchical control system would make the process more predictable at the lower control levels and would therefore reduce the complexity of the control scheme required. Intuitively, it would seem useful to distinguish between at least three hierarchical levels: the machine level, the process level and the supervisory level. Machine-level control is intended to optimise the dynamic behaviour of the machine tool itself. Process-level control is aimed at increasing productivity by maximising the technological parameters that are subject to constraints (permissible work space). The aim of supervisory-level control is to manage the process in the best possible way on the basis of an established objective function. This is, of course, not the only potential scheme for a hierarchical control system, but it is the one that we feel is the most intuitive. This new hierarchical architecture must include a powerful monitoring system that can monitor the machining process, the condition of the machine (machine condition) and the condition of the cutting tool (tool condition, which, while forming part of the process, will be dealt with separately due to its particular characteristics and huge importance).

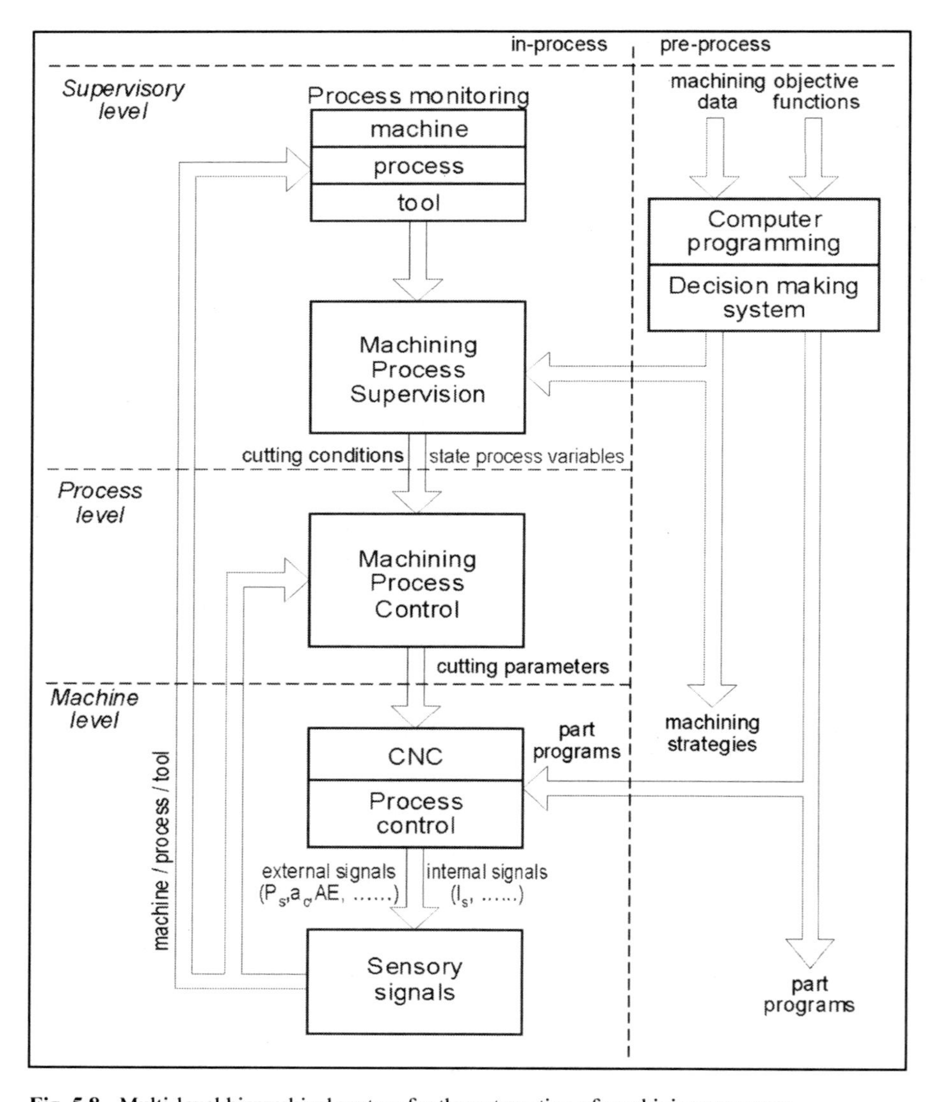

Fig. 5.8 Multi-level hierarchical system for the automation of machining processes

This proposed architecture would work as an "in-process" (*on-line*) system. It would also be necessary to provide the whole system with a powerful programming support unit that should include both geometric and, in particular, process simulators. This unit would operate "pre-process" (*off-line*) and would be responsible for supplying each machining unit with the relevant optimised part programs (*G-code*) and the machining strategies designed for the different stages in the machining of each part (including rough and end machining).

Accordingly, one possible structure for an advanced control system for a high speed, high performance machining process is illustrated in Fig. 5.8. It shows the three horizontal control levels described above and the two vertical levels, one of them pre-process and the other one in-process [11].

5.4.1 The Control of the Machining Process

The aim is to increase productivity by maximising technological parameters, particularly feed speeds and cutting speeds, subject to the actual limitations of the machine itself. At this process level, current research is focused on the control of variables that directly affect machine productivity, particularly cutting forces and torque. Designing systems to control these variables is no easy task because of the process variations, which are always of the variable-gain type and cause performance degradation and even stability problems. Fixed-gain controllers cannot be used, so adaptive control techniques must be employed. Different adaptive control techniques have been used for many years, such as variable gain methods with explicit parameter estimation. While these techniques have not been abandoned, the main thrust of today's research is directed towards the use of reference models and robust control techniques. These strategies always incorporate, in one way or another, a cutting force/torque model that, once calculated, is used to design an adaptive controller. In practice, however, teach-in techniques are still being used, particularly in the implementation of the AFC (adaptive feed rate control) function in some of today's CNCs.

The specialist literature contains different reference models in addition to the classic variable gain model. The kinds of controllers that use a reference model can be divided into five different techniques: linearisation controllers, logarithmic transformation controllers, non-linear controllers, adaptive controllers and robust controllers. Reference [17] contains a complete study of them all, with an analysis (by both simulation and experiment) of the transient response and stability robustness of all these systems.

Cutting force controllers, which are an AFC function in some current CNCs, begin with the customary representation of cutting force F:

$$F = k\, d^{\beta} s^{\gamma} f^{\alpha} \tag{5.16}$$

where k is the process gain, d is the depth of cut, s is the cutting speed, f is the feed and $\alpha, \beta,$ and γ are the coefficients that describe the non-linear relationships between F and cutting parameters d, s and f. Typically, the control variable is the feed, so the force process gain is therefore:

$$\theta = k\, d^{\beta} s^{\gamma} \tag{5.17}$$

which, as one can see, is variable and directly dependent on s and d.

Although these adaptive control strategies are not yet fully integrated into the control process, they at least seem to give some ideas for use in industrial environments. In the production of single parts, the cutting tools, parts and fluids are changed continuously, meaning that it is not economically viable to apply reference models. As a result, the best solution might be variable gain controllers with explicit parameter estimation. In mass-production environments, where operational characteristics remain constant, a logarithmic transformation controller could be a good solution, as it is the least sensitive to parameter variations and model uncertainties. In batch production, where work is done with a specific range of operational characteristics, the robust controller offers the best results, as it guarantees certain properties over a specific range of operational conditions. The robustness of these controllers is a highly valued characteristic in production processes such as these [17].

Given that uncertainty and incomplete information are inherent features of machining processes, artificial intelligence (AI) could offer some credible solutions, with the design of controllers based on AI techniques (fuzzy logic, neural networks and genetic algorithms). At present, artificial neural networks (ANNs) and fuzzy logic (FL) are probably the most widely used AI techniques in the identification (model production) and design of control systems. Feed-forward networks and particularly multi-layer *perceptron* networks are what are most frequently used in non-linear control. Furthermore, by using FL, human knowledge can be incorporated into the control system, expressed in qualitative terms (rule bases).

In the particular case of fuzzy controllers, the input variables used are usually cutting force error, ΔF, and the change in cutting force error, $\Delta^2 F$. The manipulated variable is usually the machine feed rate, f, while the cutting speed is gener-

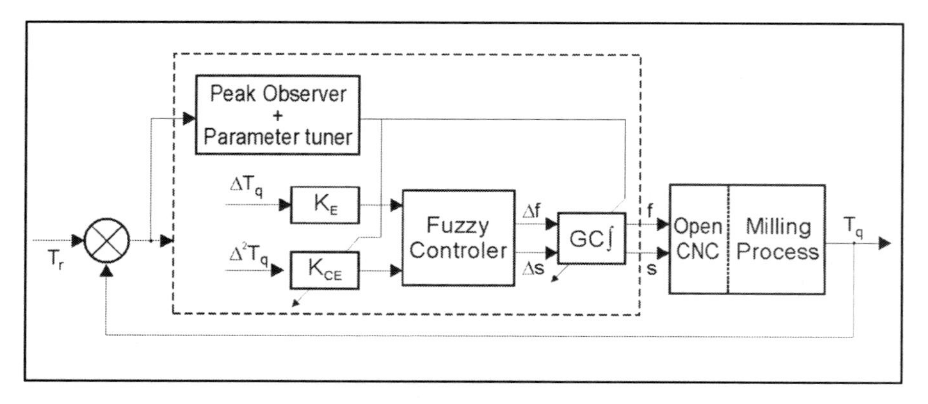

Fig. 5.9 Scheme of self-tuning fuzzy controller applied to the milling process

ally regarded as constant (MISO, multiple input – single output). However, research is currently being done into other input variables that are "easier" to measure, including cutting torque and the instantaneous power consumed by the spindle speed motor. A diagram for a self-tuning fuzzy controller for a milling process is shown in Fig. 5.9 [10], where T_r is the torque reference value and T_q is the torque measured from an open architecture CNC. As can be seen, a MIMO (multiple input-multiple output) architecture has been adopted here, with two manipulated variables, Δs and Δf (percentages of the values programmed in the CNC). The whole control scheme, including a self-tuning mechanism and the fuzzy controller, is shown in Fig. 5.9. The output-scaling factor (GC) multiplied by the crisp control action (generated at each sampling instant) provides the final actions that will be applied to the CNC:

$$\begin{aligned} f(k) &= f(k-1) + GC\,\Delta f(k) \\ s(k) &= s(k-1) + GC\,\Delta s(k) \end{aligned} \tag{5.18}$$

5.4.2 The Supervisory Control of the Machining Process: Merit Variables

As mentioned above, the aim of supervisory control is to ensure the optimum operational management of the process based on one or more merit variables and the corresponding established objective function. At this level account must be taken of all the factors that have not been specifically dealt with and compensated for in the design of the two lower levels. At this level the merit variables programmed for each individual case must be used also. Machining strategies can even be programmed that may change during the machining process for a particular part. An optimum operation such as this may depend on the tasks to be completed, such as, for example, rough, semirough or final finishing. In any case, this higher level should supply the lower levels, in real time, with the cutting conditions and process state variables, such as the cutting forces or torque values that will in turn establish the references of the relevant controllers located at the process level. The following can be used as merit variables: 1) maximising productivity; 2) minimising tool wear rates; 3) avoiding regenerative chatter; 4) minimising cost; 5) ensuring surface quality (roughness and surface integrity); and 6) ensuring dimensional quality. Depending on the machining strategy used, as represented by a certain objective function, each of these variables must be weighted to a greater or lesser degree using multi-objective optimisation techniques.

After account has been taken of maximising productivity at the process level, the most significant functions include minimisation of tool wear rates, the avoidance of regenerative chatter and the guarantee of surface quality. Remember, nowadays people do not just buy machines, or even optimised processes. They buy production, that is to say, parts; and ideas such as "right the first time" therefore become fundamentally important.

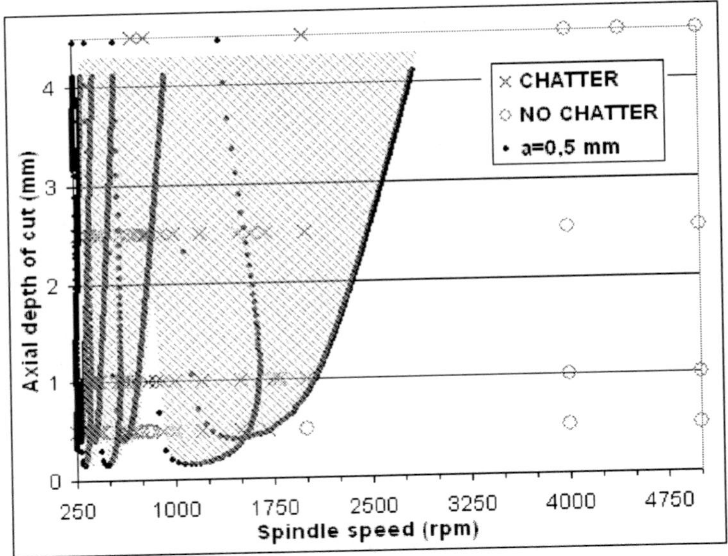

Fig. 5.10 Stability lobes diagram: depth of cut/spindle speed (white: stable zone, grey: chatter zone). Example of thin-floor machining (1 mm thick) using an axial depth of cut 0.5 mm, performed tests and diagrams at the University of the Basque Country

Merit variables should be controlled at this supervisory level, though the importance of estimating and predicting some variables must always be borne in mind. Measuring and predicting these variables will be discussed in Sect. 5.5. This section will deal only with some of the ideas concerning automatic control systems.

Current specialist literature uses two kinds of methods to address the issue of chatter avoidance; this machining problem is described in Sect. 3.5.3. The first is based on stability lobes diagram (off-line), and the second, on cutting speed modification (on-line).

Stability lobe diagrams propose two areas on the plane formed by depth of cut vs spindle speed (Fig. 5.10). The upper area is unstable, while the lower area is stable. A number of methods have been investigated for identifying the stability lobe diagrams for a particular machine and a particular tool. In general, the investigated methods display a lack of precision. There is therefore a tendency to develop analytical and experimental methods in which the transfer functions of existing systems with multiple degrees of freedom are identified by non-destructive structural dynamic tests (a piezoelectric hammer). That way the natural frequency, the stiffness of the mechanical system and the damping factor for each mode of vibration can be found through experimentation [32].

The most commonly used method for suppressing chatter automatically is to adjust the spindle speed set point. One heavily used technique is to vary the spindle speed, usually in a sinusoidal manner. While such methodologies may be promising, their main disadvantage is that there is no theory to support them. Nobody knows how to produce optimum adaptive modulation.

Another important function to implement at this supervisory level is monitoring of the surface roughness process (including surface integrity), with the imposition of cutting conditions that ensure the desired surface roughness in the final machined part. An automatic surface roughness control system could be implemented by using a sensor unit that can predict surface roughness during the process. To do so, it would be necessary to analyse the factors that affect surface roughness *Ra* and modify those values during the process in order to obtain the desired *Ra* values. As indicated in Sect. 5.5.3, there is an extensive series of factors that affect *Ra*. Of said factors, feed and cutting speed control would seem to be the most viable industrial solution. Other variables, for example, depth of cut, require the part program to be rewritten, so they rule out automatic *Ra* control. By controlling the emergence of regenerative chatter, feed rate and cutting speed, an effective automatic control process could be developed for the surface roughness of machined parts.

5.5 The Sensory System for Machining Processes

As already mentioned, if an “ideal machining unit” is ever achieved, a machining sensor system will be a key element for identifying how the machine, process and cutting tool are all behaving. Such sensors plus the post-processing units form what is identified in Fig. 5.8 as the monitoring units for the machine, the process and the tool.

The various monitoring strategies are classified as “pre-process”, “in-process” or “post-process”, depending on the time scale at which monitoring is performed, and they are divided into off-line and on-line methodologies. Some strategies are classified on the basis of the ultimate purpose of the monitoring process; these include tool wear monitoring (TWM), tool condition monitoring (TCM), the monitoring and the control of machining, surface quality prediction, tool wear detection (TWD), tool failure detection (TFD), real-time prediction, machine health monitoring, chatter stability prediction, heat generation and temperature prediction.

The following points can be made about off-line and on-line methodologies:

Off-line monitoring

- Pre-process. This type of monitoring is based more on an analytical study that attempts to predict the different parameters that will influence the process. It may be supported by an analysis of previously monitored results. Basically, what is necessary for this kind of monitoring is the appropriate IT tools to run process simulations. However, the initial set-up requires prior measurement (direct measurement) of all the elements or parts whose dimensions are known to vary. The direct measurement of these elements or parts is usually performed using instruments such as electronic or optical microscopes, optical cameras, roughness meters, hardness meters, micrometers, Vernier scales, etc.

- Post-process. Just as measurements must be taken before the process to establish the initial conditions, the final condition of the elements being studied (mainly the part and the cutting tool) must also be checked after machining is complete.

On-line monitoring

Machine status monitoring is seen as a maintenance task that is independent of the actual machining process and prepares the machine for beginning or continuing the machining process. In this case, the monitoring process will take account of factors such as the condition of the bearings and gears, the emergence of vibration in the moving parts (particularly under non-load conditions), the suitability of the temperature of machine components and the consumption of the drives.

Table 5.1 Measuring instrument type classified by machining signals

Signal type	Physical quantity (sensor type)	Location	Parameter to process	Measurement
Force	Power	Motor shaft, electric cabinet	Spindle power Feed shaft power	Cracking and tool wear Cutting power/ driving power ratio
	Torque (static, dynamic)	Table, tool	Tool torque	Tool wear Tool fracture No cutting chip condition Thread depth
	Distance/ displacement	Force transmission element	Indirect force measurement	Crack identification Pressure forces
	1–3 axis force sensor	Area under forces parallel to displacement axis	Single- or multiple-axis force	Process identification
Acoustic emission	Acoustic emission (AE)	Linked to work piece. Near signal source	Proportional cutting-force signals	High-frequency energy signals of cutting process and tool fracture
	Acoustic emission (AE) fluid sensor	Near noisy area	Proportional cutting-force signals	High-frequency strength waves of work pieces in movement or rotating components, or materials with very rough surfaces
	Rotating acoustic emission (AE) sensor	Near signal source	Proportional cutting-force signals	High-frequency energy signals of cutting process and tool fracture

Table 5.1 (continued)

Signal type	Physical quantity (sensor type)	Location	Parameter to process	Measurement
Vibration	Accelero-meters	Near signal source: spindle or work piece	Mechanical vibration of machine structure in machining process	Tool wear Give increase Collisions Excess vibrations in ball bearing and spindles
	Ultrasonic sensor	Near signal source	Ultrasonic-range vibrations	Tool wear Give increase Collisions Excess vibrations in ball bearing and spindles
	Laser barrier	Perpendicular to measurement surface	Tool fracture and wear	Thermal deformations Work-piece roughness Chatter marks Feed marks
Tempera-ture	Thermo-couple	In contact with work piece	Temperature effect on wear and machining process	Temperature gradient
	Thermo-graphic camera	Near signal source	Temperature effect on wear and machining process	Temperature gradi-ent distribution in cutting process

Although some in-depth research has already been done in this area, it is not yet widely applied, and only very simple sensors are being used, generally based on the application of constant-value upper and lower limits throughout the entire machining process.

The aim of monitoring the machining process is to detect disturbances or operating modes that have a negative effect on the process and part quality. Process monitoring can include all the phenomena that bear any correlation with the surface quality of the part being machined (e.g., vibration), the factors that result in machine process planning errors (e.g., precision in terms of dimensions) and unexpected machining conditions (e.g., collisions).

The most appropriate measuring instruments to select depend on process and computing-power constraints, as well as the variable that is to be monitored, and these depend in turn on the parameters that are to be estimated. The literature published in this field suggests the measuring instruments listed in Table 5.1, depending on the goals and constraints.

5.5.1 Correct Monitoring Conditions

Attempts to develop the appropriate monitoring instrumentation reveal that both the cutting process and the machine tool have limitations. The most significant are as follows:

- Spindle speed.
  - High-order frequencies caused by mechanical vibrations.
  - Appearance of high-order harmonics.
- Bandwidth, sensitivity, impact limit and resonant sensor frequency.
  - Bandwidth. A high bandwidth should be considered if the presence of high main frequencies and high harmonics is suspected.
  - Sensitivity. Critical when measuring signal values that are very small but very significant.
  - Impact limit. May not be so sensitive if the equipment does not involve elements that are subject to direct impact.
  - Resonant frequency. This could be considerably affected, depending on the equipment. Could become saturated unexpectedly early. The possibility of harmonics appearing above the main frequency increases with the number of moving parts that come into contact with one another at high relative speeds.
- Sampling frequency: High main frequencies require high sampling frequencies, for the acquisition of the largest amount of data possible for each tool cycle.
- Acquisition resolution: This determines the precision with which measurements are taken. The greater the resolution, the greater the precision.
- Download speed: At high acquisition speeds, the card must be able to download the data into the memory quickly. The internal memory must be large enough to be able to store all the data from the sample.
- PC requirements: In real-time monitoring and control, any acquired data must be processed as rapidly as possible. The PC processing capacity must be such that it can store the information received on its hard disk before the memory used for acquisition is entirely used up.
- Platform for communication with the PLC: This must allow the flow of data between the PC and the PLC (Programmable logic controller) to happen in real time. This means that the protocol must allow for high data transmission speeds; communication with the PLC must be as direct as possible, and any intermediate delays must be avoided; and finally the operating system must be in real time in order to guarantee the least possible loss of time in the sending and receiving of data between remote applications and the monitored system. The advantage of this type of operating system lies in the way it manages priorities. In contrast to other systems like Windows™ and Linux, these systems assign greater priority to the execution of the applications that require it.

5.5.2 Machining Characteristics and their Measurement

References [24, 25] have identified and classified the most important features that must be monitored during machining. Each feature to be monitored can be identified with one or more measuring parameters. Table 5.2 shows these machining features and their potential measurement factors, offering an idea of the flexibility with which one or more features can be monitored.

The function performed by monitoring each feature is shown in Table 5.3 [25]. Each function shown in this table can be performed using off-line and on-line strategies. Everything related with estimation forms part of the off-line (and, in this case, pre-process) strategy, while everything related with maintenance forms part of the on-line, in-process monitoring.

The general structure of a *condition monitoring system* (CMS) is shown in Fig. 5.11 [1] and contains the following functional units: 1) sensory signals, 2) signal processing, 3) extraction of sensory characteristic features, 4) decision making and classification, and 5) corrective actions. The aim in practice consists of selecting a multi-sensor system and ways of processing the resulting signals that will allow for a decision-making and classification function with minimum levels of error as regards possible process failures. As Table 5.3 shows, various measurement parameters can be used to evaluate machining characteristics.

Table 5.2 Characterisation of machining and measurement factors

Measurement parameters	Feature to monitor						
	Chatter	Cutting forces	Chip forming	Cutting temperature	Tool wear	Tool failure	Surface errors
Cutting depth	×	×	×		×		
Spindle speed	×	×	×				
Acoustic emission	×	×			×	×	
Cutting force	×	×	×		×	×	×
Feed vibrations	×	×					
Feed			×	×	×		×
Tool wear			×				
Infrared image				×			
Chip contact length				×			
Surface finish					×		
Part dimension					×		
Tool geometry					×		
Tool vibrations					×	×	
Spindle power					×		
Tool deflection							×
Cutting speed							×
Cutting temperature							×

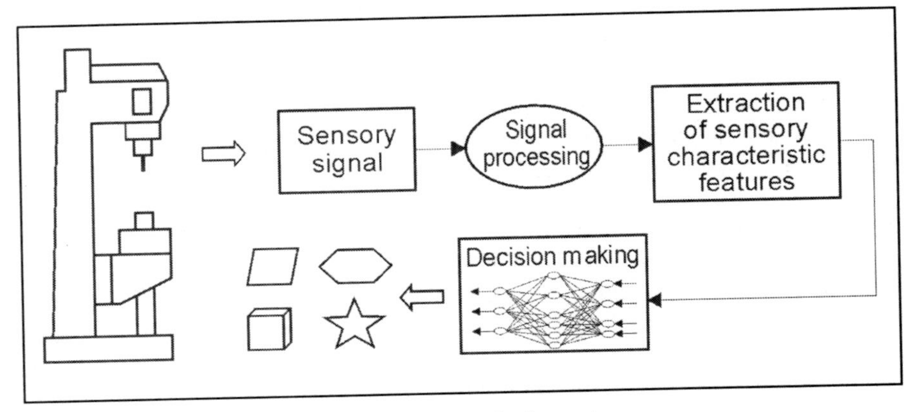

Fig. 5.11 The general structure of a condition monitoring system

Table 5.3 Functions associated with the features monitored

Feature	Function
Chatter	Estimate chatter Avoid chatter Suppress chatter
Cutting forces	Estimate cutting forces Maintain specs
Chip formation	Estimate formation Maintain specs
Cutting temperature	Estimate temperature Maintain specs
Tool wear	Estimate wear Maintain specified wear rate Compensate for wear
Tool failure	Estimate, avoid and detect failure
Surface errors	Estimate surface finish and tolerances Maintain specs

A detailed analysis of tool condition monitoring (TCM) and surface quality prediction is given in Fig. 5.11, in view of the particular importance of these two processes.

5.5.3 Two Case Studies

For many years now, a great deal of research has focused on the area of monitoring tool conditions, given that tool failure is the cause of around 20% of machine down time and tool wear has a clear impact on the quality of the parts produced, their dimensional precision, their surface quality and their surface integrity [9]. Artificial vision techniques have been successful in the laboratory and have given very precise results. Nevertheless, they have only worked in the laboratory and always in a post-process context, i.e., when the tool is no longer in contact with the part. The illumination has turned out to be a highly critical element. The majority of the developments made in the laboratory have been restricted to 2D measurements. Few results have been produced in 3D, and these only for crater wear.

The forces, both static and dynamic, that appear during machining are affected by tool wear. The radial force component would seem to be the most sensitive to wear. Sometimes the relationships between forces are utilised. The relationship between feed force and cutting force would seem to be highly sensitive to wear on the flank of the tool.

Likewise, during the cutting process, the part undergoes considerable plastic deformation, which generates acoustic signals (acoustic emissions). These high-frequency acoustic signals (50 KHz to 1 MHz) are highly sensitive to tool wear. A strong correlation has been shown between the acoustic emission RMS value and tool wear. The variance is one of the most sensitive parameters, with the highest range being shown at the end of a tool life. These laboratory results have not yet been put into industrial practice either, due to the lack of sufficient knowledge about the physical significance of acoustic emission signals and the great sensitivity of such signals to sensor location.

To date, the operation of almost all systems used to monitor tool condition has been based on very simple strategies that involve detecting when tool condition trespasses some specific pre-set limit. These strategies need to be recalibrated constantly, given the wide variation in the conditions under which machining is done. Adding greater "intelligence" to the monitoring process, along with the capacity to adapt to the production process itself, while maintaining a simplicity of operation for the end user is still the focus of research for many research teams.

From the most recent results reported, it can be concluded that the sensors most commonly used to monitor tool condition are: 1) acoustic emission sensors, 2) vibration/acceleration sensors, 3) static and dynamic cutting force sensors, and 4) cutting torque sensors. The place where sensors are located on the machine is an important factor to consider, since the system must, in all cases, be as non-invasive as possible.

Once the signals received from the sensors have been properly processed, patterns must be extracted, either in the time domain (time series models, e.g., AR, ARMA) or in the frequency domain. Once patterns have been extracted, the decision-making process must take place. For this, two types of methods may be used: statistical/stochastic methods or artificial intelligence (AI) methods. The most heavily used statistical methods are statistical analysis, time series analysis and discriminant analysis.

There is currently a wide variety of types of AI methods: 1) fuzzy logic, 2) artificial neural networks, 3) hybrid systems (e.g., neurofuzzy systems), and 4) probabilistic networks (e.g., Bayesian networks). Neural networks are currently widely used in research projects, in which both supervised networks (multi-layer perceptron, among others) and non-supervised networks are used. The latter are currently very much in fashion, with networks such as ART (adaptive resonance theory), KSOM (Kohonen maps) and RBF (radial basis function). All these neural structures are of the competitive-learning type.

Tool condition has a noticeable effect on surface roughness, surface integrity, dimensional quality, machine vibration levels and machine performance in general. It is therefore important to ensure that these systems not only account for the fresh/worn dichotomy, but also cover intermediate wear parameters that allow for the development of more effective, sophisticated tool change strategies.

Surface quality is measured in terms of what is called surface integrity, which not only describes the topology of a surface but also takes account of its mechanical and metallurgical properties that subsequently play a vital role in terms of fatigue strength, resistance to corrosion and the part's service life. Highly sophisticated laboratory techniques are generally used, such as X-ray diffraction and metallographic inspection. These are always applied post-process and in all cases cost both time and money.

Surface topology is represented by the so-called surface texture that measures several quantities related with the deviations in a part as compared with the nominal surface. Deviations are classed from first to sixth order, in accordance with international standards. First- and second-order deviations refer to form (i.e., flatness, circularity, etc.) and to waviness, respectively. Surface roughness refers to deviation from the nominal surface of the third up to sixth-orders. *Ra* is the most widely used parameter to describe surface texture due to its direct influence on friction, fatigue, electrical and thermal contact resistance, and appearance [20].

Nowadays roughness measurements are taken using modern roughness meters (profilometers), though their application is restricted to the post-process applications, with the corresponding inherent problems such as inspection times, the need for random testing (i.e., not every part produced is tested) and the production of waste, since the fact that inspection occurs post-process means that waste production cannot be avoided, with all that this entails in terms of both cost and the effect on the environment. At the present time, the quality assurance process involves two phases. The first uses Taguchi-type methods during the design stage to ensure the quality of the product being designed. This is a pre-process action, making it essential to incorporate a second intermediate stage to ensure quality during the machining process.

Ensuring the surface quality of a part during the machining process is no easy task. Vision systems have been used which involve measuring reflected light intensity in order to estimate surface roughness. However, the practical results have as yet not offered much hope, as the measurements obtained have not been very precise or reliable. Neither sensors based on artificial vision nor sensors based on acoustic emissions have actually worked on the factory floor. This has made it essential to develop sensors based on predictive models.

Predictive models can basically be separated into two main categories: models based on machining theory, which are analytical models that could be termed mechanistic, and models based on observed data, which are empirical models that could be referred to as observational. Mechanistic models tend to be more general. Empirical models are more specialised and only operate under certain specific conditions. Their properties depend heavily on the data observed (black box models), which include its abundance, integrity, completeness and timeliness. It is also possible to develop semiempirical (grey box) models.

Mechanistic models used to predict surface roughness are of the geometric type, based on the motion geometry of a metal cutting process, without taking cutting dynamics into account. Hence their main disadvantage, namely, that these mechanistic models do not include parameters related to cutting dynamics, such as speed, depth of cut or the type of material used for the work-piece. A second disadvantage is that they are purely theoretical and do not verify their models with any experimental data. As a result, in-process prediction of surface roughness tends today to involve the use of empirical models, and the design of the experiments used to obtain observed data is of extreme importance. Here, Taguchi techniques for the design of experiments (DoE) seem to be the most widespread methodologies.

Generally speaking, the compromise that must be dealt with when building empirical models is between the form and structure adopted and the way the model depends on the selected data (e.g., variance). An incorrectly parameterised model results in a biased model. An over-parameterised model will offer a high variance which fits the construction sample well, but could be of little use in validating a sampling. The bias/variance trade-off becomes particularly evident when working a small amount of data, where it is difficult to identify a smooth form due to the variability of the data. The choice between the bias and variance in a neural network model is more difficult to define.

Empirical models basically take one of two forms: There are statistical models, based on non-linear multiple regression equations, and there are artificial intelligence models, based above all on artificial neural networks (ANNs). A typical example of a non-linear multiple regression equation appears in [7], which proposes a functional relationship between surface roughness and some independent variables:

$$Ra = c\, h^{\alpha 1} f^{\alpha 2} r^{\alpha 3} d^{\alpha 4} s^{\alpha 5} \tag{5.19}$$

where Ra is surface roughness (the arithmetic average), h is the hardness of the part, f is the feed, r is the tool nose radius, d is the depth of cut and s is the cutting speed.

Because there are equations of this kind, the results can be discussed from a mechanical point of view, and the variables that are statistically more important can be identified. These models show that machining merit variables offer contrasting contributions in areas such as productivity, surface finishing, dimensional precision and machining costs, among other things. Considering only the most representative interactions between independent variables and merit variables, the following could be cited as being the most important:

- Productivity and surface roughness are contrasting figures of merit in the machining process.
- Feed has a notable influence on productivity and surface roughness. Productivity increases with increased feed, while surface quality diminishes.
- Cutting speed has a notable influence on the dynamic stability of the process and surface roughness. Surface quality increases with greater cutting speed.
- Depth of cut has a notable effect on productivity, dynamic stability and surface roughness.
- The hardness of the work-piece being produced and, above all, the radius of the tool have an effect on surface roughness. Surface quality increases as the radius of the tool is increased.

ANN-based models offer a very powerful alternative when there are no analytical models or when a low-grade polynomial is not appropriate. An essential step when using an ANN for modelling is to use real observed data to find the synaptic weight values for all neurons (learning algorithm). One of the algorithms used for adjustment is the back-propagation algorithm, which is the most often-used algorithm for modelling surface roughness.

Reference [38] proposed a model that used spindle speed, feed speed, depth of cut and vibration average per revolution as the ANN input signals. According to their results, using a 4-5-1 topology for the ANN, they could achieve an accuracy of 95.87% precision. When using a more powerful topology, with two hidden layers at 4-7-7-1, they obtained an accuracy of prediction of 99.27%. Training times were not long, and operation in real time seemed to be guaranteed.

5.6 Open-Architecture CNC Systems

Computer numerical controls have evolved at extraordinary speeds in recent years, due basically to the availability of powerful microprocessors, sophisticated software and other developments in digital technology. Many manufacturers include high performance processing blocks in their controls, with processing times of under a millisecond. Extremely high speeds can now be attained in comparison with the 70 ms that were available with classic CNCs. These processing-time specifications cannot be looked at alone, but rather in relationship with the time required to process, initiate and complete the required motion. This is the real execution time, which defines the speed and accuracy of machining

Research on integrating personal computers (PCs) with open CNC systems has continued in recent years, and this has necessarily led to superior properties in terms of flexibility. The new open architectures are more reliable, to the extent that they can operate the machine as a whole and open architecture also allows for the real-time overall management of the whole flow of information concerning a machine's operation. This avoids the instability that is inherent in a link with a PC or a connection with a PC operating system. Instability of this sort has been particularly problematic lately with the Windows operating system.

Furthermore, results on the design of complex distributed-control architecture are now available. In other words, in addition to the inclusion of high speed microprocessors in CNCs that can share control tasks in real time, there have been advances in the development of digital and "intelligent" motor actuators for each axis. Given the high speeds used, feedback loops are connected directly to the actuators instead of going to the CNC, though the CNC must still monitor the entire system as a whole.

A completely digital system, like some of those that are already commercially available (e.g., the Heidenhain TNC 620™), provides an excellent platform for achieving improved control. Some systems incorporate commands that switch to "high precision" control modes in order to increase precision and reduce the cycle time. These commands involve calculating different accelerations and decelerations in real time based on feed speeds. Elsewhere, new intelligent filtering and control strategies based on adaptive algorithms and fuzzy logic can now be used to suppress vibrations in the system within a determined range of frequencies.

5.6.1 Networked Control and Supervision

In traditional CNCs, communications architecture is by point-to-point connection. This architecture has been successfully applied in industry for many decades, though it is not suitable for systems containing a large number of devices, as the number of cables increases along with the number of elements. Nor does this kind of architecture afford the possibility of dealing with new requirements, such as modularity, decentralised control, simplicity and ease of maintenance [41].

A networked control system is one whose components (sensors, controllers, actuators, *etc.*) are distributed using some form of computer network technology. The use of this technology brings with it important advantages, such as reliability, the improved use of resources, ease of maintenance and error diagnosis and, above all, the possibility of reconfiguring the different components. Nevertheless, there are also disadvantages to this type of system: implementing closed-loop control in a communications network leads to delays that inevitably can cause instability [14].

5.6.1.1 Communications in an Industrial Environment

The real-time communications market in the industrial sector has for a long time been dominated by fieldbuses. Some of the most well-known standards are Profibus, WorldFIP and Foundation Fieldbus. At present they have been standardised but not unified. These technologies offer a number of advantages, one of which is their deterministic behaviour. However, their disadvantages include the high cost of hardware and the difficulty of integrating them with other products [23]. To solve these problems, some computer network technologies, particularly Ethernet, are being adapted for use in the industrial automation field, given their simplicity, low cost, availability and high transmission speeds.

The main technical obstacle to Ethernet in industrial environments is its non-deterministic behaviour, which in principle makes it unsuitable for applications with real-time requirements. The reason for this is that, in classic Ethernet networks, the transmission medium is shared by all stations in the sub-network. The CSMA/CD (carrier-sense multiple access with collision detection) protocol is used to access the medium. This protocol cannot guarantee that the packets sent will be received before a certain amount of time has elapsed, because a station will have to wait until the transmission medium is not being used before being able to transmit. Furthermore, if two stations on the network decide to transmit simultaneously, there will be a collision, which will mean that the message will not reach the recipient correctly. When a collision is detected, the stations must wait for a random amount of time, which is determined by the binary exponential backoff algorithm. If a station detects 16 consecutive collisions, it will cease to transmit and send an error message to the higher levels in the network protocol stack [34]. In short, Ethernet does not use up time in the bus, but collisions between packets may lead to delays in information being sent and may even cause information to be lost. It is therefore impossible to predict what delays will occur in the network.

Although it does not perform deterministically, under light traffic conditions Ethernet sends information quite quickly and with little latency [19]. However, with a greater load on the network, collisions are more frequent, thus causing delays and reduced performance. One way of improving the performance of a control system based on an Ethernet network would therefore be to reduce network traffic, with the resulting reduction in the number of collisions.

Some researchers have suggested changing the medium access protocol so as to achieve deterministic behaviour. PROFINET is one standard that does so. Although these alternatives use Ethernet hardware, they are entirely incompatible with standard Ethernet traffic. In recent times, the use of switched Ethernet (which uses switch connections) has offered further improvements. Switched Ethernet behaves differently from the classic model. Switches are intelligent devices that identify the intended recipient of the information and only retransmit the information via the required output ports. This differs from the way the older system works, via a hub that retransmitted all the information received via all its ports (Fig. 5.12). Use of the switch system on the Ethernet eliminates any potential collisions in information transmission and thus increases network efficiency.

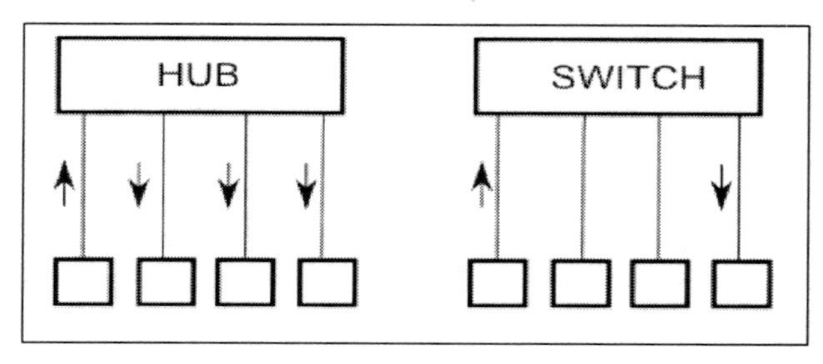

Fig. 5.12 Hub vs Switch

Because there are no collisions, the network does not become unstable under heavy traffic loads, and delays are drastically reduced. The delays that happen when switched Ethernet is used are sufficiently low for industrial networks working in real time, making switched Ethernet quite a promising alternative for networked control systems [18].

5.6.1.2 The Structure of a Networked Control and Supervision System

Figure 5.13 shows the structure for a networked control and supervision system run via switched Ethernet.

In its simplest form, the system is comprised of two stations, the first of which handles control, and the second, acquisition. The two stations are connected via a switched Ethernet network. The acquisition station is in direct contact with the process.

The system works as follows: the acquisition station, which is in constant contact with the process, acquires information from the process. At given intervals, depending on the process in question, the acquisition station sends the information it has obtained to the control station. The control station makes a decision based on a control algorithm that has been specifically implemented for the process. This information will be sent to the acquisition station so that the necessary action can be taken.

This is therefore a *client-server* architecture. The control station (*server*) receives requests from the acquisition station (*client*), makes a decision on the basis of some form of control strategy and replies to the original station. The control station is also responsible for carrying out supervision duties, and it is the point at which all the system's resources must be integrated and centralised. Its duties particularly include monitoring process variables and plant conditions and configuring devices and control systems.

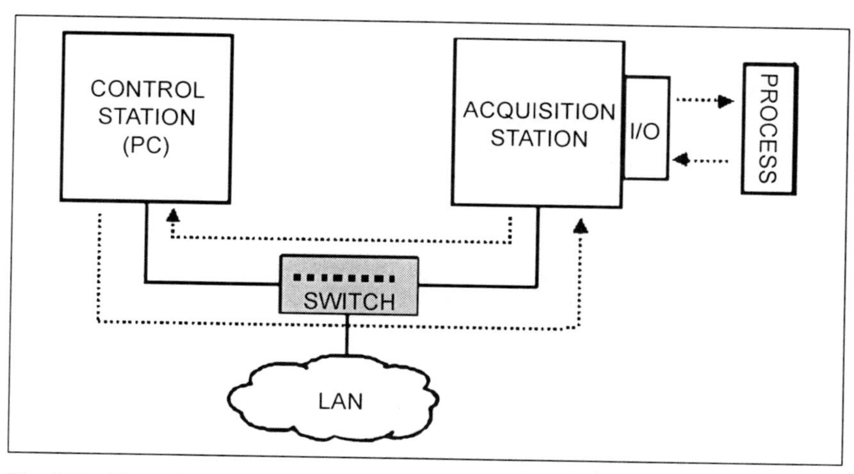

Fig. 5.13 Networked control and supervision via Ethernet

5.6.1.3 Networked Control and Supervision via Ethernet

Internal model control (IMC) is a technique that is widely used and well established in controller design. This closed-loop control scheme explicitly uses a model (G_M) of the dynamics of the plant to be controlled, which is placed in parallel to the plant in question (G_P). Furthermore, it also contains another model of the inverse dynamics of the plant (G'_M) placed in series with the process, which acts as a controller.

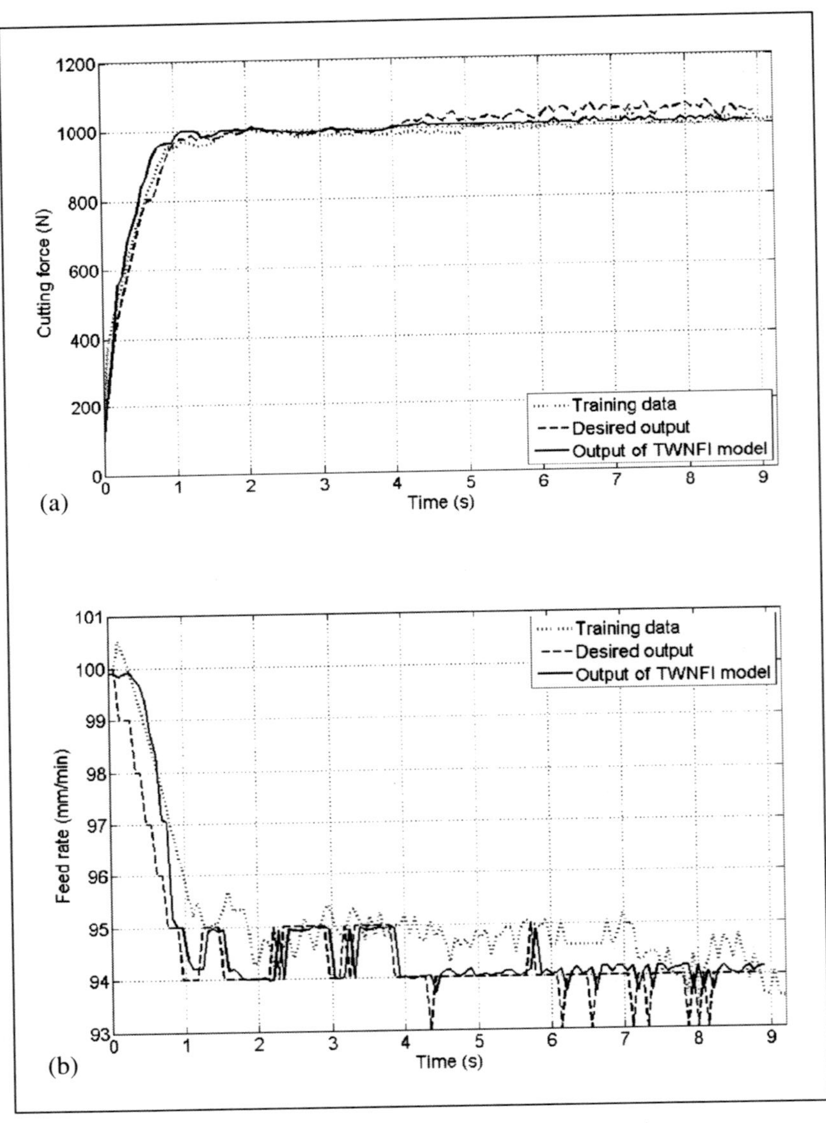

Fig. 5.14 Output of TWNFI model. **a** Direct model. **b** Inverse model

One of the advantages of this type of control lies in the fact that its stability and robustness can be guaranteed. However, the inversion of non-linear models is no easy task, and there may be no analytical solution. In certain cases, a solution may exist but be impossible to implement. Another associated problem is that the inversion of the process model could lead to unstable controllers when the system is in a non-minimal phase.

The following is a representation of the use of the TWNFI algorithm to create on-line models (direct and inverse) [33]. Both models are calculated with each new entry in the control scheme. Using this neurofuzzy inference technique, the creation of the inverse model is much easier and always offers a solution. The direct model should be trained to learn the dynamics of the process. In order to do this, a TWNFI system is used with input-output learning data in which the input data corresponds to feed speed while cutting force is used as the output variable (Fig. 5.14a).

In order to calculate the inverse model, instead of inverting the direct model obtained analytically, another TWNFI system is used in which the learning data contains cutting force values as the input and feed speed values as the output. In this way, the system is able to learn the inverse dynamics of a high performance drilling process (Fig. 5.14b).

The learning data for both the direct model and the inverse model have been obtained from real drilling operations using GGG40 material tests. This set of data does not need to be more extensive, as representative values from each area of operation are sufficient.

Inverse model G'_M and direct model G_M use prior states for the input variables in order to achieve a better approximation in the input-output relationship in the training data for the TWNFI algorithms:

$$\begin{aligned} f(k) &= G'_M(F(k), F(k-1), f(k-1)) \\ \hat{F}(k) &= G_M(f(k), f(k-1), \hat{F}(k-1)) \end{aligned} \tag{5.20}$$

where $\hat{F}(k)$ is the cutting force estimated by the model.

Once the models were laid out in a diagram, it was also decided to include a low-pass filter (G_F). This was incorporated in the control system with the aim of reducing high-frequency gain and improving the system's robustness. It also served to smooth out rapid, brusque changes in the signal, thus improving controller response. The filter takes the following form:

$$G_F(z) = \frac{1-k_2}{z-k_1} \tag{5.21}$$

where k_1 and k_2 are design parameters, and usually $k_1 = k_2$.

Figure 5.15 shows the internal model control scheme. The diagram includes the delay (both the delay introduced by the network and the delay inherent in the process) through block L. The figure also shows direct and inverse models G_M and G'_M, respectively, filter G_F and the process represented by G_P [12]:

$$G_p(s) = \frac{F(s)}{f(s)} = \frac{10.26}{0.005241s^3 + 0.09376s^2 + 0.5414s + 1} e^{-0.4s} \tag{5.22}$$

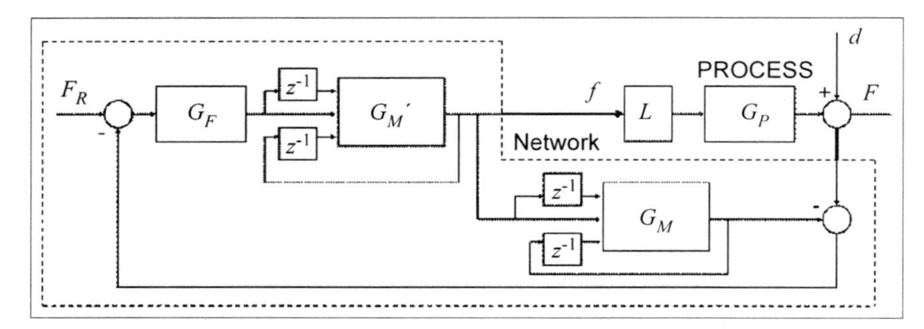

Fig. 5.15 Internal model control scheme

where *f(s)* is the feed speed, *F(s)* is the cutting force and G_p(s) is the process transfer function in the Laplace domain.

The issues concerning the delay, *L*, and the architecture of the networked control system are explained in the following section.

5.6.1.4 Networked Control Systems

All delays (both from the process itself and from the network) have been grouped together as block *L*. The use of networked control systems undoubtedly brings significant advantages in terms of reliability, better use of resources, ease of maintenance and error diagnosis and, above all, the possibility of reconfiguring the different components. Fieldbuses have become consolidated as the most commonly used communications networks in industry. Indeed, it is normal practice in the process control field for models to incorporate the delay displayed by these networks.

However, the development of information and communications technologies has meant that control and supervision systems are increasingly required for processes that cover some geographical area. For this reason, the control processes discussed here for high performance drilling use Ethernet network technology.

Ethernet is being used massively in the field of industrial automation, given its low cost, its availability and its high transmission speeds [6]. The main difficulty associated with networked control in general and control using Ethernet in particular lies in the fact that, while the control system is strong enough to withstand the design changes considered, it may not be able to tolerate unmodelled communications delays.

Ethernet's main technical disadvantage in an industrial environment is its non-deterministic behaviour, which makes it unsuitable for applications that include real-time requirements. Ethernet does not use up time in the bus, but packet collision may cause delays in information transmission and may even cause information to be lost.

However, the use of switched Ethernet (which is connected by means of switches) removes the potential for collision during data transmission, thus increas-

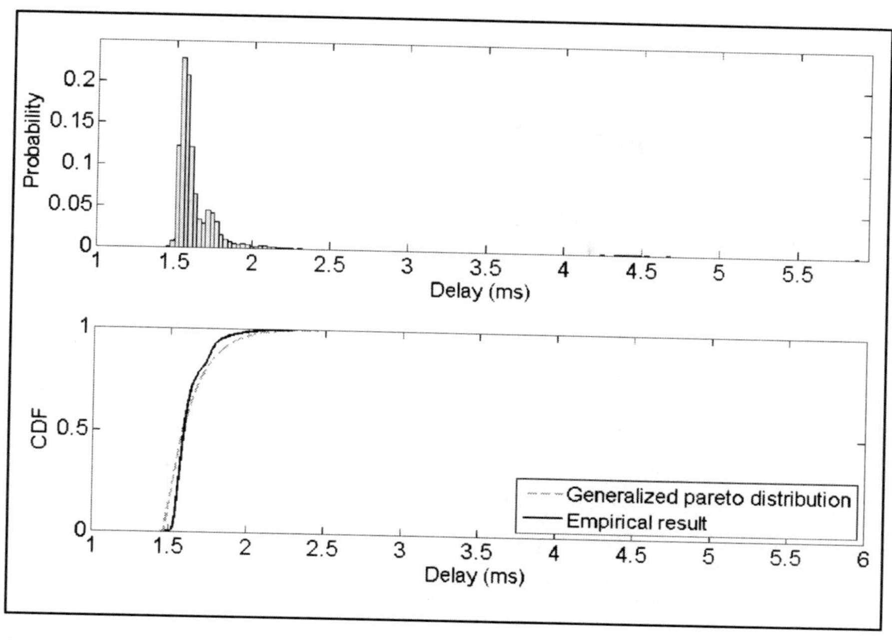

Fig. 5.16 Statistical distribution of delays

ing the network efficiency. As there are no collisions, the network does not become unstable under heavy traffic, and delays are drastically reduced. Studies have shown that switched Ethernet causes delays that are sufficiently small and deterministic for the system to be used in real-time industrial networks, and it has therefore become quite a promising alternative for networked control systems.

An in-depth study of the delays found in manufacturing processes where machine tools are connected to computers in a local network using Ethernet can be found in [21], along with some experimental results. Based on this work, as shown in Fig. 5.16, the maximum delay in an Ethernet network is established as 5×10^{-3} sec.

The architecture of the networked system is shown in Fig. 5.17. PC1 is connected using Profibus to the open-architecture CNC of the machine tool. This is where the measurement and processing of measured signal F takes place:

$$\tau_{SC} + \tau_{CA} + \tau_{drilling} = 0.4 \tag{5.23}$$

From a technical point of view, the existence of proprietary software places restrictions on connectivity to the open-architecture CNCs systems. However, it offers security in handling the CNC. PC2 uses free software (RT-Linux) and an intermediary real-time system (ACE-TAO). The internal model control system is implemented here. Assuming that the maximum delay is known (0.4 s >> 0.005 s), it is possible to implement networked control using Ethernet.

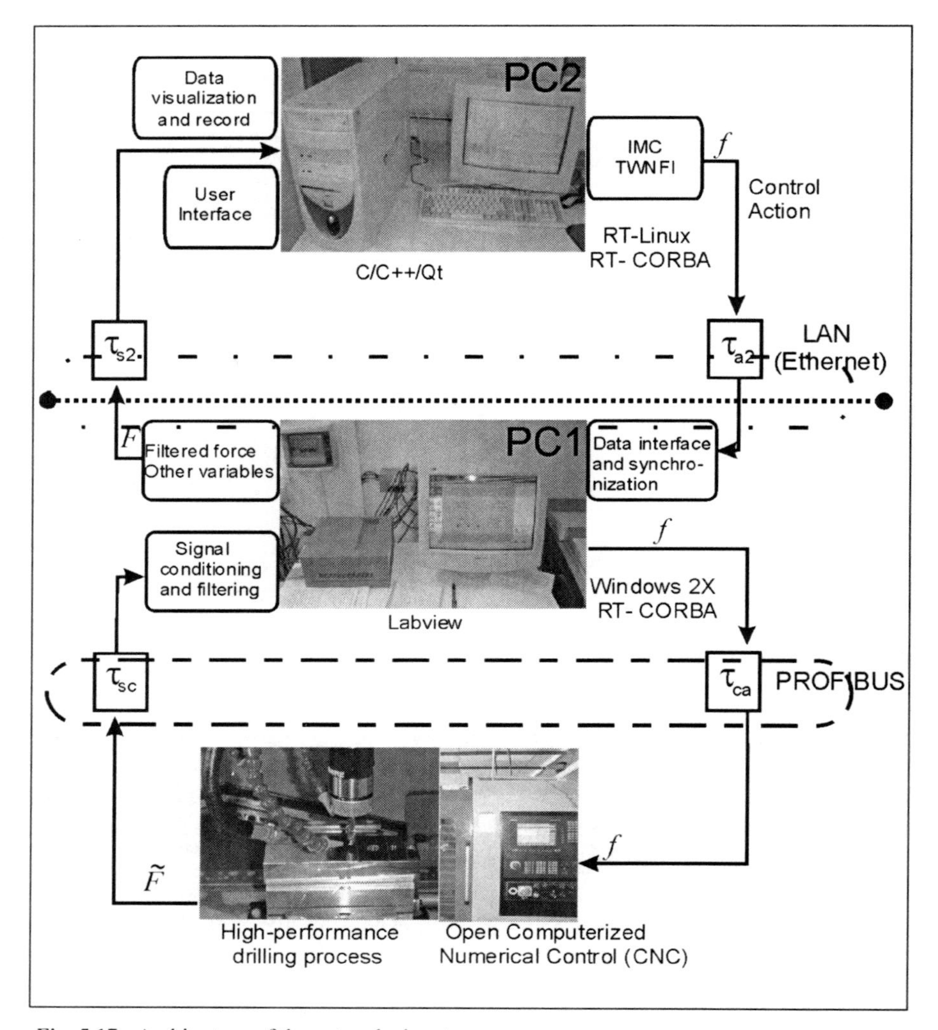

Fig. 5.17 Architecture of the networked system

5.7 Programming Support Systems: Manual Programming

Moving on from the operational function of a numerical control, the next issue is to analyse how a numerical control is programmed, what language is used and how a programmer enters all the data required for machining (i.e., the part program) into the numerical control. In the simpler cases, the programmer works from a drawing that identifies all the coordinates and the lines that connect them. However, if a complex part is being made, and particularly if the part includes free-form surfaces, the identification process is much more complicated. A solution that is frequently used in such cases is the exploration/digitalisation technique. Digitalisation is a technique by which a scale drawing is used to provide the coordinates required to prepare a numerical control program. This program is usually generated automatically by the computer using the aforementioned digital information. The computer calculates a sufficient number of discrete points to produce a satisfactory control program. All digitalisation systems offer the possibility of editing or reviewing the data collected.

In general, two different methods may be used to program a machining process, manual programming or automatic (computer assisted) programming. In the case of the former, the program is entered in machine language, with the reasoning and calculations made by the programmer himself. Programming in traditional machine language requires the end points of all segments, arcs and other geometrical data to have been previously calculated. In the latter case, a general-purpose computer such as a PC or work station is used as a programming aid to output the part program in the specific language required for the numerical controller in question.

Manual programming in machine language presents serious disadvantages, the most important being the time required to develop a program that is really operative. It requires many calculations made by hand, which lead to errors being made. These disadvantages have now been partially eliminated, since today's CNCs are able to program fixed or repetitive machining cycles, conducting a graphical simulation of the tool's movement and completing a number of mathematical operations, meaning that the CNCs do not need to know all the coordinates involved in machining a part. Nevertheless, and in spite of its disadvantages, manual programming is essential, so much so that a good knowledge of manual programming methods is essential for creating a really high quality program using the automatic programming technique. The debate between manual and computer assisted programming is not very important these days, as both are now completely essential. Nobody can be a good CNC programmer without having an in-depth knowledge of both techniques. In the case of manual programming, the programmer must complete the following operations:

- Break the machining of the part down into elemental operations and specify the technological requirements (feed, cutting speed, *etc.*), bearing in mind the tools to be used for each operation.
- Determine the preferential order for each operation.
- Define the part's curves and surfaces using curves that the controller can generate using the proper interpolation. It is important here to take account of the required level of dimensional tolerance.
- Write the program in the machine language for the specific numerical controller and its particular manual programming format.
- Enter the program into the CNC device.

Although today's CNC systems now have new features that help perform the tasks listed above, it is an unquestionable fact that they can be cumbersome and even impossible for a programmer who does not have support from a computer.

The following observations refer only to the new properties of advanced CNCs, as it is to be supposed that the reader has a good knowledge of classic manual programming. Afterwards, the discussion will turn briefly to the automatic programming languages currently used in CAD/CAM (computer-aided design/computer-aided manufacturing) systems.

A CNC program consists of a set of blocks or instructions that, when properly organised in subroutines or in the program body, provide the CNC with all the data required to machine a particular part. The CNC program may consist of several local subroutines, which are defined at the beginning of the program, along with the main body of the program. The body of the program contains a header (to indicate the start of the program), several program blocks (which contain the movements, operations, etc.) and an end-of-program instruction. An auxiliary M02/M30 function is used for the end-of-program instruction (Fagor® CNC 8070).

A subroutine is a set of blocks that may be called upon several times from the main program or from another subroutine. Subroutines can be either global or local. A global subroutine is stored in the CNC memory as an independent program. A local subroutine is defined as part of a program and is only called upon from the program where it has been defined. In modern CNCs, the blocks comprising the subroutines or the program body may be defined using ISO code commands or high-level language. Each individual block must be written in a single language, though the program may combine blocks written in both languages.

The high-level programming language allows the operator to use control commands like *$IF*, *$GOTO*, *etc*. With both types of language, the operator can use constants, mathematical parameters, variables and mathematical expressions. Programming in ISO code still involves the use of typical *N, G, F, S, T, D, M, etc.* addresses for each instruction. All of today's controllers have a look-ahead function that allows the machine to read several blocks ahead of the one currently in operation, to improve calculation of the forthcoming trajectory.

In contrast to their older counterparts, the latest CNCs incorporate logical and mathematical operations and functions. An operator can program all kinds of mathematical operations, make comparisons (e.g., greater than or equal to), perform binary operations (e.g., exclusive *OR*), use logic operators (e.g., logic *AND*), Boolean constants, trigonometric functions (e.g., arctangent), mathematical functions (e.g., Neperian logarithm) and other functions, such as "return the integer" [4].

Modern controllers also incorporate functions such as acceleration control. There is a manufacturer-defined nominal acceleration, a_0, and an acceleration a_p, which is the acceleration to be applied according to the operator. To do so, *G* functions of the following kind are used:

G130 percentage of acceleration to be applied per axis.
G131 percentage of acceleration for both axes.

Thus, the block:

G130 X50Y20

indicates an acceleration of 50% (a_0) for the X-axis and 20% for the Y-axis, and:

G131 100 X50 Y80

restores 100% of acceleration on all the axes and causes movement to point X = 50 Y = 80.

Two other features that only new CNCs have are jerk control and AC-Forward Control. The functions associated with jerk control are [4]:

G132 X20 Y50

which programs a jerk of 20% on the X-axis and 50% on the Y-axis.

The feed-forward percentage is programmed using *G134* followed by the axes and the new feed-forward percentage to be applied for each axis. Thus:

G134 X50.75 Y80 Z10

indicates a percentage of 50.75% on the X-axis, 80% on the Y-axis and 10% on the Z-axis. This percentage may be applied via machine parameters and via PLC as well as by program. The value defined by PLC will be the one with the highest priority. The maximum feed-forward value that can be applied is 120%.

Another important advantage of modern CNCs is their capacity to control high speed machine tools (HSC). As already mentioned, CAD/CAM systems provide a high number of very short blocks, from several millimetres to a few tens of microns in size. With this type of part, a CNC must be able to analyse a large number of blocks in advance and thus generate a continuous line that passes through the points defined in the program, while as far as possible maintaining the programmed feed speed and observing the limits on maximum acceleration, jerk, etc., for each axis. The HSC function offers several ways of working to optimise solutions related with contouring errors and machining speed. In such cases, the maximum contour error permitted is defined in advance, and the CNC then modifies the geometry through intelligent algorithms to eliminate any unnecessary points and automatically generate splines and polynomials in the transition between blocks. In this way, the contour is travelled at a variable feed rate, depending on the curvature and other parameters programmed (acceleration and velocity), though without exceeding the maximum level of error permitted.

The following can thus be programmed:

 # HSC ON
 # HSC ON [CONTERROR 0. 01]
 # HSC ON [CONTERROR 0. 01, CORNER 150]
 # HSC ON [CORNER 150]

indicating that the maximum contouring error is 0.01 and the maximum angle between two paths is 150°.

One of the functions that differentiate modern CNCs from their predecessors is their capacity to program spline interpolation using a series of control points. This type of machining adapts the programmed contour to a spline-type curve that passes through all the programmed points. The dashed line shows the programmed profile. The solid line shows the spline [4] (see Fig. 5.18).

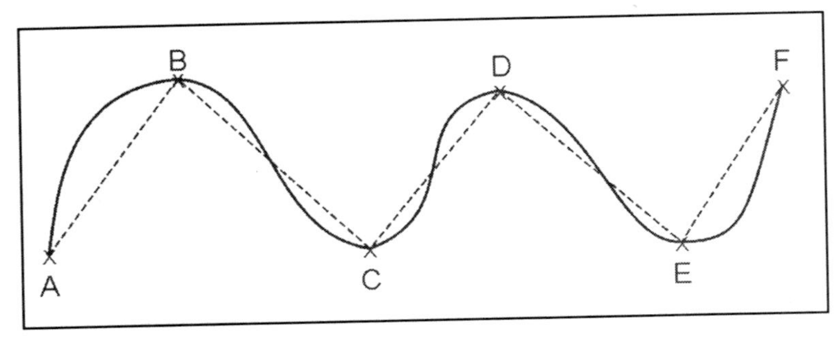

Fig. 5.18 Spline adaptation

The contour to be splined is defined with straight paths. When defining an arc, the spline is interrupted while being machined, and it is resumed on the next straight path. The transition between the arc and the spline is done tangentially.

When executing the "activate spline" adaptation, the CNC interprets that the points programmed are part of the spline and begins making the curve. The programming format is:

 # SPLINE ON

With the "instruction select"-type of tangent, the programmer sets the type of initial and final tangents of the spline that determines the transition from the previous and to the next path. Optionally, if the tangent is not defined, it is calculated automatically. The programming format is:

 # ASPLINE MODE [<initial>, <final>]

One example is [38]:

N30 # ASPLINE MODE [1,2]
N40 # SPLINE ON
N50 # XX1 YY1
N60 # XX2 YY2
N70 # XX3 YY3
…
…
N110 # XXi YYi
N120 # SPLINE OFF

where (X1, Y1) (X2, Y2) … (Xi, Yi) are the spline points and (Xi, Yi) is the last point of the spline. Value "1" indicates that the initial tangent is calculated automatically. Value "2" indicates that the final tangent is the tangent to the previous/next block (see *MODE*).

Another function that only appears in advanced CNCs is the possibility of programming polynomials of a limited degree. The instruction *#POLY* can be used to interpolate complex curves, like a parabola. With this type of interpolation, a curve given by a polynomial of up to the fourth degree can be machined [4]. The programming format is as follows:

```
# POLY [<axis> [a, b, c, d, e] ... SP <sp> EP <ep>]
```

where *<axis>* is the axis to be interpolated, *a, b, c, d, e*, are the polynomial coefficients, *<sp>* is the initial interpolation parameter and *<ep>* is the final parameter. A programming example would be:

```
G0X0Y0Z1 F100 0
G1
# POLY [X(0,60,0,0,0,] Y(1,0,3,0,0) SP0 EP60
M30
```

New CNCs also allow macros to be programmed that define a program block or part of a program block with their own names, using the format "*MacroName*" = "*CNCblock*". The Fagor CNC 8070 allows up to 50 different macros to be defined, each up to 140 characters long. Here is an example of two simple macro definitions:

```
# DEF "READY" = "GO X0 Y0 Z10"
# DEF "START" = "SP1 M3 M41"
```

In addition, the manual programming language of all modern CNCs offers the possibility of executing flow control instructions, which were once the exclusive domain of computer assisted programming languages. An example can be seen in the instructions to skip a block, such as:

```
$ GOTO N <expression>
$ GOTO [<label>]
```

These are unconditional skips, though conditional skips can also be programmed, such as:

```
$ IF <condition> ... $ ENDIF
```

The programmer can even program "*if else*" instructions:

```
$ IF <condition> ... $ ELSE ... $ ENDIF
```

or "*FOR*" instructions:

```
$ FOR <n> = <expr 1>, <expr 2>, <expr 3> ... $ ENDFOR
```

These new functions were not available in older CNCs manual programming language.

5.7.1 Computer Assisted Programming

When computer assisted CNC programming (computerised programming) is used, the programmer's role is reduced to preparing command orders, and the computer carries out all the operations that computers can complete much more quickly and with a minimum probability of error. Among other things, the computer can complete the following tasks:

- Calculate all the important machining points.
- Compose the data blocks.
- Detect and correct errors.
- Enable the definition of subprograms for use in repetitive machining operations.

The computer also uses the appropriate post-processor to adapt the results to the machine language required for each CNC.

There are two main categories of computer languages, general languages and specific languages. General languages can be used for any machine tool. A general language is also integrated software, which means it allows the programmer to use various types of machine tools. Specific languages are simpler and have been developed by machine tool manufacturers for use in their own machines. The disadvantage of specific languages is that they have their own individual vocabulary and syntax, but on the plus side there is no post-processor, and data processing can be done in a single step with a single program. In general, a specific language is dedicated software that is only supported by one type of machine, for example, lathes. Both types of languages are derived, generally speaking, from APT or Compact II, whose origin lies in the late 1950s at the Massachusetts Institute of Technology. Language improvements such as the APT Long-Range Program appeared towards the end of the 1960s.

All of these languages and their derivatives are still used, though they are increasingly being replaced by modern interactive graphical programming, where the programmer defines the geometry typically as the geometry of the tool's trajectory. Any geometrical error is detected automatically and displayed on a graphical screen for correction. This trend has progressed continually since the 1970s, with CAD/CAM systems adding the possibility of visualisation to programming processes. It is now both easy and cheap to use a CAD/CAM system in the workshop, so it is easy to develop programs using graphical interfaces. It is not even necessary to use a dedicated PC, as the same computer can also be used for other tasks. In general, a drawing is developed using CAD software (such as AutoCAD™) and is then sent to CAM software (such as MasterCAM™), which generates the control program in machine language. This includes a suitable post-processor for each CNC. The process requires an exchange of files between the different systems. The most traditional way of doing this is by IGES (initial graphics exchange specification). Another frequently used format is Autodesk's DXF. This DXF (drawing exchange format) is regarded as standard for exchanging drawings, and it was developed by Autodesk®, the owner of the popular AutoCAD program, which is the most widely used CAD program in the world for applications where highly complex geometries are not required. In general, DXF is used for the simpler cases, and IGES, for more complex geometries. For more information on computerised programming languages, the reader is directed to the relevant programming manuals. If interested in classic languages, see APT or Compact II. For modern Windows-based programming systems, see the popular, powerful MasterCAM, though there are other software packages.

5.7.2 Graphical Simulation

All computerised programming languages offer simulations of various kinds to show machining trajectories in different amounts of detail. In addition, all PC-based CAD/CAM systems offer more-than-adequate versions of this facility. Manual programming languages also now incorporate graphical simulations for choosing between different representation modes: 1) plan view, 2) view on three planes and 3) volume model. The Siemens ShopMill™ CNC, for example, offers these facilities and provides these three forms of graphical representation.

Depth is represented by colour in the plan view representation; the darker the colour, the greater the depth. A three-plane view shows the plan with two cuts, similar to a technical plan. The cut planes can be freely moved around to show hidden contours. Modern CNCs (e.g., CNC ShopMill by Siemens®) can also be used to create a 3D image in which the part is shown in three dimensions as a volume model. The volume model can then be turned on a vertical axis or cut at a desired point to show hidden contours and views. A volume model cut in cross section is a fundamental tool for verifying machining programs.

Computational geometry techniques are used to develop these simulations, with a view to representing solid objects in CAD. There are a number of different basic formats, the most well-known being the following:

1. Boundary representation (B-rep), in which the object is represented by its edges, defined as a set of vertices, borders and faces. It is necessary to know how all these elements are connected.
2. Sweep methods, in which the object is represented as a volume generated by making a curved sweep across a flat form.
3. Primitive instancing, which involves the parametric description of all possible objects.
4. Constructive solid geometry (CSG), which uses a set of primitives such as cubes, spheres and cylinders. The solid is stored as a binary tree with primitives in the main nodes and Boolean operators in the intermediate nodes.
5. Spatial enumeration, a way of representing a solid using binary volume elements of a uniform size known as *voxels*.
6. Cell decomposition, in which the object is represented as a combination of different sized cells.

The solids most commonly used in CAD are CSG and B-rep. B-rep representation is the form most used in current CAD systems, though its use to represent dynamic solids is not compatible with the speed and interactivity conditions required in a virtual simulation. A similar problem arises with the CSG format involving high computational orders, depending on the number of primitives. In order to solve the computational problems posed by simulating a CNC machining process, what is generally used is geometry approximation techniques or the spatial subdivision of the working space. Another aspect to be borne in mind is that traditional techniques do not store geometrical information during the simulation process, but merely modify the graphical output on the screen using image-based techniques. This has serious limitations, such as the fact that regenerating a different view from another point will require a recalculation of the entire simulation, and the fact that it is difficult to detect errors within realistic tolerances.

Research is currently underway into a technique known as level-based representation, or LBR. LBR objects are approximated using a set of uniformly spaced parallel levels. Each level is formed from a set of non-intersecting coplanar contours. The degree of precision obtained will be determined by the number of levels used. Boolean operations can be performed between LBR objects, thus creating an interactive 3D visualisation regardless of the point of view. The geometry can also be exported and imported [22].

All programming support systems offer powerful simulation tools, some within the CNC itself and others on an external PC, which enable the machining of a particular part to be simulated pre-process. However, all simulators are purely geometric. There should also, for example, be process simulators to show the expected cutting forces that will be generated as the machine tool manufactures the part. Such process simulators would enable much better part programs to be created, because cutting conditions could then be programmed in accordance with the objective function required.

5.8 Current CNC Architectures

For several years now, CNC systems have been evolving from closed, proprietary architectures into more open systems. This trend was first seen back in 1998 with the development of the MOSAIC system at New York University. A European initiative emerged almost at the same time in the OSACA (open system architecture for controls within automation systems) project, which in turn formed part of the ESPRIT III project. In 1994 six Japanese companies set up a working group known as the OSE (open system environment) consortium to develop open architecture through its OSEC (opens system environment for controllers) project. In the automotive sector, General Motors led another initiative to develop open modular architecture controllers known as OMACs. In practice, given the absence of any clear definition of open architecture CNCs, openness is perceived as involving modularity, portability, extendibility, interoperability and scalability [28].

There are currently a number of open system solutions, each displaying advantages and disadvantages and each offering a varying degree of openness.

5.8.1 Systems Based on Multi-microprocessor Architecture

Until a few years ago, when the incorporation of the industrial PC became inevitable, these were the most common architectures. They were proprietary architectures in terms of both hardware and software and are now no longer used.

5.8.2 The PC Front-end

This architecture involves the incorporation of a PC in a traditional CNC system. In this case, the PC is basically used to improve the interface with the operator, though all the control tasks remain rooted in the CNC.

The PC can be incorporated into the electrical control panel or the keyboard. In the former case, the PC forms an additional module to the classic CNC (see Fig. 5.19). This architecture offers a classic solution, and its main advantage is the low level of vibration from the hard disk. The significant disadvantage is that PC peripherals on the electrical control panel complicated cabling and centralised inputs/outputs.

In the second case, the PC is incorporated in the keyboard, acting as an interface with the operator, and it uses a high speed line (usually Ethernet) to communicate with a conventional numerical control located in the electrical panel (see Fig. 5.20). The advantages of this system include the fact that the PC is incorporated in the keyboard and cabling is simpler. Its most significant disadvantages are higher cost (use of Ethernet), centralised inputs/outputs and greater vibration (hard disk). This kind of configuration requires the use of two operating systems, such as, for example, MS Windows™ and a proprietary NC operating system. Use of this architecture is on the wane.

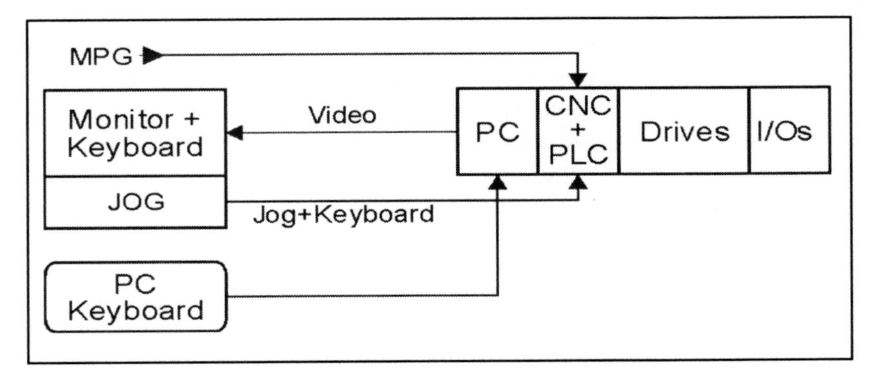

Fig. 5.19 Hardware architecture scheme I

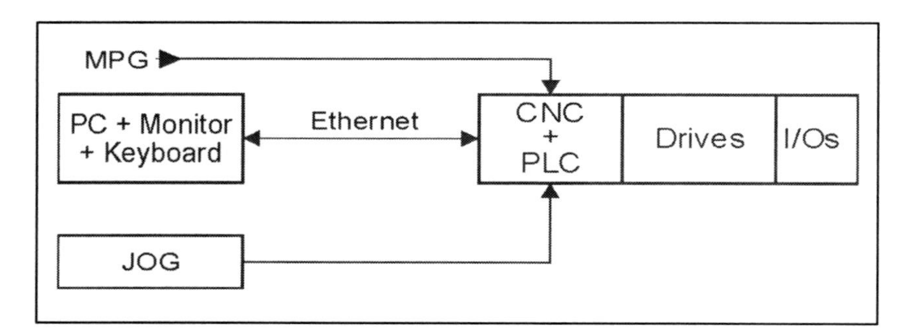

Fig. 5.20 Hardware architecture scheme II

5.8.3 The Motion Control Card with a PC

This dispenses with the use of classic CNCs altogether and instead uses a PC with tailor-made cards to control the movement of the axes. These cards, which normally use a digital signal processing system, are used to carry out the most critical tasks (real-time functions), while the PC itself carries out the non-critical operations (non-real-time functions). The two CPUs can communicate via the PC bus or via a dual RAM memory. This scheme improves the interface and offers greater flexibility to both manufacturers and operators. Its main disadvantage is the need to use additional hardware, generally a DSP motion control card. This solution is quite widely used in current CNC systems.

5.8.4 The Software-based Solution

This is the ideal solution, as all CNC functions are implemented using software run on a PC. The critical functions are implemented by software with the support of a real-time operating system (RTOS). In this case, the PC is located in the keyboard and carries out all the required tasks, both as a CNC and PLC and as an operator interface. It communicates with the inputs/outputs distributed around the machine and the digital regulators in the electrical control panel using one or more fieldbuses (see Fig. 5.21). The main advantages of this type of architecture are: 1) PC peripherals in the keyboard, 2) simpler cabling, 3) distributed inputs/outputs, and 4) facilitation of a single processor solution for low-range controls. The main disadvantage is the greater vibration (hard disk).

There are various kinds of fieldbuses: Profibus, CAN, SERCOS and others. All of them require a master card connected to the PC in order to communicate with the slave devices. Current CNC systems usually use the CAN bus for inputs/outputs and the SERCOS to activate the machines. Generally speaking, SERCOS and CAN cannot be used simultaneously as a communications interface, so one of them must be selected.

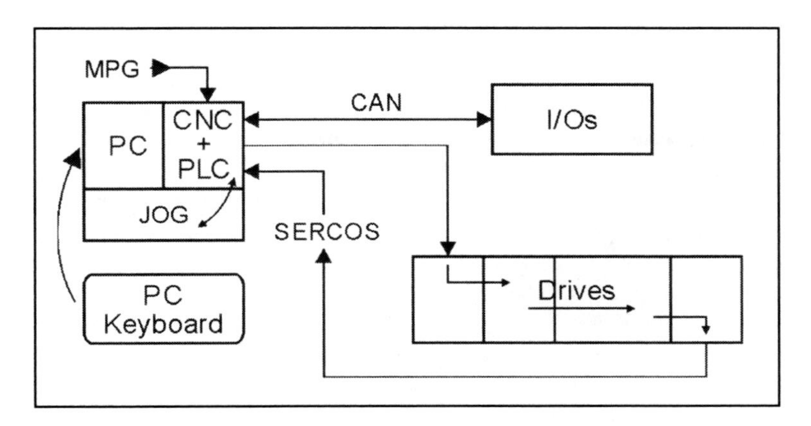

Fig. 5.21 Hardware architecture scheme III

The SERCOS interface (IEC 1491) is a digital interface for communication between systems which have to exchange information cyclically at short, fixed intervals. It is appropriate for the synchronous operation of distributed control or test equipment, e.g., connection between drives and numerical controls [31].

A SERCOS interface communication system consists of one master and several slaves. These units are connected by an optical fibre ring. This ring starts and ends at the master. The slaves regenerate and repeat their received data or send their own telegrams. By this method, the telegrams sent by the master are received by all slaves, while the master receives data telegrams from the slaves. The optical fibre ensures reliable high speed data transmission with excellent noise immunity.

The SERCOS interface controller SERCON816 is an integrated circuit for SERCOS interface communication systems. It contains all the hardware-related functions of the SERCOS interface and considerably reduces the hardware costs and the computing time requirements of the microprocessor. It is the direct link between the electro-optical receiver and the transmitter and the microprocessor that executes the control algorithms. The SERCON816 can be used for SERCOS interface masters and slaves alike.

With the SERCOS bus, the motor controllers can be connected in an optical fibre ring, usually at 16 MHz. The SERCOS communications ring offers the following functions [4], 1) It carries the speed instruction from the CNC to the controller in digital format, with greater precision and no possibility of external disturbance; 2) it carries the feedback signal from the controller to the CNC; 3) it provides information on any errors and manages the basic "controller enable" signals; and 4) it allows parameters to be adjusted, monitored and diagnosed from the CNC using simple, standardised processes. This minimises the amount of hardware required in the controller and affords greater reliability. Its open, standard structure affords compatibility between controls and drives from different manufacturers within the same machine. Work is currently underway to boost the velocity to 100 MHz via Ethernet.

The ISO 11898 CAN interface is an international standard for digital communication between controllers and machine drives using CNC. The communications protocol is CanOpen, standardised as EN 50325-4. The CAN communications bus covers the same functions as SERCOS, the only difference being that the latter is a ring connection while CAN uses a tree topology.

Position controllers can be located in the CNC or in the external regulators. The tendency with current CNCs is to use them externally, with the position loop being closed by the regulators. In this case, position references are sent to the regulator every two to four milliseconds. The regulator is equipped with an interpolator (which is cubic for the Fagor CNC 8070), which generates intermediate position references every 250 microseconds. The regulator returns the real position of the axis to the CNC purely for visualisation and diagnostic purposes, because the position loop is closed by the regulator.

In software-based solution systems, the operating system plays an important role. Real-time control applications like interpolating and servo control require RTOS to have properties of real-time determinism and low latency. For non-real-time tasks, Windows NT or Linux can be used, the latter being free-access software with open source codes that offers full protocol support, stability and low latency (with RT-Linux). Real-time tasks can be carried out using RT-Linux or RTX, a technology for Windows™ NT that offers RTOS performance and determinism.

5.8.5 Fully Digital Architectures: Towards the Intelligent Machine Tool

There is now a clear tendency for all system components to be interconnected to one another using purely digital interfaces. In some cases this involves the use of standard fieldbuses like SERCOS or CAN. In others it means using special interfaces developed by the manufacturers of the controllers themselves. For example, Heidenhain® uses its own interface, which it calls HSCI (Heidenhain serial controller interface), to connect the main computer (MC) with the machine operating panel and the machine operating panel with the CC controller units [13]. Digital interfaces are also used for connection to the PLC. The encoders are connected to the CC controllers via a bidirectional interface, also developed by Heidenhain, known as EnDat 2.2. Up to two CC controller units can be connected to control up to 14 machine axes (see Fig. 5.22).

In addition to developments in fully digital architectures, advances are also being made in the area of "intelligent machines", meaning machines that incorporate a number of "intelligent functions". Mazak®'s Integrex i-150 incorporates five intelligent functions, namely:

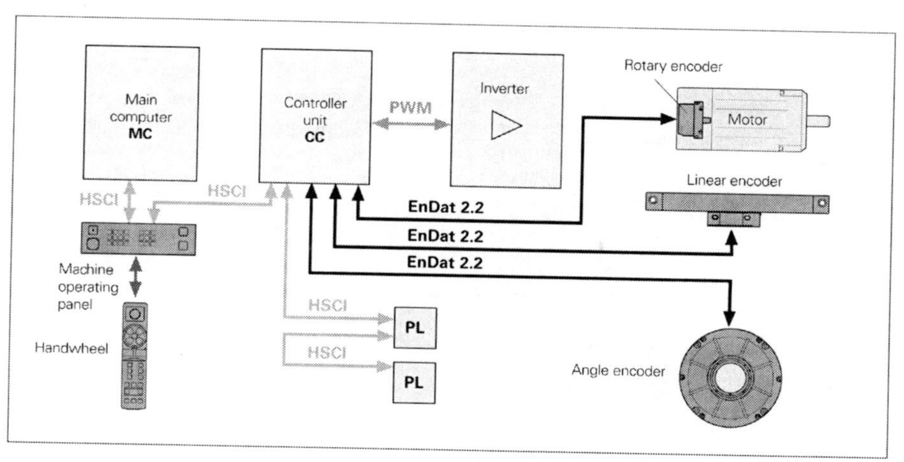

Fig. 5.22 Digital connections of the various control components (courtesy of Heidenhain®)

1. Active vibration control, which reduces the vibration caused by the movement of the axes and thus gives high accuracy positioning. This makes for high surface quality and better tool protection during machining.

2. An intelligent thermal shield. Changes in environmental temperature and heat generated during the machining process alter the machine's own precision. In order to correct this, the operator is forced to conduct the relevant thermal compensation. An intelligent thermal shield, however, offers automatic compensation for temperature changes, leading to stable, high precision machining.

3. An intelligent safety shield, which prevents errors arising from interference with the machine due to operator error. If any machine interference occurs, the machine motion immediately stops.

4. A Mazak voice adviser, by which the CNC system advises verbally of potential problems. For example, the function advises that lubricant has to be added or that the tool life is over for smooth operation.

5. An intelligent performance spindle, which uses sensors housed in the spindle to provide information on temperature, vibration and spindle displacement. This can prevent spindle-related trouble in the machine.

Mazak® has also announced that its machines now have intelligent maintenance support, which is extremely useful for machine maintenance (see Fig. 5.23).

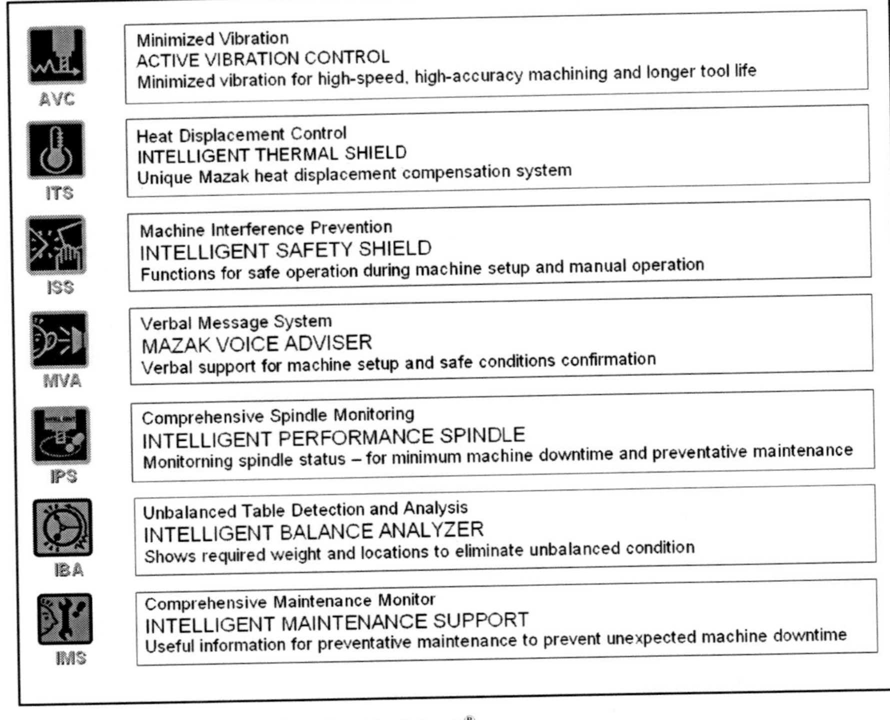

Fig. 5.23 Seven intelligent functions, by Mazak®

References

[1] Al-Habaibeh A, Gindy N (2000) A new approach for systematic design of condition monitor ing systems for milling processes. J of Mater Process Technol, 107:243–251

[2] Azpeitia JL et al. (2006) Desarrollo de nuevas estrategias de control basadas en observadores de estado para la mejora de la precisión y la dinámica en MH (in Spanish, New control strategies bases on state-observers for improving precision and machine dynamics). XIV Congreso de Máquinas Herramienta y Tecnologías de Fabricación: 457–476

[3] Boissier R et al. (2001) Enhancing numerical controllers using MMS concepts and a CORBA-based software bus. Int J Comput Integr Manuf, 14(6):560–569

[4] Fagor Automation Scoop (2007) CNC 8070 Programming Manual. Fagor

[5] Deleure C, Lannoo M (2004) Nanostructures. Theory and Modeling. NanoScience and Technology series. Springer Verlag

[6] Felser M (2005) Real-time ethernet – Industry prospective. Proceedings of the IEEE, 93:1118–1129

[7] Feng CX, Wang XF (2003) Surface roughness predictive modeling:neural networks versus regression. IIE Transactions, 35:11–27

[8] Goser K et al. (2004) Nanoelectronics and Nanosystems. Springer Verlag

[9] Haber R, Alique A (2003) Intelligent process supervision for predicting tool wear in machining processes. Journal of Mechatronics, Special Issue on Computational Intelligence in Mechatronics Systems, 13(8–9):825–849
[10] Haber R, Alique JR (2007) Fuzzy logic-based torque control system for milling process optimization. IEEE Trans on Systems, Man and Cybernetics, 37 (5):941–950
[11] Haber R et al. (1998) Toward intelligent machining: hierarchical fuzzy control for the end milling process. IEEE Trans Contr Sys Technol, 6 (2):188–199
[12] Haber R et al. (2007) A classic solution for the control of a high-performance drilling process. Intern J of Mach Tool Manufact, 47:2290–2297
[13] HEIDENHAIN (2007) Uniformly Digital. Technical Information. 6+10 Edition. Klartext.
[14] Ji K, Kim WJ (2005) Real-Time Control of Networked Control Systems via Ethernet. Int J Contr, Automat Syst, 3 (4):591–600
[15] Koren Y (1980) Cross coupled biaxial computer control for manufacturing systems. J of Dynam Syst Measure Contr, Trans of the ASME 102:265–272
[16] Koren Y, Bollinger JG (1978) Design parameters for sampled data drives for CNC machine tools. IEEE Trans on Ind Applications IA, 14(3):380–390
[17] Landers RG et al. (2004) A comparison of model-based machining force control approaches. Intern J of Mach Tool Manufact, 44:733–748
[18] Lee KC, Lee S (2002). Performance evaluation of switched Ethernet for real-time industrial communications. Computer Standards & Interfaces, 24:411–423
[19] Lian FL et al. (2001) Performance Evaluation of Control Networks: Ethernet, ControleNet, and DeviceNet. IEEE Control Systems Magazine, 21(1):66–83
[20] Liang SY et al. (2004) Machining process monitoring and control: The state of the art. J of Manufact Sci Engineer, Trans of the ASME, 126:297–310
[21] Martin D et al. (2008) Design and implementation of a networked monitoring platform based on CORBA middleware for industrial processes. Future Generation Computer Systems, in review
[22] Moreno A et al. (2006) SIMUMEK: Un sistema de simulación grafica 3D para operaciones de mecanizado CNC (in Spanish, System for the graphic simulation of 3D machining operations). XIV Congreso de Máquinas Herramienta y Tecnologías de Fabricación, 391–403
[23] Neumann P (2004) Communication in Industrial Automation – What Is Going On? Proceedings of INCOM'04 Salvador Brazil: 1332–1347
[24] Norman et al. (2006) A sophisticated platform for characterization, monitoring and control of machining. Meas Sci Technol, 17:847–854
[25] Park J et al. (1995) An open architecture real time controller for machining processes. Proc CRIP, May
[26] Poo AN, Bollinger JG (1974) Digital analog servosystem design for CNC. IEEE Ind Appli Society, 9th Annual Meeting
[27] Poo AN et al. (1972) Dynamic errors in type 1 contouring systems. IEEE Trans on Ind Applications IA-8(4):477–484
[28] Pristschow G et al. (2001) Open controller architecture – Past, present and future. Annals of CIRP, 50(2):463–470
[29] Renton D, Elbestawi MA (2000) High speed servo control of multi-axis machine tools. Inter J of Mach Tool Manufact, 40:539:559
[30] Sanz R (2003) A CORBA-based architecture for strategic process control. Annual Reviews in Control 27:15–22
[31] SERCOS Interface Controller (2000) Reference Manual 10/2000
[32] Solis E et al. (2004) A new analytical–experimental method for the identification of stability lobes in high speed milling. Inter J of Mach Tool Manufact, 44:1591–1597
[33] Song Q, Kasabov N (2006) TWNFI – a transductive neuro-fuzzy inference system with weighted data normalization for personalized modelling. Neur Netw, 19:1591–1596
[34] Tanenbaum AS (2001) Computer Networks. 3rd ed. Prentice Hall, Upple Saddle River NJ

[35] Tikhon M et al. (2004) NURBS interpolator for constant material removal rate in open NC machine tools. Inter J of Mach Tool Manufact, 44:237–241
[36] Tomizuka M (1987) Zero phase error tracking algorithm for digital control. J of Dynamic Systems, Measurement and Control, Trans of the ASME, 109:65–68
[37] Torfs D et al. (1992) Extended bandwidth zero phase error tracking control of nonminimal phase systems. J of Dynamic Systems, Measurement and Control, Trans of the ASME, 114:347–351
[38] Tsai YH et al. (1999) An in process surface recognition system based on neural networks in end milling cutting operations. Inter J of Mach Tool Manufact, 39: 583–605
[39] Waser R (2003) Nanoelectronics and Information Technology. Ed. Wiley-VCH
[40] Wecks M, Ye G (1990) Sharp corner tracking using the IKF control strategy. Annals of the CIRP, 39(1):437–441
[41] Yang TC (2006) Networked Control System: A Brief Survey, Control Theory and Applications. IEE Proceedings, 153(4):403–412

Chapter 6
Machine Tool Performance and Precision

A. Lamikiz, L. N. Lopez de Lacalle and A. Celaya

Abstract This chapter introduces main machine tool design, construction and testing aspects to achieve high precision on machined parts. Not only the machine tool but also the machining process itself is a source of errors on parts. Therefore, a holistic view involving machine and machining is required.

The design of high precision machines involves some basic principles and methodologies which are presented in depth further on. A key factor is the identification of error sources, studying their physical causes and relevance to the final uncertainty. Thus, assembly error, thermal growth, component deformations and control inaccuracy are described.

Errors in components and subassemblies are propagated along the kinematic chain of the machine, producing larger positional errors at the tool tip position and tool axis orientation. The use of homogeneous matrices is introduced as a very useful tool for estimating error propagation.

Tool deflection and machine deformation caused by cutting forces is another important error source. Unfortunately deflection is impossible to avoid although some models are now provided with machining toolpath selector implying minimal cutting forces along error sensitive directions.

After construction and just before the machine tool is delivered to the customer, the verification of precision and performance must be always performed. A description of the existing standards and procedures for this are described at the end of this chapter.

A. Lamikiz, L. N. Lopez de Lacalle and A. Celaya
Department of Mechanical Engineering, University of the Basque Country
Faculty of Engineering of Bilbao, c/Alameda de Urquijo s/n, 48013 Bilbao, Spain
{aitzol.lamikiz, norberto.lzlacalle, ainhoa.celaya}@ehu.es

L. N. López de Lacalle, A. Lamikiz, *Machine Tools for High Performance Machining*,

6.1 Introduction and Definitions

On the highly competitive market, machine tool manufacturers must build high precision machines, while at the same time keeping prices as low as possible. Machine tool history is irrevocably united to machine tool precision and consequently to the precision of parts to be produced by them, as shown in Fig. 1.5 in Chap. 1.

The improvement in machine tool precision has greatly helped to increase the added value of manufactured parts, decreasing the adjusting and finishing operations when these parts are assembled in complex systems. The final results are both technical and economic, systems are more and more reliable when machining tolerances are reduced and at the same time less manual adjusting operations are required on the final assembly. The explosion motor is a good example. Today, the reliability of this complex system is much higher than that of car engines of the past century; their production is absolutely automated using CNC machining centres, boring and honing machines, automated transfer systems and robotic assisted assembly. Although, undoubtedly, in twenty years current quality and reliability will be considered low.

In many applications the need for high precision is simultaneously associated with its usual contrary requirement, i.e., the need for high productivity rates. And even low production costs and environmental impact must be considered at the same time. A good solution where these factors are balanced may be the current concept of "high performance machining", where the eco-efficient machine tool is the basic element. This chapter is mainly focused on the precision concept, however, it will at all times be considering other aspects.

6.1.1 An Introduction to Precision Machining

Machining cost for a part rises exponentially as the required precision level increases; this is a widespread belief among many production engineers. However, this concept of precision, where the cost of the individual part is only considered, excluding any other considerations, is somewhat obsolete. Following this old idea, the 20th century criterion for setting production system accuracy level was the minimum required to avoid part rejection by customers, while at the same time maintaining item price as low as possible. Extra investments to achieve more precision were thus considered absolutely superfluous.

However, some authors [12] assess that precision machining can be much more cost effective than conventional machining if other aspects come into play. Thus, high precision machining eliminates the need for laborious finishing and adjusting operations, decreasing the final assembly cost when components are inserted. An illustrative example is a simple plastic injection element for a vehicle dashboard or bumper. In most cases these parts are designed for both aerodynamic and aesthetic functions, so high surface finish requirements (in many cases below 0.1 μm Ra) are usually required. These parts have to be assembled in a more complex system, which must meet some dimensional requirements without clearances or overlaps. In addition, these parts are assembled on mass production manufacturing lines, so roughness and dimensional tolerance must be kept throughout the entire series. Different sources of error for this simple plastic injection part can be detected:

- *Errors from the plastic injection process:* These are related to part contractions due to the solidification and cooling processes, part defects and scratches due to the flow of the plastic inside the mould, etc.
- *Errors of the plastic injection machine:* These relate to misalignment in the mould setup, thermal deviations and machine deformation due to the mould closing force.
- *Errors of the mould itself:* All mould machining errors are directly reproduced in injected parts. Mould machining must ensure a higher precision than an order of magnitude of the required injected part. This accuracy can be in the range of 0.05–0.01 mm in the entire mould. Mould errors are derived from those of machine tools and those originated by the machining process; both types will be explained in the following sections.

Taking into account the added-value of machined parts and relating it to the cost of production (as shown in Fig. 6.1), one can observe the production of high-precision parts may be much more profitable than conventional ones, not only in terms of strictly economic profit but also market consolidation for the future. As shown, when precision is near the current maximum, the balance is positive for machine users. Of course, the cost to achieve the current maximum level of accuracy is very high, however this is only required in some special applications for astronomy, lens production, metrology, physics and other instrumentation device assemblies; in these applications cost is not the key factor.

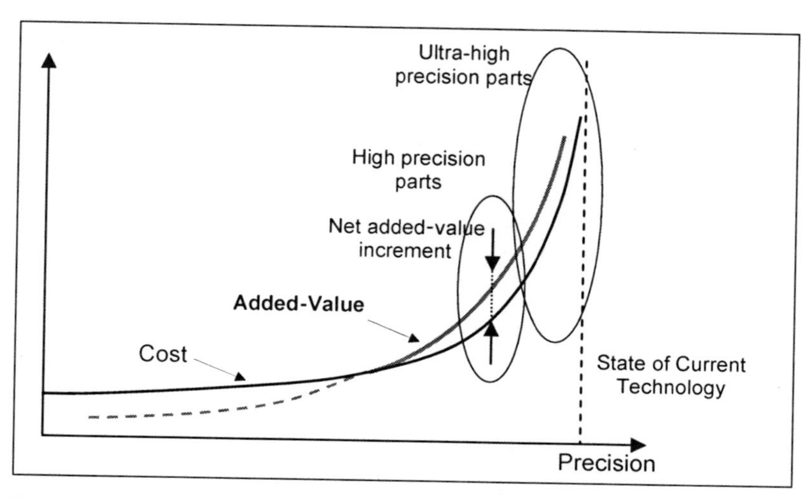

Fig. 6.1 Relation between precision, part added-value and cost

A typical definition for *high precision machining* is the production of complex components with tolerances below 10 (or even 5) µm. Below these tolerances, the phrase *ultra precision machining* is used. Obviously, the term "high precision" is continuously changing. Today machine tools are all high precision machines if we compare them with those of twenty years ago, whereas a lot of them would not be precise if compared with those constructed ten years from now.

Since the 1990s, high precision machining has become an industry in itself, including the manufacture of high-precision components for the optical, electronics or aeronautical industries. The market was about $US 60 million in 2006 and the estimated growth by 2009 is in the order of $US 100 million (growing 66% in three years!). Bearing in mind the sectors involved, the major markets are Asia (Japan, Taiwan, Korea and China), North America (USA and Canada) and Europe.

Moreover, many industries other than the above are demanding precision levels which are difficult to obtain by traditional machines [11]. Therefore some of the precision principles and methods are now spreading to sectors such as the automotive or aeronautical industries.

The greater investment for precise machine tools must be analysed considering the life span of a machine tool [12]. If we consider the required precision for machined parts increases yearly and machine precision decreases due to clearances, surface wear, etc., a high precision machine tool life span is longer than a conventional one. In Fig. 6.2 initial machine precision is presented, plus the usual decrease thereof. Therefore, the required accuracy of precision machine tools last for a longer period than that of conventional machine tools.

Precision is in fact a statistical variable. Real conditions introduce a high number of sources of error, many of them with random effects, which can affect the precision value [5, 6, 14]. Therefore, there are three basic concepts, which are introduced in the next subsection.

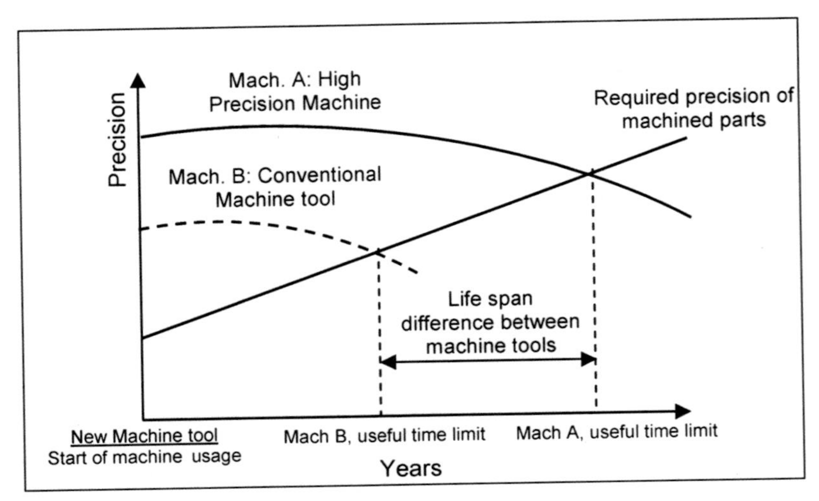

Fig. 6.2 Relationship between precision and life span of the machine tool [12]

6.1.2 Basic Definitions: Accuracy, Repeatability and Resolution

Let's take the following example. The diameter of one specific hole must be measured along a part series. The measured hole diameter is to be compared with the nominal value. Two possible results of the measurement are presented in Fig. 6.3. At the top of the figure, measurement values are represented, at the bottom the statistical function of the density is calculated

The first case implies a high "accuracy" and low "repeatability" set of measurements. Each measure presents deviation with respect to the nominal value; however, the mean value of the measurement is very close to the nominal value. In this case the machining process presents a high accuracy, but a very high dispersion too.

On the other hand, the second case represents a series of measurements whose deviation of the mean value from the nominal value is higher. However, all measurements present very close values to this mean value or, in other words, the dispersion of the measurements is much lower. It is therefore a case with lower "accuracy" but higher "repeatability".

After this example the concept of precision, and the difference among *accuracy*, *repeatability* and *resolution* must be highlighted:

- *Accuracy* is the difference between actual and nominal values. It is also referred to as "error". Statistically the accuracy is measured by the mean measured value.
- *Repeatability* is the range of deviation for the same position value due to a random source of errors. Some references refer to the repeatability as precision.

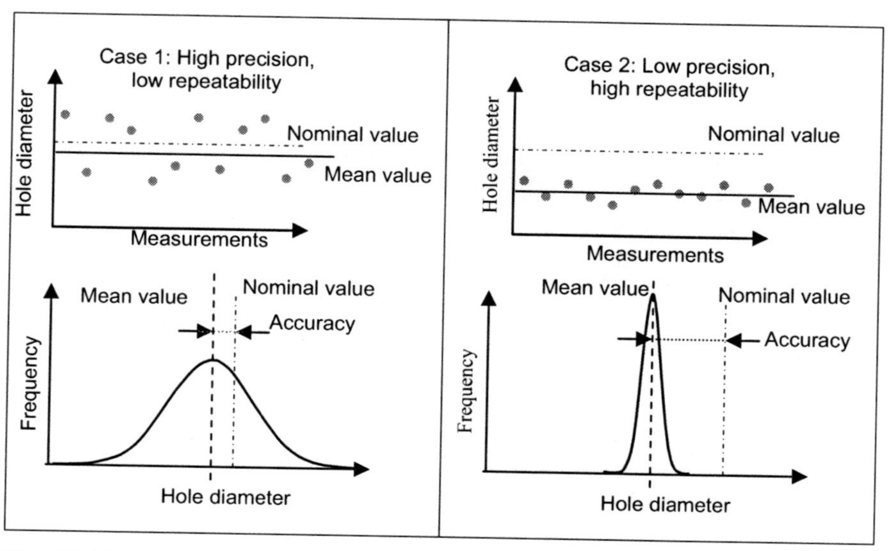

Fig. 6.3 Two extreme cases of precision and repeatability values

However, in a general sense, precision makes reference to a qualitative aspect, including both accuracy and repeatability

- *Resolution* is the smallest magnitude that can be detected with guarantee for a measurement device. Therefore it should be about one order of magnitude below the accuracy of the machine where measurement devices are installed. Today resolution is not a problem for machine precision, since for example, current machining centres incorporate 0.1 µm resolution linear encoders to close the axis position loops; to close the feed drives loops up to 16 millions of pulses per revolution encoders to ensure an absolute close speed control. With these figures, the machine tool accuracy and repeatability (measuring the deviation at the tool tip position) are higher than the resolution.

The concept of *uncertainty (U)*, as the maximum expected error of a measured value, is also important. Experience has shown that measuring the same magnitude over and over with the same instrument and under the same conditions results in different values. Therefore, it is evident every measurement contains an undetermined error, which is not always the same. The *uncertainty* is referred to as the maximum expectable error in a measurement with a probability factor, so uncertainty marks a range of error with a given probability. Thus, if a set of different measurements results in a mean value of x, it can be ensured the "true value" of the measured magnitude is within the interval $x \pm kU$ with a determined probability (usually 95%, being $k = 2$).

It is important to remember the Slocum [17] cite of Bryan's observation (Bryan is an outstanding metrologist technician of Livermore National Laboratory), which says machine tool deviations depend on a series of factors (the design itself, external/internal forces, thermal variations, …) and the probabilistic definition of the uncertainty is in fact a mathematical tool which helps to understand and evaluate a problem which may be too complex or simply unpredictable.

6.1.3 Historical Remarks and the State of the Art

One of the first references to a high precision machine is the boring machine designed and patented by John Wilkinson in 1774, cited in the first chapter of this book. This mechanical machine was capable of boring holes with a 1,250 mm diameter to an approximate error of 1 mm. It was a giant step at the time, and many references attribute the spread of the steam engine to Wilkinson, because the correct sealing of the cylinder would not have been possible without the precision of his boring machine. To make a rapid comparison, the piston/cylinder clearance of a modern motorcycle engine is now about 1.5–2 µm.

The history of machine tools is the precision history of machine tools, and should be remembered as related in Chap. 1. Since Wilkinson's patent, many factors have introduced precision improvement in machine tools. Most of them were mechanical, until the CNC introduction at the end of the 1950s that became generalized in the 1970s. This device has improved the precision interpolation from the order of 0.1 mm in the 1970s up to 0.01 µm in 2007.

To build precision machines depends not only on one single advance but also on several techniques throughout the entire machine life and use. Starting from the design, machine tool structural parts are today designed taking into account several phenomena, such as elastic deformations caused by external forces or gravity, the possible wear of moving parts, and the appearance of thermal errors or machine responses to external factors [7].

Moreover, in addition to achieving the maximum precision of each machine tool element separately, specific techniques for the manufacture and assembly of linear guides have been introduced. For example, ultra-high precision machine tool companies, estimate at more than 400 hours the work of scraping the linear guides on every machine to ensure the straightness and squareness of each axis. However, precision is required by a lot of machine tool manufacturers. Therefore the modularisation, the interchangeability of parts and easy maintenance must be also associated with precision. For these reason, nowadays the use of linear rolling guides coming just assembled and preloaded from the supplier makes the repeatability of machine assembly operations easier. Hence, with rolling guides higher precision is offered in general to the machine tool manufacturers.

The design of high precision machines requires the application of fundamental principles [5, 12], followed by the use of estimation utilities for the error budget [17]. On the other hand, spatial formulation can be introduced to define the propagation of errors along the kinematic chain. The result of the application of these principles and tools has been the increase in precision from 100 μm to 1 μm in the last thirty years, as shown in the first chapter, in Fig. 1.4.

Regarding the machine tool verification, Prof. Schlesinger in 1930 defined specification and tests for the acceptance of machine tools by users. Since then, several researchers, technicians, companies and standardisation groups worked and are working to specify metrological procedures and tests to evaluate machine tool precision (see Sect. 6.4)

Three associations promote the spread of knowledge regarding precision and its transfer to companies in three continents. The eldest in Japan since 1933 is the JSPE (Japanese Society of Precision Engineers). In the USA the ASPE (American Society of Precision Engineers) has existed since 1986, and the European EUSPEN (European Society of Precision engineers) completes the picture of precision in the world. Engineers interested in metrology, machine construction, precision devices and micro manufacturing interchange experiences through these societies. Of course those societies focused on manufacturing in several countries have also precision as a very important topic.

6.2 Basic Design Principles and an Error Budget

One of the most difficult but paramount machine design stages is the *error budget* estimation of a machine, prior to construction. The error budget is an engineering design tool that allows an a priori evaluation of the uncertainty of a complex system.

Estimation begins with each simple error, which must be described and quantified, being followed with their combination to produce a total error value for the machine tool. The main advantage from calculating the error budget is to be able to evaluate the relative relevance of each source of error on the total resulting error. The direct consequence is that efforts will be focused on those with more weight in the global uncertainty. Designers and assemblers will try to reduce them or if that is not possible, to take them into consideration when the CNC equipment is adapted to the machine.

This section introduces the main concepts involved when an error budget is carried out. The first step is logically to identify and understand the major sources of error of the design, assembly, verification and use of a machine tool. Along with this task, the main principles which inspire all precision designs are kept in mind. It is difficult to establish whether principles are prior or subsequent to analysing the error sources. Engineering principles are based on the conceptual description of mechanical solutions, nevertheless, they also consider the most common sources of error and their relative importance. For this reason we are presenting the sources first.

6.2.1 Sources of Errors in Machine Tools

The main uncertainty sources in the design and construction of machine tools are categorised as follows [17, 13]:

- *Geometric and kinematic errors*: these come from the mechanical imperfections such as misalignments of axes, slideways degradation and the wear of joints and couplings. These directly affect the relative position between tool and workpiece, producing both dimensional and shape errors on workpieces. The border to define an error as geometric or kinematic is diffuse. The machining accuracy achieved by complex machines (such as five-axis machine tools) is generally inferior to that of conventional ones. In three-axis machining centres error propagation is basically linear. Otherwise, in machines with rotary axes the error propagation along the machine kinematic chain is not linear. Errors between moveable machine elements directly affect the final dimensions of the workpiece and its shape.

  Many researchers have investigated geometric errors for 3-axis machine tools. For example, Nawara [13] has formulated the error of machine tools from 21 simple error components, proposing an error prediction and compensation algorithm. In addition to this work, there are lots of references which develop models to estimate geometrical error, implementing compensation algorithms in the CNC [16, 20]. Uncertainty in position of machine slides, gear backlash, couplings clearance, etc., is also included in this group.

  The more complex the system, the more kinematic errors in the machine. This is one reason why high precision machines present very simple structures with minimum constraints; constraints increase system stiffness although they

introduce a lot of uncertainty and require special care at the assembly stage to prevent misalignment and other hyper-static effects.
- *Thermal errors:* thermal errors have a complex non-linear nature which makes them difficult to handle. The main causes of thermal distortions are the variation of workshop temperature (for example between winter and summer, morning and afternoon), the local heating due to feed and main spindle motors, or chip heaps.
- *Stiffness error and errors addressed to the deflection of cutting tools:* machine tools are not perfectly rigid. Therefore the weight of structural components on one side and cutting forces on the other, cause large errors which are highly dependent on the machine tool position. On the other hand, cutting forces can cause important errors addressed to the deflection of cutting tools. As an illustrative figure, errors derived from tool deflection in ball-end finishing processes can exceed 40 μm [12]. These errors will be studied in depth in the next section.

Those mentioned above are the *known sources*. Their consequences are complex, but techniques to evaluate them or compensate their effects have being developed. Unfortunately in real cases there are also *unknown sources*, which can only be handed considering a maximum level of uncertainty.

6.2.2 Error Budget Estimation

In this section, typical error budgets for a conventional 3-axis high speed machining centre and an ultraprecision 3-axis micromilling centre are presented. Budgets must include all the elements affecting the final accuracy of the workpiece, i.e., machine errors (geometric errors, positioning error, etc.), machining process errors (tool runout, deflection, vibration, etc.) and errors derived from auxiliary equipment (the workpiece setup).

6.2.2.1 Machine Tool Errors

This section covers machine construction errors, errors due to the wear of parts and error due to non-optimal control of each drive unit.

- *Guideway positioning error:* The compensation applied by the NC reduces this error basically to the repeatability of the movement of each slide. For conventional high speed milling centres, the usual resolution of linear scales is below 1 μm whereas the resolution of ultraprecision micro milling centres is below 0.01 μm. These resolution values are the minimum slide positioning error; however the following errors must be added to obtain the uncertainty value of each guideway.

- *Uncertainty of the reference position.* The "machine origin" or "machine zero point" is a fixed point set by the machine tool builder. Any tool movement is measured from this point. The CNC always remembers the tool distances from the machine origin. The associate error to this action is related with the repeatability of the set of the machine zero point. This is only applicable when the machine is switched on and off, when the machine must go to its origin point (the so-called "zeroing the slides"). This value is about 3–5 μm for a conventional milling centre and 0.5 μm for an ultraprecision one.
- *Thermal expansion.* This is highly dependent on the machine tool internal heat sources (drives, friction on nuts, etc.) and temperature variations of the working room. A typical uncertainty value for conventional machine tools can be 5 μm. These errors for micro milling centres are below 0.5 μm due to the careful isolation of thermal sources in these machines.
- *Effect of reversal of linear movements.* Position measured values are different when the slide goes forward or backward. The sums of these errors are close to 3 μm on conventional machine tools and 0.1 μm for ultraprecision ones.
- *Angular errors:* The machine sliding units suffers rolling, yaw and pitch effects. The typical angular errors can be up to ±5 arcsecs/100 mm in milling centres and under ±3 arcsecs/100 mm in ultraprecision machines. On the other hand, the effect of angular errors on position and straightness depends linearly on the distance to the reference point. This length is directly related to machine size, e.g., 750–1000 mm for a small conventional milling machine and in the range of 10 mm for micro milling centres.

Applying the propagation of errors, the estimated uncertainty for a linear movement would be in the range of 12 μm for conventional machines and 0.25 μm for ultraprecision machines. This estimation includes the mechanical construction of the machine; however other uncertainties must be also considered. These are:

- *Machine trajectory errors:* This error source is negligible in micromilling because at the fastest machining feed rates (50 mm/min) the machine moves so slowly that following error with regard to the set point is minimal, even without feed-forward control installed. On the contrary, for a high speed milling centre with an average CNC, the deviation between programmed and real trajectory can be up to 5 μm due to CNC following errors in sharp trajectory deviations.
- *Errors in the spindle, including the spindle-shank, shank-collet and collet-tool interfaces:* Precision spindles guarantee runout errors of 1 μm in the spindle nose; however this error is magnified by the spindle-shank interface, the shank itself, the collet or toolholder, the toolholder-tool interface and the tool itself. The shanks and collets available on the market include simple-precision sets, with runout errors of around 10 μm, and super-precision toolholders used in ultraprecision machine tools, with runout errors of just 3 μm. The lack of tool concentricity depends on tool perfection but mostly on the correct arrangement and clamping of the tool into the collet. Today, thermal shrinking toolholders are the most accurate for conventional scale machines; usually the runout measured at the tool tip is under 5 microns.

6.2.2.2 Machining Process Errors

The main sources of uncertainness deriving from the machining process are tool deflection, tool wear and vibration of the machine, tool or part. In more detail:

- *Tool deflection:* a tool can bend considerably under the action of the cutting forces. For conventional milling, this value basically depends on the rigidity of the tool used, being given by the relationship between its length and diameter. An order of magnitude of error in the machined surface is 25–30 μm [15], considering a slender finishing endmill (Ø6 mm and 60 mm length) and applying finishing conditions on steel. In the worst cases errors could be higher than 90 μm. On the other scale, typical parameters for micromilling cause cutting forces of around 30–100 mN, which can lead to 3 μm tool deflection.
- *Tool wear:* wear increases the cutting forces and produces a variation of tool dimensions, with the consequent loss of accuracy. The corner radii of straight endmills and turning inserts become more rounded due to wear, and therefore either the tool diameter or the length becomes smaller than the nominal figure. Furthermore, ball-end milling tools lose their initial radius, especially when very hard materials are machined, as in the milling of tempered steels hardened to more than 50 HRC.
- *Vibration:* two types of vibration are common when machining. The first are the force vibrations due to the tooth passing frequency. Thus, if the natural frequency of the machine, tool or workpiece is near the characteristic frequency of the cutting force a high dynamic amplification happens. Second, autoexcited regenerative vibrations (chatter) can produce relative tool-part displacement higher than 100 μm; chatter is prevented with a good selection of spindle speed and chip section, as explained in Chap. 3. Vibrations are considered an uncertainty source only if they are impossible to eliminate with good machining expertise.

6.2.2.3 Auxiliary Alignment Systems and the Reference Set-up

Errors originating in the workpiece positioning and the subsequent "zero reference" setup do not affect workpiece precision if all operations are going to be done on a single machine with a single tool-holder. Errors in setup appear when workpieces are transferred between machines, or different tools are applied; then this is a very important error source. As a matter of fact, this is one of the main reasons for using the new multi-task machines cited in Chap. 1.

Nevertheless, there are workpiece alignment systems on the market (Erowa®, Hirschmann®, Dock-lock®, etc.) that ensure positioning repeatability below 3 μm. Thus, this error should be added to the expected repeatability of the machine tool position.

6.2.2.4 The Total Error Budget

After the above figures, common values for uncertainty estimated for a conventional and a micromilling machine are collected in Table 6.1.

Table 6.1 Uncertainties for an ultra-precision centre and a conventional machining centre

	High speed machining centre	Ultra-precision micro-milling centre
Uncertainty in machine position	2.50 μm	0.13 μm
Uncertainty in thermal expansion	5.00 μm	Less than 0.50 μm
Uncertainty in tool-holder displacement	5.00 μm	2.20 μm
Uncertainty in trajectory errors	3.50 μm	–
Uncertainty in tool due to deflection	15.00 μm	1.50 μm
Uncertainty due to workpiece alignment	3.00 μm	2.50 μm

The combined standard uncertainty (u_c) and, thus, the expanded uncertainty (U) are calculated respectively as:

$$u_c^{\ 2} = \sum_{i=1}^{N} \left| u(x_i) \right|^2 \tag{6.1}$$

$$U = 2 \cdot u_c \tag{6.2}$$

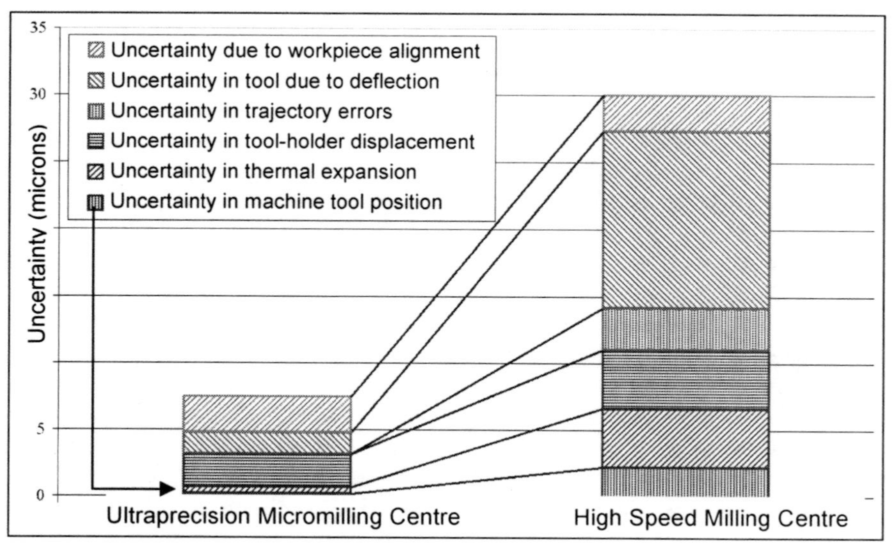

Fig. 6.4 Ultra-precision and conventional machine centre error budgets. The legend is in the same order as the columns

For a conventional high speed machining centre, uncertainty values can be around 30 µm. On the other hand, the global uncertainty for ultraprecision machine tools (in one set-up) is 5.5 µm, but if the alignment and reference errors are considered, uncertainty grows to 7.5 µm. In Fig. 6.4, the relevance of each source of error can be observed as the result of the error budget calculation.

6.2.3 Basic Principles for Precision Machine Design

After the analysis of error sources and calculation of the combined standard uncertainty (Eq. 6.1), some basic principles are deduced as a guide to minimise the positioning error and achieve the required levels of accuracy. In most cases these are common sense, easy to understand principles. However, the real and practical implementation of these principles is sometimes very complex.

Nakazawa [12] and Slocum [17] provide basic guides to establish the design and construction principles for precision in machine tools and other devices. Some of them are briefly explained in the following lines.

6.2.3.1 The Machine Tool Structure

Some principles should be applied to optimise machine tool structure. Thus, the *functional principle* states that independent frames are highly recommended in order to split the errors of each element. On the other hand, machines must be designed starting from the basic requirements, rather than improving old designs or adding new components, which is the base of the *total design principle*.

Another basic rule is the *principle of compliance*, which is directly related to the machine tool structure stiffness. The stiffness must be as high as possible, thinking not only in general terms but in local areas that will withstand great forces. Figure 6.5 presents an example of the total design principle and the princi-

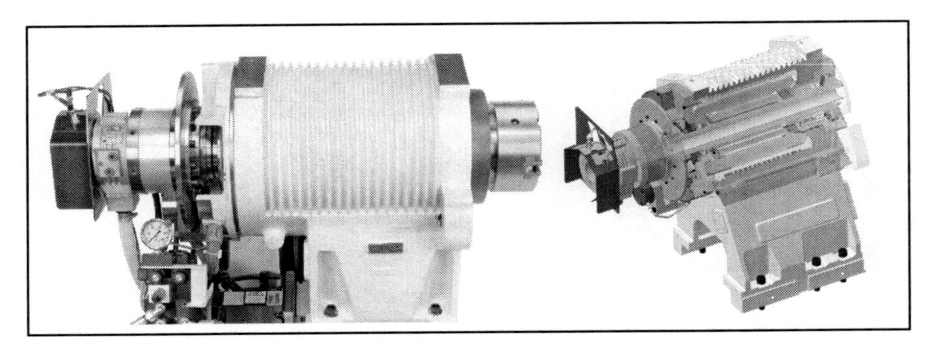

Fig. 6.5 Total design principle: built-in motor embedded in a lathe spindle, by CMZ Machinery Group®. The spindle is refrigerated with oil

ple of compliance. There, a new turning spindle designed specifically for high precision operation is presented.

Finally, the *principle of minimisation of heat deformation* refers to one of the main problems in the design of new machine tools. If possible, the machine structure should present a symmetric design, where the compensation of thermal deformations is easier. The heat sources of the machine (the main spindles, drives, etc.) should be as far as possible from the working zone and it is highly recommended to design cooling systems to avoid thermal gradients, as is the case shown in Fig. 6.5.

6.2.3.2 The Kinematic Design Principle and Smooth Motion

Whenever possible, the machine should present minimal constraints in joints, slides or other element interfaces. A *kinematic design* [17] prevents deformations induced by internal constraints and allows isolating sensitive elements from the influence of dimensionally changing supports and/or manufacturing tolerances. Kinematic design provides exactly the right number of independent, well-conditioned constraints maintained by appropriate closure forces to ensure the desired degrees of freedom between two rigid bodies.

In Fig. 6.6 an example of kinematic design for a linear axis table is shown. The guides are designed to allow motion in one direction and to constrain the possible displacements in all the other directions. To insert the minimum constraints, two different shape guides are introduced. The first one is a V-shaped linear guide, which constrains the lateral displacements of the table. The second is a flat guide which constrains only the vertical motion of the table, to increase the stiffness in that direction. If two identical linear guides with constraints in both lateral and vertical directions were introduced, the system would be overconstrained and therefore imperfections of the guides and assemblies could lead to deformations of the elements. In the kinematic design, imperfections are absorbed by the joints of the system, without additional loads. As a general rule, point constraints are less problematic than surface and line contact.

The *principle of smooth motion* states that guiding mechanisms should present minimum friction to achieve smooth motions. Friction between sliding compo-

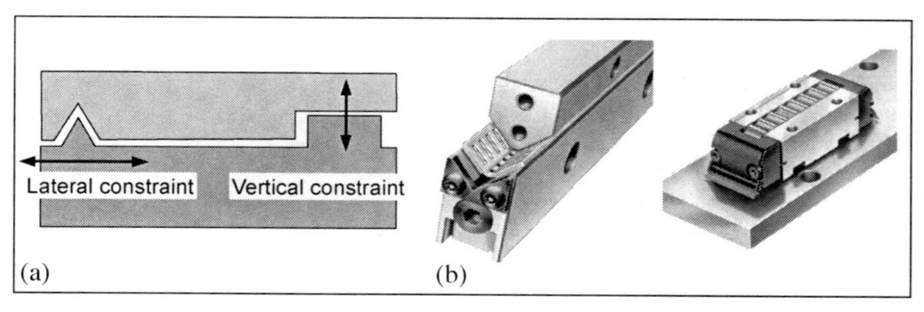

Fig. 6.6 **a** Semi kinematic design of a linear axis. **b** Details of V and flat linear guiding systems of INA – FAG®

nents (guide and counterguide) generates effects, such as *stick and slip*, that can originate difficult to control vibrations. Thus, a precise guiding system not only has to be accurate in the construction, but must also present smooth motion. Hence, the design of guiding systems should accomplish two objectives, minimum and constant friction coefficients.

6.2.3.3 Abbe's Principle

This basic principle of metrology states "the length to be measured and the measuring scale must lie on the same axis". In conventional machine tools, many linear motion systems do not satisfy Abbe's principle because the motion axis and the machine element (its reference point or its centre of gravity) do not lie on a single line.

In Fig. 6.7 two different machine tool designs are presented, Fig. 6.7a shows a C-frame with a single ball screw column. As can be observed, the tool axis line and measuring line are separated; Abbe's principle is thus not satisfied and therefore the angular deflection is translated into a displacement error at the tool. In Fig. 6.7b a double column and twin ball screw is shown. In this case, the distance between the tool axis and each ball screw axes compensate the angular deviation effect on the linear axis. This is, obviously, a more costly solution, but the machine precision is much higher.

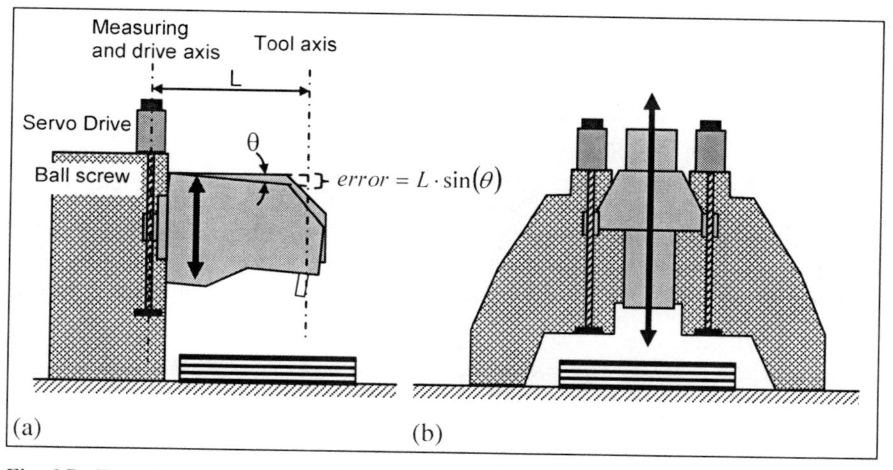

Fig. 6.7 Two different machine designs. **a** With Abbe's error. **b** With Abbe's error reduction

6.2.3.4 The Position Measurement

A direct measuring system, free of external perturbations or noises, is always better than an indirect measuring one. The latter needs some geometrical parameters that may vary due to sources of error. A classical example of this principle is the use recommendation of linear encoders instead of encoders to measure the position of linear axes.

Most of the measuring devices for machine-tools operate using the principle of photoelectric scanning, which is a contact-free method, so the measuring systems are free of wear. These systems use light scanning devices, consisting of a light source projected to two graduations with equal grating periods. One of the graduations: the *scanning reticle* is moved relative to the other: the *scale reticle*. When the light passes through a grating, light and dark surfaces are projected to a photoelectric sensor. There are two limit positions: first, when the gratings are aligned, light passes through and the photovoltaic cells generate the maximum voltage value. On the other hand, if the lines of one grating coincide with the gaps of the other, no light passes through and the photovoltaic cells signal is minimum. The resulting signal of the relative motion between the scanning and the scale reticle generate nearly sinusoidal output signals.

There are two types of photoelectric measuring systems, depending on the scale reticle. First, if the scanning reticle is a disk, the measurement device is called a *rotary encoder* or simply an *encoder*. Rotary encoders can be used for the angular measurement of rotary motions such as NC rotary axes, tilting tables or angular metrology measurements. On the other hand, if the scanning reticle is a linear scale, the measurement device is a denominated *linear encoder* or simply a *linear scale*. Both linear and rotary encoders can be divided into absolute and relative encoders. The absolute encoders use a scale graduation formed from several graduations or code tracks where each possible position creates a unique code. Thus, the position is read directly and there is no need to find a reference position. The relative encoders use a constant grating period scale, and the position is measured relative to a reference position. The resolution of the relative encoders is higher than the absolute ones and it is the most typical measuring device for high precision machine tools.

The resolution of the relative encoders depends mainly on the period of the grating scale. The smaller the period of the grating structure is the higher is the resolution. In addition, the measurement precision depends on the mounting tolerance between the scanning and the scale reticle. Actually, typical grating periods

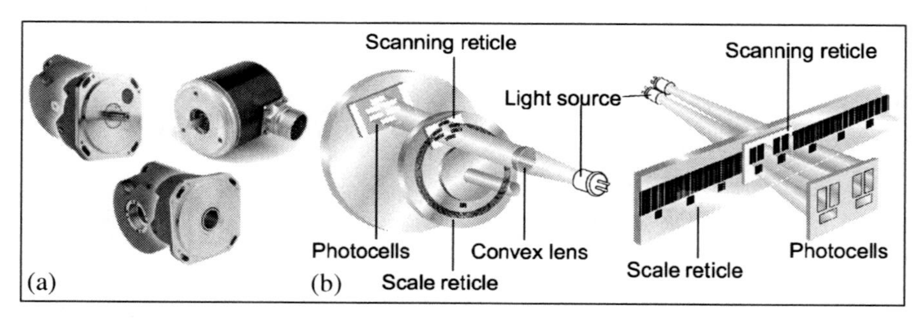

Fig. 6.8 **a** Rotary encoders. **b** Photoelectric scanning for a rotary and linear encoder. (Courtesy of Fagor Automation®)

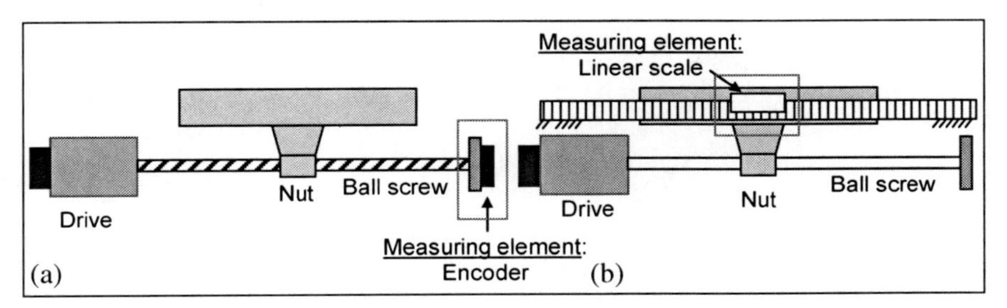

Fig. 6.9 **a** Encoder based measuring system. **b** Linear scale based measuring system

are below 20 μm and, for high precision applications, there are linear encoders with 4 μm grating periods. In addition, the scanning reticle can be divided in several scanning grids offset from each other. A photocell is assigned to each scanning field and a different sinusoidal electrical signal is obtained from each one. Thus, each grating period can be divided interpolating the different electrical signals from each photocell and the resolution of the measurement can be improved up to 0.1 μm for high precision applications. In Fig. 6.8, both a rotary and a linear encoder photoelectric scanning principle are presented.

In Fig. 6.9 two measuring solutions are presented. The solution based on the encoder directly measures the angular position of the ball screw, placed on the opposite side to the joint between the drive and the ball screw. Using this data, the position of the axis can be obtained, calculating the relationship between the linear axis motion with the ball screw rotation, which is defined by the ball screw lead. However, the measurement is affected by numerous sources of error. Thus, the encoder cannot detect the ball screw deformations due to thermal or load effects. On the other hand, backlash of the ball nut is not detected either. In other words, the encoder cannot detect what happens in those elements placed after the drive system.

Contrarily, if a linear scale is installed between the table and the machine structure, the linear movement is measured directly. This measurement is independent of the ball screw features (diameter and lead) and its deformations. Thus, it is again a more costly but very much precise solution.

6.2.3.5 The Principle of Error Correction

The global precision after assembly is usually below the required precision for the machine. This fact leads to use compensation techniques to correct and minimise those errors that can be measured by advance measuring instruments such as laser interferometers. Corrections are implemented in the machine CNC with the aim of achieving the target position for the tool centre point. This principle or method may seem an easy solution, because target positions can be obtained without the need to build a very precise machine. However, it is false because there is a limit of accuracy compensation; low-precision machines will not significantly improve the precision by compensating the errors and will never become high-precision machines.

In this respect, different types of errors can be distinguished:

- *Repetitive errors*, which can be predictable and, therefore, compensated by the machine CNC. Usually, the geometric errors are in this group. A repetitive error is an error, but it is not really an uncertainty.
- *Non-repetitive errors*, which are variable with temperature, force, vibration or other factors. These errors cannot be compensated because error value is unknown. However, a lot of research in modelling, monitoring and control techniques has been done. Through these techniques, thermal deformation or vibrations errors may become predictable and consequently compensated.
- *Random errors.* Despite the current models there are a lot of divergences between reality and design that are impossible to be estimated. Therefore only statistical treatments can be used with them.

6.2.3.6 Machine Tool Position Control

Once the machine is designed and built with the required precision, it is necessary to tune the control of its motions. The first important fact is that it is impossible to achieve a precise motion with low precision hardware.

Machine control must be designed under the *filter effect principle*, based on noise correction and obtaining precise and smooth motion. One of the noise sources is the resulting linear interpolations of complex curves. CAD files are usually defined with high order NURB curves. These curves are defined with a relatively small quantity of points; nevertheless the geometry is well defined

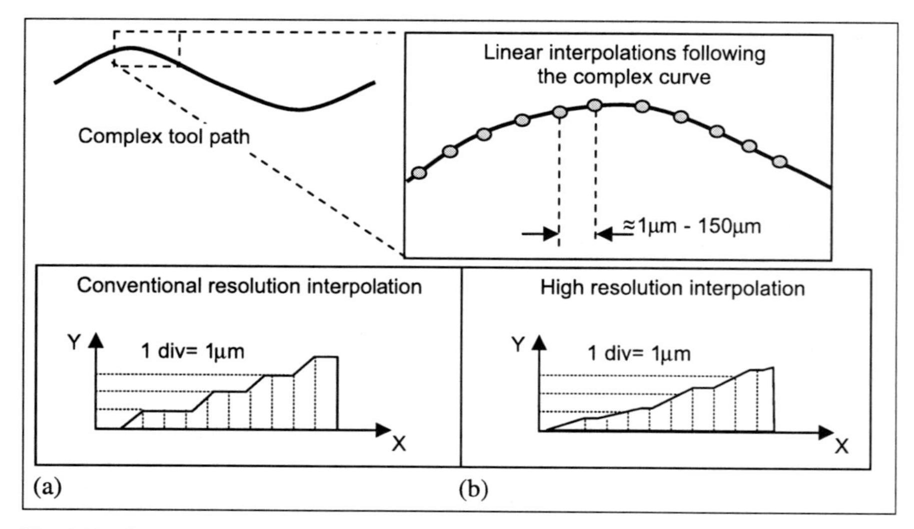

Fig. 6.10 Conventional (**a**) and high resolution interpolation (**b**) for a sequence of linear interpolations

due to the mathematical definition of the shapes. However, the NC program generated by CAM systems is usually based on small linear interpolations. As a result, a double negative effect is introduced. First, the toolpath is not exact, although the deviation between the toolpath and the exact curve can be controlled by the CAM software. Second, the sequence of lineal interpolations results in a non-smooth toolpath which can result in axis drive instability and, finally, in machining vibration.

Modern machine tool controls incorporate specific functions to minimise the discontinuity effect deriving from the linear interpolation. Thus, advanced *jerk control functions* are integrated in the CNC for controlling the feed rate at workpiece corners. Jerk control results in a smooth variation of acceleration which results in a smooth motion without any instability or shocks. On the other hand, the resolution of interpolations has been improved significantly during recent years. In Fig. 6.10 conventional and high resolution interpolation schemes are presented. A relatively rough interpolation that leads to a poor surface finish (Fig. 6.10a) is presented. In contrast to that, Fig. 6.10b shows a high resolution interpolation that enhances surface finishing.

6.2.4 Error Propagation

Machine tools are spatial mechanism with joints that can be modelled as kinematic pairs. Depending on the type of machine the kinematic model is different, and can be represented by a "bar and nodes" scheme like any mechanism. This model includes rotary and prismatic joints.

A $X_iY_iZ_i$ coordinate system is going to be fixed on each element of the mechanism; hence the position of the element will be fully defined by the $X_iY_iZ_i$ origin to the absolute reference system $X_0Y_0Z_0$ and the rotation of the X_i, Y_i and Z_i axes with respect to the $X_0Y_0Z_0$ coordinate system. To obtain the position of an element placed at the end of the mechanism (in a machine tool, the cutting tool is the element) a coordinate system must be associated to it at the tool centre point.

A known method to represent the relative spatial position of a rigid body with respect to a given coordinate system is based on a 4×4 matrix, the so-called *homogeneous transformation matrix* [10]. The first three columns of the matrix (see Fig. 6.11) represent the unit vectors of the rigid body reference system $X_iY_iZ_i$ with respect to the absolute reference system. The fourth column indicates the position of the $X_iY_iZ_i$ origin with respect to the reference system.

Using homogeneous transformation matrices, any element point refereed to the local system $X_iY_iZ_i$ can be easily obtained with respect to the $X_0Y_0Z_0$ reference system as follows:

$$\begin{Bmatrix} x_0 \\ y_0 \\ z_0 \end{Bmatrix} = [T] \begin{Bmatrix} x_i \\ y_i \\ z_i \end{Bmatrix} \tag{6.3}$$

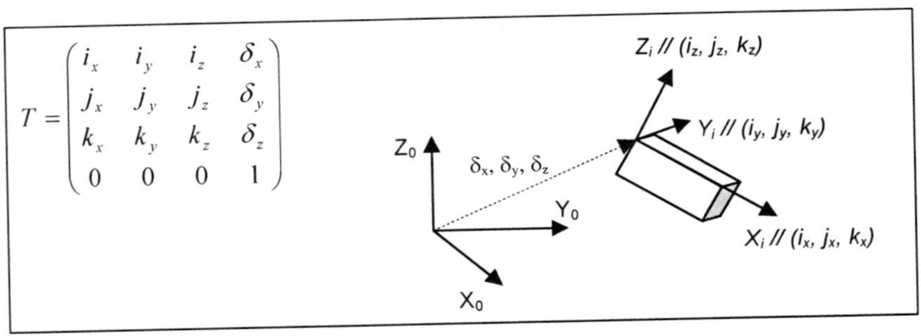

Fig. 6.11 Homogeneous transformation matrix to define the element spatial position

To represent a spatial manipulator like a machine tool, a homogeneous transformation matrix is formulated to pass from one element to the next. Elemental matrices are multiplied successively to obtain the unique homogeneous transformation matrix which can calculate the position of the tool (placed at the end of the manipulator) with respect to the reference coordinate system. Thus, the global transformation matrix T_{4x4} of the machine and the tool is obtained.

In the case of machine tools, the *tool centre point* or *TCP* is the tool reference, placed just at the tool tip, and the Z_{tcp} axis usually matches the tool axis. Therefore, the global homogeneous transformation matrix of a machine tool gives the position of the TCP system ($X_{tcp}Y_{tcp}Z_{tcp}$). This matrix is a function of the machine geometrical parameters and the position of the degrees of freedom (machine axes).

For example, in Fig. 6.12 a five-axis gantry type high speed milling centre and its kinematic model are shown. This machine presents three translational axis (X, Y and Z) in the ram, and two rotary axes (A and C) in the headstock. Once the model is designed, a reference coordinate system ($X_0Y_0Z_0$) fixed to the machine bed is specified. To obtain the homogeneous transformation matrix of the whole machine, different reference systems (not shown in the figure) are set at each structural component. Multiplying each transformation matrix from one element reference to the following one $[T_i]$, the resultant homogeneous transformation matrix $[T_{th}]$ is obtained, as is shown at the bottom of Fig. 6.12. The terms of this matrix depends on the position of the machine tool axes (Δx, Δy, Δz, θ_A and θ_C), geometrical parameters of the machine structure defined as a_l or l_4, and finally, on the tool length, defined as l_t.

In the same way, an error in any machine component can be considered in the kinematic model as a new parameter. In Fig. 6.12, the theoretical and real kinematic schemes are presented. As shown, the position of the tool tip referred to the ($X_0Y_0Z_0$) coordinate system is much more difficult to evaluate as all the errors of the machine elements and joints have to be considered.

The complexity of the homogeneous transformation approach is derived from the large number of errors that must be considered in each elemental transformation [8, 18]. For example, considering a linear guide, the transformation matrix gives the motion of a coordinate system set in the carriavge with respect to a reference system set on the guide.The ideal linear guide results in an easy translational matrix, including the constant values a, b and c dependent on the geometry of the guide and the

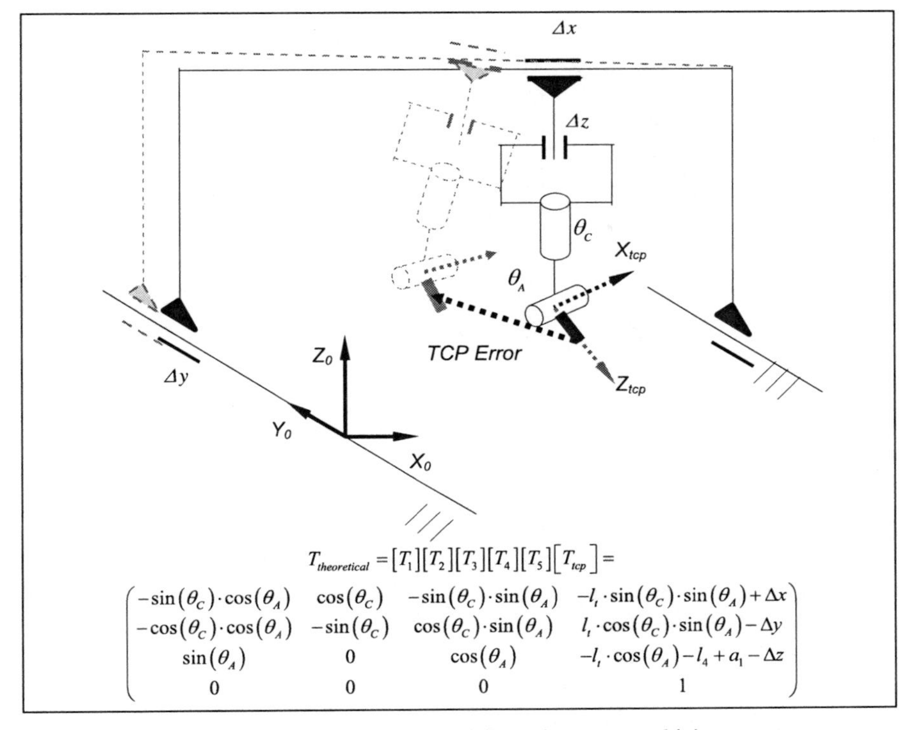

Fig. 6.12 Kinematic models of a perfect and real five axis gantry machining centre

Δx motion of the slide. But if errors are considered, the resulting matrix introduces several translation and rotation terms (see Fig. 6.13). Error values noted as δ_x, δ_y, δ_z correspond to the linear positioning errors, whereas ε_x, ε_y and ε_z are the roll, pitch and yaw errors. These errors are variables for each position of the guide and have to be measured previously. Thus, comparing ideal and general matrices, the number of parameters significantly increases.

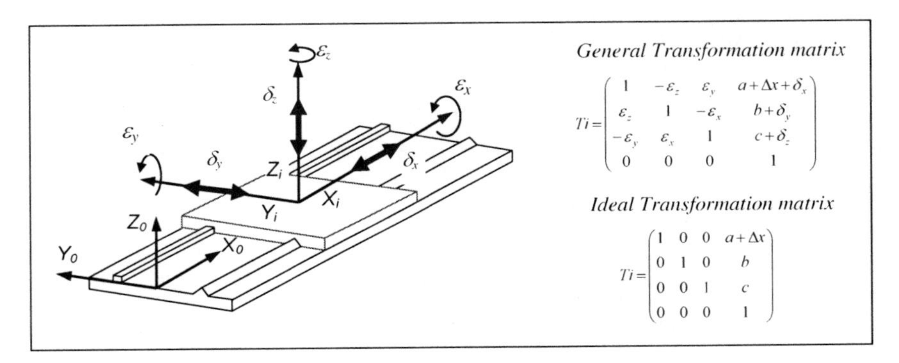

Fig. 6.13 Ideal and general homogeneous transformation matrix for a linear guide

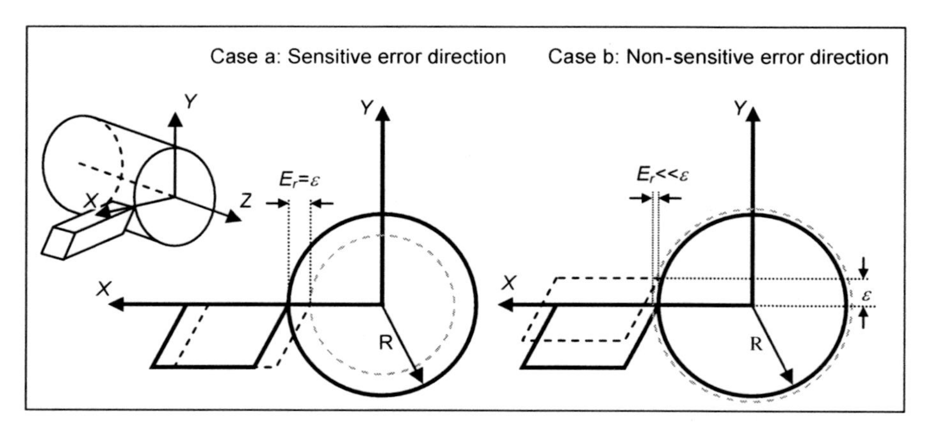

Fig. 6.14 Sensitive and non-sensitive error directions for a conventional turning operation

Despite this complexity, homogeneous transformations including errors of each element result in an outstanding tool to study and improve the machine precision. The influence of each error in the tool position could be checked ("error mapping"), concluding whether a specific error is important and worth working on to reduce it. On the other hand, it is useful to check whether errors are in a *sensitive* or a *non-sensitive* direction.

A direction is called *sensitive* when an error along it propagates directly or is even amplified at the tool tip. A *non-sensitive* direction is one where errors slightly affect at the tool tip. In Fig. 6.14 a typical example of *sensitive* and *non-sensitive* directions are presented. Here, the same error value ε is introduced in the tool for a conventional turning operation. In the first case, the error is introduced in the radial direction, resulting in a similar workpiece error E_r. However, if the same error value is introduced in the tangent direction, the resulting error E_r is much lower than ε. In this case, the radial is a *high sensitive direction* and consequently errors in that direction are the first to be reduced.

6.2.5 Thermal Errors

Thermal growth of machine tool structure and workpieces is, in general, a well-known effect. There is a general understanding that temperature variations can affect machine dimensions, consequently affecting its precision [3]. Nonetheless, over the years, few machine tools have been designed to minimise thermal growth effects. However, today some users are starting to keep the workshop temperature at 20°C commonly by air conditioning.

The size of solid elements depends on the temperature and most materials expand or contract linearly with it; graphite is an exception being practically stable for thermal variations. The increment of a linear length can be calculated as:

$$\Delta L = \Delta T \cdot \alpha \cdot L_0 \tag{6.4}$$

where ΔL is the length increment, ΔT is the temperature variation, α is the thermal expansion coefficient and L_0 is the initial length of the object. This equation considers constant thermal expansion coefficient and uniform temperature along the linear object, but it gives a reasonably good approximation of the thermal growth of a component.

In view of this formula, the variation of length can be minimised if the lengths of the element, the temperature variation or the thermal expansion coefficient are minimal. However the length of machine elements depends on the machine size and they are a design constraint. Thermal expansion coefficient is a material property, therefore the selection of component material is very important.

Granite has been widely used in coordinate measurement machines (CMMs). Likewise, several high precision machine tools include structural parts of this material; see Fig. 6.15. The main benefit is the reduction of the thermal expansion coefficient and the accuracy and flatness of the surfaces, but it is obviously a much more expensive solution than the conventional grey cast iron structure.

In Table 6.2, thermal expansion coefficients of some of the most typical materials are listed. As explained in Chap. 1, cast iron is the usual solution for conventional machine tools when several aspects are simultaneously evaluated.

In view of the values collected in Table 6.2 and using the above equation, if a workshop is not properly isolated and the temperature increases 10°C, for a 1,000 mm length cast iron component a length variation of more than 0.1 mm is caused.

A general but mistaken belief is that if temperature change is slow enough, machine and workpieces experiment similar thermal growth. Obviously it is not true because the machine tool, the fixtures and the workpiece present different thermal expansion coefficients. Furthermore, if the room temperature changes, the machine, the fixtures and the workpiece do not change their temperature at the same velocity. Again it depends on the material thermal conductivity and the size of the elements.

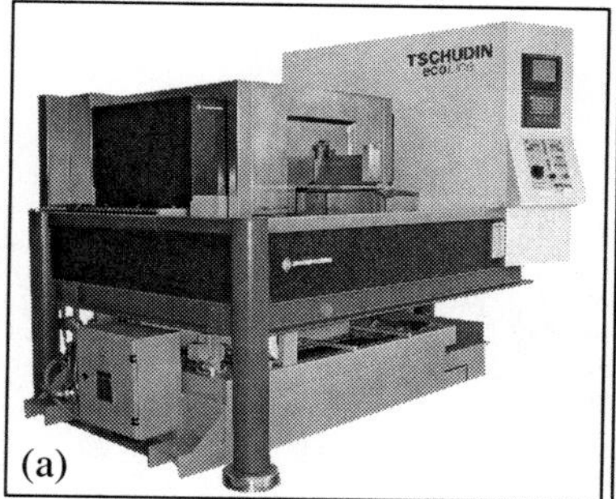

(a)

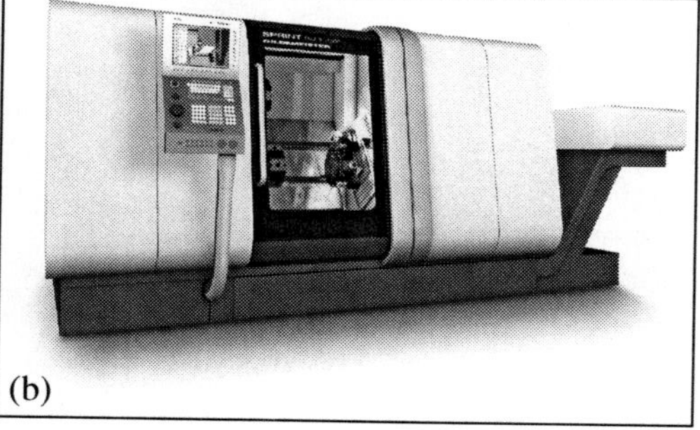
(b)

Fig. 6.15 **a** Ultraprecision centreless grinding machine of Tschudin®. **b** Advanced turning centre CTX of DMG® with granite structural parts

Table 6.2 Thermal expansion coefficients for materials used in machine tool structures and workpiece materials

Material	Coefficient of thermal expansion ($\alpha \times 10^{-6}$ K^{-1})
Grey cast iron	11.1
Carbon steel (AISI 1045)	10.8
Polymer concrete	11.5–14.0
Granite	8.0–8.5
Aluminium (7075 T6)	23.6
Titanium (Ti6Al4V)	8.6
Stainless steel (AISI 303)	17.3

Although as already mentioned, kinematic design is highly recommended, conventional machine tools are usually high-constrained structures. For this reason, thermal growths of machine components involve complex structural deformations since these growths are constrained by other elements. To minimise the effect of deformation stresses caused by thermal expansion, symmetrical machine structure is a good design practice. Thus, the expansion towards one element end is compensated with the expansion to the other side. In Fig. 6.16 a machine tool structure with a complete symmetrical structure is shown. In this machine, servos and cooling systems are placed away from the working area. This is because these elements are heat sources which can distort both the machine and the part.

Temperature control of the workshop is necessary through air conditioning systems, which usually works at 20°C as a reference temperature. As a rule of thumb, the workshop should be situated in a closed room with heavily insulated walls to

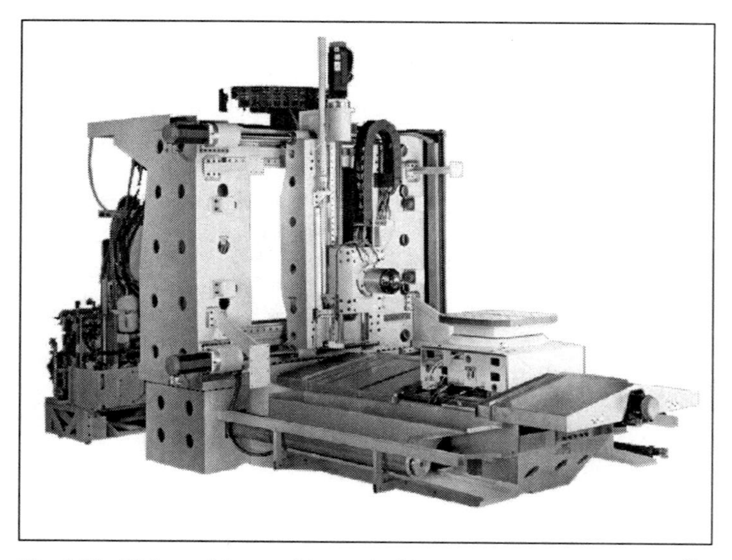

Fig. 6.16 High precision machine tool with symmetrical design, by Dixi®

prevent any entry of sunlight and ambient heat. Continuous temperature reading is also recommended at various points of the workshop every hour, to register the temperature changes.

There are some typical heat sources in the machine tool which can increase the temperature of the surrounding elements. The general recommendation is to set these heat sources as far as possible away from the working area. This is quite easy with some elements such as electrical systems, coolers or even servo drives. However, elements such as ball screws generate heat and increase its temperature.

Ball screws are used to convert rotary into linear axis motion. They are very accurate and efficient, and about 90% of the energy is transmitted [2]. However, the rest of the energy is translated into heat, which causes the thermal expansion of the ball screw being a source of positioning errors. The thermal variation of ball screws is generally unpredictable and non-uniform, so it is very difficult to compensate. A good practice is installing the shaft with a high pre-stress; in this manner the initial deformation can absorb some of the thermal dilatation without a significant reduction of shaft stiffness.

In Fig. 6.17, a real example of ball screw heating is shown, with an infrared image of the ball screw after six hours of reversing traverse at 24 m/min between two points 150 mm apart. One can observe the thermal distribution and hot area located in the screw middle area.

This effect is one of the most important errors when the position measuring system is based on the encoder of the servomotor (Fig. 6.9a). To solve this problem, there are two solutions. First, the use of linear scales for direct measuring of the axis position (Fig. 6.9b). Direct measurement will increase machine tool precision not only because the thermal expansion of the ball screw, but also other deformations which cannot be measured by the encoder. The second solution is to minimise the ball screw thermal variation by cooling the system. One technique that has been widely used in recent years is to run coolant through the machine

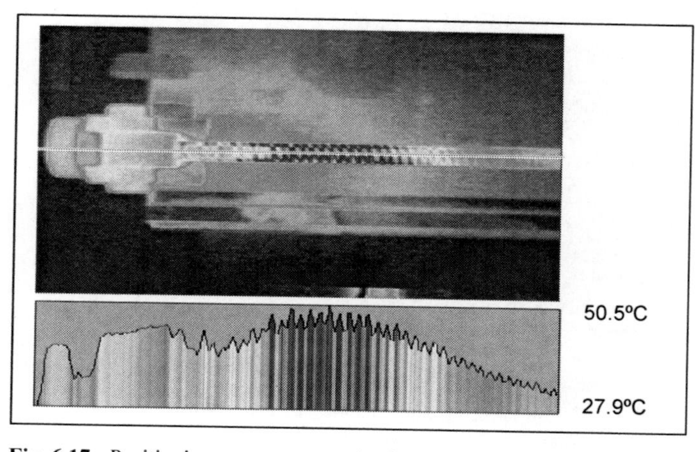

Fig. 6.17 Positioning error as a result of local temperature rise in the recirculating ball screw (Courtesy of Heidenhain®)

screws. It is a more expensive solution than the simple solid ball screws, because it is necessary to hollow out the ball screws and insert the coolant through rotary joints. Thus, not only high precision machines but also some high speed milling centres use cooled ball screws nowadays.

Finally, one of the main heat inputs is the machine main spindle. It is usually a high power electric motor and, despite being highly cooled, its temperature increases during machining. The temperature variation can lead to a thermal expansion which "pushes" the tool axially toward the workpiece. This effect is clearly shown in Sect. 3.5.4.

6.2.6 CNC Interpolation Errors

The latest generation CNCs incorporate algorithms that reduce machine velocity and acceleration in those part zones where complex interpolations have to be performed, such as corners or trajectories with very low radii of curvature. The acceleration reduction results in lower inertia forces. These forces can deform machine structural parts or deviate trajectories. Furthermore, these algorithms also provided jerk control, resulting in a smoother behaviour of the feed drives.

The CNC function for reducing the machine feed rate along complex interpolations is called *look-ahead*. Basically it is based on reading hundreds of CNC blocks prior to executing them, adjusting the feed rate depending on the trajectory complexity. This function runs in real-time, modifying the actual feed with respect to that programmed. An example of feed rate calculation is shown in Fig. 6.18, being something different for other CNC manufacturers. Here a machining toolpath obtained from the CAM system is translated into a sequence of short linear interpolations (between 0.1 mm and 1 μm length). The look-ahead algorithm evaluates each set of three consecutive points and calculates two geometrical parameters, the α angle between two successive chords and the radius of curvature R of the circle involving the three points. The CNC estimates two feed rates; the first

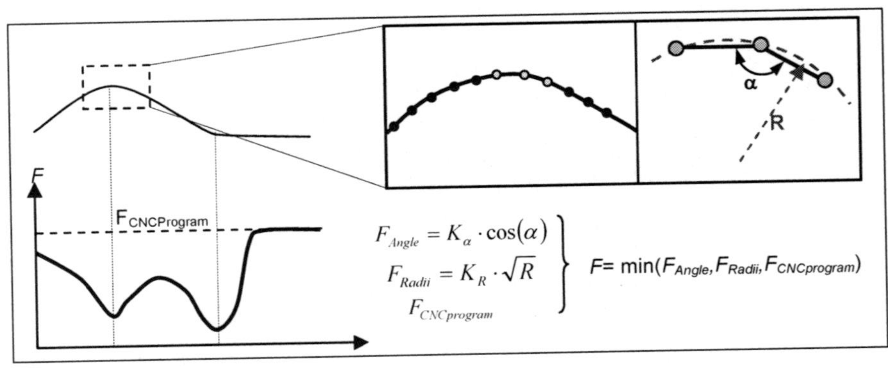

Fig. 6.18 Example of a simple look-ahead algorithm, the feed calculation for a general toolpath

is the function of the α angle, the second is the function of the R radius. Comparing these two calculated feed rates with the programmed one, the most restrictive of the three is selected and sent to the CNC axis control function.

Obviously, these algorithms imply feed reduction and consequently increase machining time; e.g., milling times can be up to 80% higher than expected using the programmed feedrate as reference. For this reason, the *look-ahead* parameters (the K factors in Fig. 6.18) can be changed along the CNC program inserting some auxiliary functions in the program. These auxiliary functions call a set of parameters stored in a CNC table. Usually, for rough operations without precision requirements but where high productivity is desired, the look-ahead parameters to be selected do not greatly reduce the programmed feed rate. In contrast, in the case of finishing operations it is important to active look-ahead parameters for reducing the feed rate and in this way the inertial forces, thus allowing a smooth motion along the full path.

6.3 Errors Originated by the Machining Process

The previous sections have described the main sources of error originated by the machine tool and its components. In addition to these errors, there are important errors that come from the machining process. Sometimes they can be greater than deviations caused by the machine itself.

Errors from the manufacturing process can start in the early stage, the CNC program generation, being followed by tool wear, tool deflection and deformation of parts due to the fixture pressure.

Of course, human error is always possible. In recent years, multi-task and multi-axis machines have extended the probability of collisions of tool against part. For this reason, specially developed software for the virtual simulation of the machining process along with machine movements are becoming interesting. These packages allow checking machining programs on the PC screen. An example is Vericut™ by CGTech®. All the same, this risk and other human errors are not considered in the error budget.

In a general point of view, errors originated by the machining process can be reduced by taking special care of the fixture of the part and the tool clamping system on one side, and making a good selection of the machining parameters on the other. Some models can be used to estimate repetitive errors caused by the cutting forces; however it is impossible to avoid others, so they have to be considered in the machine tool error budget to obtain the total error budget of the part to be machined.

6.3.1 Errors Originated in the CNC Program Generation

The CNC programs generation can be carried out in three different ways, i.e., manual programming, conversational programming or using a computer aided manufacturing (CAM) system. Manual and conversational programming are simple, direct and, assuming that no mistakes have been made, accurate because it is the user who sets the position of the tool at each point. Nonetheless they are only valid for very simple parts. For medium and high complexity parts, CAM systems must be used.

CAM systems use as inputs the part CAD geometry and a number of parameters to be defined by the user. In the 1980s and early 1990s, the calculation of the tool with respect to the part surface was based on the triangularisation of the CAD surface, which was an additional source of error. Nowadays, CAM software works with the real CAD surface to calculate tool positions. Nevertheless another discretisation is introduced, because the final CNC program is generally based on a sequence of short linear interpolations. The deviation between these interpolations and the original CAD geometry is controlled by the CAM system, using a tolerance value or, which is much more widespread, two tolerance values called *intol* and *outol* (see Fig. 6.19). Obviously, the lower are the tolerance values, the higher is the precision of the CNC program. But if *intol* and *outol* are very low, the number of linear blocks increases exponentially and consequently the size of the CNC file. Figure 6.19 shows the program size for different *intol* and *outol* values; the CNC file size is less than 500 Kb for a tolerance of 1 μm, but the part size is only $300 \times 300 \times 150$ mm and the geometry is quite simple. In the cases of large parts, such as automotive stamping dies, the CNC file size for a finishing program can be up to 700 MB or even more. In industrial applications the file size is important because it implies memory requirements for the CNC and for the mass storage systems.

Some CAM systems are able to generate toolpaths using high-order curves such as splines or NURBS. This method can obtain complex trajectories with much fewer points, so it may seem the perfect solution to obtain more precise trajectories. However, the most extended technique is still the toolpath linear discretisation due to the lack of standardisation in CNC NURB and spline programming on one side, and the higher calculation power and bigger storage capacity of the current CNCs on the other.

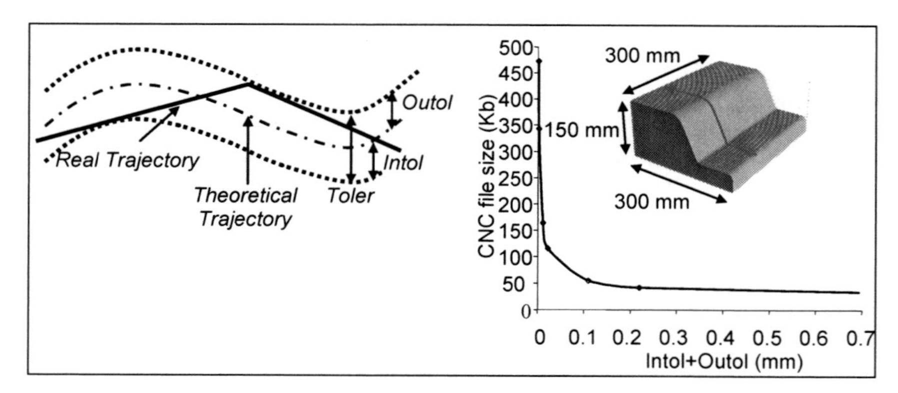

Fig. 6.19 CNC file size for a test part with different intol and outol values

Nevertheless, in a strict sense, deviation caused by CAM systems is not a conventional source of error because the maximum deviation can be controlled. However, it is necessary to take into account the value of this error, considering it within the total error budget of the part.

6.3.2 Errors Originated by the Tool Wear

During the last twenty years, new substrate materials, coatings and tool geometry have been developed to increase the cutting speed and chip load. These developments also result in a longer tool life. Unfortunately tool wear is an unavoidable effect originated by several degradation processes, such as abrasion, diffusion, plastic deformation, etc. Wear can be located on different zones of the tool edges. Hence, if wear is located on the tool rake face the cutting edge resistance is badly affected and the edge may break, but this wear type does not have a direct consequence on the workpiece dimension. On the other hand, if wear is located on the flank face, both the tool shape and dimension change, affecting the accuracy of the workpiece. Therefore, flank wear must be measured and controlled as with any other source of error, as it is shown in Fig. 6.20.

Flank wear is measured by the *VB* parameter. There is a direct relationship between *VB* and the tool dimension which depends on the tool diameter in the milling case or on the tool edge angle in the turning case. Therefore, the higher the *VB* parameter, the higher the tool tip deviation and the inaccuracy of the machined part. The mean value of *VB* must be under 0.3 mm for general applications (or 0.5 mm in some cases). However, this reference value must be lower for high added value precise parts, where a new tool can be required for each part.

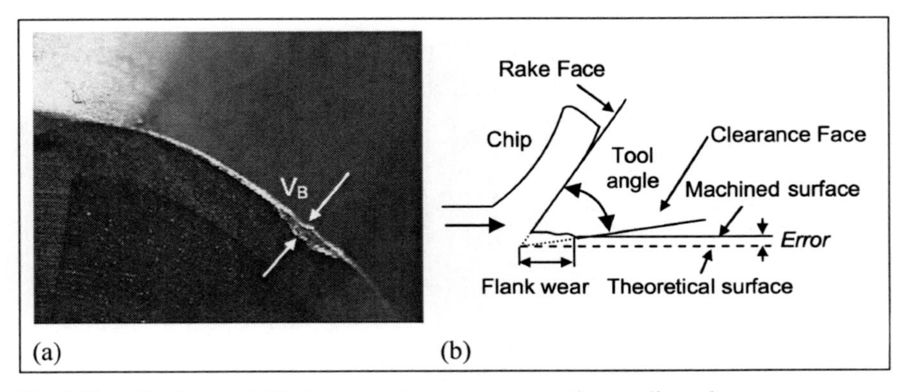

Fig. 6.20 **a** Flank wear. **b** Flank wear and consequences on the part dimension

6.3.3 Tool Deflection Error

Some machining operations require the use of slender tools in order to finish deep and narrow cavities, for example the milling of moulds for plastic components. Then there are important errors coming from the deflection of the cutting tool, or more exactly, from the distortion of the entire machine-toolholder-tool system under the action of the cutting forces [15].

There are several research works about the effect of tool deflection on machined surfaces. For example, in [9] errors of 100 μm were measured in a hardened steel mould for a 78 mm overhang and 10 mm diameter cemented carbide ball-end mill. The magnitude of this figure explains why some die and mould makers have to repeat the finishing CNC program to re-machine the stock material left over from the previous finishing, with the resultant increase in cost and production time.

The first approximation to the study of tool deflection error is considering tool as a cylindrical cantilever beam. This basic model predicts deflection as:

$$\delta = \frac{64F}{3\pi E}\frac{L_T^{\ 3}}{D^4} \tag{6.5}$$

As seen, tool deflection is a function of the following parameters: E, the Young's modulus for the tool material, $L_T^{\ 3}/D^4$, the tool slenderness parameter including D, the equivalent tool diameter, and L_T, the overhang length. Finally, F is the cutting force perpendicular to the tool axis. Considering the previous model, the use of short and high diameter mills is highly recommended, but operations in narrow cavities usually require high overhangs and low diameters.

The tool is placed at the end of a system where the machine, the shank and the toolholder also introduce flexibility. Typical values of system flexibility (meas-

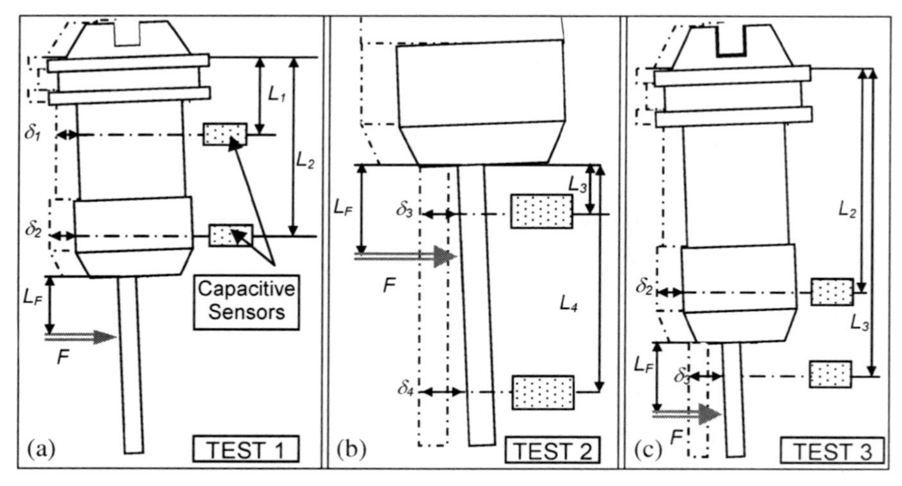

Fig. 6.21 Configuration of the tests for obtaining the shank-spindle rigidity (**a**), the tool-toolholder angular rigidity (**b**) and tool-toolholder radial rigidity (**c**)

ured at the tool tip) range from 0.4 to 1.8 μm/N; this feature depends on the type of toolholder, machine stiffness and tool slenderness.

One of the most common shanks is the HSK type (see Sect. 3.4.3) which is based in a double-face contact in the spindle nose, both in the tapered and the flanges shank faces. This feature provides excellent performance in terms of axial and radial stiffness. In [1] the radial stiffness of HSK50 shanks is measured, which ranges from 19 N/μm at 5,000 rpm to 17 N/μm at 25,000 rpm. Comparing these values with those of the tool derived from the cantilever beam model, which are about 0.15 N/μm for 6∅ mm and 4.4 N/μm for 12∅ mm; it can be noted that in the stiffer case, tool stiffness is approximately four times lesser than the shank.

On the other hand, comparing the stiffness of the tool with respect to the machine tool, in the case of the most flexible tool, machine flexibility is less than 45% of the global value measured at the tool tip, whereas it increases to 69.5% in the case of the most rigid tool.

An experimental study with HSK 63 toolholder was carried out by the authors (see Fig. 6.21), because this shank is one of the most used in high speed milling applications. Three different tests to study the stiffness of the spindle–shank and tool-toolholder interfaces were done, measuring the tool displacement δ with two capacitive sensors and applying a constant force F, measured by a dynamometer table.

In the first test, the capacitive sensors were located measuring the shank displacement at two separated points. In the second test, the sensors were both located measuring tool displacements at two tool points and finally, in the third test one of the sensors measured the toolholder displacement while the other measured the tool displacement.

The radial and angular stiffness of each element can be obtained from the measurement of δ_1, δ_2 and δ_3, combined with the position of the capacitive sensors and the force application point. The effect of the stiffness of the machine in itself can be added to the so-obtained equivalent coefficients. As can be seen in Fig. 6.22, where

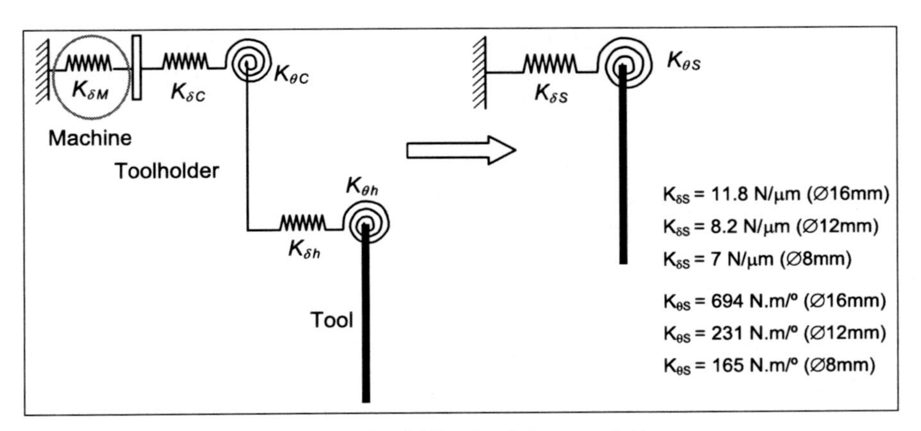

Fig. 6.22 Equivalent radial and angular rigidity for different tool diameters

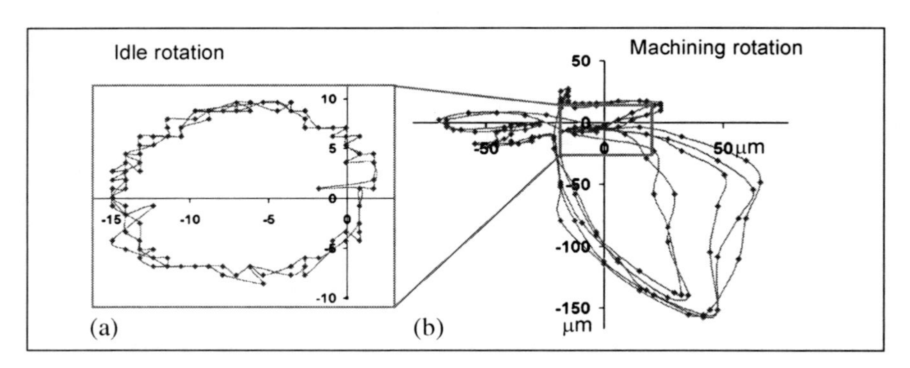

Fig. 6.23 **a** Initial runout. **b** Tool axis displacement during machining, 8 mm diameter endmill

the values of equivalent angular K_θ and radial K_δ stiffness are shown, tool-toolholder-machine stiffness depends on the tool diameter. It must be taken into account that rigidity is different for each machine axis and using the lower value implies estimating the worst stiffness case. The complete calculation of the equivalent rigidity calculation is fully developed in [15].

For rigid tools, tool deflection is lower than initial tool runout, so deflection error due to cutting forces can be ignored. Nevertheless, for slender tools cutting forces causes tool tip deflection up to 10 times the runout value. In Fig. 6.23, the real displacement of an 8 mm diameter end-mill is presented, measured with two capacitive sensors, one for the X and the other for the Y axis. In Fig. 6.23a the tool is running without machining, so displacement corresponds to the initial runout. The tool in Fig. 6.23b is machining a hard tool steel with tool deflection being more than 100 µm.

To evaluate the error generated by the tool-toolholder-machine flexibility, it is necessary to consider the magnitude of the cutting force too. Therefore we can distinguish two different cases for machining operations.

- *Turning operations:* Turning operations present constant chip load resulting in constant cutting force. Therefore, the error caused by the tool tip deflection can be predicted relatively easily, using the equivalent stiffness value of the system. In this case, it is important to distinguish between sensitive and non-sensitive errors (see Fig. 6.14).
- *Milling operations:* Error caused by the tool deflection is much more complex to evaluate. Milling involves variable cutting forces, so tool deflection is also variable along each rotation. Usually, tool deflection is a quasi-static phenomenon, so the deflection is proportional to the cutting force but time dependent. On the other hand milling is a three-dimensional operation.

In this respect, the error on the part surface caused by the tool deflection depends on the instant at which the tool cuts the theoretical surface of the workpiece, while what happens along the rest of the tool rotation will not have any effect on the part precision. For the downmilling case, the instant that must be considered is

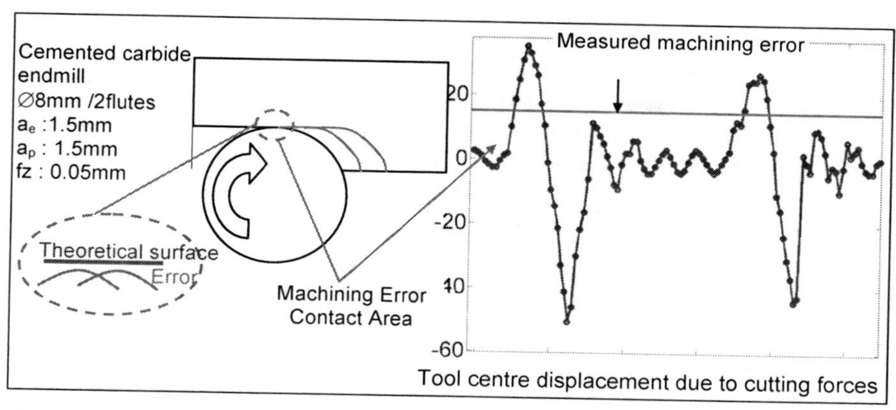

Fig. 6.24 Error caused by tool deflection for an upmilling case: the instant of the cutting edge engagement is when the tool deflection causes the machining error

when the cutting edge exits from the workpiece; for the upmilling case it is when the cutting edge engages into the workpiece, as is the case shown in Fig. 6.24. On the right of this figure, the tool axis displacement and the tool deflection error is represented. The surface error is produced when the cutting edge engages into the workpiece, and the rest of the rotation does not cause any error in the workpiece surface.

There are some research works that explain the calculation of errors due to tool deflection including some solutions [15,21] but they are difficult to apply. First, it is necessary to estimate the cutting forces at the instant when the tool contacts the theoretical surface of the part. In addition, the calculation of the cutting force must be done with the real chip load, as the tool deflection reduces the theoretical chip thickness. On the other hand it is necessary to calculate the equivalent stiffness of the complete system machine-toolholder-tool, which would be necessary to evaluate for every tool and toolholder combination. Finally, it would be necessary to compensate the estimated errors in the CAM system or directly in the CNC.

The practical solution is to use machine tools, toolholders and tools as rigid as possible. Tool rigidity depends on the slenderness so the objective is to maximise the tool diameter and minimise the overhang. Another practical solution is to reduce the chip load for finishing operations in order to minimise the cutting force, and therefore tool deflection. Again, this solution has the disadvantage of a lower productivity, but it may eliminate later finishing operations.

6.4 Verification Procedures

The objective of the machine verification is to know and measure all machine errors to understand and document its capabilities. An important purpose of verification procedures is the detection of systematic errors, providing the necessary information to compensate them. Therefore, the definition of standard procedures has been a key factor to increase machines precision.

From a general point of view, machine tool testing and verification procedures can be grouped into two types. First, geometric validations tests that measure and calibrate the machine tool in no-load operating conditions. Second, the machining of test parts, which can be performed either by the machine builder or the customer to verify machine performance during its normal function.

6.4.1 Standard Procedures for Machine Tool Validation

Geometric validation tests measure the machine tool accuracy and repeatability with a series of simple positioning tests, generally operating under no-load conditions. Standards determine how to measure, giving admissible tolerances for linear and rotary axes.

The testing procedures establish a series of target points, the accuracy and repeatability is obtained measuring several approximations to each target point. The results are the straightness, parallelism and squareness of linear axes, the concentricity errors of rotary axes, etc. These errors are measured with high precision measurement devices like laser interferometers, collimators or calibrated rules, so measurements are quite simple to perform and information is very accurate.

Many machine tool builders try to simplify the specification of the accuracy and repeatability of their machines given a numeric value, but often the user receives it without any reference about the tested conditions. In fact this is a bad practice, because the numerical value of the accuracy can vary significantly depending on the measurement procedures. For example, the accuracy is different when fewer control points are measured, or if measurements in a forward direction instead of bi-directional measurements are done. The repeatability may also vary depending on the number of measures at each control point. The fact that the accuracy and repeatability values can differ significantly makes it very difficult to compare the performance of different machine brands.

To generalise the same definition of accuracy and repeatability, some international standards have been presented. The most important ones are the ISO 230-2, JIS B6201-1993 and ASME B5.54. These standards establish both test procedures and statistical parameters to be measured, to calculate the accuracy and repeatability for linear and rotary machine motions. Nevertheless, there are some important differences among these standards, mainly in the number of target points and measurements needed to define the value of machine accuracy.

The ISO 230-2 is probably the most accepted standard in the world, including some important advantages. Testing according to the ISO 230-2 has the following bases:

- *Uniform temperature:* All tests must be performed in a controlled environment of 20° C. The temperature variations, as it has been cited in Sect. 6.2.5, can introduce thermal expansion errors which are avoided using a reference temperature for all tests.
- *Warm up cycle:* All tests include a warm up cycle that simulates the working conditions of the machine. Other standards or procedures do not include a warm up cycle, so errors caused by the machine heat sources cannot be detected.
- *Uni- and bi-directional approaches:* All tests include uni-directional and bi-directional approaches to the target points. The accuracy of unidirectional approaches seems significantly better than bi-directional results because some errors such as mechanical backslash are not included. But machine axis works in a bi-directional way, so this is the real condition and useful value for a machine user.
- *Number of target points:* Linear axes require at least 5 target points per metre and rotary axes require at least 3 target points per 90 degrees.
- *Number of measurements per target point:* Each test requires at least 5 trials per target point and per direction of approach. Therefore, there are multiple measures for each target point which allows statistical error bands calculation. The error band used in ISO 230-2 is ±2 times the standard deviation of the measurements.

The distance between target points should be different to obtain non-uniform positions. The positioning error for an *i* target point and a *j* test is defined as x_{ij}. The standard distinguishes between the unidirectional approximations in positive and negative directions defining $x_{ij}\uparrow$ and $x_{ij}\downarrow$, for each approximation. If *m* target points are selected and *n* tests are made for each target point, the mean value of the position errors is calculated as:

$$\bar{x}_i \uparrow = \frac{1}{n}\sum_{j=1}^{n} x_{ij} \uparrow \text{ and } \bar{x}_i \downarrow = \frac{1}{n}\sum_{j=1}^{n} x_{ij} \downarrow \tag{6.6}$$

The bi-directional deviation is calculated as the mean value between the unidirectional positioning errors:

$$\bar{x}_i = \frac{\bar{x}_i \uparrow + \bar{x}_i \downarrow}{2} \tag{6.7}$$

The error due to the inversion of the motion between the positive and negative approach is measured as:

$$B_i = \bar{x}_i \uparrow - \bar{x}_i \downarrow \tag{6.8}$$

On the other hand, the standard deviations for each target point can be calculated as:

$$s_i \uparrow = \frac{1}{n-1}\sum_{j=1}^{n}\left(x_{ij}\uparrow - \bar{x}_i \uparrow\right)^2 \text{ and } s_i \downarrow = \frac{1}{n-1}\sum_{j=1}^{n}\left(x_{ij}\downarrow - \bar{x}_i \downarrow\right)^2 \quad (6.9)$$

The ISO 230-2 use an error band ±2 times the standard deviation, so the repeatability for a uni-directional test for each target point is $R_i\uparrow = 4s_i\uparrow$ and $R_i\downarrow = 4s_i\downarrow$. The repeatability for the bi-directional approach is slightly more complex to calculate:

$$R_i = \max(2s_i \uparrow + 2s_i \downarrow + |B_i|; R_i \uparrow; R_i \downarrow) \quad (6.10)$$

Therefore, the repeatability of the machine tool axis, R, is calculated as the maximum repeatability of each target point:

$$\begin{aligned} R \uparrow &= \max(R_i \uparrow) \\ R \downarrow &= \max(R_i \downarrow) \\ R &= \max(R_i) \end{aligned} \quad (6.11)$$

The accuracy of the machine tool axis is also defined with an error band of 2σ and is determined by the following equations:

$$\begin{aligned} A \uparrow &= \max(\bar{x}_i \uparrow + 2s_i \uparrow) - \min(\bar{x}_i \uparrow - 2s_i \uparrow) \\ A \downarrow &= \max(\bar{x}_i \downarrow + 2s_i \downarrow) - \min(\bar{x}_i \downarrow - 2s_i \downarrow) \\ A &= \max(\bar{x}_i \uparrow + 2s_i \uparrow; \bar{x}_i \downarrow + 2s_i \downarrow) - \min(\bar{x}_i \uparrow - 2s_i \uparrow; \bar{x}_i \downarrow - 2s_i \downarrow) \end{aligned} \quad (6.12)$$

Figure 6.25 represents the accuracy and repeatability for uni-directional and bi-directional tests as defined by the ISO 230-2 standard. The target points are separated non-uniformly and measurements are made in both positive and negative directions. Once each position error has been calculated, the accuracy and repeatability for each unidirectional test is evaluated. The error band is calculated with the 2σ criterion. Finally, combining the result of both directions, the accuracy and repeatability of the machine tool axis is calculated.

The Japanese Industrial Standard JIS B6201 presents a non-statistical calculation of the accuracy and repeatability. The tests on each axis are based on three target points along the complete axis, two end points and the midpoint of the axis stroke. The approaching movement is only in one direction and the actual position of the axis is measured using a laser interferometer and compared to the positional readout of the CNC. The error is the difference between the two values. The accuracy is measured with the maximum of the three positioning errors, calculated as ±½ positioning error. In the same way, the repeatability value is measured with a series of tests for each target point. At least seven approximations for each target point must be done. Considering there are only three target points, the repeatability is calculated with 21 measurements and defined with ±½ the maximum repeatability value measured in one of the three target points.

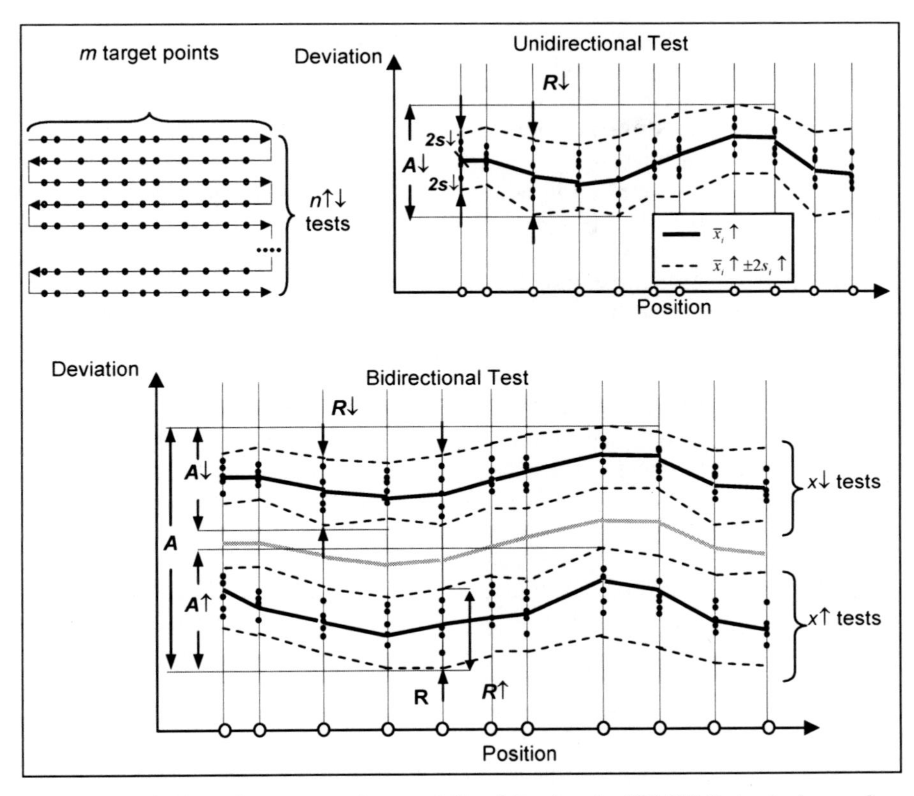

Fig. 6.25 Definition of accuracy and repeatability following the ISO 230-2 standard procedure for uni-directional and bi-directional tests

The evaluation of accuracy and repeatability with the JIS B6201 standard results in lower numerical values than the ISO 230-2, due to the lower number of target points and non-statistical data calculation. However, the ISO results are much more useful in terms of machine tool real precision (in the general term of precision) because the machine tool operating errors are used to follow statistical distribution.

ISO 230 is a general guide for several ISO standards that define more specific tests for each machine tool type. Hence, ISO 230-1 describes the general concepts of straightness, roundness, and squareness, defining the basic measuring methods. ISO 230-2, as previously described, defines the accuracy and repeatability measuring. There are different ISO standards for CNC lathes (ISO 13041), machining centres (ISO 10791), bridge-type milling machines (ISO 8636), wire EDM machines (ISO 14137), die sinking EDM machines (ISO 11090), vertical grinding machines (ISO 1985) and many other machine types. For example, Fig.6.26 shows the straightness measuring of the Y axis for a vertical machining centre, following the ISO 10791-2 guides. The straightness is measured in two perpendicular planes

(YZ and XZ) to obtain the spatial straightness. In this case, tests are carried out with a calibrated rule and dial gauge.

The proposed positioning tests measure machine axes separately, only some partial results for axes interpolation such as squareness or concentricity of rotary axis are defined, but there is no information about the contouring capabilities of the machine. For this reason, in addition to the standards, a second tests group includes some "semi-standard" (very widely used) procedures to measure the contouring interpolation errors.

One of the most extended tests for the axes interpolation testing is the *Ball-bar test*. It was first developed in the mid 1980s by Jim Bryan for the Lawrence Livermore Laboratories, as a means of testing high accuracy diamond turning lathes. Since the end of the 1980s, the ASME B5 and ISO 230 committees include this in the recommended instruments for performing verification tests.

The Ball-bar device is essentially a displacement transducer, held between two very accurate spheres. The spheres are set on the machine tool table and spindle nose. Once the Ball-bar is set up, the typical test is to execute a circular interpolation of the machine tool table with respect to the machine tool spindle, which is performed moving the table or head axes depending on the machine configuration. Theoretically, if a perfect circular motion is executed, the distance between the two spheres remains constant and the transducer would not detect any displacement. However, there are several deviations from the trajectory due to the lack of straightness and squareness between the two interpolated axes, the ball screws backlash, the CNC interpolation errors, the *stick and slip* effects on guides, etc.

The bar transducer detects the relative displacement of the spheres which is represented in an easy-to-understand polar plot. Moreover, a common Ball-bar test takes about 15 minutes to check the interpolation between two axes where many sources of error can be detected.

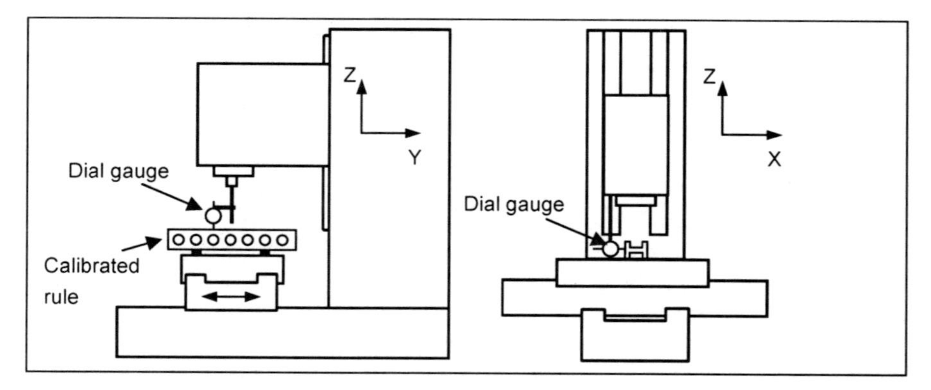

Fig. 6.26 Straightness test for the Y axis of a vertical machine centre, following the ISO 10791-2 standard

6.4.2 Test Parts

As explained above, the ISO, JIS or ASME standards define no-load tests to measure the positioning error of the machines. But these tests are far from the real machining conditions. There are some operations, such as drilling or boring, which are similar to a point-to-point interpolation, but generally, machining operations involves two or more axis interpolations. Some tests, such as the circularity tests, measure the contouring errors in no-load conditions. The information of these tests includes interpolation errors between two or more axes.

This information is valuable mainly for machine tool builders, because it is relevant data to compensate and fit the axis dynamics of the machine. However, machine tool users needs the real performance during machining operations. This is why the last verification test group includes "test parts" to check the behaviour of milling centres during real conditions. Milling implies the interpolation of three or five axes.

The best known is the NAS 979 test part, which was developed in 1969 and it is the basis of the workpiece described in the ISO 10791-7:1998 (Fig. 6.27a). The geometry of this part is a combination of a circle, a square, a diamond shape contour and a 3° angle sided quadrilateral. The standard defines two different sizes for the test part: 320 mm and 160 mm length of the external square. To obtain realistic results, the standard recommends the machining of the test part on a mid point of the machine X, Y and Z axes. In addition, the raw material and test conditions should be arranged between the machine tool builder and the final user, but the most common material is an aluminium alloy in order to maximise the feed speed with the minimum tool wear. Once the test part is machined, the standard defines a series of feature measurements such as squareness, parallelism, circularity, etc. Measurements have to be carried out in a coordinate measuring machine.

However, the ISO 10791-7 test does not include complex surfaces, while many users demand high precision in 3D contouring applications. That is why different testparts have been designed in the last years, such as the so-called Mercedes or NCG parts (the latter available at NC-Gesellschaft® association, *www.ncg.de*. Fig. 6.27b).

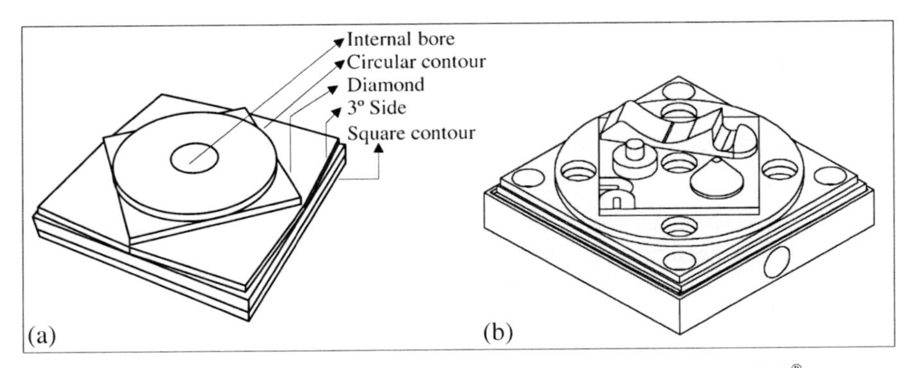

Fig. 6.27 **a** Geometry of the ISO 10791-7 test part. **b** Geometry of the NC-Gesellschaft® test part

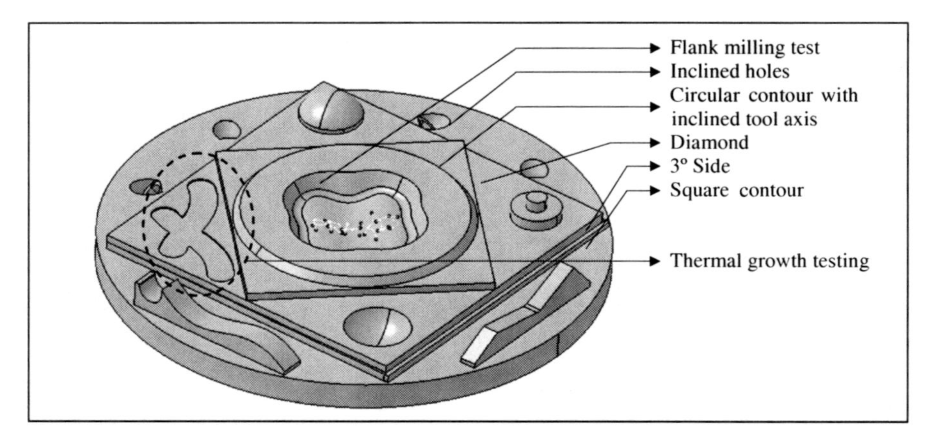

Fig. 6.28 New test part designed by Fatronik and the University of the Basque Country for five-axis milling machine centres

These test parts are widely spread due to customer demand and can be seen in many fairs or showrooms to demonstrate the machine tool capabilities, but none of them are as yet included in the ISO standard regulations.

Most of the designed testparts are for 3-axis milling machine centres, however, the demand for 5 axis milling centres is increasing rapidly and there is a lack of specific parts and validation tests for this kind of machine. In fact, ISO 230 defines some general guides to check the rotary axis positioning accuracy, but there are no tests to evaluate the interpolation error between two rotary axes or the combination of linear–rotary axes. Therefore, some research projects have been developed to complete current standards and include specific multi-axis machine testing procedures. For example, Fig. 6.28 shows the design of a specific test part for five-axis milling centres.

In this part, some features helps to evaluate the position of indexed rotary axes, while others are oriented to assessing the accuracy of multi-axis milling operations, such as flank milling. The design of new test parts and testing methods and, what it is more important, the standardisation and generalisation thereof are the only way to evaluate machine tool performance and compare machine tools.

6.5 A Brief Conclusion

The basic principles and methodologies for precision design and assembly of machine tools have been presented, along with the verification procedures to estimate the accuracy and repeatability of a machine. Precision is easy to define but difficult to numerically determine, taking into account the "true value" of a dimension is never known. The ISO 230 defines the main concepts when accuracy and precision tried to be measured.

The history of machine tools is the history of the precision of machine tools. The most precise ones are now giving errors under one micron. But surely this value will be improved in the next twenty years. In the chapter devoted to micromilling machines in this book, new concepts and application of the exposed principles are found.

Acknowledgements Thanks are addressed to Dr. J. L. Rodil for the technical support and people from Heidenhain, CMZ CNC lathes, Kondia and Ibarmia for their continuous help and advice. Thanks are also addressed to financial support of the Ministry of education and Research of Spain (DPI2007-60624).

References

[1] Aoyama T, Inasaki I (2001) Performances of HSK tool interfaces under high rotational speeds. CIRP Ann Manuf Technol 50:281–284
[2] Arnone M (1998) High performance machining. Cincinnati, USA
[3] Bryan JB (1990) International status of thermal error research. CIRP Ann-Manuf Technol. 39/2:645–656
[4] Chatterjee S (1997) An assessment of quasi-static and operational errors in NC machine tools. J Manuf. Syst, 16:59–68
[5] Dornfeld D, Dae-Eun L (2008) Machine design for precision manufacturing. In: Precision Manufacturing. Springer Verlag, USA
[6] Evans C (1989) Precision Engineering: An Evolutionary View. Cranfield Press, Bedford, UK
[7] Hocken R (Chairman) (1980) Machine tool accuracy, Report of Working Group 1 of the Machine Tool Task Force, UCRL-52960-S. Lawrence Livermore Laboratories, University of California, Livermore, CA
[8] Hsu YY, Wang SS (2007) Mapping geometry errors of five-axis machine tools using decouple method. Int Precis Technol, 1:123–132
[9] Kim GM, Kim BH, Chu CN (2003) Estimation of cutter deflection and form error in ball-end milling processes. Int J Mach Tool Manu 43:917–924
[10] Lamikiz A, López de Lacalle LN, Ocerin O, Diez D, Maidagan E (2008) The Denavit and Hartenberg approach applied to evaluate the consequences in the tool tip position of geometrical errors in five-axis milling centres, Int J Adv Manuf Technol, DOI 10.1007/s00170-007-0956-5
[11] Moriwaki T (1994) Intelligent Machine Tool: Perspective and Themes for Future Development. Manuf Sci Eng Trans ASME, 68:841–849
[12] Nakazawa H (1994) Principles of Precision Engineering. Oxford, UK
[13] Nawara L, Kowalski J, Sladek J (1989) The influence of kinematic errors on the profile shapes by means of CMM. CIRP Ann Manuf Technol. 38:511–516
[14] Reshetov DN, Portman VT (1989) Accuracy of Machine Tools. ASME Press Translations
[15] Salgado MA et al. (2005) Evaluation of the stiffness chain on the deflection of end-mills under cutting forces. Int J Mach Tool Manu 45:727–739
[16] Sartori S, Zhang GX (1995) Geometric error measurement and compensation of machines. CIRP Ann Manuf Technol. 44:99–609

[17] Slocum AH (1992) Precision Machine Design. Englewood Cliffs, USA
[18] Srivastava AK, Veldhuis SC, Elbestawi MA (1995) Modelling geometric and thermal errors in a five-axis CNC machine tool. Int J Mach Tool Manu 35:1321–1337
[19] Taniguchi N (1983) Current Status in and future of ultra-precision machining and ultrafine materials processing. CIRP Ann Manuf Techn.32:2–11
[20] Tsutsumi M, Saito A (2004) Identification of angular and positional deviations inherent to 5-axis machining centres with a tilting-rotary table by simultaneous four-axis control movements. Int J Mach Tool Manu 44:1333–1342
[21] Uriarte L et al. (2007) Error budget and stiffness chain assessment in a micromilling machine equipped with tools less than 0.3 mm in diameter. Precis Eng 31:1–12

Chapter 7
New Developments in Lathes and Turning Centres

R. Lizarralde, A. Azkarate and O. Zelaieta

Abstract This chapter describes the latest developments in lathes and related turning technologies and industry. The evolution of the industrial manufacturing tendencies and their influence on the configuration of lathes and turning centres, as well as the technological developments induced by the same, will be described. These tendencies will also be connected to the main application sectors, their particularities, specific machines and technologies devoted to them. The most significant recent developments in process and machine components will also be addressed.

7.1 Introduction

The lathe is historically the basic machine for any cylindrical part manufacturer, from the smallest shop floors to high production lines, as well as for educational purposes, in schools, institutes and universities.

From the simplest hand operated machines, made up of one workpiece head and double carriage holding a single tool, evolution can be considered as incremental over the years, through the introduction of electronics, numerical controls (NCs) and mechanical components with improved performance in terms of accuracy, speed, power and productivity, to put it in a nutshell.

However, in recent years the turning industry has moved more than one step ahead, encouraged by industrial productive demands from the main manufacturing sectors, going from pure lathes to the current, very complex state of the art turning centres capable of completely manufacturing one workpiece in one set-up, from row to final part.

For this purpose, current turning centres integrate processes from rough turning to milling, drilling, grinding and even surface mechanical or thermal treatments

R. Lizarralde, A. Azkarate and O. Zelaieta
Ideko-IK4, C/Arriaga, 2, 20560 Elgoibar, Spain
rlizarralde@ideko.es

L. N. López de Lacalle, A. Lamikiz, *Machine Tools for High Performance Machining*,

such as laser hardening or roller burnishing. Moreover, these complex processes are supported by the application of complex multi-purpose and multi-process cutting tools and assisted processes via high level technologies such as laser and ultrasonic vibrations, and likewise a high degree of intelligence and automation developed and integrated into the latest generation NCs.

From the machine configuration side, multi-purpose or multi-task lathes or turning centres are equipped with multiple workpiece heads, turrets, milling heads with one or two rotation axes, supplementary grinding devices and high speed, short time workpiece loading systems.

Brought to this level of complexity by the leading industrial sectors such as automotive, land transport, aeronautics and energy generation, turning centres have acquired a high degree of flexibility and versatility, becoming in turn a necessity for the capital goods industry and general purpose manufacturers.

7.2 Machine Configuration

This section will describe the general tendencies in machine architecture and configuration, connecting them to their main purpose and target sector demands.

The chapter will avoid general purpose "conventional" machines to focus on the newest, most advanced differentiating developments, in new horizontal high production configurations, horizontal multi-tasking lathes or turning centres, vertical high production machines, vertical large size machines, vertical turning centres, and also multi-process machines including non-conventional processes or special features and applications.

7.2.1 High Production Lathes

The most representative market for this type of machines presently is the automotive industry. The automotive industry has become the paradigm of high productivity in high accuracy applications, leading its suppliers, machine tool manufacturers among them, to the development of precise, reliable high production products. In the case of vertical and horizontal lathes, there are five basic characteristics shared by all these machines:

1. They have compact machines, for minimum floor space occupation.
2. They are equipped with one or two headstocks and two, three or four turrets for minimum process time, and maximum productivity.
3. Maximum relevance is given to the manipulation, loading and unloading of workpieces, applying a huge variety of solutions, from external conveyors integrated with the machine, to pure internal pick-up solutions.

   Vertical machines have been pioneers in solving the workpiece integrated pick-up, especially with the development of machines with upper headstocks.

Fig. 7.1 Pick-up solution in a vertical lathe from Emag®

These lathes were specially developed for this purpose, plus the advantage of easier chip evacuation. Solutions for parts such as brake discs or similar parts were developed in the 1990s by several leading manufacturers such as Emag® (Fig. 7.1), Weisser® and Hesapp® and extended later to the rest of the manufacturers.

Recently, some solutions have been developed in vertical lathes for long and slender parts, such as shafts, which were typical horizontal applications. In horizontal lathes the loading solution was performed by external elements. Lately, some remarkable integrated solutions have been developed. Among them, the patented solutions of Weisser® (Fig. 7.2), that applies the usual pick-up concept of vertical inverted lathes to a horizontal configuration, and Index®, performing the pick-up operation via tool turrets, that hold special chucks for picking the workpiece up in the two tool position.

4. They have a high speed on headstocks and high dynamics on displacement axes for minimum lead time. For this purpose, the machines are equipped with high-speed electrospindles in headstocks, linear ball or roller recirculating guides in displacement axes and, in some cases like Gildemeister®, linear motors for axe displacement.
5. Basically, they only perform turning operations, thus, with optional simple operations like drilling or basic milling, they are performed by motorised tools installed in the turret. Nevertheless, one of the latest evolutions has been the inclusion of grinding wheel heads to complete the machining processes. This tendency is being applied both in vertical and horizontal machines, as shown in Fig. 7.3.

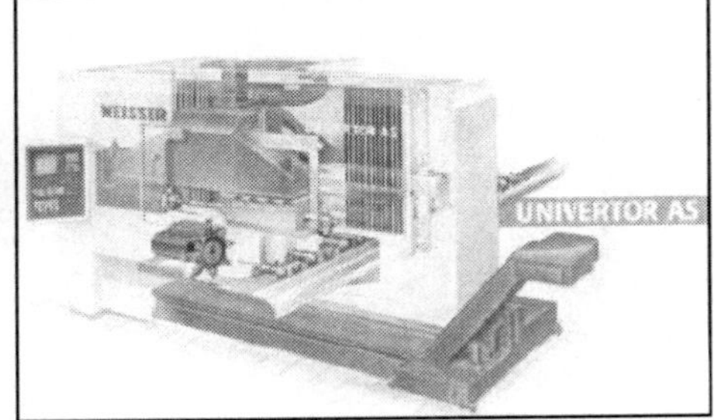

Fig. 7.2 Pick-up solution, by Weisser®

Fig. 7.3 Grinding wheels included in lathes. **a** Vertical from Index®. **b** Horizontal from Mazak®

7.2.1.1 High Production Processes: Hard Turning Versus Grinding

The high quality finishing of parts is usually a prerequisite in high production lines, in particular the automotive high production line. Additionally, hardened steel parts are very common in the automotive industry, especially in engine parts, which are subject to high precision movements, and high mechanical and contact loads.

The high production machining of these parts brought about the discussion between grinding and turning as finishing operations, especially after the development of "hard turning" technology. Basically hard turning is the turning of hardened steel using CBN (cubic boron nitride) or ceramic tools, in high-stiff and precise lathes.

Lathe manufacturers have defended their solution supported by the higher flexibility and cheaper solution, while grinding manufacturers based their arguments on the supposedly higher precision, stability and productivity of the grinding process. Additionally, the application of hard turning has been found over the years as not suitable for parts with sealing requirements, because of the helix generated in the surface by any turning process. This problem has recently been overcome with the application of special tools to eliminate that feature, as shown in Fig. 7.4.

Both lathe and grinding machine manufacturers have directed the discussion to an intermediate solution, developing mixed solutions in recent years that incorpo-

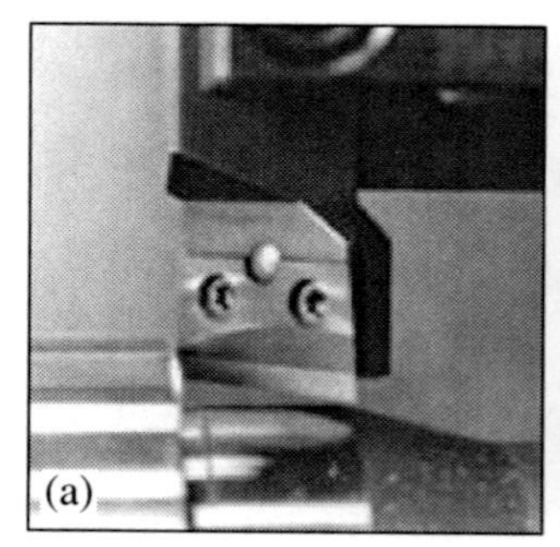

Fig. 7.4 Lead-free hard turning, **a** By Sumitomo®. **b** By Weisser®

rate the "opposite" process, as shown in Fig. 7.3. Nowadays, it is common to combine processes such as hard turning and grinding, in the same machine, depending on the parts and particular features to be machined.

7.2.2 Turning Centres: Multi-tasking Machines

Multi-tasking machines have become the most important evolution of lathes in the last ten years, holding a privileged position in the catalogue of many leading manufacturers.

The origin of turning centres or multi-tasking machines can be found in the application of additional motorized tools, which have been used in lathes for many years. The market tendency towards lower batches, higher flexibility and the reconfigurability of machines has fostered the fast extended development of solutions in this line.

It is difficult to delimit or mark the fields of applications and the configuration of multi-tasking machines, since nowadays the variations and combinations of features, devices and processes integrated are almost infinite. The imagination and development capabilities of machine tool manufacturers have made almost any combination possible, in vertical and horizontal configurations.

The first steps in turning centres consisted of conventional lathes that included a milling head instead (or in addition to) of the turret, with one or two additional axes to perform interpolated milling operations.

Multi-tasking "turning-drilling-milling" centres with the standard B, C, X, Y and Z-axes allow the complete machining of precise complex workpieces. The goal to completely machine and measure a wide spectrum of workpieces with minimal settings in only one machine has been achieved. By the interpolation of up to five axes even free form surfaces can be machined easily. To summarise, the claims for these types of machines are:

- The reduction of floor-to-floor time. The number of machines involved in the machining process is drastically reduced. Queuing up in front of machines or transport in between individual machines is reduced to a minimum. Work scheduling and planning is extremely simplified. The cost of stock piling semi-finished piece parts is reduced.
- Increased flexibility. In multi-tasking machines setup times are usually low when changing over to another clamping or piece part. Expensive, complicated fixtures are unnecessary. Due to technological capabilities, these machines offer many creative possibilities in the customer's design department.
- Increase in use. Decreasing setup times are equivalent to a significant increase in production hours. Complete machining means machining processes are not cut by manual interventions, which eliminate interruptions by breaks.
- Reduction of manpower. The reduction of machines involved in the production process automatically reduces personnel requirements for operation, inspection,

piece part transport and work scheduling. The complete machining process combines formerly individual processes in just one operation, enabling a multiple machine operation.

- Increase in quality. Due to the lack of multiple clampings as in split manufacturing not only costs for fixturing are reduced but also clamping and adjustment errors are avoided.

These machines have progressed towards higher sizes, power capabilities and the integration of different solutions and processes. Some of the latest, most significant examples will be presented below.

7.2.2.1 Large-size Machines for Heavy-duty Processes

These machines, such as the M150 of WFL® (Fig. 7.5), can hold workpieces of up to 6 metres long, with a 1 metre diameter, capable of performing heavy-roughing operations, by means of their high torque and power spindles and milling heads, reaching values up to 90–100 kW.

Fig. 7.5 Heavy-duty multi-tasking machine from WFL®

7.2.2.2 New Ideas in the Structural Machine Concept

From the original configuration of machines, usually an inclined bed lathe incorporating the milling head, manufacturers have progressed to the development of optimum configurations for multi-tasking purposes, giving the same importance to milling and turning operations. One of the best examples is the new NT family by Mori Seiki®, that integrates the milling centre concept (a vertical "box in box" configuration with the milling head located in a octagonal Z axis ram), together with the horizontal bed of a lathe, as shown in Fig. 7.6 and Fig. 1.3 in Chap. 1. Operability, precision, mechanical performance and chip evacuation are ensured in this machine concept.

Besides the machine concept, Mori Seiki® has introduced new features in these series, one of the most innovative being the tool turret that integrates a milling electrospindle, as shown in Fig. 7.7.

Fig. 7.6 Structural configuration of Mori Seiki® NT series

This example of machine configuration shortens the distance between turning and milling centres, leading to a new discussion in the machine tool industry. This discussion is more extended in the case of vertical turning and milling centres equipped with rotary tables. Vertical turning centres are the evolution of original vertical lathes towards one complete setup for part machining.

For this purpose, vertical lathes incorporate a vertical displacement axis configured with a milling ram or combination of a ram plus a moving portal. This ram can incorporate a milling head or turning tools. In the latest evolution, analogous to the horizontal turning centres, turning tools are held by the same milling head, locking the rotation of the spindle.

These vertical turning centres have found a very important market in the machining of big parts, energy generation and land transport being some of their target markets. Figure 7.8 presents one example of this type of vertical turning centres, by Pietro Carnaghi® (see also Fig. 1.21, Chap. 1). These machines hold workpieces of diameters up to 6–8 metres and the turning power can reach values from 200 to 300 kW.

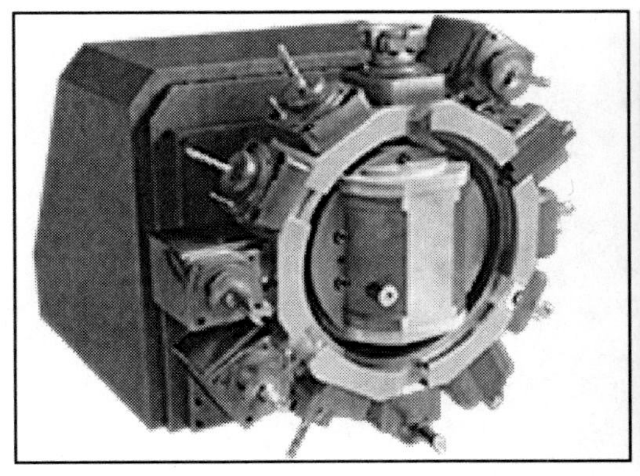

Fig. 7.7 Milling electrospindle integrated into the turning turret, by Mori Seiki®

Fig. 7.8 Heavy-duty vertical turning centre, AC 46 by Pietro Carnaghi®

From the technological side, vertical turning centres usually collect two very difficult to make compatible requirements: big workpiece sizes and narrow tolerances.

Aeronautic turbine components (rings, housings, seals ...) and wind energy generator gearboxes are typical applications of the medium size vertical turning centres, which lead to very high demands in terms of quality, precision and reliability, together with big size and high material rates to be removed. For this reason, these machines combine high level technologies from the mechanical and control sides: hydrostatic guides on all axes, including the rotation head, the monitoring and control of process and machine components, and the compensation of thermal and dynamic disturbances or online measurement are technologies usually found in these machines.

Besides the pure machine configuration evolution, these multi-tasking machines have introduced innovations in the application of processes. One innovative and significant example is the use of milling to perform operations that were usually done by turning, in a process called "turn-milling" or "spin milling" by differ-

Fig. 7.9 Turn milling operation. **a** By WFL®. **b** By Mori Seiki®

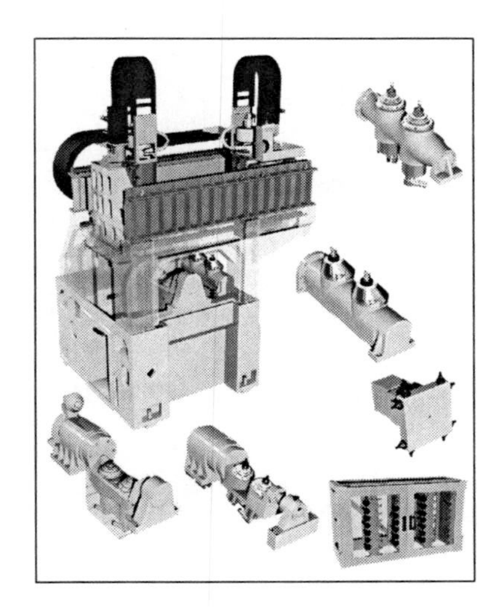

Fig. 7.10 Reconfigurable machine concept, after [1]

ent manufacturers to increases the productivity of turning by a higher material removal capability. Figure 7.9 shows some examples of this application.

Flexibility and reconfigurability associated to multi-tasking machine has also been demanded by industry, and several approaches have been developed and are currently in the process of being developed. Reconfigurability is understood in this context as the possibility of re-configuring a machine, from one purpose to another by means of substituting or adding modular components. These approaches, such as the German project "METEOR" (Fig. 7.10) seek the elimination of lead times [1], not only during the machining of one particular reference, but also in changing from one reference to a new one.

Although they cannot be considered a multi-tasking process or machine, the turn-milling or turn-broaching processes are remarkable. These are high productivity processes, mainly orientated to the rough machining of engine crankshafts, especially end journals by means of big diameter milling tools, as shown in Fig. 7.11.

Fig. 7.11 Turn-milling/turn-broaching operation, by Boehringer®

7.3 The Latest Technologies Applied to Lathes and Turning Centres

This section summarises the most important technologies developed to provide new machines with the performances required to satisfy increasing market demands.

7.3.1 General Configuration Technologies

Historically, lathes have been the most "conservative" machine tools, being mostly used as general purpose machines and, until some years ago, to perform pure roughing processes. Finishing operations were performed by grinding, honing or polishing.

Nevertheless, the latest most advanced machines, i.e., both high production lathes and turning centres, incorporate the same high level mechatronic technologies as other machines such as grinding machines or machining centres.

- The architecture of machines has been described in the previous section. There is no other machine tool that has undergone so much evolution, from the conventional lathe configuration the complex turning centres with several headstocks, turrets and milling heads, combining other tools such as laser, roller burnishing or ultrasonic sources. Machine architecture has evolved to adapt all these innovations.
- Guiding systems have progressed from the friction sliding guides of conventional lathes to a massive use of recirculating ball or roller units, pushed by the drastic increase in speed and acceleration required for faster machining, loading and unloading operations.

  In heavy-duty and high precision machines, such as vertical lathes and turning centres orientated to aeronautic and energy applications, hydrostatic guides are used in rotary and displacement axes, providing a high load capacity, high damping ratios and positioning accuracy that provide safety, precision, stability and productivity to the high added value solutions of those sectors.
- Conventional motors for main spindles and transmissions have also been substituted by integrated electrospindles for headstocks (see Fig. 6.5 in Chap. 6) and linear motors for displacement of axes when speed and acceleration requirements have become excessive for conventional transmissions.
- Numerical controls must also answer to the high demands the complexity level of these machines induces: multiple axes to be controlled and synchronized, different processes and machining cycles (turning, milling, grinding, drilling, tapping, etc.) to be programmed and controlled, and the high dynamics of axes and the implementation of measuring devices. Due to all these particularities, the latest generation controls are being applied and some manufacturers, such as Mazak®, have developed or customized their own NCs.

7.3.2 *Complementary Technologies to Improve Machine Performance*

In recent years, some complementary technologies have been integrated in lathes to improve the productivity and precision of these machines. Thus:

- The machine behaviour control and compensation of errors, in particular, vibrations and thermal effects. The "thermal friendly" concept is used by most of the leading manufacturers, with similar approaches that cover, basically, two fields: structural design, attempting to reach minimum thermal sensitivity designs, and the measurement of critical temperatures to apply compensation strategies through the NC.

  Concerning vibration prevention, there are several approaches. Some manufacturers have developed strategies to suppress chatter vibrations (Sect. 3.5.3), by the application of mechanical passive dampers or software based strategies such as the variation of the rotation speed.

  More widespread are the developments related to the balancing of parts during machining. The so-called "adaptive balancer" by Mori Seiki®, shown in Fig. 7.12, which is able to balance the potential part imbalance during turning, using two plates with small masses distributed asymmetrically, is remarkable.

  Other advanced balancing applications have been recently developed for car and train wheels machining. In these cases, the imbalance is measured between roughing and finishing, and balancing is performed via eccentric turning, eliminating more material on the overloaded side.
- Programming and simulation of processes. Three main factors make these technologies almost necessary for multi-tasking machines: 1) the inclusion of multiple different processes 2) the complex configuration of machines, with many devices and tools, and 3) the complex shape of the parts and the high number of different operations combined for a total machining in one setup.

  For this reason, specific CAM software has been developed for turning centre machines, combining features of turning and milling. These CAMs are combined with 3D simulations of the process (Fig. 7.13) to optimise the working cycles, preventing collisions and non-desired cycles. Several machine manufacturers have developed their own systems.

Fig. 7.12 "Adaptive balancer" by Mori Seiki®

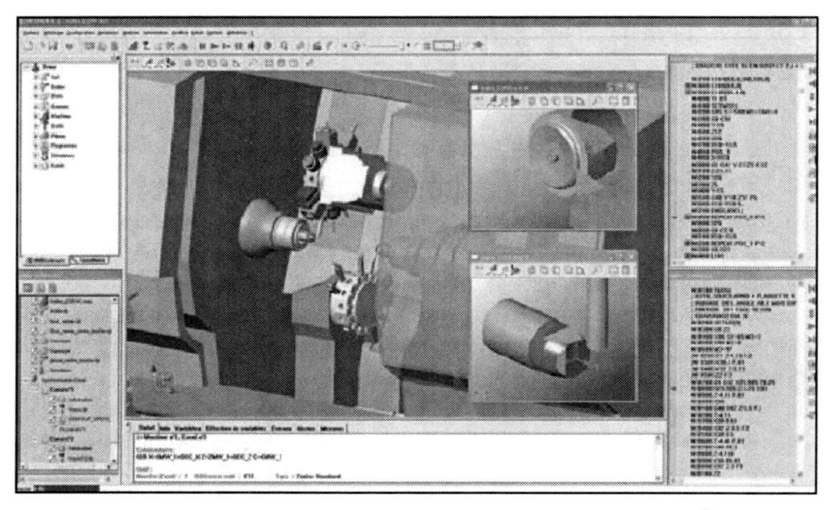

Fig. 7.13 Simulation of the multi-tasking process, by Spring Technologies®

7.4 Special Machining Processes Applied in Multi-tasking Machines

Multi-tasking machines are now integrating different machining processes, for example laser processes, cold forming treatments and others. Current developments are explained in the following sections.

7.4.1 The Laser Application

Lasers have some potential applications in combination with other processes in multi-tasking machines. Laser hardening (Fig. 7.14) is the most applied. Several manufacturers integrate a laser head as an additional tool, in different configurations.

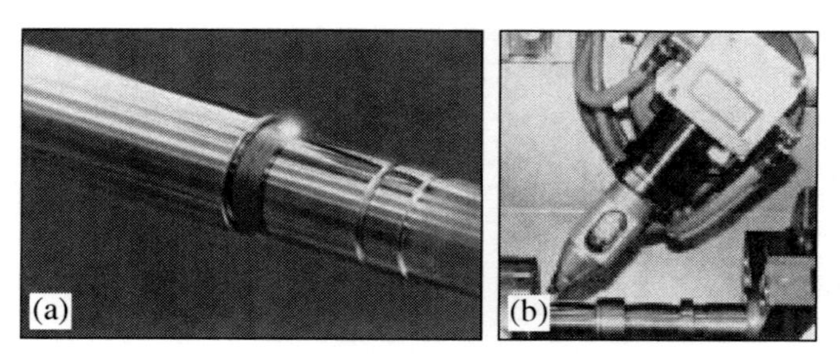

Fig. 7.14 **a** Laser hardening process integrated in the turning centre. **b** System by Boehringer®

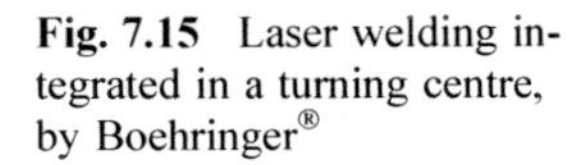

Fig. 7.15 Laser welding integrated in a turning centre, by Boehringer®

The laser head can be manipulated and controlled in the same way as a milling tool, with two displacement axes and two rotaries as needed; thus, it is possible to apply them to complex shaped parts. The advantage of laser hardening against other more productive methods such as induction hardening is its flexibility, and the possibility of applying to details in one workpiece, likewise difficult to reach points.

Laser welding has also been incorporated by some turning centre manufacturers (Fig. 7.15). In this case laser is applied for welding parts in an intermediate operation; furthermore, it can be used for hardening and even for finishing some part details.

Laser-assisted turning has been also investigated in the last ten to fifteen years for hard to machine materials. This application is based on the idea that continuous laser beam induced heating of the workpiece provides a local softening of the material, enabling the material to be machined using a defined cutting edge.

Several research works have reported positive results from laser-assisted machining for machining ceramics, whereas in the case of metal alloys such as nickel and cobalt based stainless steel or titanium alloys, the results are quite contradictory [2, 8]. In industrial terms, the application of laser-assisted machining is almost negligible.

7.4.2 Roller Burnishing and Deep Rolling

Roller burnishing is a cold rolling process without actual metal removal. It is a new concept in finishing components.

The surface of metal parts worked through turning, reaming or boring operations is a succession of peaks and valleys when microscopically examined. The roller burnishing operation (Fig. 7.16) compresses the peaks into valleys (Fig. 7.17), thus forming a smooth mirror finished surface [4].

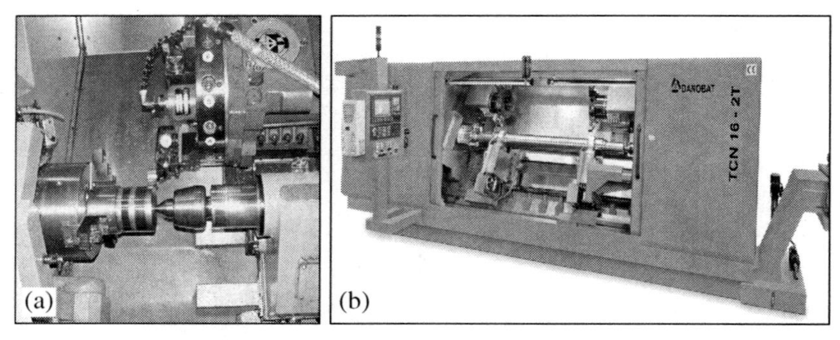

Fig. 7.16 **a** Roller burnishing. **b** Lathe ware rolling is applied to train axles, by Lealde®-Danobat

One step ahead, deep rolling is one of the most valuable mechanical processes to improve the fatigue strength of dynamically loaded components. It eliminates or at least reduces fatigue, especially on notches like fillets and shoulders which can lead to fatigue cracks. The deep rolling process works similarly to roller burnishing; however, the task is different. To ensure equal component quality all process parameters, in particular the burnishing force, must be controlled during the process. This control is performed mechanically or, in more advanced systems, hydraulically, controlling the pressure applied to the roller or ball tool.

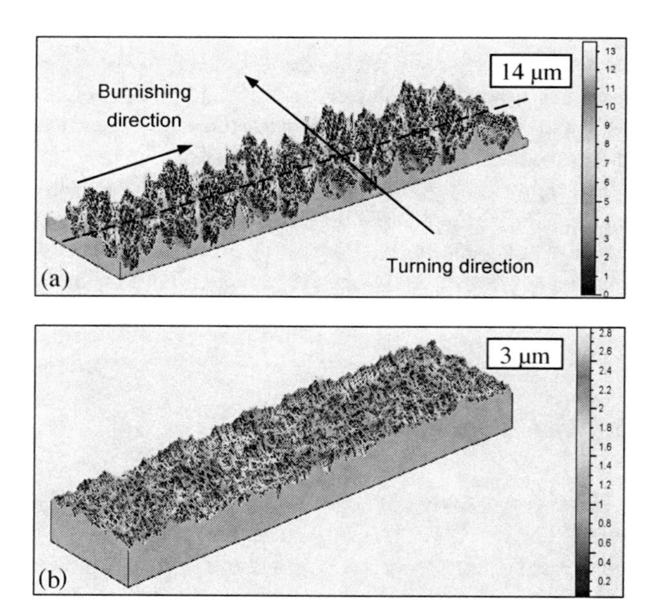

Fig. 7.17 **a** Workpiece topography after turning. **b** Workpiece topography after turning and burnishing

Fig. 7.18 Deep rolling of turbine blades, by Ecoroll®

Deep rolling is based on the combination of three simultaneously working physical effects:

- A deep layer of residual compressive stress on a component surface.
- A strength increase through cold working.
- The elimination of micro notches and the improvement of surface roughness quality.

Compressive stress, generated on the surface layer during the rolling process, remains to a high extent after the rolling process is finished. The compressive stresses in the axial direction are most important for the improved fatigue strength.

In the past, deep rolling was only applied to special machines; however, nowadays, due to the evolution in the design and configuration of deep rolling devices, it can be incorporated in machines such as lathes and turning centres. The deep rolling device can be incorporated in the lathe turret as with any other tool.

Several successful applications of combined turning and deep rolling processes have been addressed by researchers and industrial manufacturers, from automotive engine parts, such as shafts of valves, through railway applications, such as wheel axes, to aeronautic parts such as turbine discs, blades (Fig. 7.18) and shafts or landing gears.

7.4.3 Ultrasonic Assisted Turning

In difficult-to-cut and brittle materials several research approaches have been carried out for the purpose of modifying the mechanics of chip formation to improve process capabilities, in terms of chip breakage, decrease in forces and friction, and thus an increased tool life.

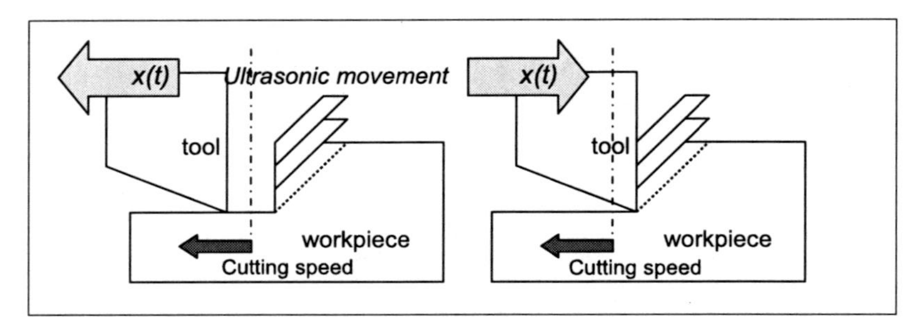

Fig. 7.19 Principle of ultrasonic vibration, where *x(t)* is the ultrasonic movement (20–40 kHz)

One of the most researched technologies is the application of ultrasonic vibrations on the tool, overlapped to feed movements. The principle behind this technology, as shown in Fig. 7.19, is the application of a high frequency (usually between 20 and 40 kHz) low amplitude (5–20 μm) linear or elliptical vibration on the tool.

Ultrasonic vibration is introduced by means of a generator into an intermediate component, the so-called "sonotrode," designed to have a *normal mode* at the required frequency and with the required *mode shape* to generate vibrations on the defined plane.

Several authors [3, 7] refer to good results in terms of chip breakage, load reduction and tool life increase for very diverse materials, such as titanium alloys, copper, or stainless steel and glass. Nevertheless, industrial applications are mostly only reported in the machining of brittle materials such as glass for optics, and ceramics.

7.4.4 Cryogenic Gas Assisted Turning

This process consists of the application of cryogenic gas (nitrogen usually), to assist the turning process in hard to machine materials such as hardened steels. The application of this cryogenic gas is claimed to provide some significant advantages:

- Increasing cutting speeds and material removal rates by as much as 200% and reducing the overall cycle time.
- Alternatively, increasing tool life up to 250% and reducing machine downtime for tool changeover.
- Enhancing the removal of chips from the cutting area and eliminating oily residues on machines, parts and chips.
- Improving the reliability of cutting with ceramic inserts while maintaining desired dimensional tolerances.
- Improving the quality of produced parts by preventing mechanical and chemical degradation of the machined surface.

Fig. 7.20 Icefly® industrial application, by Air Products®

Some authors [9] have researched the application of this process to hard turning, allowing the use of ceramic tools instead of CBN, and eliminating the need for conventional coolant fluid. The advantages, besides the above, are related to an environmentally lower impact and a better surface integrity achieved by this alternative process.

Concerning the application in machine, it must be mentioned that Company Air Products® have patented the technology, under the Icefly® brand, collaborating with some machine manufacturers, like Hardinge®, for the development of industrial applications, as shown in Fig. 7.20.

7.4.5 High-pressure Coolant Assisted Machining

High-pressure coolant jet assisted machining is being researched and applied in industry for difficult-to-machine materials, such as titanium and nickel-based alloys. It is well known the poor machinability of these materials is due to several factors: their low conductivity, chemical reactivity with tool materials, shearing mechanism during chip formation and the strain hardening during machining.

These factors lead to very high tool wear forcing the use of very low cutting parameters, especially a very low cutting speed.

To overcome these limitations, the application of coolant jet at high pressures (up to 350 bar), properly aimed at the interface between the tool tip and chip [5] generates a effect of the hydraulic wedge that significantly eases the chip formation and cutting process, via these factors:

- A significant increase of cutting speed and feed rate. The coolant jet facilitates chip breaking, decreasing the friction and forces generated on the tool. Additionally, the coolant effect is improved, resulting in the possibility of increasing cutting parameters.
- An increase in tool life, due to the decrease of cutting loads and temperatures.

- An improvement in surface finishing, via the easiest selection of optimum cutting conditions and elimination of long chips, a decrease in friction and thermal effects on surface.
- The final effect, i.e., or sum of all the above is the decrease in cycle times and cost per part.

Aeronautical part manufacturing is the leading sector in the application of this technology, due to the extensive use of related materials and critical quality and productivity process parameters.

Acknowledgements Gratitude is expressed to all cited companies for the pictures, and Prof. Abele for the Fig. 7.10.

References

[1] Abele E, Wörn A (2005) Economic Production with Reconfigurable Manufacturing Systems (RMS) Production Engineering, XII/1:189–192
[2] Chryssolouries G, Anifantis N, Karagianni S (1997) Laser Assisted Machining: An Overview. Journal of Manufacturing Science & Engineering. Transactions of the ASME, 119: 766
[3] Kumar J, Khamba JS, Mohapatra SK (2008) An investigation into the machining characteristics of titanium using ultrasonic machining. Internat J Machin Machinabil Mater, 3:1–2
[4] López de Lacalle LN, Lamikiz A, Muñoa J, Sánchez JA (2005) Quality improvement of ball-end milled sculptured surfaces by ball burnishing, Int J Mach Tool Manufact, 45(15): 1659–1668
[5] López de Lacalle LN, Perez Bilbatúa J, Sánchez J A, Llorente J I, Gutierrez A, Albóniga J (2000) Using high pressure coolant in the drilling and turning of low machinability alloys. Internat J Adv Manufact Technol, 16/2:85–91
[6] Moriwaki T, Shamoto E (1995) Ultrasonic Elliptical Vibration Cutting. Annals of the CIRP, 44/1:31–34
[7] Moriwaki T, Shamoto E (1991) Ultraprecision Diamond Turning of Stainless Steel by Applying Ultrasonic Vibration, Annals of the CIRP, 40/1:559–562
[8] Rajagopal S et al. (1982) Machining aerospace alloys with the aid of a 15 kW laser. J Appl Metalwork, 2:170
[9] Zurecki Z, Ghosh R, Frey JH (2003) International Conference on Powder Metallurgy and Particulate Materials

Chapter 8
High Performance Grinding Machines

R. Lizarralde, J. A. Marañón, A. Mendikute and H. Urreta

Abstract This chapter deals with the latest developments in technology, and describes the current state of the art in machine configuration, critical components, and control technologies applied specifically in grinding machines. Moreover, this description of latest developments is structured into the different machine types: cylindrical, surface, vertical, centreless and their variants, and will also reference special-purpose machines, oriented to particular but technologically significant applications and markets, such as the aeronautic or automotive markets. The principal machine components and peripherals will also be revised, showing the tendencies in the last years and the newest and most innovative solutions.

8.1 Introduction

Grinding is a machining process that removes material from a workpiece by means of an abrasive grinding wheel that rotates at high speed. Grinding machines vary from very simple "surface machines" with a reciprocating table to very sophisticated machines incorporating 5 axes, several grinding wheels and complex devices for dressing, coolant application and workpiece on-line measurement, complemented with last generation control units that introduce "intelligence" to the machine.

In the last years, the industry is demanding process performances that were unbelievable 20 years ago. Productivity, reliability and accuracy demands have been increased 10 and 100 times in some cases. Industry demands shorter cycle times,

R. Lizarralde, J. A. Marañón, A. Mendikute and H. Urreta
Ideko-IK4, c/Arriaga, 2 20560 Elgoibar, Spain
rlizarralde@ideko.es

L. N. López de Lacalle, A. Lamikiz, *Machine Tools for High Performance Machining*,

minimum human actuation, minimum set-ups, the minimum number of machines and workshop space, and finally "intelligent and autonomous" machines. In the 1990s, high production machines for high batches were demanded. Nowadays, the shorter life time of most products (the automotive industry is a representative example) has shortened the batch sizes, adding the requirement of flexibility and reconfigurability to machines.

These industrial demands have been answered by grinding machine manufacturers by means of developments in several areas, which can be summarised in the following list:

- Multi-purpose machines. Machines that include several wheels for external and internal grinding, combining conventional abrasives with superabrasives.
- High speed and high dynamics machines. Wheel rotation speeds have increased 10 times reaching values up to 300 m/s. Machine dynamics allow the introduction of new high production strategies.
- New developments in grinding wheels. New abrasives, such as new ceramics and CBN (cubic boron nitride), in combination with machine advances allow material removal rates up to 1,000 $mm^3/mm/s$ in processes such as "creep feed" grinding, "high efficiency deep grinding" or "speed stroke" grinding (see Sect. 8.3).
- New numerical controls, with higher processing speeds, the integration of sensors for monitoring and control of the process, faster and easier communication, and integration of measuring devices to close the control loop.

8.2 The Machine Configuration

This section will describe the main general trends for the configuration of grinding machines. It will cover the principal machine elements from the general architecture, axes configuration, material applied for the construction of the structural parts, driving units, guideways, main spindles, dressing systems and cooling fluid application systems.

Although the general configuration of grinders could be considered as "conservative", with few variations in many years, a very significant number of modifications and improvements have been performed in the last decades, beyond the introduction of CNC, which is always mentioned as the greatest innovation in machine tool industry in the last 40 years.

Besides the non-discussed relevance of the electronics and software in the evolution of machine tools, many technologies related to the mechanical parts of the machine have been developed, and translated from books and laboratories into the shopfloors and the production facilities. Grinding technology is not an exception in this tendency, trying to answer to the demands of the 21st century society and industry, under the target parameters of productivity, accuracy and the reliability of machines and productive processes.

8.2.1 The Machine Architecture

The architecture of grinding machines will be analysed separately for the typical configurations: cylindrical, surface, centreless and vertical machines, since each one of these configurations presents a complete range of applications and, therefore, a complete range of configurations. The separation in the analysis of architectures will simplify the understanding.

8.2.1.1 Cylindrical Machines

The basic process of cylindrical grinding consists in a grinding wheel rotating in parallel to the workpiece, which is supported between the working head and the tailstock.

Concerning the general architecture of the machine and the axes distribution, there are two main tendencies: the "traverse table" and the "traverse wheel".

The traverse table is the most conventional and widely applied configuration. The advantages of this configuration that make it the most used are related to the accuracy and stability of the process, in particular from the wheel side. This configuration structurally separates the infeed movement, in the grinding wheel side, from the traverse movement, in the table. The machine is more compact in the grinding wheel side with only one guideway; this provides better static and dynamic behaviour. Additionally, the manufacturing and assembly of this side is significantly less complex.

On the other side, the major disadvantage on the traverse table is the larger space required by the table and, consequently, by the machine. The space required is the sum of the table size (depending on the maximum workpiece to be grounded) plus twice the traverse stroke. For this reason, limitations in the footprint and big workpieces to be machined are situations in which the traverse grinding wheel is selected.

The grinding wheelhead is also subject of several different configurations, depending on the requirements of flexibility, the adaptability to different applications and workpieces and tendencies towards one set-up machining, to avoid errors produced by workpiece manipulations.

Figure 8.1 is representative of several configurations that are common among the grinding wheel manufacturers.

Position 1 presents the most basic straight fixed head. It is the least flexible, applied in simple machines, extending the principle of ancient manual machines to general-purpose and training-purposes grinders.

Position 2 shows the oriented head disposition. These heads present a higher flexibility, since they can not only perform simple plunge operations but also can grind faces and shoulders. This type of head presents also some variants: from two fixed positions in minimum and maximum angles, up to a continuous controlled rotary axe.

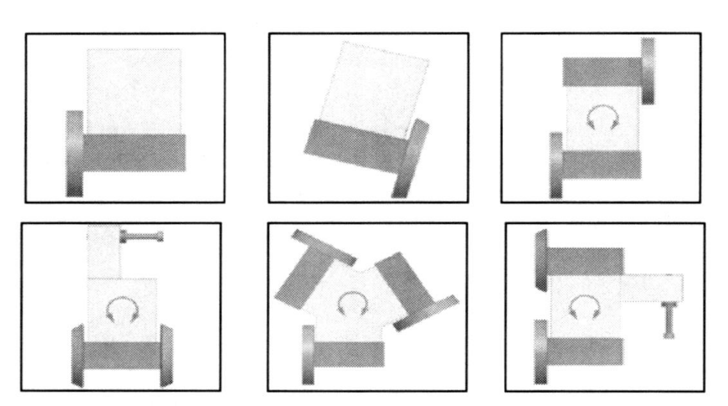

Fig. 8.1 Configurations of grinding wheel heads (*top row*: 1,2,3, *bottom row*: 4,5,6)

The increasing productivity requirements have promoted the development of more complex solutions, implementing several (2, 3 up to 4) wheels in a rotary head, as presented in positions 3, 4, 5 and 6 in Fig. 8.1. These multi-wheel heads can include external and internal grinding wheels (positions 4 and 6), or conventional wheels combined with high speed superabrasive wheels, or wheels for special processes such as peel grinding. The applications of these configurations are wide: from general-purpose shopfloors, that manufacture a wide range of parts, to high production facilities that produce complex parts with several different features that require different wheel geometries or process specifications.

Finally, in these multi-wheel configurations, the search of maximum operability with minimum floor space required has led to configurations where the wheels are located in not perpendicular faces, as shown in Fig. 8.2.

Fig. 8.2 Two non-perpendicular high production wheel configuration

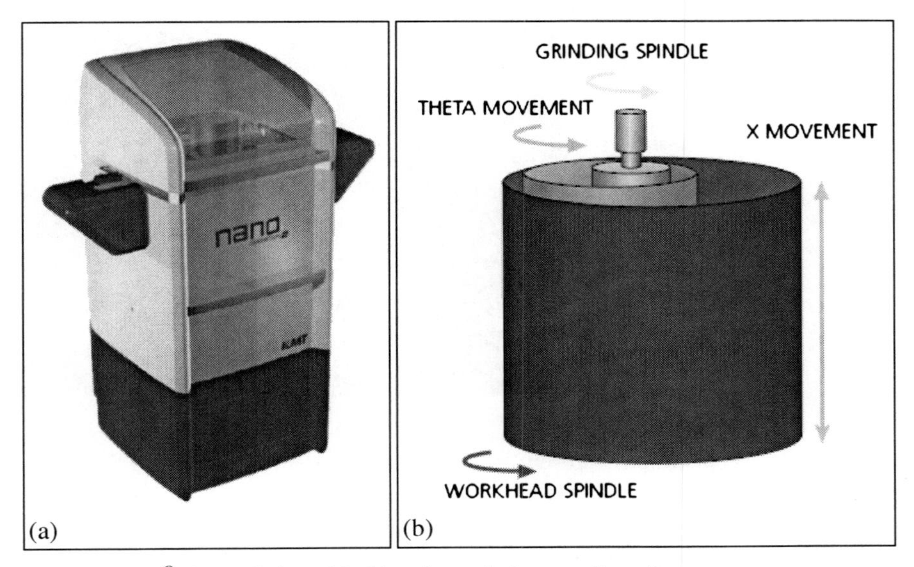

Fig. 8.3 KMT® Nano grinder. **a** Machine picture. **b** Axes configuration

A very special new machine architecture was presented in 2005 by the company KMT Lidköping®. The new concept is a very compact small machine, called the Nano Grinder, in which the conventional combination of the rotating workpiece and the longitudinal penetration displacement of the wheel is substituted by two eccentric rotary axes, as shown in Fig. 8.3b. The machine is particularly oriented to the external and internal grinding of bearing rings.

With this axes configuration the machine can perform internal and external cylindrical grinding, of parts up to an 80 mm external diameter, with a very small footprint. High accuracy is also claimed by means of the new architecture, supported by the use of linear motors and aerostatic guides.

8.2.1.2 Surface Grinders

In general-purpose surface grinders two general configurations are possible: "reciprocating column" and "reciprocating table". Similar to cylindrical, reciprocating table is the preferred configuration except in cases of high length or weight workpieces, where reciprocating column is selected.

For both configurations a very common architecture, used in case of high volume (height and width) parts, is the bridge-type machine. The grinding of machine tool structures and components such as guideways is a typical application for bridge-type grinders.

Among surface grinders, "creep feed grinding" machines must be remarked upon. These machines, applied in very high material removal processes, in particular for difficult-to-machine materials such as titanium and nickel-based alloys

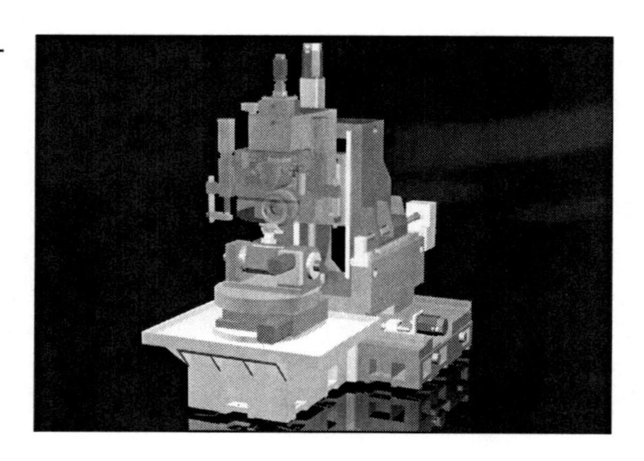

Fig. 8.4 Creep feed grinding machine with continuous dressing device

for aeronautic applications, present very compact and stiff architectures to overcome the high grinding loads generated by such processes. Traverse column is the most extensively used architecture, incorporating high power main spindles, and special devices such as continuous dressing and very special coolant application systems. An example of this configuration is presented in Fig. 8.4.

8.2.1.3 Vertical Cylindrical Grinding Machines

Vertical machines are a variation of cylindrical ones with some architectural similarities with surface grinders. These machines are applied for the finishing operation of big size parts, in particular big diameters with lower lengths, gaining positions in the last years in the aeronautic and energy generation markets. The typical configuration of vertical machines is presented in Fig. 8.5, a rotary table supporting the workpiece, and a vertical column holding X and Z-axis and the grinding wheelhead.

Following the tendencies of the high productivity and flexibility of cylindrical machines, vertical ones usually present multi-wheel heads, for different operations

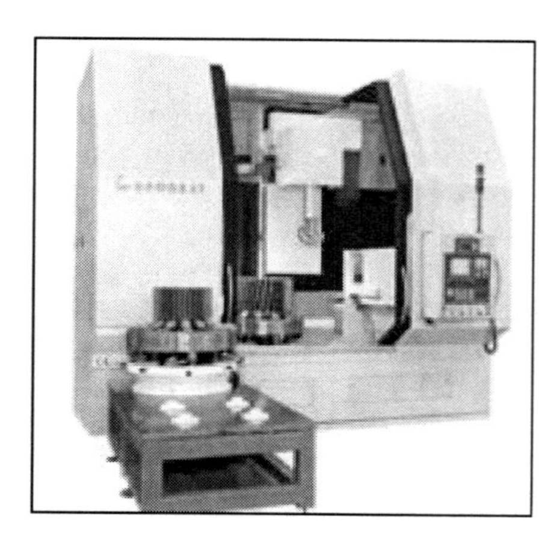

Fig. 8.5 Vertical grinding machine, by Danobat®

Fig. 8.6 Detail of vertical grinding machine with three tools: surface grinding, internal grinding and turning, by Danobat®

to be performed, including external and internal cylindrical grinding, surface grinding and even turning in the most recent applications, as presented in Fig. 8.6.

8.2.1.4 Centreless Machines

In centreless machines there are two main configurations, "fixed work centre" and "mobile work centre", applied depending on the type of workpieces to be machined. In the latter case, the grinding wheel is fixed on the machine base, and the regulating wheel and work-rest move to allow different workpiece diameters, different set-ups and to perform the grinding infeed movement.

This configuration presents some variants in the movement of regulating wheel and work-rest:

- Two displacement axes, one in the low carriage that holds the work-rest and a second one between this carriage and the regulating wheel.
- In some machines the driving unit is unique and the two carriages can be clamped or released, thus moving both together, to set up the workpiece, or only the regulating wheel, for the infeed movement.

Fixed work centre configuration is mainly used for the grinding of long parts and especially in *through-feed* grinding, in which the movement of the work-rest makes very complicated the setting up of the workpiece loading system. The grinding of long bars and pipes is a typical application of fixed work centre machines.

An unusual configuration in centreless machines is the one presented by company Nomoco®, with the grinding wheel located above the regulating wheel, as presented in Fig. 8.7.

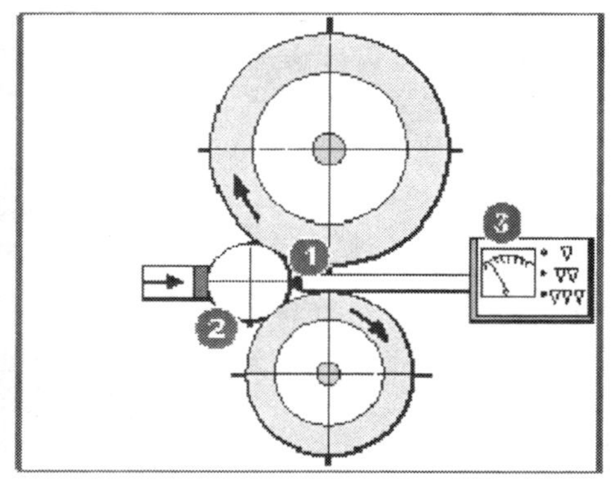

Fig. 8.7 Nomoco® centreless vertical wheel configuration with gauge

With this configuration the workpiece setting up can be simpler and the possibility of applying a measuring gauge in the workpiece area is also claimed, as shown in Fig. 8.7b.

8.2.1.5 Combined Configurations Integrating Grinding and Turning

The discussion about the frontiers and overlapping between turning and grinding was launched in the 1980s supported in the development of "hard turning" processes (see Sect. 7.2.1.1) and more accurate lathes that were able to compete in some applications with grinding, and at the same time, the development of superabrasives for grinding wheels boosted the development of high removal rate grinding processes, allowing the complete machining of parts, roughing and finishing, in grinding.

This discussion, together with the growing tendency of the industry towards the complete machining of parts in one set-up has promoted the development of lathes that integrate grinding devices and grinding machines that incorporate turning tools, drifting the discussions towards the question of which solution is more suitable. Some of the most important grinding machine manufacturers have materialised this tendency into new models such as the Studer® S242, shown in Chap. 1, Fig. 1.4, besides the above mentioned vertical machines that usually integrate turning tools.

8.2.2 Materials Applied in Structural Parts

The main characteristic that differentiates grinding machines from any other machine tool is the required high accuracy. Specifically, this is the accuracy represented by accurate positioning, accurate movements, and the avoidance of any

Table 8.1 Comparative values: polymer concrete vs cast iron and steel

Characteristic	Polymer concrete	Cast iron	Steel
Density (kg/dm^3)	2.3	7.15	7.85
E-Modulus (kN/mm^3)	30–40	80–140	210
Tensile strength (N/mm^2)	10–15	150–400	400–1,600
Compressive strength (N/mm^2)	110–125	600–1,000	250–1,200
Bending strength (N/mm^2)	25–35	250–490	–
Relative damping	0.02–0.03	0.003	0.002
Thermal conductivity (W/K·m)	1.3–2	50	50
Specific effective heat capacity (KJ/kg·K)	1	0.5	0.5
Thermal coefficient of expansion	14–16	11	12

perturbance: vibration, static or thermal deflections, stability and repeatability. For this purpose, special care is taken in the selection of the components and the design of the machine structure, searching for compact architectures, with high stiffness and dynamic stability.

The same accuracy and stability criteria have been historically applied to the selection of the constructive materials. Cast iron is the clear leader, applied in more than 80% of the machines, as explained in Chaps. 1 and 2.

Although some manufacturers decided to use steel instead of cast iron the better damping characteristics of this material is the principal motivation for its massive use in respect to steel.

Nevertheless, grinding manufacturers are the most active among the machine tool manufacturers in the search for alternate materials. Polymer concrete, in several slight variants, and natural granite are the two materials mostly used after cast iron.

Polymer concrete (also known as mineral casting, in particular in Germany) is a material composed by a polymeric resin, usually epoxy based, filled with different

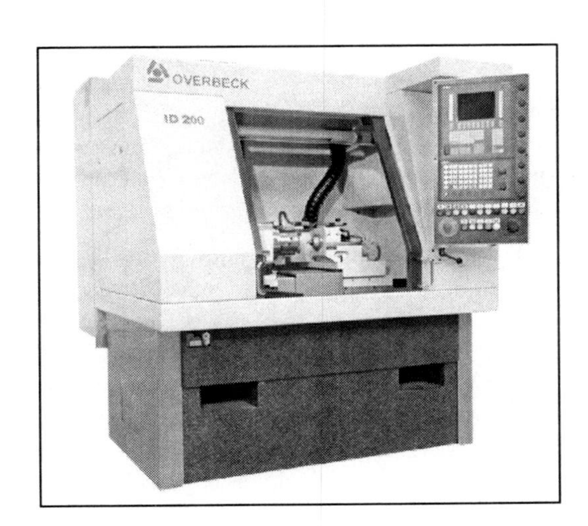

Fig. 8.8 Cylindrical grinding machine with pure granite bed, by Overbeck®-Danobat

sized stone particles in specific percentages. Variations in this distribution lead to different material denominations, such as the Granitan®, registered by the company Studer®. The advantages of polymer concrete in comparison with cast iron are:

- Technically, a much better damping ratio, a better thermal response and similar or better stiffness achievable in the parts. Table 8.1 presents the comparative values between polymer concrete, cast iron and steel.
- From the manufacturing side, a proper design of the structural parts allows to include into the structural elements components such as guideways, or fluid and electric conductions, obtaining better assembly and set-up times.

Natural granite is less extended than polymer concrete, but in the last years it has grown in applications, especially in very high precision machines. The main virtue supporting its use is the stability along time, besides the mechanical characteristics achievable. An example of a cylindrical high precision machine bed made by granite is presented in Fig. 8.8 and in Chap. 6, Fig. 6.15a.

8.2.3 Main Components

This section will focus on the components that have a significant influence on the machine behaviour, in particular driving systems and guideways.

8.2.3.1 Driving Systems

Since the strong development of electronics, electric motors and controls, the most expanded driving system in machine tools and, consequently in grinding, has been that composed by a ball screw driven, by means of a belt transmission directly operated by an electric motor. This is the most standard configuration, applied in horizontal and vertical axes, to drive the workpiece and wheelhead.

Nevertheless, this basic configuration has experienced several significant developments, oriented to improve the performance of the system in concordance with the market increasing demands. Beyond the general improvements achieved, in terms of higher speeds available, obtained through new lower-friction materials, surface coatings applied and new ball screw designs, the most remarkable concerning grinding process requirements, are:

- Hydrostatic ball screws. One of the disadvantages of ball screws concerning their use in grinding is the metallic contact between the two moving parts connected by the ball screw. On the one side, this is because any geometric or surface error in the system can be directly translated to the contact between part and grinding wheel, and finally "copied" into the ground surface. On the other side, this is due to the absolute lack of damping introduced by that metallic contact. Hydrostatic technology has been developed for ball screws using the experience and principles applied over the years in hydrostatic guides. The

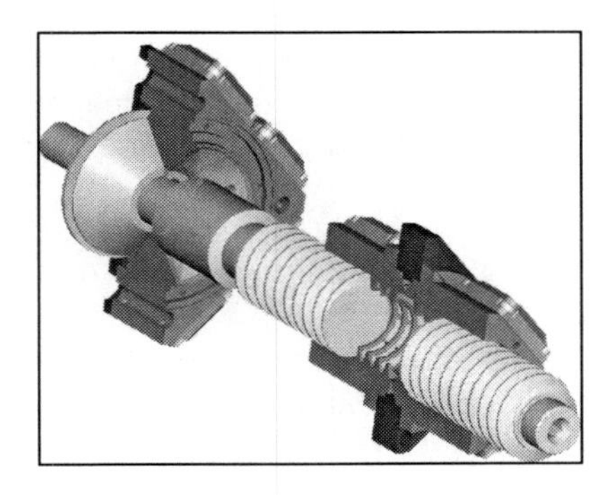

Fig. 8.9 Scheme of hydrostatic screw (company Hyprostatik®)

balls between screw and nut are replaced by a pressurised oil film, avoiding the metallic contact between both elements, as shown in Fig. 8.9.

Hydrostatic ball screws are mainly applied in the grinding of eccentric parts, like cams and crankshafts. In these eccentric parts the process is performed by means of the synchronisation of the part rotation with the forward and backward displacement of the grinding wheel. Hydrostatics improves the quality and smoothness of the grinding wheel movement and also presents a significantly better life for the system, due to the lack of contact wear.

- Isolation of the driven component from the ball screw system in the traverse axes. This solution is performed to avoid the transmission of perturbances from the grinding wheel side to the workpiece or vice versa. By means of this system, the axial performance of the ball screw and the whole system remains invariable, but the traverse directions are decoupled. There are several different solutions, by means of sliding units or fluid chambers, the solution patented by Toyoda® being one of the most advanced.

Linear motors were dedicated in their beginnings to machine tools in high speed milling, supported by their ability to reach significantly higher speeds and accelerations, compared to traditional transmissions. The evolution of linear motors towards smaller sizes, more efficient constructions and ironless motors which minimise the cogging effect, have opened their application range, and in the last years many grinding machine manufacturers are selecting them for their high

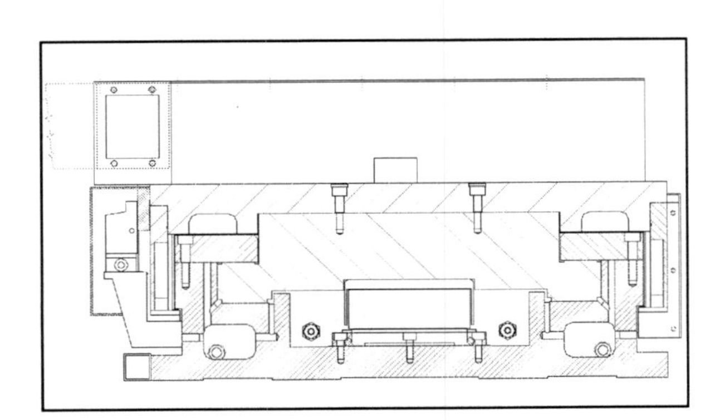

Fig. 8.10 High precision driving solution: hydrostatic guideways plus a linear motor

precision applications, combining them with hydrostatic guides to configure ultra-high precision machines without any mechanical contact and high dynamic response machines, as presented in Fig. 8.10.

8.2.3.2 Guideways

The demands for guideways in grinding machines are: smooth movement, high repeatability and positioning accuracy and providing of static stiffness and damping for a stable, vibration free process.

Historically, slide guideways have been the most widely applied system. The advantages of slide guideways are the large contact surface, that provides very good stiffness and good damping. Hand scraped surfaces and special low friction materials such as Turcite® are among the machine elements that have been the subject of maximum care for the manufacturers. On the other side, their main disadvantage is their relatively high friction that penalises the positioning accuracy and repeatability, in particular in reversing movements due to stick-slip.

Nowadays, slide guideways are still the most used system, although the competitors are increasing.

Commercial linear guideways have gained their market share in the last 10 years, due to the great efforts of the manufacturers towards the minimisation of their limitations. The main advantages of these systems are the low friction that allows good positioning and repeatability, even in short stroke and reversal movements, the good stiffness achievable and the easy assembly and maintenance operations. On the negative side, the low friction and purely metallic contact between parts rebounds in very poor damping values.

Several configurations of linear guideways are being applied, form the compact recirculating units applied in general-purpose cylindrical and surface machines, to

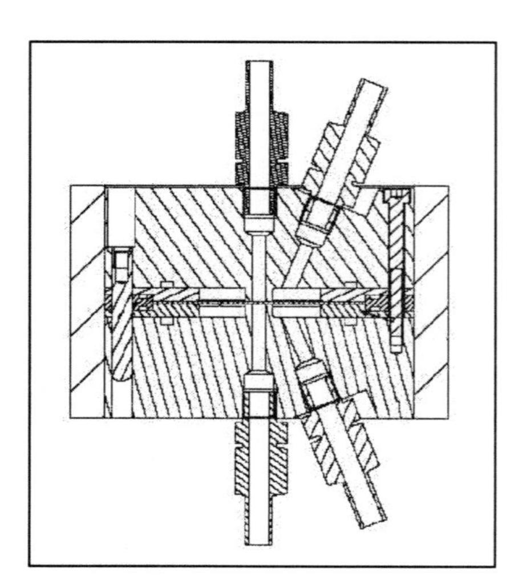

Fig. 8.11 Self-compensating membrane valve

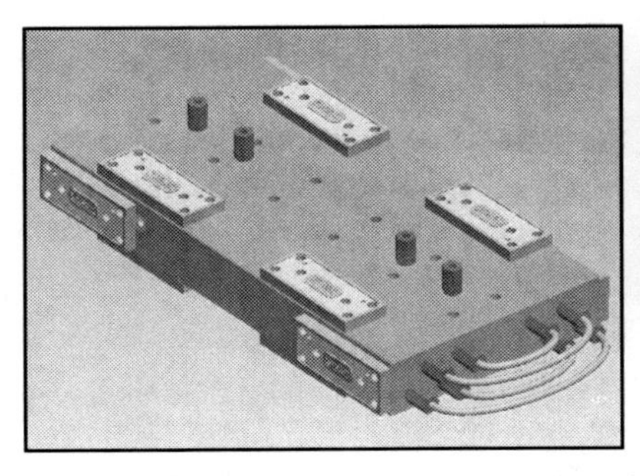

Fig. 8.12 Hydrostatic self-compensated coupled bearings

the needle cages applied in high precision machines, centreless machines, dressing units and heavy-duty creep feed surface grinding machines.

Finally, non-contact systems are applied in high precision machines. Hydrostatic or hydrodynamic systems collect the benefit of the other systems, in terms of very smooth and accurate positioning, high stiffness and high damping, and long operating life, with the sole disadvantage of the high cost of manufacturing and operation.

Hydrostatic systems are the subject of much research in several ways, which searches for more controllable, more flexible, adaptable and easier to manufacture elements. Among the various research lines are the following:

- *Self-compensating valves.* These systems replace the typical capillary restrictors that provide a fixed pressure drop, by valves that are able to absorb and compensate variations in the pocket pressure, stabilising the system for the new conditions. The scheme of these valves can be seen in Fig. 8.11.
- *Self-compensating coupled bearings*. In this type, coupled bearings (on one side and the opposite side) are connected to each other, allowing the system compensation and avoiding the use of any external elements, such as capillaries, to generate the necessary pressure drop. This solution is much easier to manufacture and set up, requiring also lower maintenance. A scheme and a prototype of this type of systems are presented in Fig. 8.12.

Very high precision machines and also heavy weight machines configure the main market of hydrostatic guideways.

8.2.4 Wheel Dressing Systems

Wheel dressing is the process that regenerates grinding wheel geometry and topography, removing ground material chips that penetrate in the cavities of the porous bonding, removing also the most worn abrasive grits and sharpening the rest of the grits, renewing the cutting capabilities of the wheel.

8.2.4.1 Conventional Dressing

This process is performed by means of the interaction between diamond grits with the grinding wheel surface, grits and bonding. These diamond grits are structured in different distributions and the interaction is performed by different strategies and devices, configuring the various dressing systems existing in the market:

- The original and most basic dressing is the single-point process, performed by a simple diamond tip. It can provide a very accurate profiling of the wheel, and is the simplest system, not requiring any special device. On the negative side, the dresser has a short life due to the small size of the tip.

  Variants of this single-point dresser are the long diamond blade and needle diamond dresser tools. Both are also simple static elements, and their difference in relation to the single-point is that they present several diamond grits, or a bigger diamond plate, providing in both cases a longer life for the dresser.

  Single-point dressers are nowadays used mostly in very cheap, simple, general-purpose machines or to obtain complex profiles in low to medium production applications.
- The roller disc dresser (Fig. 8.13). A metallic disc with diamond grits plated on it. This is the next step in terms of productivity and it is necessary for dressing harder abrasives such as CBN. Besides the pure advantage of the longer life provided by the higher surface of the diamond, the roller allows variants in terms of dressing strategies, playing with the disc diameter and the combination of wheel and dresser speeds, searching for the optimum condition for each application.

  Concerning the configuration of the system, the dresser disc is a simple device conformed by a motor driving a precise spindle that holds the disc. The whole dresser is usually fixed in the workpiece side of the machine (workhead or tailstock body in case of cylindrical grinding and table in surface grinding) to compensate any geometric or thermal error between the wheel side and workpiece side.

Fig. 8.13 Disc dresser. **a** Vertical machine. **b** Disc integrated in workhead

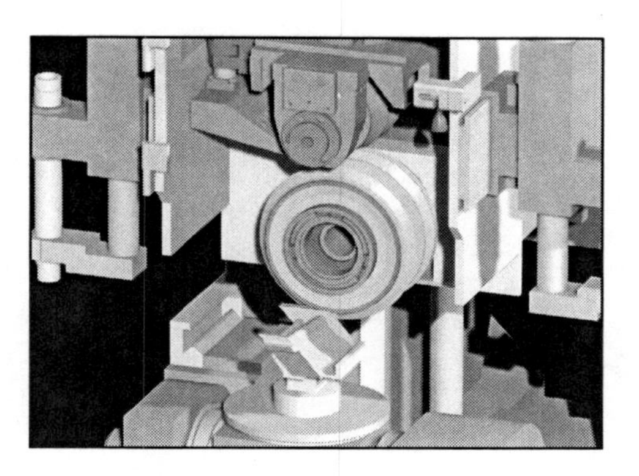

Fig. 8.14 Creep feed grinding with dressing roll tool

One recent tendency in cylindrical grinding is the use of the headstock spindle to hold the dresser disc, as presented in Fig. 8.13b.

- Some high removal rate processes, such as creep feed grinding, or processes with complex wheel profiles require the use of diamond roller dressers. In these devices, the diamond roller profile is shaped with the profile that must be obtained in the grinding wheel and the dressing is performed in a plunge process, as shown in Fig. 8.14.
- A special process that applies the diamond roller is continuous dressing, performed in very high material removal rate grinding processes, that requires a continuous sharpening of the grinding wheel surface to ensure the maintenance of abrasive capabilities. Continuous dressing maintains constant pressure between the dressing roll and the grinding wheel, which is adjustable to ensure the proper area of the wheel is exposed during each revolution. This is a process called microprofiling, as only a micron or less of wheel is removed with each revolution. This ensures that fresh, sharp abrasive crystals are always properly supported in the bonding material of the wheel.

8.2.4.2 Non-conventional Dressing Processes

The development of new grinding wheel compounds with superabrasives to grind hard and difficult to cut materials has led to the development of new dressing technologies that substitute the conventional diamond dresser by alternate principles, by means of electrochemical and laser tools. Most of these processes are still in research stage, but some few applications are present in industry.

ELID. Electrolytic in-process Dressing

This is an electrolytic method that was introduced in 1985 by Murata [8] and optimised to its current configuration by Ohmori [9]. Grinding wheel bonding removal is produced by means of a chemical reaction chain that occurs due to the

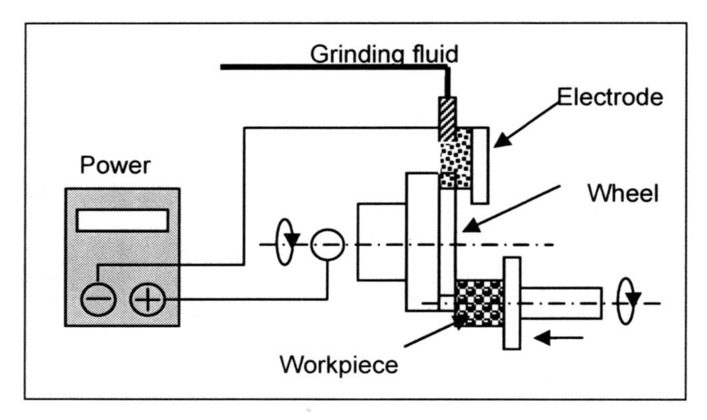

Fig. 8.15 Principle of ELID grinding

application of a voltage between the grinding wheel and one electrode, following the scheme presented in Fig. 8.15.

The applied voltage between wheel and electrode generates chemical reactions between them that depend on the bonding material that must be conductive. The fine control of the electric parameters allows a very accurate control of the material removed from the wheel and, consequently, a very precise dressing. On the other hand, it is a low removal rate method, limiting the application to fine grain wheels or to be combined with other dressing methods.

These two characteristics, high precision and low removal capabilities, have oriented the application of ELID method to superfinishing operations, obtaining mirror-finishing surfaces. Industrial examples of this technology are superfinishing and ultraprecision grinders of manufacturers such as Jung®, Toyoda Koki® and Nachi Fujikoshi®. Reportedly [12], very good results can be obtained in the grinding of brittle materials, such as glass, ceramics, hardened steels, and in some cases substituting final lapping or polishing operations.

EDD, Elecro-discharge Dressing

This is a widely known non-conventional dressing process for superabrasive wheels. Wheel material removal occurs very much as in the EDM process: electrical discharges occur within a locally ionised dielectric medium between the electrode and the conductive material on the wheel surface (either a metallic bond material, or a work metal loaded into wheel pores), as shown in Fig. 8.16. Local temperatures may rise up to several thousand degrees, removing material in the form of craters. Dielectric ionisation is achieved using a generator that applies a voltage (known as open-circuit voltage) between the wheel and workpiece [4]. Discharge duration is in the order of tenths or hundredths of microseconds. An electrode can be made from copper or graphite and the dielectric can be the same emulsion used as the grinding fluid, simplifying the application of the technology.

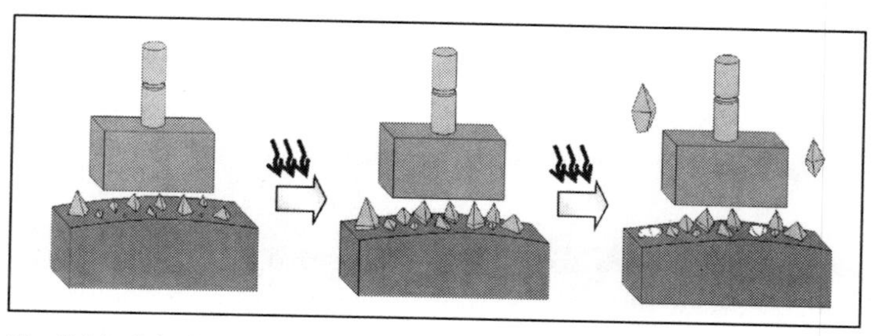

Fig. 8.16 Principle of EDD dressing

The material removing capability is limited and the process is being used for fine grain wheels and superfinishing processes.

The latest research activities [10, 11] are oriented to the truing-dressing of metallic bonded CBN wheels, providing high removal rates and improving the wheel life of conventional high removal wheels such as electroplated CBN ones. In this application, the EDD process is also used to prepare the grinding wheel for its use, eliminating the run-out of the wheel, providing a very accurate surface.

ECDD, Electro-contact Discharge Dressing

This is a truing-dressing method for metallic bonded wheels, introduced by Y. A. Pachalin in 1987. The application of a voltage generates discharges between the wheel and the chips of the electrode in continuous contact whit the wheel (Fig. 8.17). These discharges generate a thermal process that removes bonding material in the wheel.

Several authors [1] have demonstrated the efficiency of the method for small grain size wheels, obtaining significant reduction in cutting forces and the thermal

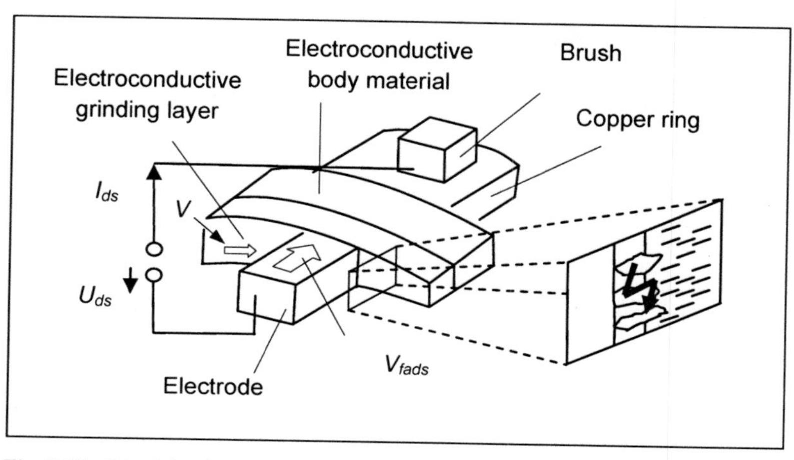

Fig. 8.17 Principle of ECDD, after [1]

affection of the workpiece surface. Ceramic grinding is one of the main fields of application for this technology.

Laser Assisted Dressing

This is a new technique in which a laser beam is proposed as the non-contact thermal dressing tools. Several researchers have studied the process applied for vitrified and resin bonded conventional wheels [5, 13] and also superabrasive metal-bonded diamond wheels [2, 3, 6].

The basic principle of the process is supported in the melting or vaporisation of the bonding material by the heat generated by the laser beam (Fig. 8.18), resulting in the elimination of bonding material.

The use of a pulsed laser with a controlled operation prevents the thermal damage of grains and the remaining wheel bonding, and usually the system is assisted by an air jet to remove the melted material.

Like other alternative dressing methods, laser is applied only to very specific processes such as ceramics ultra precision and super finishing grinding.

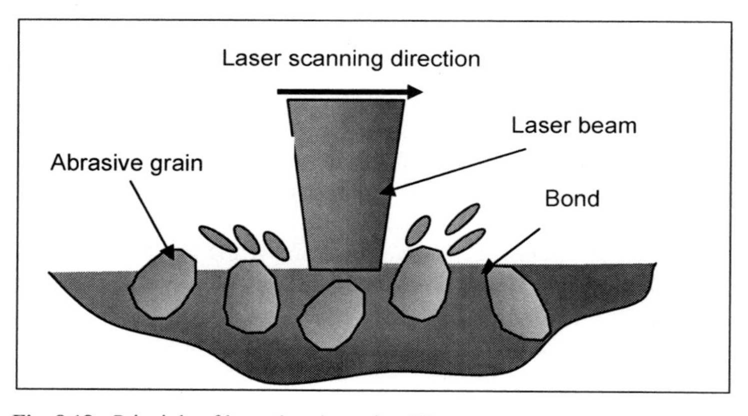

Fig. 8.18 Principle of laser dressing, after [3]

8.2.5 Process Lubrication and Cooling

The grinding process is the machining process where the cutting fluid is more important, due to the high heat quantities generated. The cutting fluid must provide lubrication and cooling to the process for correct process behaviour and to avoid thermal damages to the workpiece.

Most of the applications and machines are equipped with coolant systems that provide high volumes of cutting fluid (neat oil or emulsion) applied at low pressures by very simple nozzles, assuming that the high fluid volume is enough to cool the process.

Fig. 8.19 VIPER grinding set-up, by Makino®

Nevertheless, many approaches have been done to design more advanced and efficient systems that optimise the use of fluid, studying the optimum combination of volume, pressure and nozzle design to reach to the contact area between grinding wheel and workpiece. Most of these approaches have been induced by grinding processes with particularly critical coolant requirements, due to the severe process conditions.

This is the case of high speed grinding with CBN wheels. This process collects three particularities: a high tangential speed, up to 150 m/s in industrial applications, that creates an air film around the grinding wheel, which is a barrier for the fluid to reach the contact area, the high heat generated by the grinding process, and the high cost of the wheels that gives importance to the grinding wheel cleaning to minimise the number of dressings, increasing life. Due to these particularities, high speed CBN grinding machines are usually equipped with high pressure coolant systems with special very precise nozzles.

With these systems coolant is applied to clean the grinding wheel, penetrating in the cavities, removing workpiece clogged chips, and also is able to reach the contact area between the workpiece and the grinding wheel.

Creep feed grinding is also a process with high coolant demands. While most machines apply high volumes of coolant, Rolls Royce® developed, and patented, a technology, called "Viper" that optimises coolant by means of a controlled adjustable nozzle that ensures the application of the coolant exactly in the correct point, in each condition and for any grinding wheel and workpiece configuration. Machine manufacturers such as Makino® and Bridgeport® apply this system, by means of a license form Rolls Royce®. The system can be seen in Fig. 8.19.

8.2.6 Integrated Measuring Devices

In cylindrical grinding, measurement devices based on touch probes have been used for a long time to control the process and stop the cycle once the controlled

Fig. 8.20 Laser measurement in tip grinding of turbine blades, by Danobat®

measure reaches the programmed values. There are many commercial systems available in the market, manufactured by specialised companies.

Beyond these applications, that usually present some limitations, the most important is the short range of measurement of each sensor, that forces the system to use different sensors when different diameters must be measured, grinding machine manufacturers are developing special devices, for special applications and for wider range measurements. Market higher production and efficiency requirements have also encouraged these developments.

The following are some of the most representative special measurement devices developed in the last years by grinding machine manufacturers, providing them in the associated applications:

- Laser based measurement for airplanes turbine blade tip grinding. In this process, represented in Fig. 8.20, the workpiece must rotate at operational speed (around 3,000 rpm) and each blade must be measured in each lap, to control the outer diameter of the workpiece. For this purpose, machine tool manufacturers, such as Danobat®, have developed non-contact laser systems, able to perform this precise and high speed process.
- Measurement of crankshaft journals by means of a tracking interpolation device. Most crankshaft grinding specialised manufacturers have opted by grinding journals and pins in the same setup, performing the journals grinding by interpolation rotation with wheel head translation. Measuring of the journals diameter is a complex operation in this configuration, and has been successfully solved by manufacturers such as Naxos-Union® by means of a very special device that follows the journal rotation, maintaining a stable relative position of the probe on the journal.
- Multi-diameter cylindrical parts measurement. These systems, developed by several grinding machine manufacturers, allow the measurement of diameters, roundness and even axial measurement, which permits taper measurement, in multi-diameter parts, avoiding the use of one element for each diameter, as it has been done for many years, increasing machine operability, saving space and gaining flexibility.

Fig. 8.21 Multi-diameter measuring device, by Danobat®

These systems, as presented in Fig. 8.21, are based on two touch probes whose position is controlled by the relative movement of two axes, obtained by different solutions, depending on each manufacturer. The control and mechanics of the systems must be very accurate and stable, ensuring the repeatability of measurements. Latest developments include non-contact technologies based on linear motors and aerostatic guides, special thermally stable materials and temperature control, following the rules applied in coordinate measurement machines.

8.3 Special Grinding Processes

Most of the systems and technologies described in this chapter are applied to any type of grinding machine and process, except some particularly mentioned cases. Beyond the general-purpose machines and the evolutionary developments, there are some processes and machines that present interesting particularities and have gained an important place in the market, as new technologies that have introduced productive advantages based on new ideas. In some cases, these new processes have become the "signature technology" of their inventor companies.

8.3.1 Peel Grinding–Quick Point

Peel grinding can be considered as the answer of grinding manufacturers to the "interference" of hard turning in historical grinding applications, in particular in strategic sectors, such as the automotive.

Peel grinders, as shown in Fig. 8.22, basically collect the flexibility of turning with the precision and quality of grinding. Supported in the use of a very narrow

Fig. 8.22 Peel grinding, by Danobat®

CBN wheel, working at very high speed (up to 200 m/s), and high workpiece rotation speeds (up to 10,000 rpm), is able to grind, in one clamping, complex shapes, multi-diameter parts, faces, grooves, tapers, or any other feature.

Due to the small contact length between wheel and workpiece and the high operational speeds, cutting forces and thermal effects are dramatically decreased, the process is optimum for hardened parts. Besides this, the low forces simplify the workpiece clamping, eliminating the need of special chucking devices and, specially, the need of steady rests in slender parts. Coolant application is also simpler, due to the small contact region to be covered.

Because of these advantages, car transmission shafts and cutting tools (drills, mills), even carbide ones, are some of the most successful applications of peel grinding.

Peel grinders must consider some particular features to guarantee their competitiveness. The effectiveness and productivity of the high speed CBN grinding wheel must be ensured by a proper wheel cleaning, by means of high pressure nozzles, and the wheel spindle must be very accurately and continuously balanced. The high workpiece rotation speed requires a precise spindle, preferably an electrospindle. And the machine structure must be dynamically stable.

One step ahead in the development of peel grinding was presented and patented in 1985 by Junker®, under the denomination of Quickpoint™. This process is a variation of peel grinding based on a tilting wheel head, which converts, by means of a very small inclination of the wheel, the contact between workpiece and wheel from a line to a single-point. Therefore, the advantages of peel grinding in terms of flexibility, reduction of forces and the thermal effects and increase of wheel life are maximised.

8.3.2 Speed Stroke Grinding

The *speed stroke grinding* process was first investigated by Japanese researchers (Inasaki, Yuji, and Akinori), in the late 1980s and oriented to the machining of brittle materials, in particular ceramics, diminishing thermal affection in a high material removal process.

The principle of the process is a combination of very high table speed with very low grinding depth. Table speed reached 80–100 m/min in the early years and up to 200–250 m/min nowadays, due to the application of high dynamics driving systems, based on linear motors. The depth of cut is maintained around 1 micron, or even below.

This combination of very high speed and very low cutting depth generates a high volume of material removed with very low forces, very specific energy, and consequently very low thermal affection to the workpiece.

These process characteristics opened the range of application of speed stroke from ceramics to other parts that required high material removal and were sensible to thermal affection, presenting an alternative to creep feed grinding. The German company Blohm® collaborated with RWTH Aachen University to develop a process and machine for the speed stroke grinding of titanium alloys for aeronautic applications. This machine presents some interesting solutions to achieve table speeds up to 200 m/min and accelerations up to 50 m/s^2 compensating the high dynamic loads induced in the machine structure. The avoidance of the effect that dynamic forces generate on the machine is the major challenge for machine manufacturers. For this purpose different speed and acceleration-deceleration profiles and strategies have been investigated.

8.3.3 Creep Feed Grinding

Creep feed grinding is a surface grinding process characterised by a very high infeed rate or depth of cut (in the range between 0.5 to 30 mm) and low feed rates (0.1–50 mm/s). The process is performed in very few passes, even one single pass in some applications, in contrast to conventional reciprocating grinding, characterised by multiple very low infeed passes at higher feed speeds.

With the special combination of parameters of creep feed grinding, the chip thickness and therefore the cutting force for each grain are smaller which allows those high removal rates with less affection to the wheel integrity, since the grains are easily held in the bonding. The contact length is much higher, but the number of cutting edges (grains) involved simultaneously is much higher and because the feed rate is much lower, the achieved surface roughness is much better than in reciprocating grinding.

On the other hand, the total cutting force is much higher and the thermal effects as well. The thermal negative effect is increased by the large contact length, which makes difficult the application of the cutting fluid to the whole contact zone. The chip length is also larger than the generated in reciprocating grinding.

These characteristics define the design of the machines for creep feed grinding. Higher loads derive from a very stiff machine and components (guides, ball screws and spindles) design. Thermal effects recommend the measurement and control of the deflections, especially the axial deflection of the wheel spindle. The coolant requirement is more critical and its application is more complex, requiring special high-pressure nozzles with very accurate design to cover the maximum of the contact zone. The correct application of the coolant jet is critical, thus adjustable nozzles are also recommended. The previously mentioned Viper is one of these special systems. Conditioning of the grinding wheel is also very important, and continuous dressing is very usual in creep feed processes.

Creep feed typical applications are large batch parts with special complex profiles. Creep feed provides high productivity, very good surface roughness and very precise shapes. The aeronautics, energy generation and automotive industry are target sectors for creep feed machine manufacturers. In aeronautics and energy generation, blades and rotor parts with complex slots and profiles in low machinability materials and in automotive parts such as steering racks are typical parts in which creep feed is much more productive than reciprocating grinding.

8.3.4 High Efficiency Deep Grinding

HEDG, *high efficiency deep grinding* (or *high performance grinding*) is an improvement of creep feed grinding that introduces high cutting speed and higher feed rates (closer to reciprocating grinding, see Table 8.2).

HEDG provides the advantages of creep feed grinding, in terms of surface quality and profile accuracy, and reduces some of its limitations with a better thermal balance and a significantly increased material removal rate.

Machine requirements and applications are analogous to those related to creep feed grinding.

Table 8.2 Comparative parameters between conventional surface grinding, creep feed and HEDG

Process	Infeed	Feed rate	Cutting speed	Specific material removal rate
Reciprocating grinding	0.001–0.05 mm	1–30 m/min	20–60 m/s	0.1–10 $mm^2/mm/s$
Creep feed grinding	0.1–30 mm	0.05–0.5 m/min	20–60 m/s	0.1–15 $mm^2/mm/s$
High efficiency deep grinding	0.1–30 mm	0.5–10 m/min	80–200 m/s	50–2,000 $mm^2/mm/s$

8.4 Machine and Process Monitoring and Control

Monitoring and control strategies become a very powerful tool to ensure an optimum performance in grinding process both from the machine and process sides. The grinding process is characterised by a high number of cutting edges interacting with the workpiece, in a non-uniform distribution, making difficult a pure modelling to understand and control the process. This particularity of grinding gives more relevance to monitoring as a practical tool to understand and analyse the process.

Current and future monitoring systems are characterised by their high processing capability, by means of the integration of more powerful hardware and software. Faster and more efficient DPs, the miniaturisation of microelectronic components, and the development of hybrid and more accessible components, together with the extensive use of information technologies has lead to true adaptive, self-learning, knowledge based systems.

The latest monitoring systems integrate sensors of different natures to collect information about the workpiece, the machine, the grinding wheel, the dressing system and the process condition and behaviour. Force, power, temperature, vibrations and acoustic emission sensors are among the most applied, in single and multi-sensor monitoring strategies.

Great importance has to be conceded to the new generation "open numerical controls" (see Sect. 5.8), which offer the possibility to access the internal signals of the drives, allowing the acquisition of very valuable signals without any external sensor.

8.4.1 Monitored Parameters and Applied Sensors

Direct measurement of the workpiece geometry is the first approach to grinding monitoring and control. The use of contact gauges that measure continuously the geometry of the part that is being ground and defines the end of the cycle when the programmed dimension has been achieved is widely extended, in particular in external cylindrical grinding, and has evolved to special devices and technologies, described in Sect. 8.3.

Non-contact measurement of the workpiece geometry has also been investigated, by means of laser or optical sensors, but all these systems come up against the coolant fluid in the measuring area, generating non-controllable errors that make difficult their practical application.

Monitoring of the process behaviour is made by means of force, power, acoustic emission, vibration (accelerometers) and thermal sensors mainly. These are non-direct measurement strategies to detect deviations that can have different origins such as wheel wear, coolant fluid deficiencies, workpiece geometrical variations, programming errors or dressing process variations.

Grinding forces measurement is based on the determination of a displacement. The first approaches used strain gauge systems, located into the force flux. The main limitation of these systems is the reduction of stiffness in the machine, necessarily induced by the implementation of the measurement device.

This limitation was partially overcome with the apparition of piezoelectric sensors, which present a much higher stiffness and measurement range. These sensors are widely used for research purposes in many different configurations, from commercial load cells to homemade configurations. But their use in industrial applications is not extensive due to the complications and cost of implementation.

In industry, force measurement is performed by means of grinding wheel spindle power consumption measurement. Although this is not a direct measurement and the ratio between tangential and normal forces depends on each process configuration, the measurement of power is a valuable parameter to monitor the evolution of a process and detect any perturbance or variation occurred. With the inclusion of digital drives and open numerical controls the use of external sensors is unnecessary, simplifying the system. Collision detection is also performed by means of this technique.

Acoustic emission (AE) sensors have a broad application in grinding process monitoring. In industry they are being applied to detect the contact between the wheel and workpiece (gap detection), collision detection and dressing and grinding process control to minimise cycle times. This is one of the most successful applications, being used by most of the companies that supply monitoring systems, such as Marposs®, GT Electronics® and Dittel®. Nevertheless, AE sensors present some limitations, and especially, their location in machine must be very carefully studied, to avoid signal disturbances generated by non-desired sources such as bearings.

Accelerometers are applied to detect vibrations from different sources: the simplest and widely used for wheel automatic balancing, extensively applied in high speed wheels such as CBN. Chatter vibrations also are detected by means of accelerometers since they are associated to frequencies that are more in the range of accelerometers than AE sensors.

Thermal balance is a fundamental aspect in grinding process and the control of the heat transferred into the workpiece is one of the most critical limitations for process productivity, since workpiece thermal affection can generate unacceptable microcracks and residual stresses is the ground surface. Heat conduction is the main principle used for these measurements, performed by means of thermocouples. This technique is widely used in research activities, providing essential information to understand the physics of grinding process. Nevertheless, the industrial application is almost non-existent, due to the difficulties to implement an accurate and reliable system under industrial conditions.

8.4.2 Control Strategies

Adaptive control systems are implemented in grinding machines for the on-line regulation of the set-up parameters and the reaction against any contingency, stabilising the process towards the target quality or productivity values. The first investigations in adaptive control strategies were carried out in the 1970s and 1980s, and the great advance in hardware (sensors, microprocessors) and the new open CNCs have allowed an important development in the last years. Two main AC strategies can be distinguished:

- Adaptive control constraint (ACC). The control signal (force, power, AE, *etc.*) is set up under a limit value and maintained constant bellow that value by means of on-line regulation of machine parameters (speed, feed, *etc.*).
- Adaptive control optimization (ACO). The process is driven towards the optimum selection of parameters. The optimisation criteria can be the final workpiece result or grinding time or other process output. This is a step behind ACC because it provides for the modification of the pre-defined control parameters.

Successful strategies of AC control of the normal cutting force that provides a significant optimisation of the whole grinding cycle, increasing as well the stability of the dimensional results in the final workpiece, can be seen in many research approaches.

Acknowledgements Thanks are addressed to all companies cited in the pictures.

References

[1] Denkena B, Becker JC, Van der Meer M (2004) Potential of The Electro Contact Discharge Dressing Method in Truing and Sharpening Super Abrasive Grinding Wheels. Key Engineering Materials, 257–258:353–358

[2] Hoffmeister HW, Timmer JH (2000) Laser Conditioning of Superabrasive Grinding Wheels. Industrial Diamond Review, 60/7:209–218

[3] Hosokawa A, Ueda T, Yunoki T (2006) Laser Dressing of Metal Bonded Diamond Wheel. Annals of the CIRP, 55/1:333–336

[4] Iwai M, Ichinose M, Qun H.B, Takeuchi K, Uematsu T (2001) Suzuki K, Application of fluid-free EDM to on-machine trueing/dressing for superabrasive grinding wheels, Proc Thirteenth Int Symp for Electro-Machining, ISEM-13:371–380

[5] Jackson MJ, Robinson NB, Khangar A, Moss R (2003) Laser Dressing of Vitrified Aluminium Oxide Grinding Wheels. British Ceramic Transactions, 102/6:238–245

[6] Kang R, Yuan JT, Zhang YP, Ren J.X (2001) Truing of Diamond Wheels by Laser. Key Engineering Materials, 202–203:137–142

[7] Marshall ER, Shaw MC (1952) Forces in Dry Grinding. Trans of ASME, Jan:51–59

[8] Murata R, Okano K, and Tsutsumi C, (1985) Grinding of structural ceramics. Milton C. shaw grinding, Symposium PED, 16: 261–272

[9] Ohmori H, Qian J, and Lin W, (2001) Internal mirror grinding with a metal/metal-resin bonded abrasive wheel. Int J of Machine Tools and Manufacture, 41:193–208)

[10] Ortega N, Sánchez J.A, Aranceta J, Marañon J.A, Maidagan X (2004) Optimisation of grit protrusion in the electro-discharge dressing process of large grit size CBN grinding wheels, J Mat Proc Tech, 149:524–529

[11] Sanchez JA, Ortega N, Lopez de Lacalle LN, Lamikiz A, Marañón JA (2006) Analysis of the electro discharge dressing (EDD) process of large-grit size cBN grinding wheels, Int J Adv Manuf Tech, 29: 688–694

[12] Spanu C, Marinescu I (2002) Effectiveness of ELID grinding and polishing, International Manufacturing Conference, IMTS, Chicago

[13] Westkämper E (1995) Grinding Assisted by Nd:Yag Lasers. Annals of the CIRP, 44/1:317–320

Chapter 9
Wire Electrical Discharge Machines

J. A. Sánchez and N. Ortega

Abstract Amongst the non-conventional machining processes, electrical discharge machining has no doubt gained a deserved reputation and popularity in the second half of the 20th century. Wire electrical discharge machines appeared some 30 years ago, and since then the development of new applications in the field of very hard material precision machining, and consequently, the market share of technology is growing continuously. In this chapter the main aspects related to the technology and equipment for wire electrical discharge machining are addressed: the first research that led to the process as it is known today, machine components and intelligence, industrial and academic research and the latest advances in the field of thin-wire EDM are discussed in the following sections.

9.1 Introduction

The EDM (electrical discharge machining) process is, by far, the most popular amongst the non-conventional machining processes that can be found in industry nowadays. As with other non-conventional processes, the specific removal energy involved in an EDM operation is very high. In other words, partial material removal can only be achieved by using a large amount of energy. In this sense, EDM cannot compete with conventional removal processes such as turning, milling or drilling in terms of removal rates in low hardness materials such as carbon steels, for instance. However, the feature making EDM unique with respect to those conventional processes is that the removal mechanism is not related to

---

J. A. Sánchez and N. Ortega
Department of Mechanical Engineering, University of the Basque Country
Faculty of Engineering of Bilbao, c/Alameda de Urquijo s/n, 48013 Bilbao, Spain
joseantonio.sanchez@ehu.es

L. N. López de Lacalle, A. Lamikiz, *Machine Tools for High Performance Machining*,

mechanical contact between the tool (electrode) and the part. In short, during the EDM process, a series of discrete electrical discharges occur between the electrode and the workpiece (which must obviously be electrically conductive) in a dielectric medium, which can be deionised water or oil depending upon the application. Sparks require the existence of a distance between electrode and workpiece, which is called the gap, and is filled with dielectric. Discharge duration can be variable; however, it can be measured in terms of microseconds. During the application of each discharge local temperature rises by several thousand degrees (ranging between 10,000°C–20,000°C). Consequently, part of the material melts and vaporises generating craters on the surface of the workpiece, which is removed in the form of debris by dielectric flushing. This is the core of the phenomenon involved, although scientists still argue on some points of this explanation. The result is an EDM'ed surface whose roughness depends mainly on electrical parameters, where the craters produced are responsible for a non-directional surface finish, as shown in Fig. 9.1.

The implications of the aforementioned removal mechanism are clear. Part material does not depend upon the mechanical properties of the workpiece such as hardness, brittleness or abrasiveness. Static and dynamic problems related to machining forces disappear. It can be noticed the process works best when conventional machining cannot give an optimum solution. This fact can be illustrated with examples such as the machining of complex tooling with a very smooth surface finish on very hard materials, such as heat-treated tool steels or tungsten carbide for the manufacturing of drawing dies, sets die-punch, injection moulds, etc.; machining of difficult-to-machine alloys such as those used in aerospace applications; machining of very small parts, with a growing demand for micromachining applications, *etc*.

More than 60 years have already gone by since the Russian scientists B. and N. Lazarenko [7], with the help of a young researcher, B. Zolotykh, used the effect of electrical discharge to carry out controlled removal of material from metallic parts. Their work was the basis for the first EDM generator, a simple relaxation-type circuit, which was the core of the first industrial EDM machine: the structure of a milling machine equipped with a pulse generator. Of course, there had been

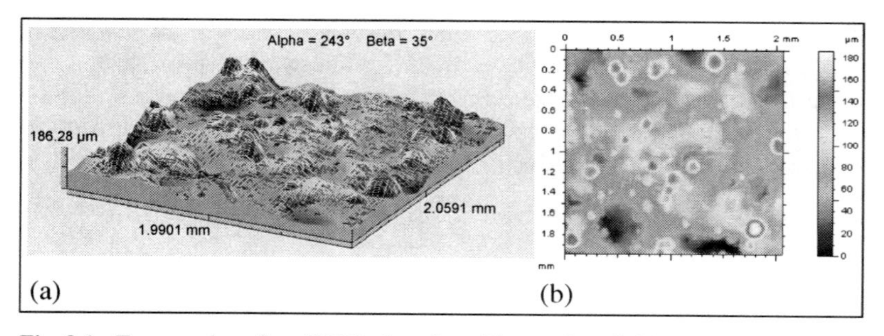

Fig. 9.1 Topography of an EDM'ed surface. The surface finish measured with a contact profilometer is $Sa = 15.8$ µm

previous experiences in the 18th and the 19th centuries, continued during the first half of the 20th; however, the work of the Lazarenkos was the basis of a completely new technology which would open the door to the machining of extremely hard materials. Thus, the sinking electrical discharge machining (SEDM) process was born. According to this principle a complex cavity could be created in a hard metallic material using a shaped electrode, whose form is mirrored in the workpiece. Relaxation-type circuits were soon replaced by transistor-based circuits, which are the basis for modern EDM machines. The introduction of numerical control in the 1980s was critical for the improvement of technology efficiency and ease-of-use, which is widely accepted nowadays among users.

Based on the same principle other techniques have appeared in the last decades, such as electrical discharge grinding (EDG), the electrical discharge dressing/truing of metal-bond grinding wheels, and some others as well as some hybrid processes and machines. However, it was the advent of the wire electrical discharge machining (WEDM) technology that brought about a radically new solution for the cutting of complex geometries in very hard materials. There are references of wire-cut machines from the Soviet Union in the 1960s, but the first industrial machine was the Agie DEM-15 (Fig. 9.2), which was the basis for future developments. But the popularity of the new technology would only be boosted by the numerical control, which allowed off-line programming of wire path and on-line control of the process. Today, the market of WEDM machines is continuously growing and manufacturers are putting most of their effort into this technology [3].

The rate of technological advance and implementation thereof in industrial solutions can only be explained by looking at the collaborative research effort carried out by both the machine manufacturing industry and academic groups over the years. Research results can be found in some prestigious international journals and congresses, among which the International Symposium on Electro Machining (ISEM) deserves a special mention. The ISEM, which has worldwide recognition as a high-level meeting, is held every three years as a common forum for industry

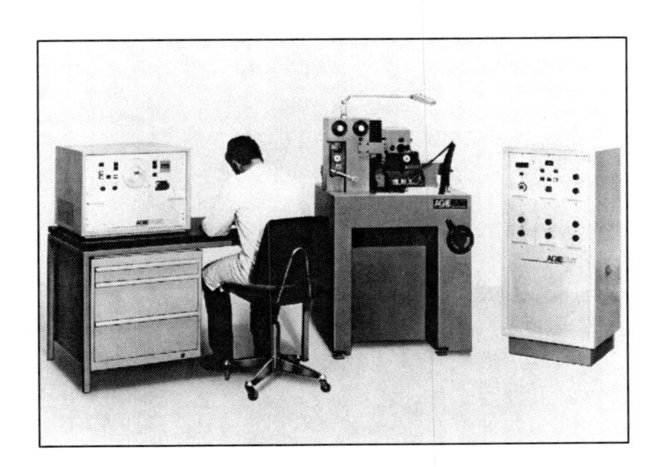

Fig. 9.2 The Agie® DEM-15, the first WEDM machine

Fig. 9.3 A historic meeting: Professor B. Zolotykh (*left*) and Mr. K. Onandia (*right*) during ISEM XIII held in Bilbao in 2001

and researchers. The last venues were Bilbao in 2001 (ISEM XIII), Edinburgh in 2004 (ISEM XIV) and Pittsburgh in 2007 (ISEM XV) and hopefully, ISEM XVI will take place in Shanghai in 2010. Figure 9.3 shows a historic meeting: Professor Zolotykh, one of the fathers of the technology, with K. Onandia, founder in 1952 of ONA-Electroerosion® S.A. during ISEM XIII in Bilbao.

9.2 The WEDM Process

In the WEDM process a small diameter wire is used like an electrode to cut a narrow channel in the work. Wire diameter ranges from a maximum of 0.33 mm to 0.1 mm in conventional WEDM; however, it may be as low as 0.020 mm in micromachining applications (see Sect. 9.6). The workpiece is fed continuously and slowly past the wire to achieve the desired cutting path. Numerical control is used to

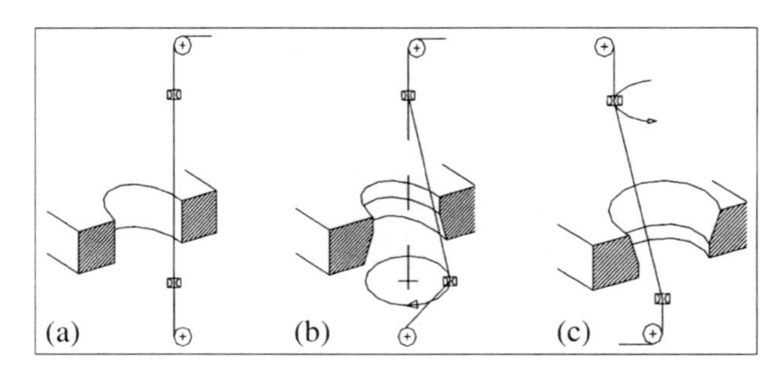

Fig. 9.4 Schema of the WEDM process. **a** Vertical cutting. **b** and **c** Taper cutting

control the relative motion between wire and workpiece during cutting. As it cuts, the wire is continuously advanced between spools to present a constant-diameter electrode to the work, as shown in Fig. 9.4. Therefore, wire wear is not a primary concern, since it is continuously renewed; nevertheless in precision applications and especially when part thickness is high, this is a point to be considered.

As described in Sect. 9.1, discharges occur within a dielectric medium. In most WEDM machines, deionised water is used. A water jet is introduced by the upper and lower nozzles through the gap between wire and workpiece, as shown in Fig. 9.5. Clear flushing is by no means easy, and the difficulty to adequately remove the debris generated and cool the wire electrode increases with part thickness. Therefore, most machine work is done in submerged mode, where the machine tank is completely filled with deionised water.

Precision tooling is probably the best known application field of WEDM. Stamping and high speed stamping, extrusion, wire drawing, amongst others, are processes that benefit from this interesting manufacturing technique. Industrial sectors such as automotive, aerospace and medical are good examples of the technology users. Examples of this application are shown in Fig. 9.6. In the photograph, we can see a blade for the aeronautical industry and a hardened steel part, both machined by WEDM and characterised by their large thickness (up to 400 mm in Fig. 9.6b).

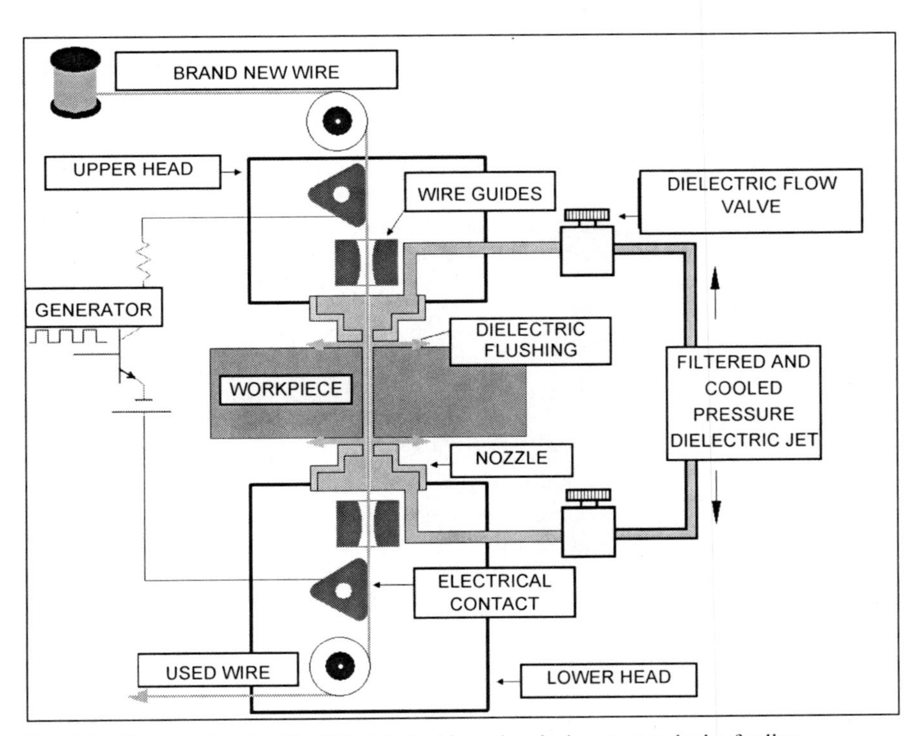

Fig. 9.5 Elements involved in dielectric flushing, electrical contact and wire feeding

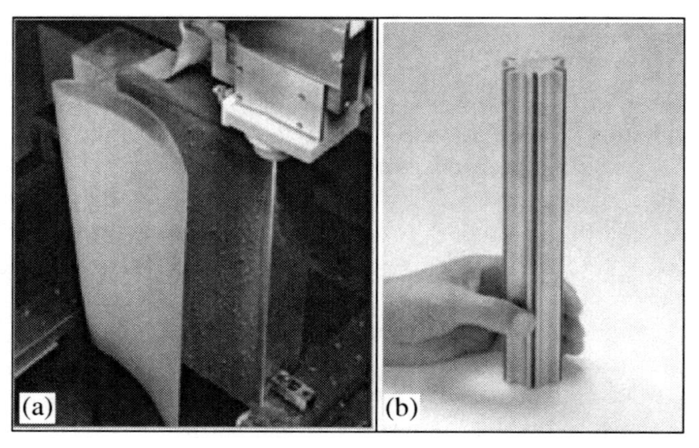

Fig. 9.6 Applications of the WEDM. In this case, parts of high thickness in difficult-to-machine materials. **a** Aeronautical alloy. **b** Hardened steel

9.2.1 Accuracy and Speed

For most users WEDM is a matter of speed and accuracy. Cutting speed is of primary importance during the first cut. Although logically speed should be expressed in terms of length divided by time, this is not a practical measure of material removal rate in WEDM since the latter is strongly dependent upon part thickness. This is the reason why it is commonly accepted for cutting speed to be expressed

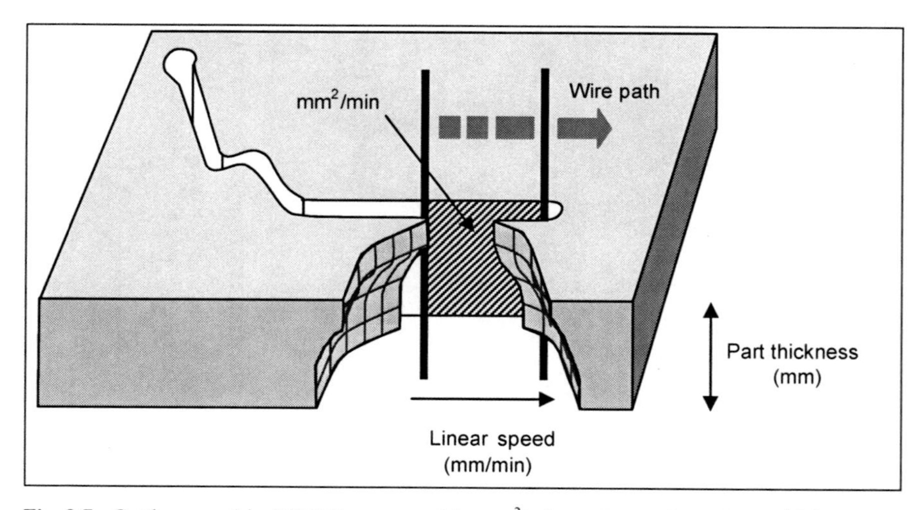

Fig. 9.7 Cutting speed in WEDM, expressed in mm^2/min as the product of part thickness and linear speed

in mm^2/min. The value is obtained by multiplying the linear cutting speed (in mm/min) times part thickness (in mm). Figure 9.7 illustrates this concept.

Probably, it would be even more accurate to define the material removal rate as the volume of part material removed per unit time. This definition includes aspects such as wire diameter and gap width which in fact define the actual quantity of material removed by the process. In practice this concept is sometimes found among scientists and academics though rarely in industry, which directly accepts the dependency of cutting speed with part thickness. This relation has a maximum at a certain value of the thickness, normally at some point within the range 50–100 mm.

The maximum values of cutting speed provided by the most important machine manufacturers in their catalogues can reach even 500 mm^2/min when cutting steel in part thickness below 100 mm. This must be seen as an upper-limit value that can only be achieved under ideal conditions such as very high dielectric pressure, simple part geometry and large diameter coated wire. Ideal conditions are not always possible in the everyday workshop job; therefore, real cutting speed values below this should not surprise the machine user. For instance, for the above conditions, a maximum cutting speed in the first cut could be about 450 mm^2/min if accuracy requirements are not placed and part geometry is very simple. When further trim cuts are programmed, or numerous changes of direction in wire path are present, the first cut must be carried out at a lower speed. It must also be taken into account that those cutting speed values correspond to machines equipped with transistor-type generators enabling much higher values than those provided by previous machines.

Cutting speed is directly related to machine parameters. Mention has already been made of dielectric pressure. Small variations in this variable may induce important reductions in cutting speed. Provided pressure is kept at an optimum value, cutting speed is directly related to discharge current, pulse on/off-time, and that kept for the gap servo. The higher the energy input to the process, the higher the cutting speed. However discharge energy is limited by the wire resistance. If one tries to increase discharge energy over a certain value wire breakage will occur. Wire breakage and process stability are closely related, and both are strongly dependent upon flushing conditions likewise to the presence of debris in the gap that had been impossible to evacuate. It can be easily understood that gap width, which is governed by the servo of gap, also plays a determining role in the evacuation of debris.

When accuracy is demanded trim cuts must be programmed after the first cut, and this latter must be designed taking into account the requirement for further finishing cuts. Trim cuts improve both part tolerances and surface finish. Precision levels are not identical for all the users, therefore machine manufacturers equip their machines with technologies of one, two and even more trim cuts after the first cut, to provide an optimal answer to the balance between productivity and accuracy. For some important machine manufacturers the trend to reduce the number of trim cuts while keeping accuracy is evident, in view of the ever-growing demands of users.

Again, surface finish is related to discharge energy. In short, a high-energy input into a single discharge produces the melting and evaporation of a large crater on the workpiece; furthermore, the superposition of large craters deteriorates surface finish. Therefore, trim cuts are commonly associated with low energy sys-

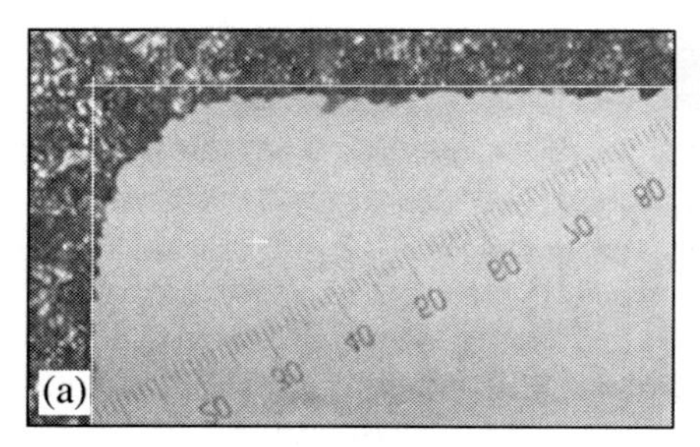

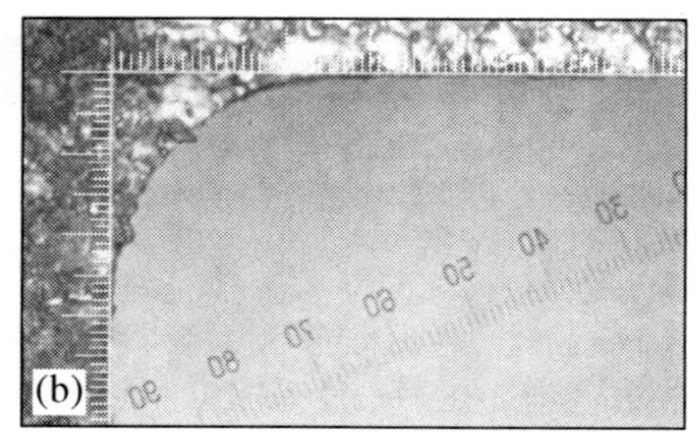

Fig. 9.8 Geometry of the error in corner cutting. **a** It can be observed that after the first cut, the amount of material along the corner is not constant. **b** The situation is clearly improved after trim cuts

tems, i.e., low discharge current and short pulse on-time. Roughness values for the first cut in tool steel are about *Ra* 2.80 μm. This value can be improved down to *Ra* 0.8 μm after two trim cuts. Surface finish is also highly dependent on part material. It is common for machine manufacturers to specify minimum values of surface roughness for hard metal, where using special modules with values below *Ra* 0.10 μm can be achieved.

Different aspects affect precision during WEDM cutting. The wire, subjected by the machine to tensile stress, from a mechanical viewpoint behaves as a beam on which deformation is induced by the forces acting: dielectric pressure, electrostatic force, electrodynamic force and electromagnetic force. Modelling the above system of forces is a classical topic in WEDM research ([1], [2] and many others). Under these forces, the wire suffers flexing and vibration that make it lose its vertical equilibrium position. As a result, wire geometry is not exactly aligned with machine

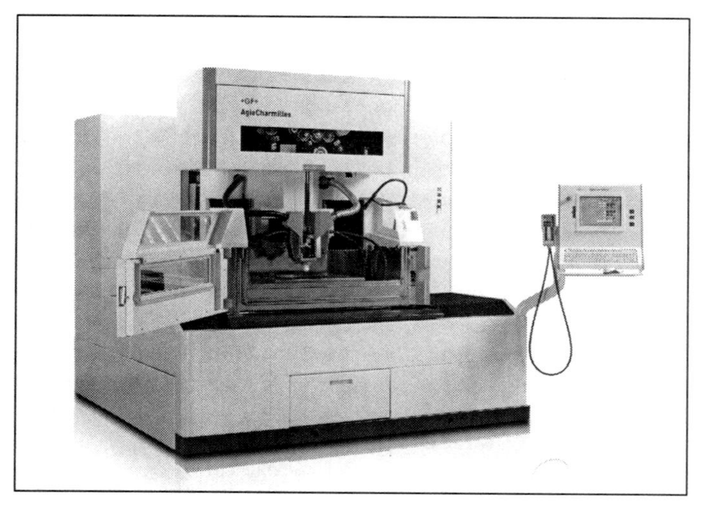

Fig. 9.9 Twin-Wire technology of the Swiss company GF AgieCharmilles®. The photograph shows the FI2050TW machine [20]

guides (which are the points actually programmed for wire path), and errors are transferred to the machined part. Deviations in part accuracy include the verticality of faces, the corner geometry (the so-called "back-wheel effect") and angular errors in taper cutting. As already mentioned, accuracy can be greatly improved using trim cuts where low dielectric pressure and low energy systems are used. For vertical cuts, verticality can be maintained below 5 μm after two trim cuts in a conventional WEDM machine. Figure 9.8 shows the geometry of a corner with internal radius 0.20 mm after the first cut (a) and after two trim cuts (b). The geometry is sounder, the maximum error has been reduced down to 7 μm, and the surface finish has been clearly improved.

Low-diameter wires are commonly used for machining intricate details and small internal corner radius. This type of wire is used in the concept of twin-wire machine developed by the Swiss company GF-AgieCharmilles® [20]. Twin-wire technology incorporates automatic wire change to reduce total machine time and therefore increase productivity. The first cut is carried out using a low cost wire, and then trim cuts are performed with wires with diameters as low as 50 μm. This is a very precise machine, in which excellent roughness, surface finish and dimensional tolerances can be achieved due to the use of thermal stabilisation systems. Figure 9.9 shows a photograph of the FI2050TW model of this company.

9.3 WEDM Machines

A WEDM machine shares many elements in common with other machine tools. Structural elements and CNC, for instance, do not differ too much from those of other manufacturing equipment. However, some systems can only be found in these machines, and thus they will be separately analysed in the following paragraphs. These elements can be identified in Fig. 9.10.

- Wire transport and wire thread devices.
- A working tank.
- A spark generator.
- A filtering system.

Structural components, mechanical transmission, guiding and measuring devices do not differ too much from those used in other machine tools. Machine bead is normally built in iron casting, although proposals of using polymer concrete can be found both in industry and in literature. In machines where precision demands are very strict, ceramic components and thermal stabilisation of the working area can be found.

The guiding system is responsible for the accuracy and straightness of linear axis movement. Linear guides of hardened steel, held to machine bead, are commonly used. This technology increases element wear resistance, although damping is committed. Movement is transmitted using preloaded linear recirculating ball bearings mounted on the moving element.

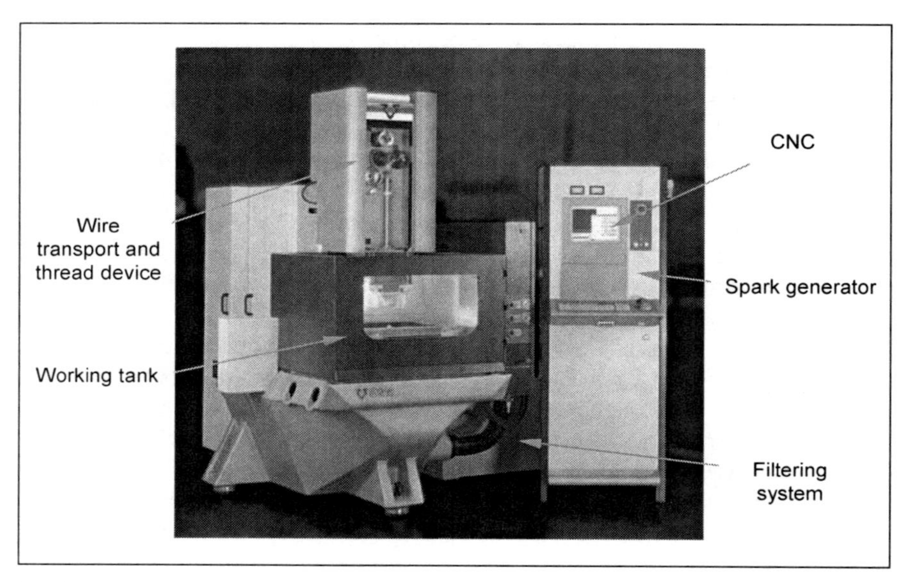

Fig. 9.10 Main systems of a WEDM machine. The photograph shows an ONA® AX300 wire EDM machine [21]

In most machines, linear axis movement is generated by a transmission engine-ball screw. The dynamic response of ball screws has increased dramatically in recent years at the sight of the competence of linear motors. In fact, some manufacturers use linear motors for their linear axes, although this is not the most common option on the market nowadays. Linear motors, due to their extremely high dynamic response, and lack of backlash, provide a good solution for improving flushing (the so-called natural flushing) in the sinking the electrical discharge machining of difficult geometries (for instance, high-depth and low-width slots). A good example of application of linear motors to the WEDM technology is given by the AQ300L machine by Sodick® [22]. Measuring devices involve precision glass scales mounted on the axes, and in some cases encoders.

Mechanics of wire electrodischarge machines are, in general, more complex than those of sinking electrodischarge machines. The machine structure commonly has 5 axes amongst which 4 are interpolated. WEDM machines can be classified on the basis of the maximum admissible workpiece size. Figure 9.11 shows the basic mechanical structure of a WEDM machine.

Workpiece height is critical since stability and cutting speed are dependent on it. Moreover, the WEDM process is the only one on the market capable of precision cutting of hard parts of high thickness (Fig. 9.12). Currently, there are machines capable of machining workpieces with the dimensions $2300 \times 1300 \times 600$ mm and a weight of 10,000 kg. In these large machines, cutting must be carried out in a submerged mode due to fact that the dielectric has a difficult access to the machining area. As far as taper cutting is concerned, and depending on the machine,

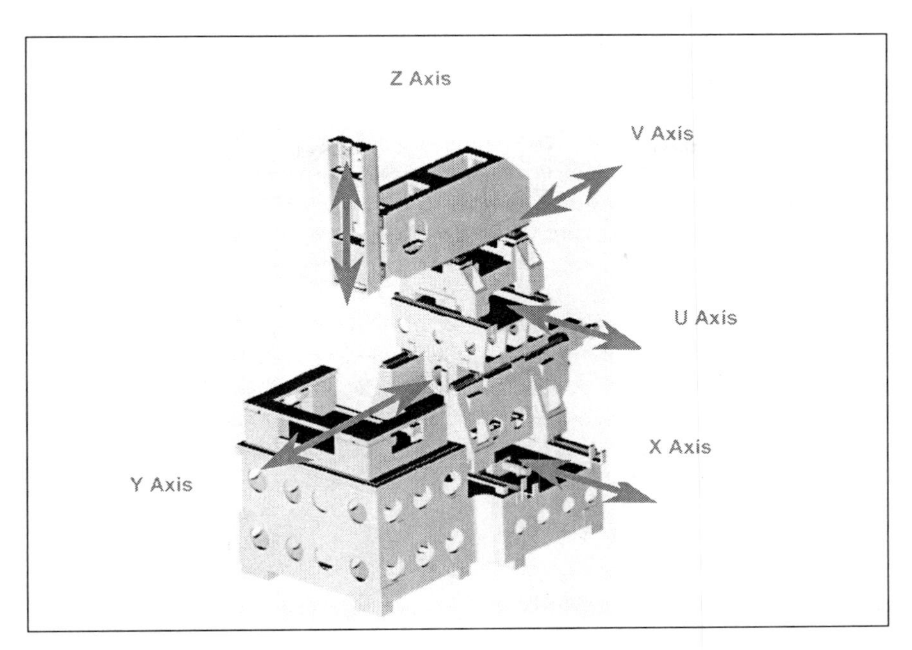

Fig. 9.11 Basic diagram of a WEDM machine

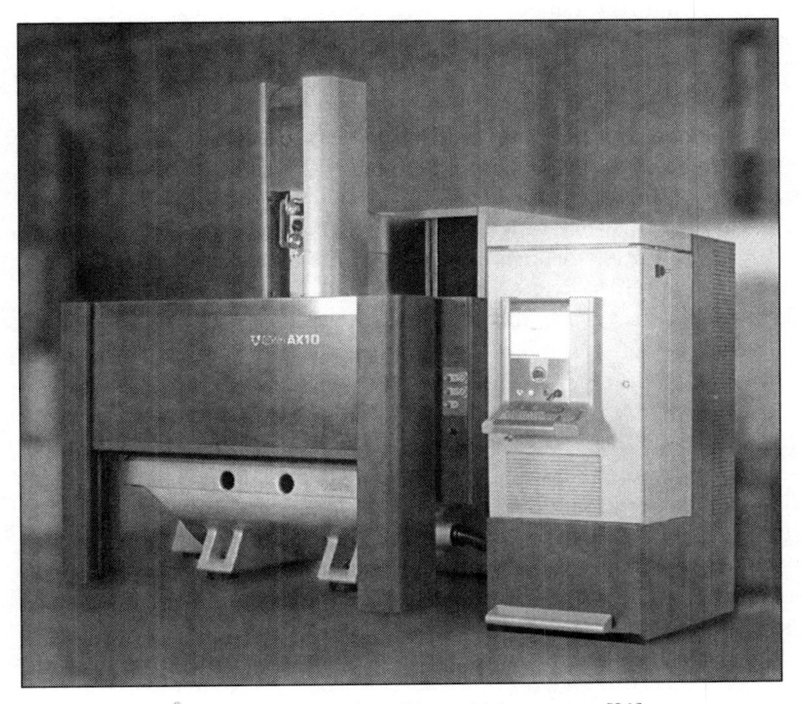

Fig. 9.12 ONA® AX-10 for the cutting of large thickness parts [21]

angles as high as ±30° can be cut in a 400 mm thick workpiece, but the angle is normally limited for higher part thicknesses.

The numerical control of a WEDM machine incorporates some important differences with respect to other conventional machine tools, such as machining centres and turning centres, for instance. In a WEDM machine the servo system not only closes loops for position and velocity, but it is also in charge of keeping a constant gap between the wire electrode and the workpiece. Solutions to this problem involve acquiring and analysing signals related to gap voltage and/or to delay time. Moreover, "machine intelligence" is of primary importance in WEDM machines. Manufacturers include in the NC of the machine technological data for optimum cutting of different materials and thicknesses, strategies for improvement of accuracy in corner cutting, the intelligent selection of EDM parameters for situations of degraded erosion (such as stepped parts, large thickness parts, taper cutting, etc.), and systems for avoidance of wire breakage, amongst others. Some of these will be addressed in Sect. 9.3.3.

9.3.1 Wire Transport and Wire Thread Devices

As said before, in most cases wire wear is not considered in WEDM since wire is continuously fed between pulleys. The wire supplying spool can contain from 1.6 to 45 kg with 3,700 m and 105,000 m respectively, and in the case of wire diameter 0.25 m. With the aim of assuring wire position, different pulley configurations have been designed by manufacturers providing the wire with the programmed feed rate and axial tension. Prior to and after the machining zone, the wire is driven through two guides (the upper and lower guide) corresponding to the nominal position programmed.

The guides are wear resistant accuracy devices; hence, they are made from sapphire or diamond. The guide diameter depends on the wire diameter used. There are, however, several wire thread systems based on wire drawing which allow using the same guides for a range of wire diameters.

There are two types of wire guides used by manufacturers on their wire EDM machines. Round or toroidal shaped wire guides are used by a number of EDM manufacturers, which may provide a slight advantage when machining larger tapers. A round wire guiding system may help to produce a slightly better finish in larger taper angles (greater than 15'). A round guide requires some clearance (~5 μm) to thread the wire through the guide. However, a number of manufacturers use V-type wire guides due to their reliability for automatic wire threading.

The nozzles drive the upper and lower flushing pressure jet. It becomes apparent that the location of the nozzles, especially in the lower arm, can be affected by these higher flushing pressures. Hence, a more rigid mechanical structure can withstand higher flushing pressures better. Nozzle geometry varies also in the taper cutting to accommodate the wire deformation shape.

Once the wire has passed through the lower guide, it enters the second part of the transport system called the evacuation system. The wire passes through the evacuation tube to the pinch roller after which the wire is cut to store.

Axial force is imposed on the wire to ensure straightness during cutting. Wire straightness is critical for precision applications. As explained in Sect. 9.2.2, the forces exerted during the process, due to the discharge and dielectric jet, tend to deform the wire. As a result accuracy is lost, since the points where discharges occur do not exactly match the position of the guides (i.e., the position programmed in the NC path). The value of the axial force imposed by the machine is limited by the mechanical strength of the wire and its diameter. Thermal load on the wire produced by continuous discharges also affects its resistance.

Nowadays, apart from these devices, most of the wire EDM machines have some kind of automatic wire thread (AWT) system which automatically provides thread or re-thread through the slot or start hole, with nearly 100% reliability. Automatic wire threading is much simpler and more reliable than manual threading, primarily when wires with low tensile stress resistance are used. Most AWT systems heat, draw and guide the wire by high pressure flushing water or air. AWT systems provide the guides with much longer life and therefore lower cost per hour and downtime.

9.3.2 Machine Automation

Automating EDM processes is a key aspect in the way to optimise productivity and work throughput, running machines continuously and unattended. Thus, major EDM machines manufacturers have developed new devices with the aim of minimising downtimes. In this section, various devices recently established commercially are presented.

The concept of cylindrical wire EDM is based on a rotary axis added to a conventional two-axis wire EDM machine to enable the generation of a cylindrical form. As conventional 2D machining, the electrically charged wire is controlled by the X and Y slides to remove the work material and generate the desired cylindrical form. The original idea of using wire EDM to machine cylindrical parts was first reported by Prof. Masuzawa's research group at the University of Tokyo in 1985 [9], and this configuration has eventually been industrialised. The Mitsubishi Electric® BA8 submerged WEDM machine, equipped with rotating spindle for EDM-turning and EDM-grinding applications, and the CNC ONA-W64 (by ONA-Electroerosion® S.A.) capable of controlling 6 axes are examples of this application. Of course, special machines can be customised. The configuration of 6 controlled axes was first aimed at the manufacture of small-diameter pins and shafts to be used as tools for 3D micro-EDM applications [13]. Nowadays, new fields of application are emerging, such as tool sharpening, diesel engine injector plungers, gear wheels with integrated shafts for easy gear assembly, conductive bonded grinding wheel dressing, parts for medical industries and aerospace parts industries.

No doubt, one of the critical concerns in WEDM application is wire breakage. The demands for a very high cutting speed and overnight cutting without attendance require wire breakage prevention since this would result in unacceptable increases of machining times, a decrease in machining accuracy, and the deterioration of the machined surface. A great amount of research effort into wire breakage prevention has been carried out which concluded wire breakage could be identified by symptoms such as high gap voltage, a sudden rise in the total sparking frequency, and an excessive instantaneous energy rate.

Control strategies to prevent wire breakage are therefore based on pulse discrimination according to the characteristics of voltage waveform during machining for the purpose of improving machining stability and efficiency. To do so, the control system analyses each discharge and evaluates ignition delay and frequency of the pulse current as an indication of arcing. Once instability is detected, the system corrects gap distance recovering a stable state. Modern digital EDM generators include any pulse monitoring system which means a wire breakage prevention system in the case of WEDM. Adaptive control systems provide effective control of the cuttings process, automatically modifying the programmed parameters of the power supply, ensuring optimum machine performance at each stage.

There are a number of other topics dealing with productivity and accuracy which can obtain enormous benefit from adaptive control systems. For instance, complex pieces like staggered components, parts with wide taper cut angles etc., can be EDM'ed resulting in improvements of as much as 30% in the cutting speed when compared with conventional control systems. Industrial examples of such systems are the AI (artificial intelligence) workpiece thickness adaptive control developed by Fanuc®, the Expert Erosion System by ONA Electroerosión®, the AutoMagic by Mitsubishi Electric®, and the Pilot Expert 3 by Agie-Charmillies®. Advanced research work in this field includes the use of knowledge-based control systems [16], explicit mathematical models [4], neural networks [8], and fuzzy logic controllers [15].

Intelligence is also introduced in the setting of EDM parameters after restarting a terminated process. In machines equipped with AWT systems (see Sect. 9.3.2) if the same parameter setting as that before wire breakage is used, the process tends

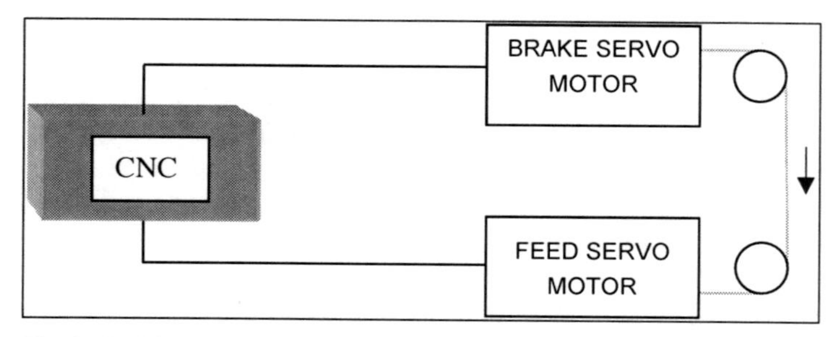

Fig. 9.13 Diagram of Twin Servo Wire Tension Control (patent pending) developed by Fanuc® Co.

to experience breakage again after restart. Thus an "intelligent" system is usually called up to change the parameter setting after each wire breakage. It is actually off-line knowledge about wire breakage.

As already explained, strict control of the axial force imposed by the machine is critical to ensure wire straightness. A good example of machine intelligence applied to the process is the Twin-Servo Wire Tension Control system developed by Fanuc, which uses servo motors to control wire tension and wire feeding. By applying Fanuc's digital servo technology to the wire feeding system, variation in tension is reduced to less than one fourth of previous systems, resulting in stable high speed and high precision machining. This function, represented in Fig. 9.13, is available in the high speed and precision Wire-Cut EDM Fanuc Robocut α-iD series [19].

9.3.3 Workpiece Fixturing Systems

Palletising is very closely related to automation. The link between mechanical interfaces – such as those between a workpiece and a machine tool – that have interchangeability, modularity and flexibility, is a critical first step toward eliminating the downtime associated with workholding. This became even more important in the case of EDM machines due to the long machining times associated to the process. Many workpiece holders' manufacturers have developed new palletising systems compatible among machines with very different configurations. For instance, this is the case of WEDM machines. In these machines, the work table is formed by two stainless steel guides with plenty of threaded holes and whose work surface is in a specific position related to the XY-plane and which is arranged in the direction of movement of the tool. It is necessary to transport the chuck-mounted work piece from one processing location to the next without changing workpiece alignment, e.g., in an orthogonal system.

Standardisation provides a stable reference system. However, it is modularity which gives a method of dealing with various applications. Modularity is the key to processing different sizes and shapes of workpiece blanks.

Erowa®, Hirschmann® and System 3R® are probably the best known manufacturers for workpiece palletising systems (Fig. 9.14). There are countless possible combinations in the product range adapted to any single application set up by reference elements, mounting heads, chuck adapters, rulers, adapter elements, supporting elements, and presetting/inspection devices. These devices allow the user to set up the machine quickly with high-quality repeatability specifications. A repetitive accuracy (consistency) for these systems is quoted in all the cases at 0.002 mm.

Growing demands on the manufacturing industry must be met with increased flexibility, increased quality and increased productivity. So, workpiece fixturing systems' manufacturers also provide pallet handling robots as effective with one-off manufacturing as with mass production. Examples of robots such as the WorkMaster by System 3R®, the Robot Multi ERM by Erowa®, and the Erobot

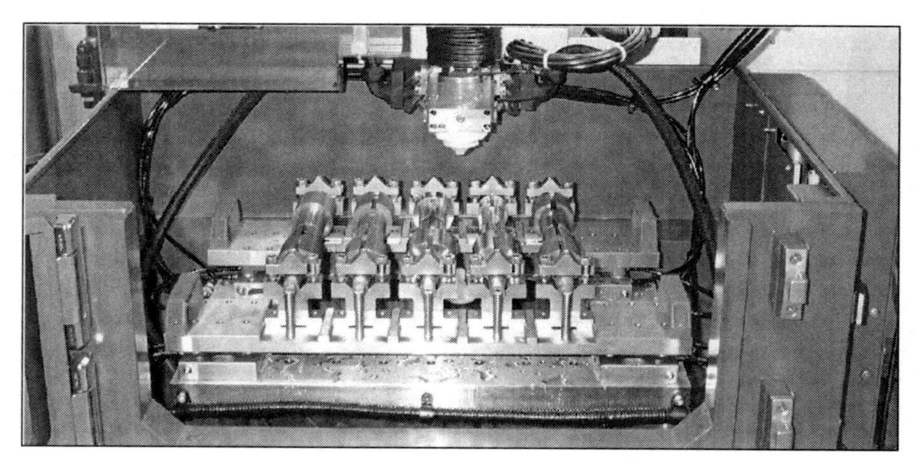

Fig. 9.14 Example of duplication of frame systems by System 3R® [23]

4018 Workpiece Changer by Hirschmann® can be cited as examples that provide a significant impact on productivity, profitability and competitiveness.

9.3.4 Filtering Systems

The filtering system is responsible for supplying the dielectric with suitable conditions into the erosion area. In most applications tap water conductivity is reduced to 106 Ω·cm, although some new materials such as hard metal must be machined with less conductivity prevent corrosion. Once tap water is electrically conditioned, it is used as a dielectric to cool and evacuate the debris from the machining which increases dielectric conductivity. No matter which filtering system is used, electrical conductivity is reduced employing special resins. If accuracy demands must be met, cooling capacity must be ensured using an external cooling system.

There are two main types of filtering systems: paper cartridge filters and mineral filters. Paper cartridge filters are replaceable or consumable which can be used for 200–300 working hours depending on the EDM process, water volume, *etc.* The filtering media consists of pleated impregnated cellulose base material which is cured and retained within the inner and outer screens by means of hot melt glue. The construction feature prevents excessive movement of the cellulose fibre in humid conditions and maintains an even pleat-spacing thus ensuring the most effective usage of the available media surface area. Nowadays, filters assure a consistent EDM fluid quality, and uniform process condition to avoid blockage of rinsing nozzles, increase resin consumption, sediment in the cooling and supply system, and to increase corrosion deposits. Flow and particle size are kept constant (particles over 1–3 μm are kept out) throughout the entire machining process with repeatable quality results. In some cases, WEDM machine manufacturers also manufacture filtering systems.

Mineral filter systems for wire EDMs do not require filter media replacement. It is a stand-alone unit which takes dirty water from the machine filtering it to a 3-micron cleanliness level to supply clean water for machining. When the filter vessel reaches its cleaning capacity, as detected by a pressure switch, a backwashing cycle starts and clean filtered water is supplied to the machine dielectric tank. With this feature the machine always has a supply of clean water.

9.4 Wires for WEDM

The correct choice of the wire for WEDM (Fig. 9.15) is a critical point in optimising the process. For each type of wire the user must be aware of the physical and chemical properties which influence the cutting process, together with the economic aspects that surrounding it. The result of a proper choice is a stable cut, economical and at maximum speed. It is obvious that the cheapest wire is not always the optimum solution for a certain application.

The first wires ever used were made of copper. Previous experience in SEDM machines using this material, and the availability of the technology for wire production in diameters 0.2 mm and 0.25 mm made copper the first option. However, the poor

Fig. 9.15 Wires for WEDM

mechanical and thermo-physical properties of this material led to the rapid development of new alternatives such as brass and coated wires. Mechanical behaviour is critical during the WEDM process. Axial force is imposed on the wire by the machine itself to minimise deformations which can affect accuracy; however excessive axial force may produce wire breakage, and this is why the ultimate tensile strength is considered a critical property. Tensile strength is 245 N/mm^2 for copper, 900 N/mm^2 for "hard" brass, and may reach as high as 1,930 N/mm^2 for molybdenum. Of course, these values depend not only on the material itself, but also on the operations (successive drawing operations and intermediate heat treatments) which take place during the manufacturing process. Wires of low ultimate tensile strength may exhibit high elongation values of well over 20%. These are considered "soft" wires. However, elongation is much more reduced (below 2%) in "harder" wires. This is especially important in taper-cutting, particularly when the part angle exceeds 7°. In this case, a "soft" wire fits the geometry of the guides better, resulting in a more accurate angle. Better fitting also ensures a higher stiffness of the system, and therefore deformations induced by the forces exerted by the process will also be smaller.

The explanation is only complete when one looks at the temperatures of melting and evaporation, properties that depend upon material composition and coating. A low melting point is associated with the generation of debris from the wire that facilitates ignition during discharge. A low evaporation temperature ensures most of the heat generated is transferred to the workpiece, which is associated to higher process efficiency.

Brass wire was introduced into the market in 1977. Immediately, cutting speed was accelerated up to values unknown so far: on a 50 mm thick part cutting speed was dramatically increased from 12 mm^2/min to 25 mm^2/min, values still very far away from current speeds (up to 500 mm^2/min, as stated in Sect. 9.2.2). The reason for this improvement is the presence of zinc in the gap. Current basic brass wires contain 63% copper and 37% zinc. Melting temperature of the latter is 420°C, well below 1,080°C of copper. Zinc evaporates during cutting, because its evaporation temperature is 1,000°C. After coming into contact with a cool dielectric, zinc re-solidifies in the form of debris, favouring ignition in the next discharge. Following this, higher zinc contents in the wire composition would be beneficial for the process. However, the phase-diagram for brass shows that over 40% content of zinc the α structure of CuZn37 results in a fragile ($\alpha+\beta$) structure which causes very important problems in the production process for small wire diameters.

Brass wire can be found in different tensile strengths, from 490 N/mm^2 to 900 N/mm^2, depending upon the production process. An alternative used in Japan is the CuZn33Al2. Addition of 2% of Al enables tensile strength increase in wire of up to 1,200 N/mm^2.

Due to the above-mentioned impossibility to manufacture high-zinc content brass, coated wires were developed in the 1970s. The first trials were made using a zinc coating and copper as a base material, although the low tensile strength and large dimensional variations forced the industry to move into other alternatives. The current industrial solution involves using CuZn20 or CuZn30 as a base material, coated with 50% Cu and 50% Zn, or even CuZn37 coated with zinc, which

Table 9.1 Recommendations of use of different types of wire for WEDM

(Base) coating	Tensile strength (N/mm^2)	Elongation (%)	Cutting speed	Cutting accuracy	Taper-cutting (> 7°)	Automatic threading
CuZn37	900	1	×	✓✓	×	✓✓
CuZn37	500	15	×	✓	✓	×
CuZn37	400	25	×	✓	✓✓	×
(Cu) CuZn50	520	1	✓✓	✓	×	×
(CuZn20) CuZn50	430	30	✓✓	✓	✓✓	×
(CuZn20) CuZn50	800	1	✓✓	✓✓	×	✓
(CuMg) Zn	600	3	✓	✓✓	✓	✓
(CuZn37) Zn	900	1	✓✓	✓✓	×	✓
(CuZn37) Zn	500	15	✓✓	✓✓	✓✓	×

×: not recommended ✓: recommended ✓✓: strongly recommended

becomes zinc oxide during the manufacturing process. The coating is compressed against the base material during the manufacturing process. The result is a wire that can be easily drawn and produces very important improvements in cutting speed. Tensile strength of coated wires may range from 480 N/mm^2 to 800 N/mm^2 depending upon the application. Coated wires are more expensive; however, they result in a more stable process and a higher cutting speed of about 30% with respect to brass wires. Optimum choice between coated and brass wires involves analysing the following aspects:

- Eroding time, in cutting hours per day.
- Non-eroding times (maintenance, part loading/unloading, *etc.*), in hours per day.
- Wire wasted in non-eroding times.
- Cost of the wire, per kilogram.
- Wire feeding speed.
- Machine cost, per hour.

Where economic aspects play a determining role, technical recommendations must also be observed. Therefore, Table 9.1 includes technical recommendations of use of different wires for WEDM.

Molybdenum wires are a high-cost choice for parts with intricate details such as small internal fillet radii. In this case, tensile strength is over 1,900 N/mm^2, and therefore a wire diameter as low as 50 μm can be used confidently. Molybdenum wires are not very popular due to their cost, and the fact that the high melting and evaporation temperatures do not particularly benefit the cutting process. The main field of application is in μWEDM, in the range of "high" wire diameters (around 50 μm). Other alternatives for micromachining applications are multi-layer wires (steel core, Cu, CuZn50 and Ag), tungsten and molybdenum carbide.

9.5 The Wire EDM of Advanced Materials

Nowadays, new industrial products involve the development of new materials with improved properties. This causes the need to obtain optimum machining parameters and complete analysis of material behaviour during the electrodischarge process. This new situation has become more and more regular, since aeronautic, electronic, and machining tools industries demand exotic materials. Aeronautic alloys, sintered carbides, and PCD (polycrystalline diamond) are some examples of the most widely used new materials.

9.5.1 Aeronautical Alloys

Aeronautic alloys, also called superalloys, are metallic alloys for elevated temperature service, usually based on group VIII elements of the periodic table, and are generally suitable for elevated temperature applications where resistance to deformation and stability are primary requirements. The common superalloys are iron, cobalt, titanium and nickel-based, the latter being best suited for aero-engine applications. Due to their exceptional properties the conformability and the machining of these allows have become very difficult tasks. For instance, the low thermal conductivity, plastic deformation resistance, mechanical hardness, and chemical affinity with tools raise the temperature during conventional machining. Regarding the acceptance of workpieces, the final user decision is actually prompted by the apparition of a modified microstructure, cracking extending from the recast to the bulk material, a pervading tensile stress regime, and others which seriously affect material fatigue resistance. Such is the case of some aerospace industry components. These types of parts are normally subjected to strict tolerances and very complex geometry. To prevent this situation, conservative machining conditions must be applied, reducing the productivity of the process.

Both wire and sinking EDM can provide an optimum solution for the machining of these materials. When applying processes with thermal removal mechanisms, surface integrity is a significant problem due to the formation of a heat affected zone. This is actually the main limitation of EDM, since subsequent manual operations are compulsory to eliminate the recast layer. In fact, a number of industries ensure their results in machining superalloys are excellent but no reference to the surface integrity damage is made. The most important EDM machines manufacturers have actually developed anti-electrolytic generators which minimise the surface damage of the parts.

One of the most popular aeronautic alloys is Inconel 718 whose applications are widely extended in aerospace and nuclear industries. This is an example of an alloy which can be easily machined by EDM, once the machining conditions are properly developed with the aim of preventing thermal surface damage. One of the most common standard tests carried out to analyse the process influence on the component life service is the fatigue test under constant stress range (as specified by ASME E 466).

9.5.2 Tungsten Carbide

Tungsten carbide is an important material for tool and die manufacturing because of its high hardness, strength, superior wear and corrosion resistance over a wide range of temperatures. Due to these extreme properties tungsten carbide cannot be easily machined by conventional machining techniques. However, electrical conductivity is high enough to make EDM possible. In particular, wire EDM provides sufficient accuracy for the manufacturing of precision tooling, such as that used, for instance, in "high speed stamping" applications.

The major drawback when EDM'ing tungsten carbide is the poor surface integrity obtained due to the electrolytic and powerful thermal effects which can cause machine surface cracking, pitting or flaking. These phenomena can drive the machined part to premature failure [17]. The electrolytic effect is generated during the phase prior to ignition and when high voltage is applied. The local chemical property of water on the surface of cemented carbide may be dominated by the dissolved cobalt ions, dislodging WC grains during the EDM process. Thus, the amount or concentration of WC grains decreases from the internal structure of the workpiece to the top surface layer. To sum up, electrochemical effects appear as a consequence of the type of dielectric [18] and electrical parameters applied [5].

Regarding dielectric fluids, deionised water is commonly used in conventional wire EDM applications. However, like most metals, cobalt is electrochemically very active in aqueous media. To minimise this, wire EDM machines manufacturers recommend users remove the workpiece from the tank immediately after machining and then dry it in an oven. When using water as a dielectric fluid, two pieces of advice should be taken into account. The first is related to the rise of electrical conductivity of water over 5 μS; the second is corrosion prevention using a sacrifice metal (usually zinc) [12]. The pH factor of water must also be carefully controlled.

Much research has been done using alternative dielectric fluids [18]. In these studies, the material removal rate obtained using oil and oxygen gas as dielectric fluids is compared. For high accuracy applications, and especially in thin wire EDM (that will be addressed in Sect. 9.6), oil can be used as dielectric fluid. In this case, the surface integrity of hard metal is greatly improved.

When it comes to machining parameters, high frequency and very low energy pulses are recommended to obtain better surface integrity. Most new machines incorporate anti-electrolysis circuits. Basically, this type of circuits keeps the average voltage close to zero by working in alternate current mode. The latest developments [17] include anti-electrolysis circuitry and CPLD-based pulse control with a frequency of 500 kHz, discharge duration as short as 150 ns and peak current as low as 0.7 A. The experiments demonstrate the fine-finish power supply can reduce the recast layer, reduce cobalt depletion with respect to that produced by a standard DC power supply.

9.5.3 Advanced Ceramics and PCD

The limitations imposed to conventional machining of advanced ceramics are widely known. Most ceramic components are manufactured by sintering at a shape close to the final geometry ("near-net shaping") and then ground using diamond tools. However, since EDM is not dependent upon part hardness, abrasiveness or brittleness, it can provide a feasible alternative to the manufacturing of ceramic parts, if the electrical resistivity of the workpiece material is lower than 100 Ω/cm. This is why some research has been done about EDM engineering ceramics. Konig et al. [6] classified it as non-conductor, natural conductor and conductor, which is a result of doping non-conductors with conductive elements. Much research has been devoted to the addition of conductive particles with the aim of increasing electrical conductivity. Following this methodology, Matsuo et al. [10] investigated the EDM'ing of ZrO_2 and Al_2O_3 doped with quantities of TiC, NbC, and Cr_3C_2. From their investigations it can be deduced that the composition is a very important factor, there being an optimum carbide content that optimises the removal rate and surface finish.

Due to their inherent properties, industrial interest has focused on materials such as boron carbide (B_4C) and silicon infiltrated silicon carbide (SiSiC). Boron carbide has long been known as an excellent abrasive powder; however it is only in recent years that its optimal properties have been applied to different industrial fields. It must be taken into account that this is the third hardest material known, just after diamond and cubic boron nitride, and is currently the hardest material produced in tonnage quantities. Common applications of B_4C are armour and wear protections, neutron absorbers in the nuclear industry, bearings, nozzles and turbines.

Silicon infiltrated silicon carbide is manufactured by the infiltration of Si into a matrix of SiC (Fig. 9.16). Its thermal conductivity is four times that of steel, and has a low thermal expansion coefficient, which is why it is commonly used in manufacturing high temperature heat exchangers. Other applications are high-efficiency gas turbines, seals, components in pumps, bearings, shot blasting nozzles, burner nozzles, *etc*. Sanchez et al. [14] provided a literature survey of EDM advanced ceramics studying the feasibility of electrical discharge machining B_4C and SiSiC.

An innovative method of overcoming the technological limitation of requiring a minimum electrical resistivity has eventually permitted insulating ceramics to be machined by EDM [11]. Both wire EDM and sinking EDM have been applied to the machining of several insulating ceramics materials. In the case of WEDM oil was used as dielectric fluid combined with long discharge duration and low discharge current. To prevent wire breakage at a very low tension down to ¼ of the

Fig. 9.16 Heat exchanger cut by WEDM. The material component is silicon infiltrated silicon carbide

tension usually applied, affecting the accuracy. Shot peening after machining is compulsory to remove the conductive layer left on the generated surface.

Polycrystalline diamond (PCD) products usually consist of diamond powder (1–100 μm grain size) treated at high temperature (1,400–1,600°C) and pressure (5.0–6.5 GPa) on top of a cemented tungsten carbide backing. PCD products are widely used in abrasive applications such as cutting tools and rock drill bits. The introduction of polycrystalline diamond (PCD) on a tungsten carbide substrate has greatly increased cutting efficiency.

The cobalt in PCD does not act as a binder, but rather as a catalyst for the diamond crystals. In addition, the electrical conductivity of the cobalt allows PCD to be EDM'ed. When PCD is EDM'ed, only the cobalt between the diamonds crystals is being EDM'ed. This sometime generates a gap between the carbide substrate and PCD because of the higher cobalt concentration in the intermediate layer. EDM'ing PCD, like EDM'ing carbide, is much slower than cutting steel. Cutting speed for PCD depends upon the amount of cobalt sintered with the diamond crystals and the PCD particle size. As a general recommendation, large particles of PCD require very high open circuit voltage.

Machine manufacturers have adapted EDM machines to allow quick precise profiling of PCD cutting tools. The PCD Edge System is the current Fanuc proposal for wire EDM machines. Available from EDM Methods, the system proposed consists of programming/probing/cutting software, a micro-finish power supply for cutting PCD, a Hirschmann® rotary axis, a Renishaw® probe (mounted on the EDM) and a tool-clamping system. On the other hand, Walter® has developed the Helitronic product family opening up a gateway into the future market of sharpening tools. These machines have two functions eroding and grinding. Within the product family Helitronic Power Diamond offers the possibility to erode rotationally symmetrical PDC and CBN and to sharpen carbide and HSS tools. Other important manufacturers such as GF AgieCharmilles® or ONA-Electroerosión® S.A. also incorporate technologies for machining PCD in their machines.

9.6 Thin-wire EDM

The origin of microtechnologies, as we know them today, can be found in the techniques of manufacturing of microsystems developed in the 1990s. Technologies such as thin-film techniques, photolithography and others became very popular for the silicon industry; however, they are unable to meet the requirements imposed by other applications related to micromechanics. To satisfy this growing demand processes such as micro-milling, laser and micro-EDM have been developed and launched on the market in recent years.

The micro-EDM role has become prominent due to its intrinsic advantages in the production world of very small components. Since there is no contact between tool and part, deformations due to cutting forces and dynamic effects disappear. Micro-EDM is profitable for short runs and even for the production of single parts, and it is the ideal partner for other technologies, for instance, in the production of plastic injection micro-moulds or micro-punches for sheet metal forming.

It is usual to reserve the term micro-EDM for sinking operations with very low diameter electrodes, while thin-wire EDM is commonly used for micro-WEDM. Under this concept the whole technology of cutting using wires of diameter ranging

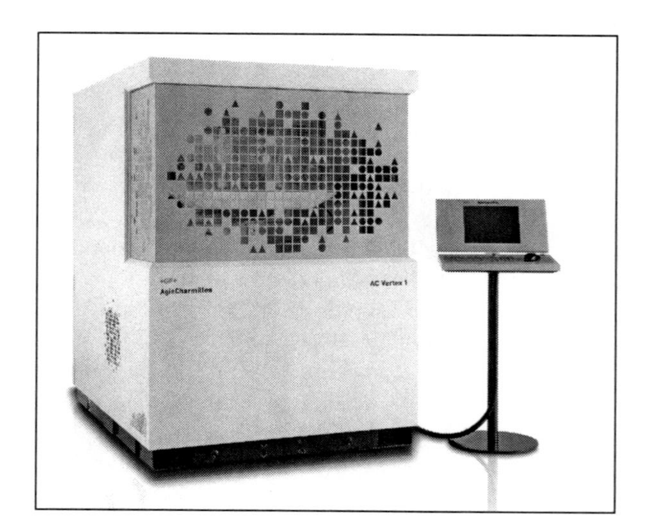

Fig. 9.17 The AC Vertex model of GF AgieCharmilles® for fine wire machining jobs [20]

from 15 to 50 μm is included. Thin-wire EDM may be conceptually similar to conventional WEDM in the removal mechanisms, however the technology is radically different and machine manufacturers have developed absolutely new solutions for machines to meet the extreme requirements imposed. In fact, absolute tolerances of about 1 μm, a surface finish as low as Ra 0.05 μm, a minimum wall thickness of 10 mm and slots of width 40 μm are usual values in this type of product. Of course, cutting speed is no longer a critical issue, ranging in values of about 1 mm^2/min. Part thickness is usually limited to about 3 mm with wire of 30 μm in diameter, increasing up to 5 mm when 50 μm diameter wire is used. Typical sectors of application are the watchmaking and medical industries, specifically in the manufacture of surgical components. Figure 9.17 shows a popular model of a thin-wire EDM machine for micromachining applications, the AC Vertex of GF AgieCharmilles®.

The first point at which one must look is the ratio between wire diameter and gap width. This ratio is about 10:1 in the case of conventional WEDM, and is reduced to 5:1 in the case of thin-wire EDM. Implications of this fact include a lower ability of the wire to stand thermal loads, a lower wire rigidity, and a lower wire strength to stand variations of axial load. Moreover, aspects such as machine precision, and the effect and compensation of heat sources are also very important. Some current solutions to these problems proposed by manufacturers will be analysed in the following paragraphs.

A new family of discharge generators has been developed for thin-wire EDM. Discharge energy must be limited to a minimum in order to prevent wire breakage due to thermal load and wire deformation. Minimum values of discharge energy in conventional WEDM machines equipped with transistor-type (FET) generators are about 1.5 μJ, value imposed by pulse on-time, the discharge current and the voltage, which is excessive for micromachining applications. The alternative is the use of relaxation-type circuits, in which the energy is controlled by a condenser. In this case, minimum theoretical values of a discharge energy of 3.5 nJ can be obtained.

Manipulation of such a thin wire is also a matter of critical concern. Axial tension imposed by the machine, guiding and threading must be carefully studied in these machines. Conventional systems for application of the axial force on the wire are usually based on the opposite action of an engine and an electromagnetic brake. This is a good choice in WEDM, where load variations of load of 20 grams do not affect process performance. However, in thin-wire EDM this value is unacceptable since, for instance, in a wire of 20 μm diameter the maximum load it can stand is about 70 grams. Controlled solutions that ensure fluctuations below 5 grams are used in thin-wire EDM. Important research work is currently being carried out by machine manufacturers, since this is obviously critical in the occurrence of wire breakage. Threading is another aspect requiring improvement. While in WEDM automatic threading has been completely solved, with reliability over 95%, only a small number of thin-wire EDM machines include this option. Manual threading is cost- and time-consuming, and can be improved by the use of tungsten wires due to their higher stiffness. However, this is yet another important field of industrial research.

Machine manufacturing includes using high precision components, high-stability structures and systems for thermal compensation. Most manufacturers use preloaded linear guides, although cross-rollers guides are also used. Precision preloaded ball screws are used for the linear axes, however linear motors are also the choice of other manufacturers. These latter provide a high dynamic response and the suppression of backlash, nevertheless the generation of large quantities of heat must be seriously considered. Measurement is carried out using high-resolution (about 5 nm) glass scales. Isostatic structures of a high stability using polymer concrete and including vibration absorbers can be used. The Monoblock configuration patented by GF AgieCharmilles® is a good example.

Compensation of external conditions is critical in thin-wire EDM machines. Thermal isolation of the working volume can be combined with intelligent heat generation to compensate for internal heat sources and external temperature variations. The cooling of critical elements, including structural components, is compulsory. These solutions combine with the use of ceramic materials with a very low thermal expansion coefficient for these elements. Thermal expansion coefficient in ceramics is two times smaller than that of graphite, three times that of castings used in precision beads, and five times that of stainless steel.

Acknowledgements The authors wish to thank the University of the Basque Country, the Basque Government and the Spanish Ministry of Education for their continued support to the EDM Research Group of the Faculty of Engineering of Bilbao. Special thanks are also given to ONA-Electroerosion® S.A. for their interest and support to our activities during the last 15 years. Finally, we would like to thank the companies GF AgieCharmilles®, Sodick®, Fanuc®, System 3R® and Bedra® for the information and pictures provided for this chapter.

References

[1] Dauw DF, Beltrami I (1994) High-precision wire EDM by on-line wire position control. Ann CIRP 43(1):193–197
[2] Dekeyser WL, Snoeys R (1989) Geometrical accuracy of wire EDM. Proc ISEM IX: 226–232
[3] Ho KH, Newman ST, Rahimifard S et al. (2004) State of the art in wire electrical discharge machining (WEDM). Int J of Mach Tool Manufact, 44:1247–1259
[4] Huang YH, Zhao GG, Zhang ZR et al. (1986) The identification and its means of servo feed adaptive control system in WEDM. Ann CIRP, 35(1):121–123
[5] Juhr H, Schulze HP, Wollenberg G et al. (2004) Improved cemented carbide properties after wire-EDM by pulse shaping. J of Mater Process Technol, 149:178–183
[6] Koenig W, Dauw DF, Levy G et al. (1988) EDM – future steps towards the machining of ceramics. Ann CIRP 37 (2):623–631
[7] Lazarenko B, Lazarenko N (1943) About the inversion of metal erosion and methods to fight ravage of electric contacts. Moscow WEI-Institute
[8] Liao YS, Yan MT, Chang CC (2002) A neural network approach foro n-line estimation of workpiece height in WEDM. J Mater Process Technol 121:252–258
[9] Masuzawa T, Fujino M, Kobayashi K et al. (1985) Study on Micro-Hole Drilling by EDM. Bull. Japan Soc of Pre Eng, 20(2):117–120

[10] Matsuo T, Oshima E (1992) Investigation on the Optimum Carbide Content and Machining Condition for Wire EDM of Zirconia Ceramics. Ann. CIRP 41 (1):231–234
[11] Mohri N, Fukuzawa Y, Tani T et al. (1996) Assisting electrode method for machining insulting ceramics. Ann CIRP, 45(1):201–204
[12] Obara H, Satou H, Hatano M (2004) Fundamental study on corrosion of cemented carbide during wire EDM. J of Mater Process Technol, 149:370–375
[13] Rajurkar KP, Yu ZY (2000) 3D Micro-EDM Using CAD/CAM. Ann CIRP 49:127–130
[14] Sanchez JA, Cabanes I, Lopez de Lacalle LN et al. (2001) Development of optimum electrodischarge machining technology for advanced ceramics. Inter J Adv Manuf Technol, 18(12):897–905
[15] Sato M (2002) Adaptive control technology for wire-cut EDM. Mitsubishi Electrical Corporation, R&D Progress Report
[16] Snoeys R, Dekeyser W, Tricarico C (1998) Knowlegde-based system for wire EDM. Ann CIRP, 37(1):197–202
[17] Yan MT, Lai YP (2007) Surface quality improvement of wire-EDM using a fine-finish power supply. Int J of Mach Tool Manufact, 47:1686–1694
[18] Yu Z, Takahashi J, Kunieda M (2004) Dry electrical discharge machining of cemented carbide. J of Mater Process Technol, 149:353–357
[19] www.fanuc.co.jp/en. Accessed 2007-12-10
[20] www.gfac.com. Accessed 2007-12-10
[21] www.ona-electroerosion.com. Accessed 2007-12-10
[22] www.sodick.com. Accessed 2007-12-10
[23] www.system3r.com. Accessed 2007-12-10

Chapter 10
Parallel Kinematics for Machine Tools

O. Altuzarra, A. Hernández, Y. San Martín and J. Larranaga

Abstract Parallel kinematics is a branch of mechanics that focusses on manipulators formed by closed kinematic chains, i.e., mechanisms that have an end-effector joined to the fixed frame by several limbs. Such a kinematic structure provides some advantages regarding stiffness, acceleration and weight, but has some drawbacks due to mechanical complexity and limited workspaces. In the field of machining, there have been several applications of such mechanisms to machine tools. Earlier designs based on hexapods did not fulfil expectations but new topologies are promising. In this chapter, there is first a description of the evolution of parallel kinematics in the manufacturing industry. Second, the authors expose a design methodology giving some hints on the main problems to overcome. Third, there is a study on calibration processes that can be applied to these machines. And, at the end, there is a description of control issues.

10.1 Introduction

Spatial mechanisms for *parallel kinematics* have a special magnetism for many researchers. In the beginnings of the nineteenth century, authors such as Cauchy and Lebesgue had already studied some interesting problems related to this kinematic structure. One of them was the analysis of singular postures, an issue that

O. Altuzarra and A. Hernández
Department of Mechanical Engineering, University of the Basque Country
Escuela Técnica Superior de Ingeniería, c/Alameda de Urquijo s/n, 48013 Bilbao, Spain
{oscar.altuzarra, a.hernandez}@ehu.es

Y. San Martín and J. Larranaga
Fundación Fatronik – Tecnalia. Paseo Mikeletegi, 7 – Parque Tecnológico
20009 Donostia-San Sebastián, Spain
ysanmartin@fatronik.com

L. N. López de Lacalle, A. Lamikiz, *Machine Tools for High Performance Machining*,

has attracted a lot of attention in the last decade. However, it is evident that the technology available until the second half of the twentieth century did not allow practical applications of this kinematic structure. The first industrial product was a tyre testing machine by Gough (1957) [17]. In 1965 Stewart [49] invented a very similar device for a completely different task: flight simulators. This later application, whose functional requirement is the acceleration of the platform, was the reference for parallel kinematics in the 1970s. Parallel structures were introduced in the field of robotics in the 1980s, mainly for pick and place tasks. The most popular designs were the Delta™ robot [7] and the Tricept™ [37]. In the 1990s, parallel robots were applied to diverse tasks such as precise positioning and haptic devices. Also, in this decade the first industrial prototypes of parallel kinematics machine tools were released. A lack of technological know-how in the field of

Table 10.1 Parallel kinematic machines (PKMs), their kinematic models, and applications

Parallel robot	Kinematic notation	Kinematic model	Applications
Rotobot™ 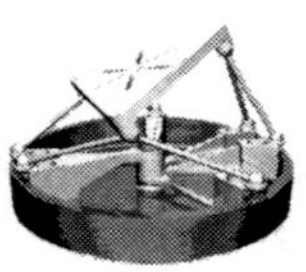	6-RSS 6 degrees of freedom (DOF)		High precision positioning
ABB IRB340 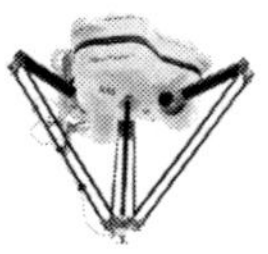	3-RRPaR 3 translational DOF	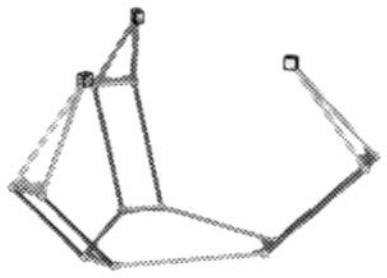	Pick & place
Hexapode CMW300 	6-SPS 6 DOF		High speed machining
HCMM 	5-SPS 5 DOF		Hexapod for metrology, co-ordinate measuring machine
Hexaglide™ 	6-PSS 6 DOF		Machining and manipulation

control had prevented this step into industrialisation until then. Now, sice these limitations have been overcome, and with the outcome of new parallel morphologies different from the hexapods, parallel kinematic machines are gaining new positions for the future in the field of machining. In Table 10.1 there are some examples of existing parallel manipulators.

10.2 Main Characteristics of the Parallel Kinematic Machines

A way to analyse the features of parallel kinematic machines is to place them against the characteristics of their dualities, i.e., serial machines. In the latter, the kinematic structure is an open kinematic chain (an arm of an anthropomorphic shape), in which the end-effector is at the end of the chain to perform a task. On the contrary, parallel robots are closed kinematic chains whose end-effector is joined to the fixed frame by several kinematic chains (articulated limbs).

Serial-robot technology is well known, and its application in most industrial domains is also well established. Therefore, before starting a competition against serial architectures, a deep analysis of the capabilities of parallel manipulators has to be done. There is not a general hegemony of one type of mechanism over the others. The morphology of the machine has to be defined for each specific application, and its dimensions have to be optimised for that task. If the comparisons are made to these fundamentals, it will be possible to know in which applications parallel manipulators are actually more competitive than serial machines.

Advantages and weaknesses for both types of mechanisms come from the kind of kinematic morphology. Serial machines may reach a wider workspace because of their anthropomorphic shape. This open architecture is also responsible for the uncoupling between orientating and positioning freedoms. This simplifies the solution of the position problems as well as calibration issues; in fact this allows an independent compensation for each axis. Parallel machines have a better stiffness because of their multiple connections to the ground. Hence, their load to weight ratio is much higher than that of serial machines. For the same reason, given the same errors on individual joint variables, parallel machines will produce a lower positioning error than those of serial ones. Dynamic response is also better in parallel structures, as well as the capability for velocity, acceleration and jerk. In conclusion, the potential of parallel machines in modern machining is better than those of serial ones if the correct morphology and the optimum dimensions are chosen for a specific task. The last condition requires multi-objective optimisation methods because the characteristics to improve are often opposed.

There is a great variety of tasks that parallel manipulators can achieve. Some of the most popular were already mentioned, such as pick and place or flight simulators. Other possible applications are precision surgical operations, the assembly of electronic components, and micromanipulators that can perform motions measured in nanometres. Regarding machining, the fields of application with more possibilities are aeronautical and automotive; in both cases it is mandatory to get cheaper and better quality parts. For example, new designs in airframes require the machining of 80% of the material of monolithic slabs to get complex geometries in one piece. In the automotive industry, machining of power trains are leading the research for better machining centres. A solution can be found in parallel kinematic machines, but in order to avoid the errors of the past, it must be born in mind that there are many topologies of parallel machines apart from hexapods.

10.3 A Classification of the Parallel Kinematic Machines

Parallel manipulators can be classified into two categories: those of complete mobility (six degrees of freedom, or DOF), and the so-called lower mobility manipulators (five or less DOFs). The latter are subclassified into two groups. The first one gathers all manipulators with motion capabilities that are constant in their workspace. Some typical examples are: translational manipulators [15, 50], spherical mechanisms [16], and SCARA type robots [40, 43]. However, other tasks may require that the tool is placed in variable orientations and positions. In such cases, the manipulator must have a type of motion that does not fit into the first subgroup; these manipulators are said to have mixed freedoms. They have rotational and translational freedoms that are coupled. The end-effector has a screw motion, variable with the posture, and in general it is not easy to get pure rotations. Translations have to be added to compensate the screw displacement and generate pure rotations.

Most modern applications in the field of machining require five axis operations, i.e., a motion with three translational and two rotational freedoms. Serial machines are used extensively for five axis machining. In the bigger machines, hybrid architectures with a moving machine head and a movable working table are often used. On occasions, rotational freedoms and a vertical translation are placed in the machine head while the working table has two translational freedoms. But the inverse architecture is also possible; the machine head is only translational with three Cartesian axes while a tilting table is employed to hold the piece. Sometimes, one of these translations is moved to the working table. For the machining of large moulds and matrices, and in the machining of outsized aeronautical components the working table has translational freedoms. The main flaws of five axis machines have been their limitations regarding dynamics due to gear transmissions, and their limited static stiffness.

In the beginning, the first alternative from the field of parallel kinematics was the use of architectures based in the Gough-Stewart's platform. Examples of such parallel kinematic machines are: Giddings&Lewis®' Variax [13], Ingersoll®'s Octahedral Hexapod HOH-600 [25], Mikromat®'s 6X [34], or Okuma®'s Cosmo Center PM-600 [38]. They had better stiffness and dynamic behaviour than that of comparable serial machines. Nevertheless, they had a considerable limitation on the rotation range that prevented the machining on five faces. Another important limitation was their overall size in relation with the reachable workspace. And regarding mechanical complexity, the use of their six degrees of freedom provided a good dexterity but at a high cost in components and control.

In order to improve the tilting range, other parallel architectures have been used different from the Gough-Stewart's; for example, Daeyoung Machinery®'s Eclipse-RP [10] or Metrom®'s P 800 [47]. However, these machines still have the problem of an excessive ratio of machine size over workspace volume.

One way to reduce this ratio is to combine in chain serial modules with parallel modules to create hybrid machines. This is the case of Neos Robotics®' Tricept [37], where such parallel-serial architecture increases tilting ranges as well as reduces its footprint. But it still lacks a good dynamic and stiffness response in rotations due to the fact that it uses a tilting head very similar to traditional machine heads. Another hybrid parallel kinematic machine is Fatronik's Space 5H [9]. This machine has a parallel module (the so-called Hermes machine head) in a serial arrangement with a frame that has two translations. This machine is designed for the machining and drilling on large aeronautical components and has very good qualities but it does not allow a complete machining of five faces. A very similar device is the Ecospeed [18] with a machine head on a parallel module with two rotational and one translational freedom called Sprint Z3 from DS Technology®.

There is still another alternative that can increase tilting angles as well as reducing the size of the machine. In this case, the machine is divided into two modules with three and two freedoms respectively, mounted on the fixed frame and working in cooperation. For example, a first module is a parallel manipulator with three translational DOF, where the machine head is placed. The second one is a tilting table with two rotational freedoms. The relative motion between both modules generates the five axis motion to machine effectively five faces. Other possibilities combine parallel modules with mixed freedoms (for example two translational and one rotational) with tilting tables or mixed tables (one rotation and one translation). In the following section the authors go over the design of such machines.

10.4 A Design Methodology for Parallel Kinematic Machines

Parallel kinematic machines can be designed following a procedure that has the following steps. The first one is the detailed definition of the requirements for the application. The second one is the type synthesis of the manipulator where the links and joints of the mechanism are determined. The third is a kinematic analysis to evaluate the potential of the mechanism chosen. The fourth is a detailed analysis carried out to define dimensions and so on. The fifth is a rigid body dynamic analysis performed to define materials, sections, drives and control. The sixth is an elastic body dynamic analysis which can be done to redefine and optimise dimensions for vibration and stiffness. At the end, the final design of the first prototype is ready and experimental studies will be done to calibrate. In this section the authors point out some specific issues to be considered in designing parallel kinematic machines.

10.4.1 The Motion Pattern

Regarding the requirements for the application, first, one must identify clearly which is the motion pattern needed at the end-effector. By motion pattern it is understood the type of motion that the tool must perform, given by a set of continuous poses reachable inside the workspace. This notion includes not only the number of DOF of the machine, but also the type of freedoms required. Typically, conventional machining centres make use of three translational freedoms and up to two rotational ones. Sometimes the piecework is fixed and the tool has the type of motion required, and other times a Cartesian or a tilting working table are used, being the motion obtained from the relative motion between tool and piece. Parallel kinematic machines can be designed following the same kinematic structures. It is common to combine a moving frame in a series with a parallel module, or to make a parallel machine head work on a movable working table. A fully parallel five axis machine will be compromised by a complex mechanical structure reducing stiffness and accuracy.

Regarding rotational freedoms, these are implemented in traditional machines using tilting tables or rotating machine heads; in both cases these axes are physical and visible. However, in PKMs this is not usually the case. Moreover, in some occasions these rotations are not pure but a screw motion needing a compensation motion on translational freedoms to accomplish pure rotations. These rotations are attained on instantaneous screw axes obtained in the kinematic analysis and not always evident. In Fig. 10.1 there is a photo of the Hermes [19] machine head that has two rotational and one translational freedom. The kinematic model used to find the possible screw axes on that pose, called the screw system, shows that these are on a plane perpendicular to the linear actuators.

Such a motion pattern is more effective when the Hermes module is mounted in a series on a translational frame, or working on a tilting table, in order to compensate screw displacements and get pure rotations when needed, as shown in Fig. 10.2. Another example of such a parallel kinematic structure is the Z3 machine head from DS Technologie® GmbH (DST) [24] (see Fig. 10.3) that again

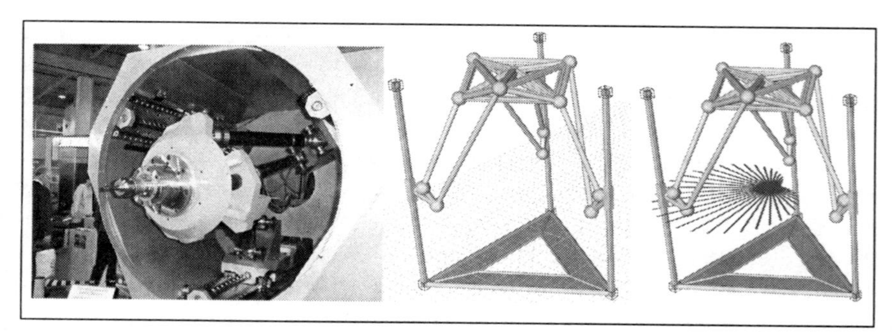

Fig. 10.1 Hermes machine head, a kinematic model and a screw system

Fig. 10.2 Hermes machine head on a translational frame and in a hybrid machine

Fig. 10.3 Z3 Sprint Head from DS Technologie® GmbH and a screw system

can create A and B axes of motion, this time with possible axes of pure rotation on a plane. Both devices are designed for the high speed milling of aerospace structural components.

Note that in these examples, a common approach in successful machines is to employ a parallel mechanism of mobility lower than the six DOF of the hexapods. These types of parallel mechanisms can accomplish their tasks with the expectations of better stiffness and accuracy while keeping their mechanical complexity and kinematic problems under control. However, the type of motion generated, i.e., their screw systems, can be very tricky and require a careful study.

10.4.2 The Type Synthesis

Once the designer has chosen the kinematic structure for the machine, and the mobility and motion pattern required at the parallel module, the following step is to find the mechanical morphology of this mechanism. The "type synthesis" is the process of finding all possible architectures of parallel manipulators whose end-effectors generate the specified motion pattern, and then to choose the most appropriate one. Bear in mind that for serial kinematic arrangements there is a limited variety, while for parallel manipulators the possibilities are many more. Parallel mechanisms with industrial applications have been largely synthesised using intuition and inventiveness, and sometimes even serendipity. Systematic procedures date only from the late 1990s, being that this an academic topic that is quickly evolving, and new classes of parallel mechanisms are continuously issued.

Existing procedures for the type synthesis are based on the same principles, but differ in the type of mathematical tools that they use, and argue about the limitations that each encounters. The starting point is the motion pattern required and the number of limbs of the parallel manipulator, usually equal to the DOF needed. First, it is to find which motion patterns may be generated by each limb so that in assembling them to a single end-effector the resulting motion pattern is the one desired. This is the first source of multiplicity in the synthesis process. It is obvious that several limbs generating exactly the same motion at the end-effector can be assembled together and keep it. But it is also true that if one of the limbs has a motion pattern of a higher dimension and this does not constrain the other limbs' motions, the resultant assembly is also valid. Therefore it is possible to assemble different motion patterns as far as they do not constrain the motion required.

Second, it is to get the different kinematic chains that can be used as limbs to produce the motion pattern defined in the previous step. Again, here there is another source of multiplicity in the synthesis. This time, the number of different possibilities is even higher and the design is determined by many restrictions related to mechanical complexity, reliability, type of actuators, workspace and so on.

Regarding the mathematical tools used in these systematic procedures, "screw theory" and "displacement group theory" are the most relevant ones. Screw theory [3] is a very powerful tool to deal with spatial kinematics. It uses Plücker's coordinates, that define a line through the medium of its own proper coordinates without regard to specific points on it or planes that contain it. In this way Ball extended this concept to the coordinates of a screw, which must be understood as a geometrical element. This can be applied to instantaneous kinematics on the one hand and statics on the other, these two being parallel concepts (a kind of duality). In the field of instantaneous kinematics, the screw serves to the definition of the velocity state of a rigid body as a single entity, namely the twist:

$$\$(h) = \begin{Bmatrix} \boldsymbol{\omega} \\ \mathbf{r} \times \boldsymbol{\omega} + h\boldsymbol{\omega} \end{Bmatrix} = \omega \begin{Bmatrix} \mathbf{s} \\ \mathbf{r} \times \mathbf{s} + h\mathbf{s} \end{Bmatrix} \tag{10.1}$$

where $\boldsymbol{\omega}$ is the vector of angular velocity of the body with a module ω and a direction $\mathbf{s}$, $\mathbf{r}$ is the vector that locates a point on the screw axis with respect to the origin of the fixed coordinate frame, and h is the pitch of the helicoidal motion (null for pure rotations and infinite for translations).

In multi-freedom mechanics, the motion of a body is given by the linear combination of n linearly independent screws defining the basis of an n-system. The motion restricted is constrained by a $6-n$ wrench system, and this is found with a specific operation called a *reciprocal product*. If in a manipulator, the end-effector has three DOF; hence, a 3-system of screws expresses its motion. The screw system can be showed as the possible locations of the instantaneous screw axes for any combination of inputs. In Fig. 10.4. the screw system of this open kinematic chain indicates that any motion generated is a pure rotation through a unique point, therefore a spherical motion. Using the reciprocal product, the wrench system is found to be any force through the same unique point as shown in Fig. 10.4. This latter meaning that such a force is sustained by the joints themselves and not by torques applied.

In type synthesis, the motion pattern required determines the n-screw system of the end-effector and the use of the reciprocal product gives the wrench needed [27]. For example, if the parallel module has the motion pattern of a five axis machine, the screw system has a dimension 5 and the wrench is a couple that constrains the rotation in one direction. This latter wrench has to be in common to the wrenches of each unassembled limb, being the first source of multiplicity in the synthesis procedure. In the example, any limb with a 6-screw system, i.e., no wrench system as in the hexapods (S$\underline{\text{P}}$S); or with a 5-screw system whose wrench is the same couple as the one to be constrained (S$\underline{\text{P}}$U) can be used. Then, the following step is to define kinematic chains that accomplish with those screw and wrench systems, another source of multiplicity. There is no room in this text for a deeper analysis of the synthesis process of the legs but the result is a huge variety of designs. As the reader can hint, for mechanisms with lower DOF the variety rises. Finally, bear in mind that the screw theory deals with instantaneous kinematics, and for finite displacements the designer should check his results in many different poses.

Concerning "displacement group theory", this method uses the Lie group properties that the set of all possible rigid body displacements have. The reader can refer to publications such as [20, 48] to learn about these issues; here only essential concepts are presented. The rigid body displacement group $\{\mathcal{D}\}$ has a dimen-

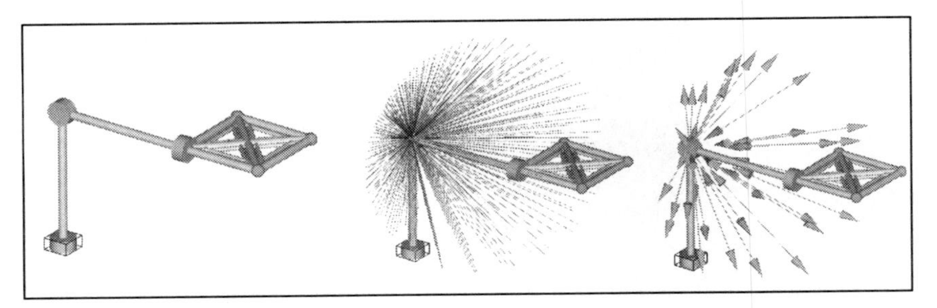

Fig. 10.4 3 DOF open kinematic chain, 3-screw system (rotation axes) and 3-wrench system (forces constrained)

sion 6, and eleven subgroups can be depicted inside it. These are: the null displacement subgroup $\{\mathcal{I}\}$ for dimension 0; translation in a constant direction **u** $\{\mathcal{T}_{\mathbf{u}}\}$, rotation about axis A $\{\mathcal{R}_{A}\}$, and screw displacement about an axis A with a pitch p $\{\mathcal{H}_{A_p}\}$ for dimension 1; translation on a plane defined by **u** and **v** $\{\mathcal{T}_{\mathbf{u,v}}\}$, and cylindrical motion about axis A $\{\mathcal{C}_{A}\}$ for dimension 2; translation in any direction $\{\mathcal{T}_3\}$, planar motions $\{\mathcal{F}_{\mathbf{u,v}}\}$, spherical motions about point O $\{\mathcal{S}_O\}$, and planar translations in combination with possible screw motions in a perpendicular direction to that plane $\{\mathcal{Y}_{\mathbf{u}_p}\}$ for dimension 3; and Schönflies motion $\{\mathcal{X}_{\mathbf{e}}\}$ consisting on translations in any direction and a rotation about a fixed direction **e** for dimension 4. There are other types of displacements but they do not have the group structure. Any mechanical system that produces a displacement subgroup is called a *motion generator*. Two main operations are performed with these groups, the product and the intersection. With the product one can analyse which is the resultant displacement in a serial chain, while the intersection provides the resultant allowed displacement in a closed chain assembly.

In the type synthesis, the first step is to find all sets of n displacement subgroups whose intersection is the specified motion pattern. Then, the most suitable must be chosen and finally each of the n displacement subgroups is replaced by the appropriate motion generator that will constitute each limb (kinematic chain).

The Verne machining centre from Fatronik (Fig. 10.5) is a five axis hybrid parallel kinematic machine; whose machine head is on a 3 DOF parallel module and the piecework is placed on a tilting table (see Fig. 10.5). The tilting table has a displacement subset of dimension 2 that is produced by the product of two independent rotations about axis A and C $\{\mathcal{R}_{A}\}\cdot\{\mathcal{R}_{C}\}$. In order to get the displacement subset of dimension 5 needed, namely $\{\mathcal{T}_3\}\cdot\{\mathcal{R}_{A}\}\cdot\{\mathcal{R}_{C}\}$, the parallel module may have several displacement subsets of dimension 3. The choice is a $\{\mathcal{T}_{\mathbf{u}}\}\cdot\{\mathcal{T}_{\mathbf{v}}\}\cdot\{\mathcal{R}_{A}\}$ displacement subset, that on this side will be obtained with the intersection of 3 displacement subsets generated by each of the limbs. The set of possible intersections is wide; the chosen ones are shown in Fig. 10.6.

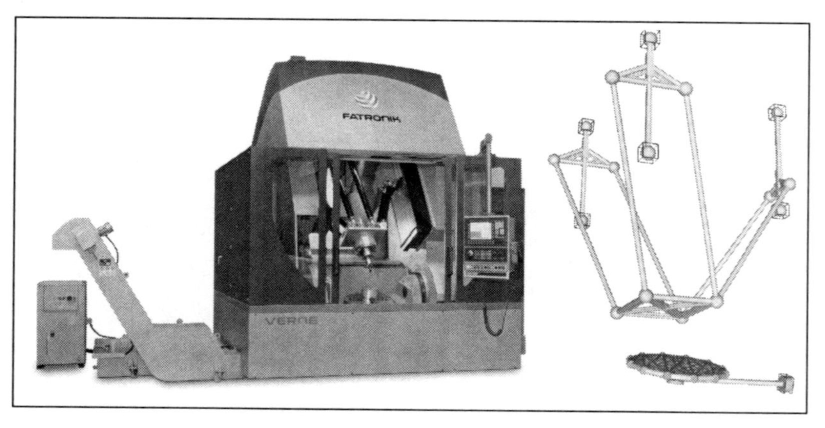

Fig. 10.5 Verne™ PKM, a 5 axis machining centre from Fatronik®, and its kinematic model

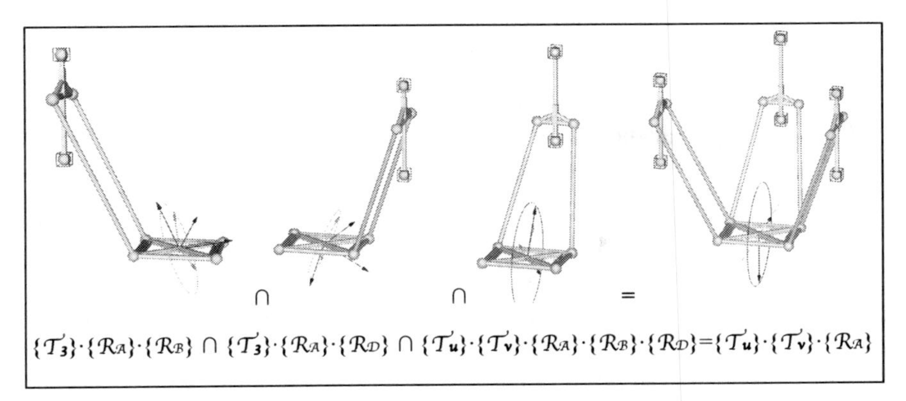

Fig. 10.6 Displacement subsets intersected to get the motion desired in model Verne™

In the last paragraph the reader may have noticed that the term *subsets* is used instead of subgroups; this is because those displacements do not fulfil some property of the Lie group. This issue imposes some limitations on the method regarding the difficulty to find all possible motion generators of the displacement desired and its validity for finite displacements.

10.4.3 The Position Analysis

Once a mechanism has been synthesised, the following step is to state the equations that locate the pose of the end-effector for given inputs, called the *direct kinematics* problem; and vice versa, those equations to provide the inputs needed for a given pose, called *inverse kinematics* problem. These equations are used in the position control, either to monitor the pose or to generate orders to the actuators. As opposed to serial mechanisms, while the inverse kinematics problem is trivial for many parallel mechanisms, the direct problem is in most occasions very challenging.

In order to get the system of position equations, methods such as Denavit-Hartenberg's used in serial kinematics are complicated by the existence of multiple closed loops. Therefore, it is very common to employ the so-called "geometric method". In this approach, a vector-loop equation $\mathbf{g}(\mathbf{x},\mathbf{q},\mathbf{p})=\mathbf{0}$ is written for each limb and constraint equations are added for lower mobility parallel manipulators. These equations relate the pose of the end-effector $\mathbf{x}$, the inputs $\mathbf{q}$, and the passive variables $\mathbf{p}$. Then mathematical manipulation tries to get rid of passive variables to arrive to a reduced system of equations $\mathbf{f}(\mathbf{x},\mathbf{q})=\mathbf{0}$ relating only input and output variables. In this system $\mathbf{f}$ are n-dimensional implicit functions of n-dimensional vectors $\mathbf{x}$ and $\mathbf{q}$ (n being the number of DOF).

Unfortunately, the resultant system of equations is a set of non-linear equations often including sine-cosine polynomials. As most research into non-linear equations solving is on algebraic polynomials, it is also common to transform sine and cosine into variables. These systems have generally multiple solutions, some real and some imaginary, being not always able to be solved analytically. Closed form solutions are rare in parallel machines; at best an univariate polynomial is obtained and then solved numerically, and often the multivariate polynomial system has to be solved numerically from the very beginning. On the one hand, when looking for all real solutions to the problem, some methods deserve a mention: polynomial continuation and homotopy continuation methods, elimination methods, Gröbner basis technique, and resultant methods [33]. On the other hand, if only one solution is sought and this is properly cornered by a closed approximation obtained with external data, mechanical constraints or other means, it is very common to use a Newton-Raphson's procedure. This procedure is quick enough in most cases as it has a quadratic convergence when starting with a good approximation. Moreover, there are several techniques to test its convergence, like Kantorovich's, which help in giving enough reliability to this method for use in the control. The last method is commonly applied in parallel kinematic machines to generate the kinematic transformation needed in the CNC. A more detailed analysis of these methods is beyond the scope of this book.

On the one hand, the position problem is used as mentioned above. Due to the parallel configuration of the PKMs, there is a non-linear relation between the position values of the drives and the position value and the tilt angles of the platform that carries the spindle. From the control point of view, this relationship is represented by a mathematical model that simplifies the mechanical system to a computable degree. The implementation of this model inside the numerical control is called *kinematic transformation*. The model describes only the rules of the relation between the ingoing and outgoing variables of the system.

On the other hand, for design purposes, a very useful and practiced strategy in the industry is to use commercially available CAD packages to perform the position analysis [4]. The main advantage of these packages is that there is no need to write the equations as described previously. It is enough to introduce the joints, as determined in the synthesis stage, in the correct position and the software is able to determine the DOF.

The limitation of these packages is that they only provide the number of DOF, but not their nature (rotation or translation), so a prior knowledge of the type of the DOF is needed. However, if the number and type of DOF is known, a complete position analysis can be performed.

One of the objectives of the position analysis is to obtain the working volume maintaining the restriction of the joints (rotation limits) and avoiding collisions between elements. For this latter purpose, CAD software packages are very useful because they usually include a collision detection algorithm. Figure 10.7 depicts a PRO/Engineer model of the Verne machine.

A position analysis was performed in this machine, in order to determine the working volume of the machine. Usually the size of the working volume is a requirement, so the machine dimensions are changed iteratively in order to obtain the working volume considering the body interferences and the joint limits. Also,

Fig. 10.7 PRO/Engineer model of Verne™ machine (by Fatronik®)

the working travel on the axes, i.e., maximum and minimum values, can be found with the same technique.

A differentiation has to be made between the working volume and the workspace of the machine. The first one is the volume reachable by the tool, while the second one is a more complete concept. It refers to the positions reachable and the range of rotation attainable for each position. This space is not easy to represent. The knowledge of the work space is very useful for the further analysis of velocity, accelerations, singularities or manipulability, above all if an iterative process is to be used.

10.4.4 Velocity Analysis, Singularities and Dynamics

Analogously to position problems, there are two types of velocity coordination problems: the *direct velocity problem*, that solves the velocity state of the end-effector given the input rates, and the *inverse velocity problem*, that does vice versa. In order to get the velocity equations, there are several approaches. One is the differentiation of the position equations, i.e., loop-closure equations and constraint equations. Another is the use of screw theory to get the velocity equation. Both methods generate an input-output velocity equation characterised by the so-called Jacobian matrices.

The analysis of the singularity of these Jacobians is a way to detect the singularities of these manipulators [33]. Several classifications of singularities have been issued [14, 53]. The common point is that the singularity is a posture where the control will have a problem to control the end-effector and must be avoided. Mathematically, singularities arise at certain specific configurations of the mechanism, but in practice, due to backslash and assembly errors, the singularity will occur in an area around that specific posture. Therefore, it is very useful to get indicators of proximity to singular configurations and restrict the motion to areas with acceptable values of such indicators. These numerical parameters are sometimes tricky and require deep mathematical analysis [33].

Acceleration and jerk analysis can be performed in successive derivations of velocity equations. In relation to them, dynamic analyses are also performed. To do this, classical dynamic methods can be used for ad-hoc analysis of a certain manipulator while multi-body methods can be used for CAE evaluation of virtual designs. Such an analysis will provide the tools to determine the material and sections of links, as well as the drives needed for acceleration requirements.

Again, a practical approach is the use of the latter CAE software packages that implement multi-body methods to get kinematic and dynamic equations. Once the dimensions of the machine are defined and the desired working volume is reachable, iterative analysis of speeds, accelerations and singularities can be performed.

For this analysis, it is possible to use the same packages used to perform the position analysis. In the Verne machine of Fatronik, different trajectories have been analysed, like linear motions, circular interpolations and angular rotations, obtaining velocities and accelerations values in the motors.

Singularities can also be analysed with the commercial CAD packages but only to a certain extent. Like the situation with the mobility, they do not provide information about the type of singularity, but just some warning messages.

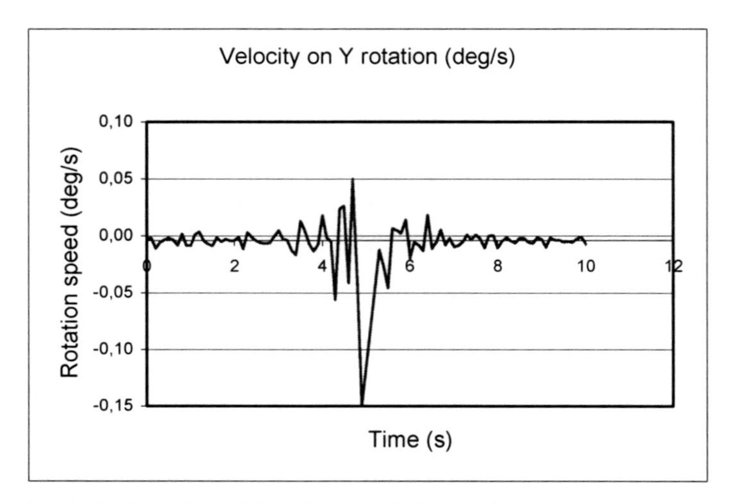

Fig. 10.8 Velocity profile in the analysis for singularity

For example, if a PRO/E representation of the mechanism is used for the analysis of singularities, and the mechanism has an inverse kinematic singularity, i.e., several inputs become dependent. The velocity profile of some axis will be altered in the proximities of the singular point as seen in Fig. 10.8. Also, the program reports a problem assembling the mechanism only when the position is close to the singular configuration. As shown no specific information on the singular points is provided; however it helps to detect possible problems in the mechanism.

10.4.5 The Optimisation

Previously to develop the design further, an optimisation analysis can be made. The objective of this study can be very different: minimising the torques on the drives or reducing the size of the machine. Commercial packages implement optimization processes with maximisation or minimisation objectives. This objective can usually be defined as a function, depending on geometrical parameters, or on measures (position, torques, speed, and so on), giving a high flexibility in order to design an optimal mechanism.

These optimisation tools have been applied to the design of the Verne machine to obtain the desired working volume with the minimal machine footprint. The optimisation has been made taking into account only geometrical parameters:

- The position of joints in the supporting frame. In order to maintain the DOF a relationship between some of the parameters is fixed.
- The length of the struts.
- The distance of the first axis of the tilting table (A axis) to the origin of the base coordinate system.

The objective of the minimisation is defined as a function, in this case the shop floor used surface. This function is defined based on two basic features of Pro/Engineer, the distance between two planes. The product of these two distances gives the surface of the machine, and this value has been set as the objective function.

Multi-objective optimisation is more difficult; it seeks the simultaneous optimisation of several functions that sometimes are opposite. There are very few works on this subject and it is an open problem; an approach could be to adapt techniques used in structural analysis such as the Pareto front based methods.

10.5 The Kinematic Calibration of PKMs

After the design process, the first prototype can be manufactured and several experimental tests should be carried out to check the correct assembly. After that, the control that implements the theoretical kinematic transformation has to be adjusted to the actual dimensions and assembly of the machine. For that purpose a calibration procedure has to be applied.

Because of the kinematically non-linear behaviour of PKMs, common methods for calibration like the adjustment of axes and setting the zero point cannot be applied. Parallel kinematics' axes can not be adjusted one by one, but need to be looked at as a complete system of coupled ingoing and outgoing values. This can be compared to the parameter-identification of any complicated system, for example a controlled system, which has to be identified. The system parameters that have to be identified here are the zero points, beam lengths and numerous different geometrical parameters that characterise the kinematical behaviour of the machine. To make the model fit as closely as possible to reality, we have to adjust these parameters; this adjustment is called *kinematic calibration.*

Though these values are all specified in the design drawings of the machine we still need to determine them by calibration. The manufacturing and especially the assembling tolerances of the parts of the motion mechanism can never be made small enough to spare calibration.

Calibration is always the determination of parameters of a model that describes a real system of any physical nature. The goal of this procedure is always to make the model behave as closely as possible to the real system itself. The principle of each calibration is to record the input/output-behaviour of the system and based on these recorded values fit the model to reality so that it simulates the input/output-performance of the modelled system.

Calibration is a well known problem for serial robots and conventional machine tools, and now it is a well-treated problem. It may be thought that the calibration of parallel robots may rely on the methods developed for serial morphologies but unfortunately this is not exactly the case. Indeed there is a major difference between both mechanisms: for serial ones small errors on the geometrical parameters induce large errors on the positioning of the end-effector, while for parallel machines these errors will also be small. Simulation for calibration is essential: it allows determining how much a calibration method is sensitive to noise in the measurements and to numerical errors. It allows for example to show that methods directly adapted from the calibration of serial robots may lead to results that are worse than the initial guess as soon as the simulated measurement noise is realistic.

There are two types of calibration methods:

- *External*: an external measurement device is used to determine (completely or partially) the real position of the platform for different desired configurations of the platform. The differences between the measured pose and the desired pose give an error signal that is used for the calibration.
- *Self-calibration*: the platform has extra sensors (sensors that can be used to solve other problems like the *forward kinematics*) and only the information of these extra sensors is used for the calibration.

The procedures of the first type tend to be difficult and tedious to use in practice but may give good results. The second method may be less accurate but is easy to use and has also the advantages that it can be fully automatised.

Calibration has two important aspects. On the one hand, we have the mathematical approach: (i) a selection of parameters for calibration, (ii) measuring machine configurations and (iii) a method for estimating the new machine parameters. On the other hand, it is the determination of (i) the external measurement artefact or (ii) the number and type of sensors (self-calibration).

10.5.1 A Mathematical Approach

Concerning the selection of parameters for calibration, the most common approach is to select those controlling the kinematic transformation of the machine. As the kinematic transformations are very different depending on the kinematic structure of the machine one may conclude that the calibration models are very different depending on the machine. However, reality shows that most of the parameters that rule the kinematic model are the same for most machines, such as the position of joints, the lengths of struts and the encoder offsets.

It is typical that models comprise simplifications. The mathematical model that the control uses to transform positions into drive values assumes that the motors move along absolutely straight lines. It is relatively difficult to model the movement in a more appropriate way, so we have to be aware of the fact that we might introduce errors here. Moreover we disregard the fact that universal joints perform a motion that might differ considerably from moving around an infinitely small point. Elasticity, clearance and elongation due to varying temperatures are not represented at all in the models.

Though it is partly possible to measure certain parameters directly, this is not necessary and not even reasonable. Because of all the simplifications the model contains, the values of the parameters will not be assigned in exact accordance to their physical values. Instead they will be adjusted by the algorithm in a way that makes the model as a whole fit reality as closely as possible. So the values often differ significantly from what would be determined by direct measurement.

The calibration minimises the difference between the output of the model and the output of the real machine. The input is the same for both. It is important to realize that the result of this optimisation is optimal in respect to the measurement values that have been gathered. To ensure that they are good in respect to the whole workspace and applicable under common working conditions, certain preconditions have to be kept. As a logical consequence of this it is clear that the platform poses that are measured have to be chosen deliberately.

The idea of the pose selection is to extract the points of the working volume that result in the most accurate estimates. For some poses, the parameters of the kinematic model do not influence the measurements much: the effects of measurement noise and un-modelled sources of error dominate over the effect of the variation of the kinematic parameters of the structure. As a result, the calibrated parameters obtained will not be reliable.

Researchers have proposed a variety of observability indices to quantify the goodness of pose selection; these indexes area based on the singular value decomposition (SVD) of the Jacobian matrix of the differential kinematics. Menq and Born [32] proposed an observability index related to the product of all singular values. Driels and and Pathre [12] proposed the condition number; Schroer et al. [46] stated that a condition number below 100 is required for reliable results. Nahvi et al. [35] proposed the minimum singular value. Hollerbach and Nahvi [36] proposed the noise amplification index, defined as the ratio of maximum singular value to condition number. Daney et al. [11] proposed a method based on the identification Jacobian, combined with a meta-heuristic approach to decrease the sensibility to local minima. The last method applies also to Gough platform calibration.

Other issues that sometimes appear are related to the presence of redundant parameters during the calibration. This term applies to the fact that the error of two different parameters of the kinematic model translate into the same measured error, causing a poor estimation in the parameters. Meggiolaro and Dubowsky [31] proposed an analytical method to eliminate the redundant parameters in calibration. The method is based on the non-singular of the Identification Jacobian matrix using the D-H parameters.

From the mathematical point of view, kinematic calibration reduces to obtaining the unknown parameters of set of non-linear functions (the kinematic transformation) evaluated at a discrete number of points (measuring poses) with a very good estimation of the function parameters (theoretical machine configuration). This problem is already solved and is known as the non-linear least squares problem.

However there is an important aspect to be considered in this approach: there is noise in the measurement. If this noise is too high, the resulting fitting problem can be useless, and the estimated parameters very bad.

Another aspect to take into account during calibration is the presence of errors due to the reduced DOF of the kinematic structure. In the case of the Gough/ Stewart platform or 6 DOF parallel kinematic machines, if other sources of error are neglected (thermal effects, deformations due to gravity, measuring errors ...) the calibration will obtain zero error in position and orientation. This is true since the number of controlled DOF is the same as the possible sources of error of the platform: three translations and three rotations.

Parallel kinematic machines with reduced DOF use passive constraints to eliminate the non-desired DOF. Due to machining and assembly errors, it is very likely that these passive elements induce some error in the same DOF that restrict, or in the worst case in a controlled DOF. For example a three DOF Cartesian machine with passive elements for angular rotations will have some deviations between the desired fixed orientation and the real orientation.

The problem is that the kinematic model cannot compensate for these angular errors, since the orientation is uncontrolled, and as a consequence calibration is not possible, so a remaining error is present. Even worse, if the rotation error induces a displacement in the X direction for example this error will affect the calibration, producing in some cases strange results in the parameters. Passive elements must be carefully controlled in precision, reducing the errors to the minimum during the assembly process, or measuring the magnitude of the errors and taking into account the calibration process.

10.5.2 Measuring on External Methods

The measuring stage of the calibration is not 100% independent from the already common aspects of the mathematical approach. The pose selection and adjusted parameters can limit the number of applicable measuring methods. Another aspect is that with the commercially available systems, it is not possible to measure at the same time the six coordinates. In fact, the lower the number of simultaneous measurements, the better the measuring precision. Some typical TCP measuring devices are:

- The laser tracker.
- Photogrammetry, vision.
- The double ball bar.
- The grid encoders.
- Dial gauges plus extra fixtures.
- The reference piece plus touch probes.
- Others (inclinometers, accelerometers).

The choice of the measuring apparatus must be done taking into account the following criteria:

- The precision of the measuring device (above all).
- The time needed for measuring (degree of automation).
- The qualification required for the measuring device user.
- The limitation on measuring range.

Figure 10.9 shows the calibration method for the Verne machine [44]. The calibration strategy uses a double ball bar from Renishaw®, a commercially available element. Combining this element with a calibrator provided also by Renishaw, it is possible to measure the absolute distance between the centres of two calibrated spheres.

The calibration procedure measures the distance (150 mm theoretical) between three spheres placed in a known position in the tilting table (measured with a MMT) and a sphere placed in the centre of the tool holder. The machine moves along a set of points that covers the entire working volume in position and rotation. Two tool lengths are used to measure the distance between the spheres, so 6 sets of measurements are obtained, making a total of 1944 measured points.

The upper part of Fig. 10.10 shows the errors of the programmed 150 mm distance on the calibration process. As shown there is a maximum error of 1 mm. The lower part shows the verification of the machine precision of the calibration. The error is now less than 5% of the starting error in most of the points, with a maximum of 10% in some extreme positions.

Fig. 10.9 Calibration test bed for Verne™ PKM

A similar method to the described is used to calibrate the Space-5H machine. Once the programs files are created and the procedure standardised, a complete machine calibration can be performed in 5h by a machine user. The main limitation of this method is the range of measure of the ball bar, limited to ±1 mm. However, this problem only appears in the first calibration after assembly, but not in the recalibration process, so this is not a problem for the end user.

This method is an example of external calibration methods. Another calibration strategy that uses a ball bar is found in [39]. In this case it is applied to the calibra-

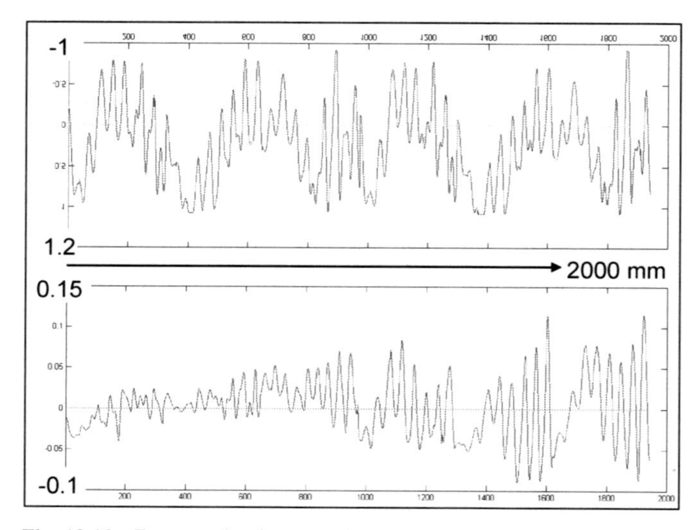

Fig. 10.10 Error results (in mm) of the calibration in the Verne™ PKM

Fig. 10.11 HexaM™ machine, by Toyoda®

tion of a fully parallel structure, the HexaM™ (Fig. 10.11). The Cartesian positioning error is reduced from ±200 μm to less than 50 μm.

Figure 10.12 shows the calibration procedure of a 4 DOF parallel robot using a 3D camera [2]. The camera "sees" a target of 4 points and obtains the position of the target in the camera coordinate system.

The precision of the camera system is 0.25 mm, and a measuring range of 2 m^3 is available. The main problem of this method is determining the position of the camera coordinate system with respect to the machine coordinate system, but it can be solved using transformations between coordinate systems and some additional measurements. A calibration is performed using this methodology and the obtained results improve the accuracy by 80%, giving precision values close to the precision of the measuring artefact. A similar strategy is used to calibrate the Tricept™ PKM.

There is yet another method widely used to calibrate based on constraining the motion of the TCP in one dimension by the use of test parts and probes

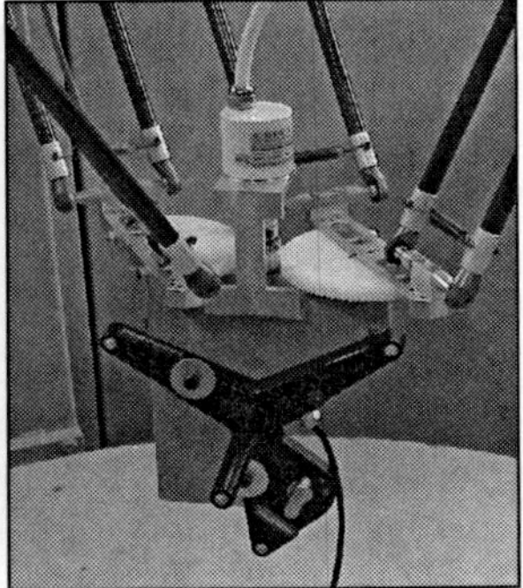

Fig. 10.12 Calibration of a 4-DOF PKM robot [2], the camera and target

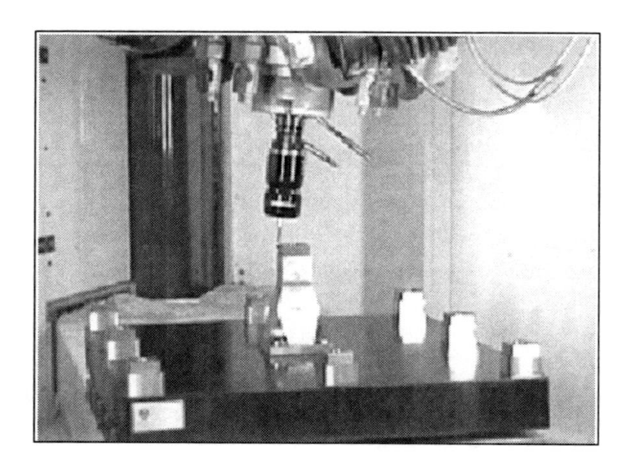

Fig. 10.13 Calibration with a sampling plate

(Fig. 10.13). The principle of the calibration procedure can be explained as follows: a number of defined points are approached with a probe and a special NC program on a calibration plate with some special towers whose sizes and position relatively to the plate centre-point are exactly known (measured with a CMM). In this procedure, the co-ordinates of these points and the corresponding strut length values are determined by the machine control and stored into a file. This file is the initial basis for calibration itself.

Other approach for calibration consists on machining a part in the machine to be calibrated. Then the deviation between the real part and the theoretical part is measured, usually in a CMM. From this information a calibration is performed.

The approach from the ISW Institute [38] consists in a test-work piece, a parallelepiped with an array of spherical holes (Fig. 10.14). Holes are arranged in a grid of 20×20 mm. On a surface of 300×400 m this yields 273 holes. The workspace is actually bigger than 300×400 mm, but this choice simplifies the handling of the work piece. As the kinematical properties of the machine differ only slightly with height (because the vertical and parallel direction of all drives), it is sufficient to arrange all holes in one horizontal plane.

The holes have a diameter of 10 to 16 mm. This is chosen in order to ensure good measurability with the CMM. The depth of the holes should be about half their diameter. It is better to choose a bit less than that, because the depth may vary in case of inaccuracies and it is more secure to be rather too high than too deep. In case one mills too deep, the shaft of the tool might be damaged or deformed. Even elastic deformation without damaging the tool is unwanted, because it leads to a wrong measurement.

The holes are machined with line wise varying spindle orientation (see Fig. 10.14). The changing of spindle orientation improves the value of the measurement information. The more the different poses (a pose is a combination of position and orientation) vary from each other, the better. This rule is proven by practice and has a clear theoretical background. The matrix that is used for the method of steepest descent is better conditioned then. When all poses are almost

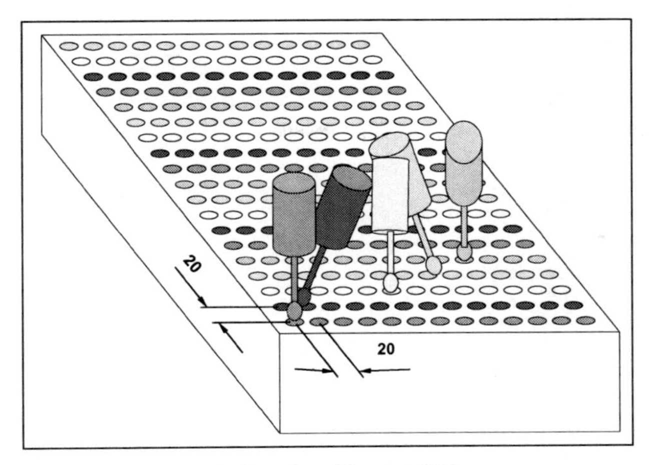

Fig. 10.14 Test work piece for calibration [38]

the same, we will have a rank deficiency in the Jacobian matrix. As soon as the rank is lower than the number of parameters we want to identify, the method fails.

The holes will be machined with a spherical cutter with changeable cutting plates. Feed while milling will always be in z-direction. The revolution speed of the spindle will be 1000 U/min at a feed rate of only 0.5 mm/min while finishing. Coolant will be used to achieve a surface that is as smooth as possible and to reduce forces that would affect the position of the TCP.

Other research on this topic has been done by Chanal et al. [5, 6] with the Verne machine.

A special combination of calibration strategies is being developed by Siemens to calibrate the SKM 400 machine, a three translational axes horizontal machine [45]. The calibration is done in two stages. The first one uses the laser tracker LTD500 by Leica®. With this laser, the three Cartesian positions of the TCP are obtained. With this information, a conventional calibration is performed, resulting in a global machine precision of 35 μm.

Taking into account that the precision of the measuring system is around 15 μm, the remaining 20 μm are probably due to limitations on the error model (elasticity, geometry). In that case two solutions are possible: (i) to improve the error model (to take into account other effects or to improve the quality) or (ii) to develop an additional compensation strategy.

This is being done by Siemens adding special functionality to the NC by means of the space error compensation (SEC) strategy. With SEC the space errors can be compensated simultaneously in the three spatial directions x, y, and z. The basis of this system is measuring instruments that record the three coordinates of a points simultaneously, such as the 3D-Laser Tracker.

First the NC moves the TCP to a defined target position in the working area, making the assumption that the point defines ideal positions. Then, the actual positions are determined at these discrete points by means of the Laser Tracker. By comparing the actual position and the target position, the error can be determined for each measuring point. The resulting 3D error table is then fed to the NC. This table enables compensating the error in real time, based on a spatial, linear interpolation process.

By using this method the space error of the machine was reduced to a value of 15 μm, the same value as the external measuring system

10.5.3 Self-calibration Strategies

Concerning the self-calibration strategies, it is known that two or more redundant information from sensors is enough to perform a self-calibration. Self-calibration has the potential of (a) removing the dependence on any external pose sensing information; (b) producing high accuracy data over the entire workspace of the system, (c) being automated and completely non-invasive, (d) facilitating on-line accuracy compensation and (e) being cost effective.

Researches have been working on self-calibration for a number of years, due to the advantages over conventional calibration. Various self-calibration systems are presented in [30]. All this approaches use the strategy of blocking joints and use information of sensors to perform calibration

This approach, although permitting the calibration of the parallel structure, allows bad automation, and requires sometimes the manufacturing of parts that can give precision problems on the clamping procedure. This strategy of using redundant sensor placed on universal joints is used in [52]. Using this approach all the advantages of the self-calibration strategy can be obtained.

10.6 The Control of Parallel Kinematic Machines

The usual strategy nowadays is to use conventional strategies applied to PKMs. The most common scheme is the "inverse kinematics" one. The general principle is summarised in Fig. 10.15: the trajectory generation module produces desired Cartesian positions which are transformed into desired joint positions by the inverse kinematics module. Then a classical low level control loop is in charge of servoing the joint positions.

There are two main reasons for this quasi-systematic use of such schemes:

- It is possible to use a different sampling time for the control loop (which remains very simple here: it is motor control) and the computation of the inputs (which needs in many cases a lot of floating point operations). These two tasks can even be done by two different systems (e.g., amplifiers can take in charge the low level position control while a computer does the *inverse kinematics*).
- For most parallel mechanisms, *inverse kinematics* is much easier to compute than *forward kinematics*.

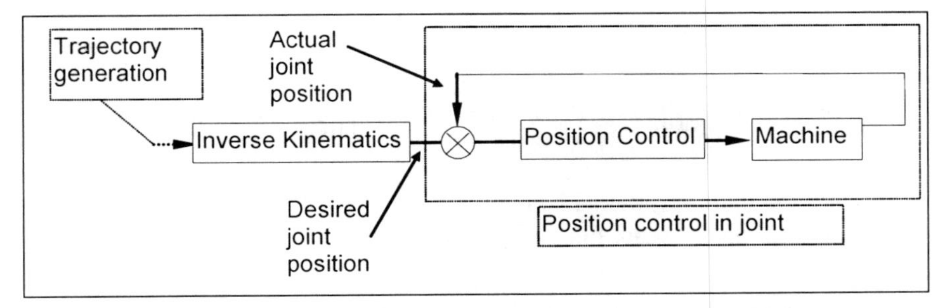

Fig. 10.15 Classical control schemes with position control in joint space and trajectory generation in Cartesian space

One possible drawback of the method comes from the joint level control: it is difficult to obtain a good accuracy in Cartesian space (which is often required) with a control independent of the Cartesian configuration.

The second strategy is based on *forward kinematics*, as shown in Fig. 10.16. The control loop is closed at the Cartesian level: the servoed variables are the one really important for the user which is very appealing to get good accuracy in Cartesian space.

However, the drawback comes from the necessary computation of the forward kinematics and the inverse of the Jacobian inside the control loop (that is: at a high sampling rate). So, for a given control hardware, the sampling time will be bigger in Cartesian control than in joint control.

Moreover, in the case of parallel robots, the computation of *forward kinematics* cannot be expressed as an algebraic equation, in general (note that some mechanisms have this characteristic). The most common way to compute the forward kinematic is based on numerical iterative techniques generally described above.

In the case of redundant machines, the control schemes should support (i) more motors than DOF, or (ii) more sensors than DOF, and will probably be controlled with schemes requiring you to get information from both the joint level (as in schemes based on inverse kinematics) and the Cartesian level (as in schemes based on forward kinematics) at the same control frequency.

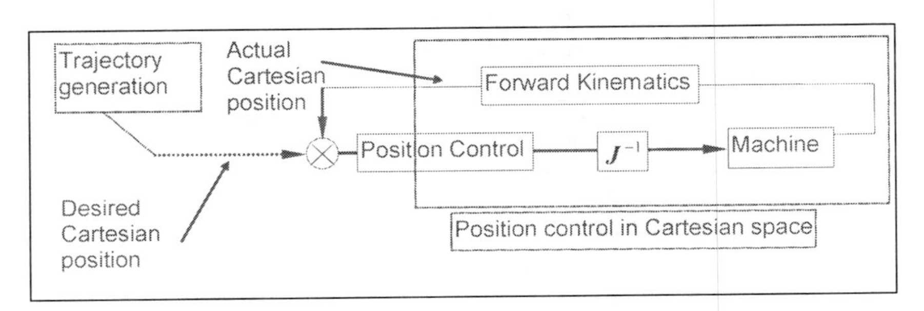

Fig. 10.16 Control in Cartesian space

10.6.1 Models Specific to Parallel Kinematics Machines

By definition, the mechanical architecture of a PKM differs from a serial one. Even if control schemes developed for serial architectures are usable for PKMs, they are far from giving good results.

To reach good accuracy and a high speed, some particularities have to be taken into account during the controller design. The main difference between classical serial architectures and parallel ones are recalled:

- The synchronisation of the axes has to be as perfect as possible. For example, machining a straight line with a classical machine tool is very easy, but for a PKM, the straightness of the machined line is the result of the quality of the synchronisation between all actuators.
- The direct model, in general, does not exist in an analytical form. It has to be computed in an iterative way. The inverse models often exist in an analytical form. The control schemes used for PKMs take this consideration into account. The open controller to be developed has to handle such considerations and also the mathematical functions to compute these models.
- As intrinsically PKMs are able to reach higher speeds and accelerations, the control must be able to handle these performances and to do so, to offer the possibility to implement advanced controls like dynamic or predictive.

There are some specific problems to PKMs that influence the requirements of control systems:

- The inverse kinematic model: for many simple PKMs (e.g., those based on a Delta™-like principle, or Hexapods™, or HexaM™, or Hexaglide™, and so on), the inverse kinematic model is rather easy to derive and to implement. Such models, as no iteration processes appear during their resolution, are extremely easy to compute and should give no problem to control systems.
- Velocity relationships: the control system has to support linear algebra relationships.
- Singular configurations: as already mentioned, singular configurations are detected analysing the Jacobian matrix of the mechanism, so the control system has to support linear algebra calculus.
- The forward kinematic model: writing the position relationships for a general PKM gives usually a set of polynomial equations that are (in general) not solvable in a closed form. Usually, it is possible to resort to an iterative scheme to work on the problem.
- Redundant machines: specific control strategies are required.
- Other cases: when inverse kinematics is not solvable without iteration (Space-5H, Verne) or singular configurations are not visible resorting only to the Jacobian matrix.
- The inverse dynamic model for advanced control: the use of this type of control requires the computation of the inverse dynamic model of the machine. The aim is to calculate, in an analytical way, the torques applied by the motors, as a func-

tion of the position, velocity and acceleration of the motors. First of all, the Jacobian matrix of the mechanism has to be calculated. This matrix permits having a linear relationship between the velocity of the motors, and the velocity of the end effector. In addition, the transpose of this matrix gives a linear relationship between efforts exhorted on the end effector and the torques applied by the motors.

There are two types of control suites for the PKM structure: the "dynamic controller" and the model-based predictive controller, although both can be used also for conventional machines (serial).

10.6.2 The Dynamic Controller

Some tasks performed by machine tools require fast motions and high dynamic accuracy. In that case, the performance of the control has to be improved by taking into account dynamic interaction torques. This control is named *computed torque control* or *inverse dynamic control*, and is based on the use of the dynamic modelling of the machine. This control permits the linearisation and the decoupling of the equations of the model, permitting a uniform behaviour in the whole workspace of the machine. Thus, in order to implement such a control, the inverse dynamic model of the machine has to be calculated and implemented in the control.

Although the dynamic control is suitable for conventional serial machines and parallel kinematic machines, it becomes almost essential in the second category. Usually, and due to the spatial distribution of the elements, the drives experience a big variation in the commanded inertia. The traditional PID control structure can manage limited variations in this inertia, and usually, the variations on the parallel kinematic machines are bigger than these limits. This gives a situation of bad trajectory follow-up, resulting in machining inaccuracies and instability problems.

Another problem related to parallel kinematic machines concerns the drive tuning. The usual strategy is to tune the drives in the "worst" position. This position is usually characterised by a lower stiffness, big axes inertia, *etc*. Sometimes, the drive tuning is performed in such a way that a compromise in the entire working volume is achieved. In both situations, the result is that the machine is optimised in a usually small area of the working volume. In the rest of the position of the working volume, the drive parameters could be increased or adjusted to obtain a better machine performance.

Both problems can be minimised using the dynamic control. As a result, the machine will have the same behaviour on the entire working volume, with increased and optimised dynamic control and optimal machining precision.

The inverse dynamic control consists of transforming a non-linear control problem into a linear one by using an appropriate feedback law (see Fig. 10.17). The dynamic model can be written using two vectors directly used in the control. The block diagram of the control is presented in the following figure.

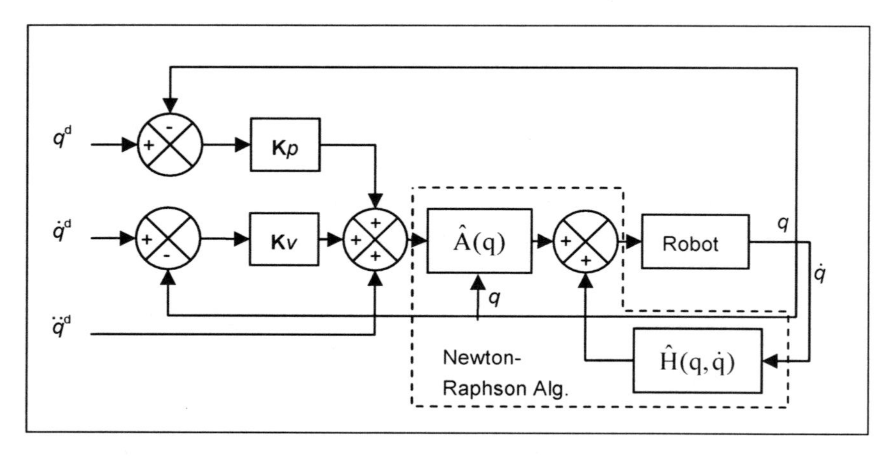

Fig. 10.17 Block diagram of the inverse dynamic control

In the ideal case of a perfect modelling and in the absence of a disturbance, the problem consists in a linear control of decoupled double integrators, considering that w(t) is the input control vector:

$$\ddot{q} = w(t) \tag{10.2}$$

The use of this control requires on-line calculation of the inverse dynamic model, and a good knowledge of inertial, friction and mass parameters. The CPU has to be sufficiently powerful to be able to make this computation. In the classical case, the needs of such a control, in terms of the controller, are drives having current loop, i.e., drives able to control the torques applied by the actuators, and encoders. The velocity and acceleration is calculated by the control itself using the position information.

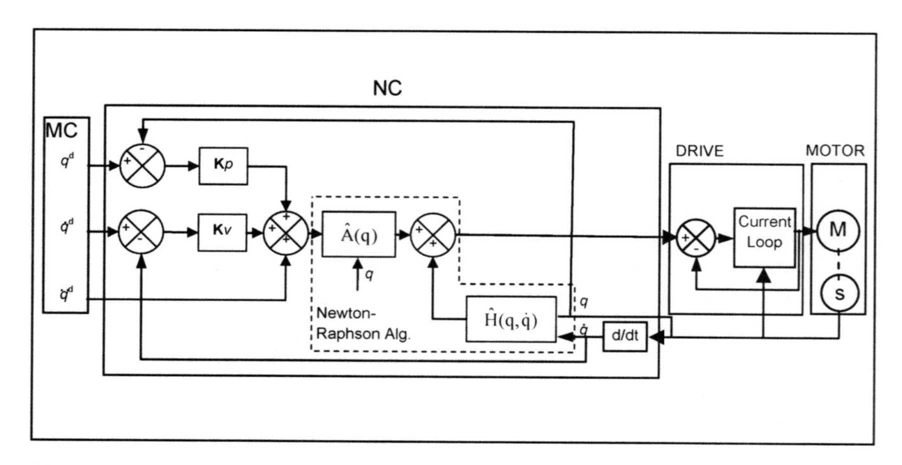

Fig. 10.18 Implementation of the inverse dynamic control in the controller

In order to improve the behaviour of the control, some additional sensors can be added. For example, an accelerometer or a force sensor can be placed on the effector of the machine. The implementation of this inverse dynamic control in the controller can be done as presented in Fig. 10.18.

This figure shows that all the calculations are done in the NC. It needs an input that corresponds to the encoders of the motors, and an output allowing the control of the torque of the motor through the current loop of the drive. As explained previously, in order to extend the control, some additional inputs should be added, as accelerometers or force sensors.

10.6.3 The Model-based Predictive Controller

Advanced controllers proposed to improve performances of production machines will be based on the use of the dynamic model of the machine encountering for inertial parameters. Efficient modelling can also take into account friction parameters and elasticity. Model-based controllers are then implemented after the identification phase, which consists of estimating the model parameters.

The inverse dynamic model of machine tools composed of n moving links calculates the motor torque vector $\boldsymbol{\tau}$ (the control input) as a function of the generalised coordinates (the state vector and its derivative). It can be obtained from the Lagrangian or Newton Euler equation:

$$\boldsymbol{\tau}=\mathbf{A}(\mathbf{q})\,\ddot{\mathbf{q}}+\mathbf{H}(\mathbf{q},\dot{\mathbf{q}}) \tag{10.3}$$

where $\boldsymbol{\tau}$ is the $(nx1)$ motor torque vector, $(\mathbf{q},\dot{\mathbf{q}},\ddot{\mathbf{q}})$ are the (nx1) vectors of generalised joint positions, velocities and accelerations respectively, $\mathbf{A(q)}$ is the (nxn) inertia matrix, and $\mathbf{H}(\mathbf{q},\dot{\mathbf{q}})$ is the $(nx1)$ vector of the centrifugal, Coriolis, gravitational and friction torques.

Equation 10.3 can be rewritten as a linear relation to a set of standard dynamic parameters $\boldsymbol{\chi}_s$:

$$\boldsymbol{\tau}=\mathbf{D}_S(\mathbf{q},\dot{\mathbf{q}},\ddot{\mathbf{q}})\,\boldsymbol{\chi}_S \tag{10.4}$$

$\boldsymbol{\chi}_s$ is the $(13nx1)$ vector of standard dynamic parameters:

$$\boldsymbol{\chi}_S^j=\begin{bmatrix}XX_j & XY_j & XZ_j & YY_j & YZ_j & ZZ_j & MX_j & MY_j & MZ_j & M_j & Ia_j & F_{Vj} & F_{Sj}\end{bmatrix}^T \tag{10.5}$$

It is composed for each link j of (XX_j, XY_j, XZ_j, YY_j, YZ_j, ZZ_j) the 6 components of the inertia tensor, (MX_j, MY_j, MZ_j) the 3 components of the first moment, (M_j) the mass, (Ia_j) the total inertia moment for the rotor actuator and gears, (F_{Vj}, F_{Sj}) the Coulomb and the viscous friction parameters.

For the estimation of the parameters of the dynamic model, usually $\boldsymbol{\chi}$ is estimated as the least squares (WLS) solution of an over determined $(r\ x\ p)$ linear system obtained from sampling and filtering the dynamic model along a trajectory $(\mathbf{q}(t), \dot{\mathbf{q}}(t), \ddot{\mathbf{q}}(t))$:

$$\mathbf{y}(\boldsymbol{\tau}) = \mathbf{W}(\mathbf{q}, \dot{\mathbf{q}}, \ddot{\mathbf{q}})\,\boldsymbol{\chi} + \boldsymbol{\rho} \quad (10.6)$$

$\mathbf{y}$ is the $(r \times 1)$ measurement vector.
$\mathbf{W}$ is the $(r \times p)$ observation matrix.
$\boldsymbol{\rho}$ is the $(r \times 1)$ vector of errors.
The next steps can be divided in two stages:

1. Identification procedure:
   - exciting trajectories,
   - signal acquisition,
   - filtering,
   - derivative estimation,
   - weighted least computation.
2. Control strategy:
   - model parameters,
   - controller parameters.

Figure 10.19 illustrates the model-based predictive control scheme.

Recent researchers are working on control algorithms without a model. Control strategies based on dynamic models fit very well to parallel kinematic machine tools. However, this approach depends mainly in the precise fitting of the dynamic model with the machine. During the machine tool life, small variations in the operating conditions occur, such as ambient conditions, wear, and different load conditions that lower the precision of the dynamic model. In this case, alternative algorithm models based on dynamic models with different precision levels can be used.

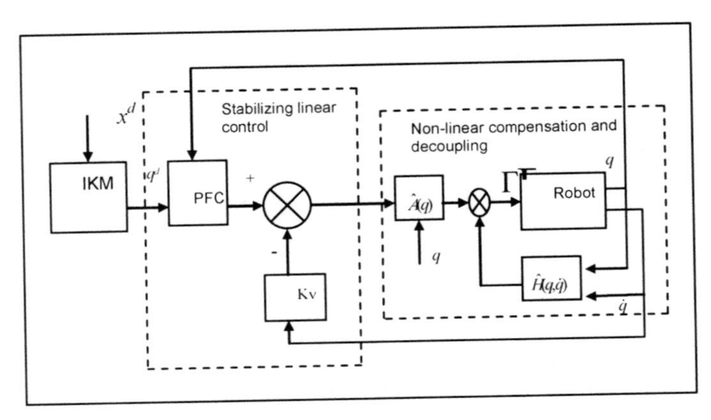

Fig. 10.19 Model-based control loop architecture

One of these approaches can be a dynamic model with almost linear independent kinematic chains. The control structure can be a standard PID controller for each kinematic chain that includes a self-tuning procedure of the regulation loops. Although this is not a multi-variable controller, variations in machine conditions are considered. In the same working line adaptive systems can be considered. These adaptive systems perform measures in real time over the machine but considering the multi-variable aspect and offer good perspectives for parallel kinematic machines.

Adaptive systems use rough dynamic models, but as they obtain different measures from the machine, they are more subject to external disturbances. Their industrial application is therefore more realistic than the systems based on models or systems with independent control loops. With the objective of industrial applications, a robust control can ensure the robustness of the system and control laws, in order to guarantee the correct trajectory follow-up. Robust control is effective if small and bounded disturbances over the model appear.

Finally, controllers based on neural networks do not consider any dynamic model, and are based on a learning process and an adaptation process based on measures available throughout the machine. All controllers except the neural controller are based on the correct choice of the dynamic model. The main disadvantage of the controllers based on neural networks, is that they are very difficult to adjust if many DOF are considered.

10.7 Conclusions and Future Trends

Parallel kinematic machines are promising devices in the field of high speed machining. In order to get the full potential of such kinematic structures it is mandatory to do designs for specific tasks. These designs require a systematic procedure of synthesis, analysis and optimisation, which needs a deep theoretical background. Due to implicit geometrical restrictions their workspaces are limited and that is the reason for the prevalence of hybrid morphologies of parallel modules and movable working tables.

Calibration is a key part in the design of such machines. Although the calibration methodology more suitable for industrialisation is the self-calibration methodology, the reality is that calibration with external devices is more widely used in the industry. Sometimes, adding sensors for self-calibration is not easy. For example, some industrial machines include universal joints that do not allow the integration of such sensors. But the main problem is that each machine configuration needs a deep analysis of the number and the type of redundant sensors to be integrated. And sometimes the required sensors are against the construction and functionality of the machine. Another problem is that adding sensors needs a calibration of the sensors offsets, and sometimes the required sensor precision is bigger than the state of the art.

Acknowledgements The authors would like to acknowledge the grants received from the Ministerio de Educación y Ciencia (Project DPI-2005-02207) and the FEDER funds of the European Union, as well as from the Universidad del País Vasco-Euskal Herriko Unibertsitatea (Project GIU05/46).

References

[1] An CH, Atkeson CG, Hollerbach JM (1986) Experimental determination of the effect of feedforward control on trajectory tracking errors. Proc of IEEE Int Conf on Robotics and Automation, 55–60
[2] Andreff N, Renaud P, Pierrot F, Martinet P (2004) Vision-Based Kinematic Calibration of an H4 Parallel Mechanism: Practical Accuracies. Industrial Robot Int J, 31(3):273–283
[3] Ball RS (1900) A treatise on the theory of screws. Cambridge University Press, Cambridge
[4] Bianchi G, Fassi I, Molinari Tosatti LA (2001) Virtual Prototyping Environment for Parallel Kinematic Machine Analysis and Design 32nd Int Symposium on Robotics, Korea
[5] Chanal H, Duc E, Ray P, Hascoët J-Y (2007) A new approach for the geometrical calibration of parallel kinematics machines tools based on the machining of a dedicated part. Int J of Mach Tool aManufact, 47(7–8):1151–1163
[6] Chanal H, Duc E, Ray P, Hascoët J-Y (2006) Design of a dedicated machined part for calibrating a parallel kinematics machine tool, 5 congrès Int Usinage à Grande Vitesse, Metz, 317–328
[7] Clavel R (1991) Conception d'un robot parallèle rapide à 4 degrés de liberté. PhD Thesis, EPFL, Lausanne, Switzerland
[8] Codourey A (1998) Dynamic modeling of parallel robots for computed-torque control implementation. Int J of Robotics Research, 17(12):1325–1336
[9] Collado V, Herranz S (2004) Space 5H – A New Machine Concept for 5-Axis Milling of Aeronautical Structural Components. PKS 2004 Conf Proc
[10] Daeyoung Machinery Eclipse-RP. US Patent No. 6,135,683
[11] Daney D, Papegay Y, Madeline B (2005) Choosing Measurement Poses for Robot Calibration with the Local Convergence Method and Tabu. Int J of Robotics Research. 24(6):501–518
[12] Driels MR, Pathre US (1990) Significance of observation strategy on the design of robot calibration experiments. J of Robotic Systems, 7(2):197–223
[13] Giddings&Lewis Variax. US Patent No. 5,388,935
[14] Gosselin CM, Angeles J (1990) Singularity analysis of closed loop kinematic chains. IEEE Trans on Robotics and Automation. 6(3): 281–290
[15] Gosselin CM, Kong X (2004) Cartesian Paralle manipulators. Patent US 6729202
[16] Gosselin CM, Pierre ES, Gagné M (1996) On the Development of the Agile Eye. IEEE Robotics and Automation Magazine
[17] Gough V, Whitehall S (1962) Universal tyre test machine. Proc of the FISITA ninth Int technical congress. 117–137
[18] Hennes N, Staimer D (2004) Application of PKM in Aerospace Manufacturing – High Performance Machining Centers ECOSPEED, ECOSPEED-F and ECOLINER. PKS 2004 Conference Proceedings.
[19] Hermes Patent EP1 245 349
[20] Hervé J M (1999) The Lie Group of Rigid Body Displacements, a Fundamental Tool for Mechanism Design. Mechanism and Machine Theory. 34(5): 719–730
[21] Hesselbach J, Kerle H, Frindt M, Pietsch (2001) I PORTYS – A machine concept with parallel structure for precise pick and place operations at high-speed. In: Proc of RAAD Int Workshop on Robotics in Alpe-Adria-Danube Region

[22] Hesselbach J, Becker O, Krefft M, Pietsch I, Plitea N (2002) Dynamic modelling of plane parallel robots for control purposes. In: Proc. of the Chemnitz Parallel Kinematics Seminar (PKS):391–409
[23] Honegger M, Brega R, Schweiter G (2000) Application of a nonlinear adaptive controller to a 6 DOF parallel manipulator. In Robotics and Automation, Proc. ICRA '00. IEEE Int Conference 2:1930–1935
[24] http://www.ds-technologie.de
[25] Ingersoll Octahedral Hexapod HOH-600. US Patent No. 5,392,663
[26] Kim DH, Kang JY, Lee KI (2000) Robust tracking control design for a 6 DOF parallel manipulator. J of Robotic Systems,17:527–547
[27] Kong X, Gosselin CM (2007) Type synthesis of parallel mechanisms. Springer, Dordrecht
[28] Loreto H, Garrido R (2005) Stable neural PD controller for redundantly actuated parallel manipulators with uncertain kinematics. In: Decision and Control, 2005 and 2005 European Control Conf CDC-ECC '05. 2035–2040
[29] Li CG, Ding HS, Wu PD (2003) Application of mrac to a 6-DOF parallel machine tool. In: Machine Learning and Cybernetics, 2003 Int Conf, 4:2164–2167
[30] Maurine P, Abe K, Uchiyama M (1999) Toward more accurate parallel robots. In: 15th World Congress of Int. Measurement Confederation, Osaka
[31] Meggiolaro MA, Dubowsky S (2000) An Analytical Method to Eliminate the Redundant Parameters in Robot Calibration. Proceedings of the 2000 IEEE Int Conference on Robotics and Automation, 3609–3615
[32] Menq CH, Borm JH, Lai JZ (1989) Identification and observability measure of a basis set of error parameters in robot calibration. J of Mechanisms, Transmissions and Automation in Design, 11:513–518
[33] Merlet JP (2006) Parallel Robots. Springer, Dordrecht
[34] Mikromat 6X. WO9943463A1: Hexapod Machining Centre
[35] Nahvi A, Hollerbach JM, Hayward V (1994) Calibration of a parallel robot using multiple kinematic closed loops. Proc IEEE Int Conf Robotics and Automation, 407–412
[36] Nahvi A, Hollerbach JM (1996) The noise amplification index for optimal pose selection in robot calibration. In: IEEE Int Conf on Robotics and Automation, 647–654
[37] Neumann KE (1988) Robot. US Patent No. 4732525
[38] Okuma Cosmo Center PM-600. US Patent No. 6,203,254
[39] Ota H, Shibukawa T, Tooyama T, Uchiyama M (2002) Forward Kinematic Calibration Method for Parallel Mechanism Using Pose Data Measured by a Double Ball Bar System. In: 1st Korea Japan Conf Positioning Technol 2002,113–118
[40] Pierrot F, Company O et al. (2001) Four degrees of freedom parallel robot. Patent EP 1084802
[41] Pierrot F, Marquet F, Company O, Gil T (2001) H4 parallel robot: Modeling, design and preliminary experiments, ICRA '01: Int Conf on Robotics and Automation, 3256–3261
[42] Pritschow G, Eppler C, Garber T (2002) Influence of the dynamic stiffness on the accuracy of PKM. In: Proc of The 3rd Chemnitz Parallel Kinematics Seminar, PKS 2002, 313–333
[43] Salgado O, Altuzarra O et al. (2007) Robot paralelo con cuatro grados de libertad. Spanish Patent P200702793
[44] San Martin Y, Gimenez M, Rauch M, Hascoët J-Y (2006) Verne – a new 5 axes hybrid architecture machining centre. In: Proc of the 5th Chemnitz Parallel Kinematics Seminar, PKS 2006
[45] Schoppe E, Pönish A, Maier V, Puchtler T, Ihlenfeldt S (2002) Tripod Machine SKM 400 Design, Calibration and Practical Application. In: Proc of the 3rdh Chemnitz Parallel Kinematics Seminar, PKS 2002, 579–594
[46] Schroer K (1993) Theory of kinematic modelling and numerical procedures for robot calibration. Chapman & Hall, London
[47] Schwaar M, Jaehnert T, Ihlenfeldt S (2002) Mechatronic Design, Experimental Property Analysis and Machining Strategies for a 5-Strut-PKM. PKS 2002 Conference Proceedings

[48] Selig JM (2005) Geometric Fundamentals of Robotics. Springer, Dordrecht
[49] Stewart D (1965) A platform with six degrees of freedom. Proc of the IMechE, 180(1):371–385
[50] Tsai L, Joshi S (2000) Kinematics and optimization of a spatial 3UPU parallel manipulator. ASME J of Mechanical Design, 122(4):439–446
[51] Vivas A, Poignet P (2005) Predictive functional control of a parallel robot machine. Control Eng Practice, 13(7):863–874
[52] Zhuang H, Liu L (1996) Self-calibration of a class of parallel manipulators. Proceedings of The IEEE Int Conf on Robotics and Automation, 2(2):994–999
[53] Zlatanov D et al. (1998) Identification and classification of the singular configurations of mechanisms. Mechanism and Machine Theory, 33(6):743–760

Chapter 11
Micromilling Machines

L. Uriarte, J. Eguia and F. Egaña

Abstract In this chapter micromilling machines are presented, starting with a clear differentiation between ultraprecision milling machines and micromachines. The main characteristics of the micromilling process are also summarised, including size effects affecting milling at the micro scale and more usual applications. Furthermore, a more detailed description of the different subsystems and components specific for micromilling machines is shown: guideways, structural materials, drives, measuring systems, spindles and additional equipment. Finally some commercial micromilling machines from well-known machine tool builders are presented.

11.1 Introduction and Definitions

Miniaturisation has become widespread, with a broad range of applications in many fields including microelectronics (mobile phones and sensors), motor vehicles, medicine (implants, microdosage), biomedicine, the chemical industry and watch-making. However, there is a clear imbalance between the ease with which batch-fabricated microcomponents can be produced in silicon compared to the difficulties and the cost associated with their manufacture in other materials.

In micromanufacturing, machining equipment and techniques can be roughly grouped into "ultraprecision processes", including micromilling, and processes based on silicon machining techniques – typically found in the electronics industry and known more precisely as "micromachining processes" – which require

L. Uriarte, J. Eguia and F. Egaña
Department of Mechatronics and Precision Engineering, Foundation Tekniker-IK4
Fundación Tekniker-IK4, Avda. Otaloa 20, 20600 Eibar, Spain
luriarte@tekniker.es

L. N. López de Lacalle, A. Lamikiz, *Machine Tools for High Performance Machining*,

clean-room facilities. The first group [1, 13] is based upon conventional machining processes, but in which the critical dimensions require submicrometric geometrical precision. On the other hand, in spite of their relatively high capacity in terms of resolution and precision, micromachining processes are currently limited to 2½D geometries and to a relatively small number of materials, and high production volumes are required to make them profitable [11].

So, the most usual meaning of a micromilling machine refers to ultraprecision milling machines, with submicron accuracies, that is, accuracies under 1 micron or less, usually one tenth of a micron. Machine tools capable of such extreme accuracy may be applied to microscopic workpieces (micromachining), but they are more typically applied to workpieces with features and details measurable in submicron increments or even in the mesoscale. This meaning of a micromilling machine (as shown in Fig. 11.1) is likely to have many, or all, of the following features:

- Ultraprecision linear drives in the major axes.
- Friction-free guideways.
- Extremely good thermal design plus some means to compensate for or eliminate sources of heat.
- Very high speed spindles (40,000 rpm and higher).
- A granite base and/or column.
- An ultraprecision position feedback system.
- A CNC unit capable of processing and displaying nanometre units.
- Additional equipment for non-contact cutting tool detection or a measurement system, and a high amplification vision system.

Also, it is typical to find in the literature cases of micromachines, meaning in that case an extremely small machine comprising several millimetres or less, yet highly sophisticated functional elements that allow it to perform delicate and complicated tasks. The miniature machine should be no larger than 2–10 times the size

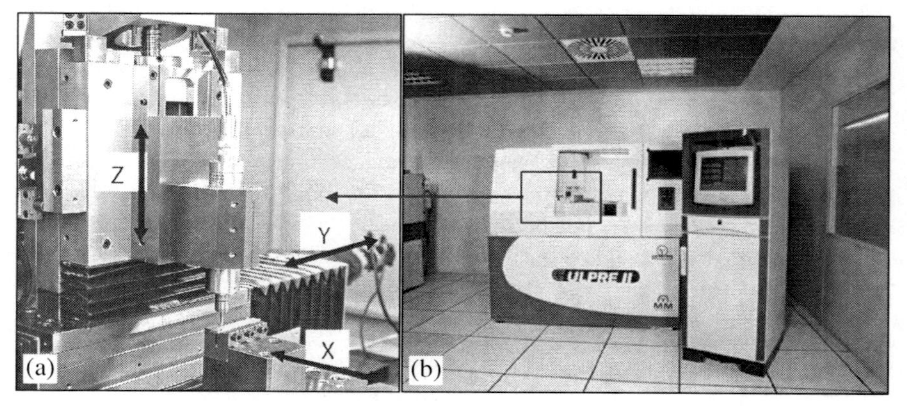

Fig. 11.1 **a** Detail of the working area. **b** Ultraprecision milling machine ULPRE II [9], by Tekniker®

of the product. This manufacturing trend has been deeply investigated at Japan, where in January 1992 the “Micromachine Centre” initiative was started to promote the engineering for such kind of machines.

11.2 The Micromilling Process

The current trend towards product miniaturisation is leading to a major increase in microtechnologies, including micromilling. Although this technique is highly similar to conventional scale milling, the great reduction in dimensions (a scale of around 40/1) means that cutting phenomena and mechanisms appear that are hardly ever encountered on a conventional scale. This scale reduction can be seen in some of the usual parameters of micromilling: a feed per tooth less than 1 μm, a depth of cut 2–15 μm, a spindle rotational speed more than 50,000 rpm and a tool diameter less than 0.3 mm. The milling machine itself must also be specific to this application and designed and built to ultraprecision requirements, with positioning accuracies on the order of 0.1 μm.

The two main advantages of micromilling in relation with other microtechnologies are its apparent similarity with conventional milling – which enables user to tackle the process from a position of in-depth knowledge – and the fact that it enables intricate parts with 3D forms to be machined (moulds, electrodes, *etc.*) in a large range of materials.

Its main drawbacks lie in the tool, which is critical in terms of size, wear, deflection [17] and accuracy and in the limitations inherent in a cutting process, including first and foremost the production of burrs that are hard to eliminate (Fig. 11.2). It is well understood that vibration is another enemy of extremely small cutting tools. Preventing vibrations that originate outside the machine from reaching the setup is one strategy. Eliminating vibrations that originate within the machine is another. Collets, toolholders and spindle interface systems that minimise

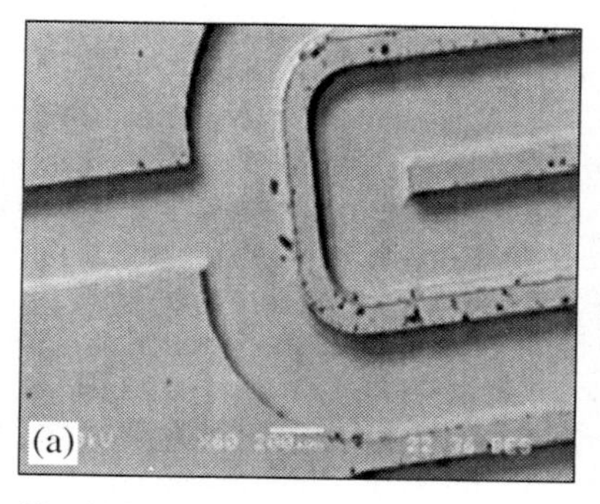

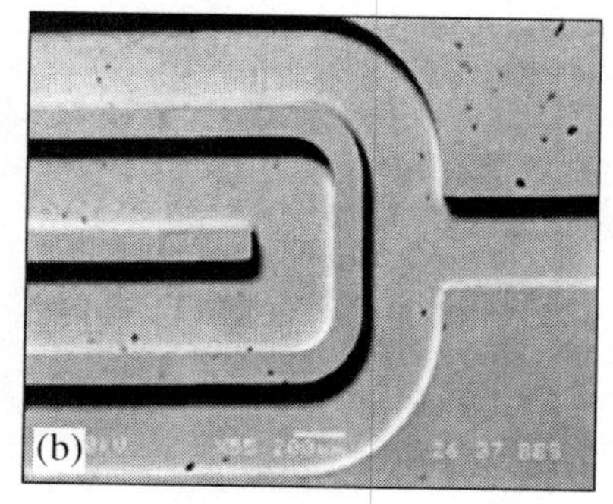

Fig. 11.2 Minimisation of burrs on edges. **a** Before. **b** After a final cleaning pass (micromilling tool ∅ 0.2 mm)

tool runout are proving to be mandatory. Several studies have looked at the validity and problems of the micromilling of microcomponents [7, 14, 21, and 22].

There are several phenomena in micromilling that prevent the results of conventional milling from being applied to it directly. Speeds, feeds, depths of cut, coolant application and chip evacuation are issues that require attention, just as they do in any machining operation. However, because micromilling applications represent a new frontier, there is no accepted body of knowledge about setting the machining parameters for tools this small. Users are largely compelled to find what works by a process of "trial and error", and this know-how then becomes a closely guarded secret. The solution should come from the development of specific cutting models that consider the three basic differences that arise from the drastic reduction in size:

- It cannot be assumed that the microstructure of the workpiece material is homogeneous [3, 19, and 20]; as tool size is becoming smaller its effect becomes more important.
- The effect of the cutting edge radius is not negligible [2, 12, and 23]: it affects the chip forming mechanism. Minimum chip thickness is a function of this parameter, and determines the transition between two cutting conditions, where chips are produced and where ploughing takes place [10].
- As a result of high tool compliance [4] and the relative size of the cutting edge radius the associated dynamic effects, i.e., forced vibration and regenerative chatter, differ from conventional milling [10].

11.2.1 Micromilling Tools

The end mills and twist drills used in micromilling applications are proportionately tiny. A number of cutting tool manufacturers offer cutting tools in diameters as small as 0.1 mm and specials as small as 0.01 mm, or even smaller. The growing importance of machining with very small cutting tools is evidenced by the number of companies specialising in small cutting tools are emerging Although this small cutting tools are almost invisible to the naked eye, correct procedures for their effective use must be ensured.

Tool market is very active in the development of new tools for micromilling. Two different cutting tool materials are generally used: diamond and tungsten carbide. There are diamond tools with almost atomic cutting edge sharpness which avoids the "cutting edge radius effect", but they are restricted to non-ferrous materials, and consequently they are not suitable for micromoulds manufacturing in tool steels. The offer of sintered tungsten carbide tools includes spherical and straight two-flutes mills made with different coatings: TiAl, TiAlN, DLC, *etc*. In any case, the offer is limited and there are not different geometries for different materials, so there is no option to choose face angle, helix angle, rake angle or other geometrical parameters. Besides the geometry of the tools presents a high

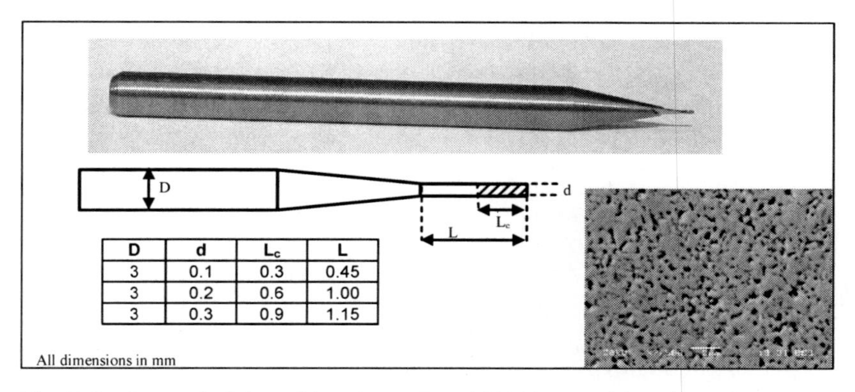

D	d	Lc	L
3	0.1	0.3	0.45
3	0.2	0.6	1.00
3	0.3	0.9	1.15

Fig. 11.3 Geometrical data of the microtools and SEM image of the micrograin structure

dispersion, being this issue important when changing the tool to continue a long machining process.

Carbide micrograin grade end mills, in this case by M.A. Ford™ (Fig. 11.3) are the most usual case. It is important to indicate that grain size and homogeneity has a big influence on the tool performance, and limits the minimum achievable edge radius, typically between 2 to 4 μm for a new tool. Figure 11.4 shows a microtool composed of cobalt (binder) 7.9% and tungsten 92.1%. The grain size varies between 0.25 to 2 μm.

Tool wear during micromilling of tool steel is quite high and that is the reason why it is usual to use two or more micromills per operation, one for rough machining and other for finishing. Tool change is a critical operation because the tool runout, tool height and collet runout are modified, thus it must be performed carefully cleaning all the shank cone, collet, tool and nut and applying controlled torques. Tool wear modifies the cutting edge profile, and consequently the accuracy and roughness of the micropart are affected, cutting forces and vibrations are increased, and burr generation is induced; this last being effect specially detrimental because postprocessing for burrs removing is not always allowed in micromanufacturing.

Figure 11.4 shows the evolution of tool wear during the machining of tool steel AISI H13 of 54 HRC. The process was slotting with a two-flutes tungsten carbide tool of Ø0.3 mm, coated with TiAlN. Tool degradation is shown after removing

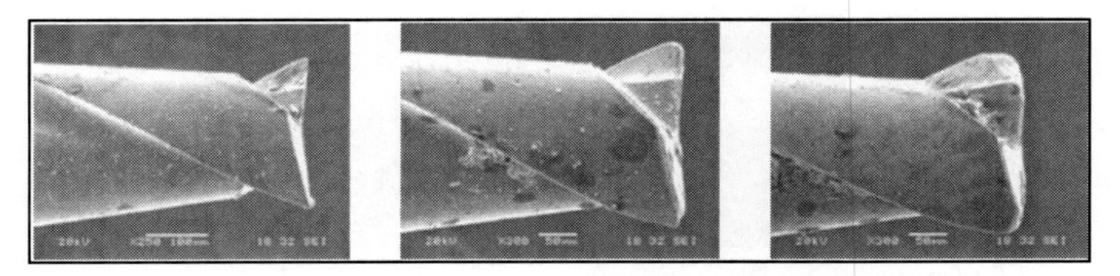

Fig. 11.4 Wear evolution in a micromilling tool (Ø0.2 mm) machining hardened steel (60HRC)

a volume of 0.2 mm^3, from (a) to (b), and after removing an additional volume of 1 mm^3, from (b) to (c). Cutting conditions were a rotational speed of 60,000 rpm, depth of cut 10 µm, and a feed per tooth of 0.4 µm. In some micromilling cases can also be noticed that wearing appears not only in the cutting edge area, but also in the opposite zone, due to the high tool deflection.

11.2.2 Applications

The applications of micromilling are multiple [15] and gradually with greater input in the industrial market, which is multiplying the efforts devoted to its study from all areas: theoretical, experimental, markets, equipment, and so on. In general, and after this outstanding ability for machining complex 3D shapes in multiple materials, we can see that the favourite demonstration of micromilling was a finely detailed mould cavity for plastic injection of a miniature device. Workpieces featuring complex arrays of holes (as small as 0.01 mm) were also conspicuous. Mirror-like finishes produced solely by milling were also prominently displayed.

For such case, the limit of current micromilling technology for steel moulds is a minimum dimension in the order of 100 µm. Below this size is really complicated for machine geometries with good repeatability. Certainly it is not usual to manufacture moulds whose maximum size is of this order of magnitude; in fact most of the industrial applications identified relate to moulds of small size, short of the category of "micro" with few details or features at the microscale. We can affirm that industrial applications are more focussed on the mesoscale.

Table 11.1 Micromilling capabilities in tool steels

Micromilling data sheet	
Mould materials	Tool steel up to HRC 62, Al7075, Cu
Cutting tool	Tungsten carbide end mill up to Ø0.1 mm
Machine	Ultra-precision milling machine 3–5 axes
Removal rate	1–3 mm3/h
Machining of channels & ribs	
Minimum width	100–110 µm
Aspect ratio	10–15
Accuracy	5 µm
Roughness	0.3 µm Ra
Machining of holes and pins	
Minimum diameter	110–150 µm
Aspect ratio	5
Accuracy	5 µm
Is a 3D freeform surface possible?	Yes

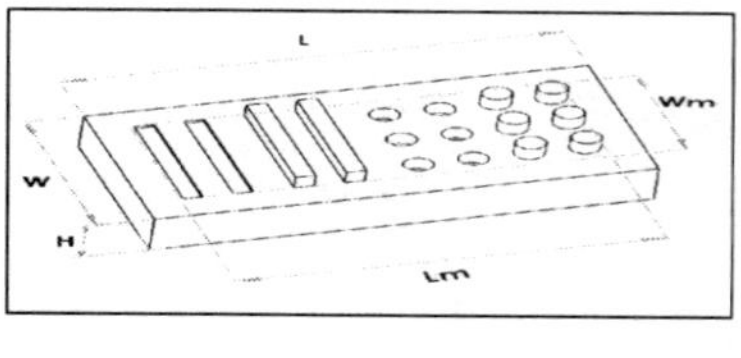

Table 11.2 Application cases of micromilled moulds

Workpiece	Mould for a surgical dosifier	Mould for dental brackets
Photo		
Material	W.Nr 1.2343 (AISI H13) 55 HRC	W.Nr. 1243 (AISI H13) 55 HRC
Tool	Cylindrical micromill, z 2, Ø0.1 y Ø0.2 mm	
Cutting conditions	N 40–60,000 rpm; F 15 mm/min; a_p 3–10 μm	
Geometric figures	Mould microchannels of straight rectangular section 0.1 × 0.1 mm and 0.2 × 0.2 mm	3D complex geometry with geometrical details in the order of 0.2–0.3 mm
Result	0.1 μm Ra, geometrical accuracy +2 μm, no burrs allowed	0.03–0.08 μm Ra, geometrical accuracy +3 μm, no burrs allowed

Table 11.1 shows the current industrial capabilities of micromilling [18] for certain simple geometries such as channels and ribs of straight rectangular sections, and circular holes and pillars in steel for moulds. Table 11.2 shows some industrial cases of micromilled moulds.

In ultraprecision cutting with tools of monocrystalline diamond (it is extremely hard and can be given a cutting edge of almost atomic-level sharpness) the production of very sharp cutting edges is therefore one of the most important tasks specified for microcutting. Among the ultraprecision-machining components are moulding tools made of non-ferrous metals – for example, for Fresnel lenses or for roughness standards. Ultraprecision milling is more flexible than ultraprecision

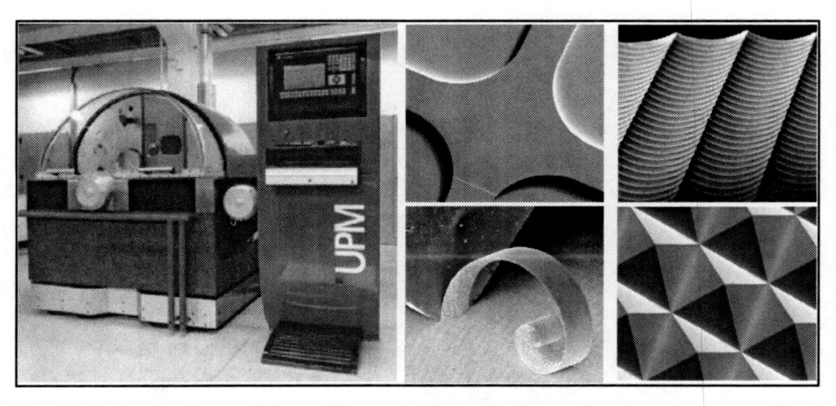

Fig. 11.5 3 axis ultraprecision milling machine with diamond tools and some application cases (courtesy of Fraunhofer IPT)

turning. For instance, single-point diamond fly-cutters allow the production of grooves that can be crossed in corresponding angles to create columnar or pyramidal structures. Materials for diamond-cutting are mainly aluminium, copper, brass, nickel silver and nickel (see Fig. 11.5).

11.3 Miniaturised Machine Tools

In the literature we can find similar terms that can lead us to misunderstanding: *ultraprecision machines*, *micromachines*, *miniature machines* and *desktop machines* among others. Up to now and in further sections we will be focussed on *ultraprecision machines*, conventional size machines with submicron accuracy, which can be employed to produce large parts with submicron accuracy and optical quality; for this purpose diamond tools are generally used, or to produce small meso-micro parts; for this purpose carbide micromills are generally used.

The term *micromachine* has a more general meaning [8] than production equipment, and it refers to any kind of microdevice able to do an action in a sense closer to MEMS (micro electro mechanical systems).

Despite the vast volume of work and developments in the machine tool industry for macro-machining, fairly limited research has been carried out in the field of *miniature machines*. In general they are specially designed and built miniature machine tools, integrated in a single platform (Fig. 11.6), which usually also integrates assembly and inspection capabilities. Thus, it is more usual to find microfactories than independent miniaturised machine tools. A worldwide effort is currently underway to bring such microfactories to fruition. Such miniature machine tools, however, remain unavailable commercially to-date. Most of the current micromachining is done using the conventional-size machine tools, so this chapter is devoted to those so-called ultraprecision machines.

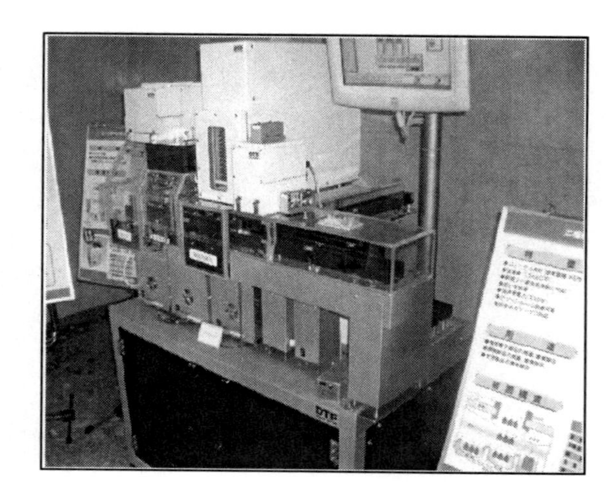

Fig. 11.6 A series of miniature factory units at Sankyo Seiki®

On the other hand there is a large market of *desktop milling machines*, which in their external size could be similar to the previous miniaturised machine tools; however they are specifically suited for meso-machining with low accuracy (0.1 mm per slide as the standard). They are applied for printed circuit boards, jewellery, *etc*.

11.4 Machine Drives

Drives for ultraprecision milling machines should be designed according to the principles of precision engineering, the first one being *deterministic design*. Any precision drive should fulfil the principles of minimising the Abbe error (see Sect. 6.2), minimise the number of restrictions (kinematic *or semi-kinematic design*), and to act at any centre of action (centre of gravity, centre of inertia, etc.) minimising the perturbations at the guiding system, a measuring system centred in the goal position. The main objective to be achieved is the repeatability of the movement, and by means of a further calibration, the final goal of precision.

Practical applications need a balance between contradictory requirements, but the goal should be to move away from the minimum from the theoretical precision principles. The next pages show examples of drives of existing micro-milling machines.

11.4.1 Conventional Ball Screw Configuration

Transmitting the movement to a linear carriage trough a ball screw is without question the most common system in machine tools, just as in precision machines. Ball screws enables a repeatability of 1 μm with ease, and even submicron repeatability in the case of ultra high precision ball screws. In spite of this, no matter how high the manufacturing accuracy of the screw is, the intrinsic nature of such a system, in the end based on rolling elements, creates unavoidable perturbations which can be clearly identified, even if they are submicrometric. This intrinsic sources or error derive from the roundness and the uniformity of the balls, the accuracy of the lead of the screw, the system for the recirculation of the balls and the preload. Moreover, errors coming from the manufacturing and the assembly of the machines must be added, which are the misalignment between the spindle and the guideways, the capability of the design to cope with thermal expansions, the bending of the spindle itself and several others. All of them affect the accuracy of the movement of the linear carriage.

11.4.1.1 A Ball Screw Drive with a Floating Nut

In precision machines required for micromilling, where the final precision is the main design criteria, it is necessary to isolate the movement of the carriage, defined

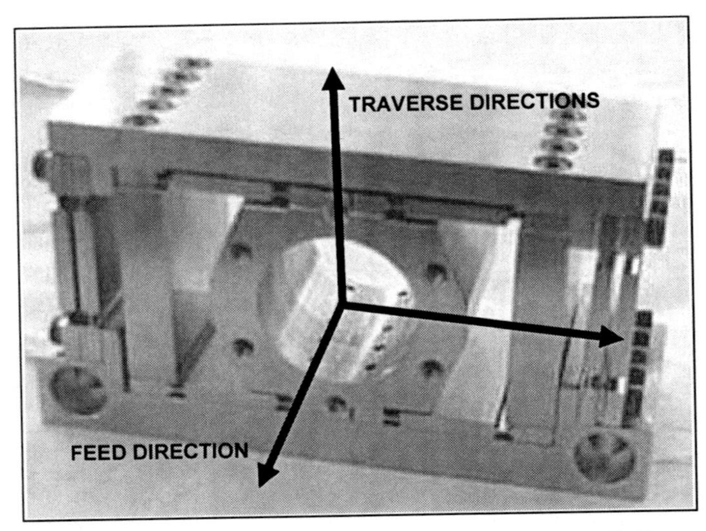

Fig. 11.7 "Floating nut" system to minimise the transmission of ball screw perturbations

by its guides, from the errors and perturbations due to the drive. The ideal situation is that where the element in charge of transmitting the movement to the carriage, in this case a ball screw system, creates no perturbation to the guiding elements of such carriage, so that the guides become the elements that exclusively define the trajectory of the moving elements.

One of the existing design solutions to achieve such objectives is the use of a *floating nut system*. The principle behind the "floating" nut is that the flange that joins the nut to the carriage is not completely coupled to said carriage; instead, the flange is designed to be very stiff in the axial direction (the direction of the movement) and under torsion, and very flexible in the other two directions perpendicular to the movement (Fig. 11.7).

Besides this, to improve its effect, the spindle has been one with a minimum preload configuration, fixed on one side and in a cantilever configuration on the other side. It is essential that the dimensions are correctly set during the design phase to ensure that the stiffness in each direction is appropriate in order to avoid affecting the dynamic response behaviour in comparison with the traditional solution with a nut tightly joined to the carriage.

11.4.1.2 A Ball Screw Drive with Intermediate Joints

In drives where there are space restrictions that avoid the use of the floating nut system, errors of the spindle can be isolated by means of a ball screw with intermediate joints. It consists of a traditional ball screw that is divided into several sections, by purposely reducing its diameter in two points, so that a relative-joint-effect is created. The points with reduced diameter are out of the range of the

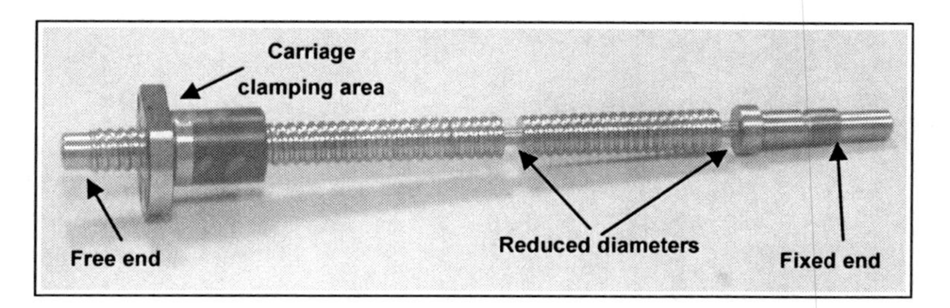

Fig. 11.8 "Intermediate joints" ball screw system

movement. Figure 11.8 shows the "ball screw with joints" concept. A cantilever-fixed preloaded precision ball screw has been thus divided. It is also worth saying that in order to maximise the "joint effect" the nut is fixed to the carriage in the area close to the free end of the screw.

The correct dimensioning of the system is critical in this case as well, so that the loose stiffness, which is obviously produced when reducing the diameter, is minimised and does not affect the performance of the carriage. Concisely, the stiffness under bending loads is reduced while, at the same time, trying to minimise the loose of rigidity under axial and torsion loads. For example, in the application case shown in Fig. 11.8, the bending stiffness has been reduced from $K_{bending} = 160$ N/mm (original screw) to $K_{bending} = 30$ N/mm (reduced screw), which means a reduction of 81% from the original figure. Meanwhile, the axial stiffness has changed from $K_{axial} = 160$ N/μm (original screw) to $K_{axial} = 130$ N/μm (reduced screw).

11.4.2 Friction Drives

Friction drives have been commonly used in ultra high precision machines such as coordinate measuring machines (CMMs). Their main advantage is the reduction of the error sources such as clearances, lead errors, ball-recirculation errors *etc*. The only "death area" of this particular drive comes from the deformation of the elements in the transmission, which can be regarded as non-existent if the system is designed properly. On the other hand it is a pretty simple drive from the concept point of view, and although commercially available solutions are uncommon, both their design and their assembly are not challenging at all.

Nevertheless, it shows important drawbacks, namely low stiffness, low damping, low load capacity and low capability for reduction. If one wishes to increase the pushing force, the preload between the wheel and the pushing rod must be considerably increased, thus creating problems dealing with deformations and stress in the contact between elements.

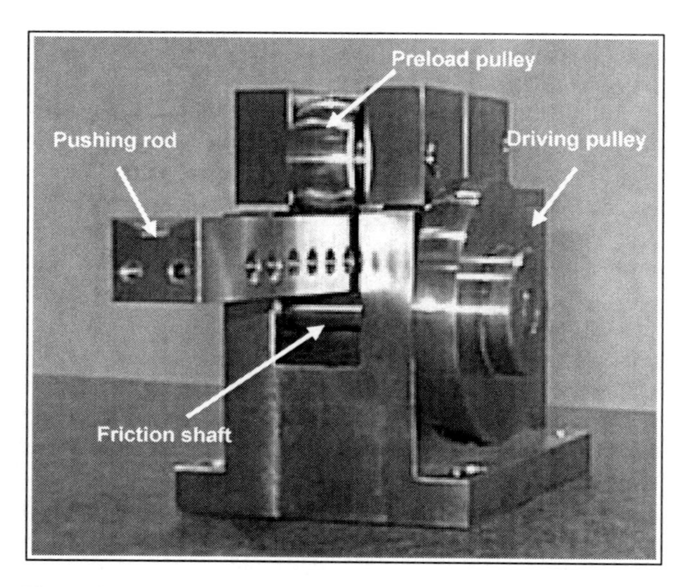

Fig. 11.9 Friction drive for a micromilling machine

Swing to this, their application is limited to movements requiring extremely high precision and softness, low pushing force and low speed, as in the micromilling machines.

In a friction drive (Fig. 11.9), the rotation motion of the servomotor is transformed into the lineal advance by means of the interaction between a friction wheel and a pushing rod. The rotation of the servomotor is transferred to the friction wheel either directly or through a belt and pulley system. The pushing rod transmits the lineal movement to the carriage trough, a flexible coupling which absorbs the misalignment between rod and carriage. The contact and rolling between the rod and the friction wheel are ensured by means of an adjustable preload system, which acts over the pushing rod through the preload pulley.

11.4.3 The Linear Motor

There are many advantages of linear motors that make them suitable for precision applications:

- The accuracy, resolution, and repeatability of the system to be moved is determined by the feedback measurement system.
- Regarding the stiffness on the movement direction, the advantage of the linear motors is that they do not introduce additional flexibilities to the system because

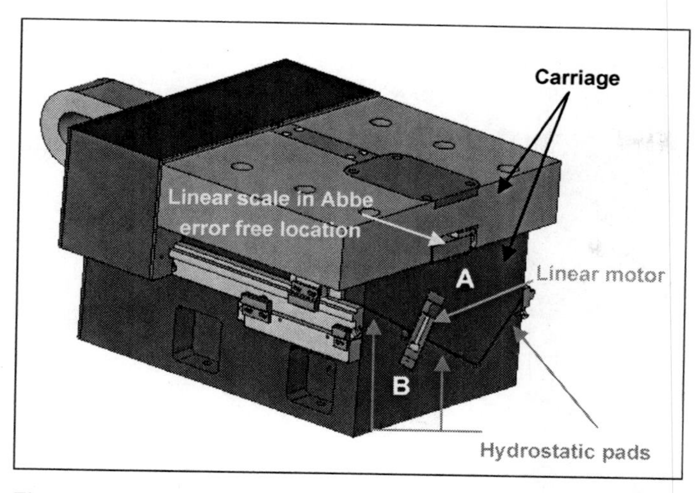

Fig. 11.10 Precision drive by linear motor preloaded by the carriage weight

the transmission is direct and no additional components are needed as in the case of rotary motion that "transmission elements are needed" to convert rotary motion to a linear one. Then on the movement direction the achievable stiffness of an axis driven by a linear motor is dependent on the mechanical structure to be moved and the control strategy implemented.

- Without mechanical transmission components, there is no backlash and also there is no wear.
- For ultraprecision applications the best option is an air core or ironless linear motor. The winding itself has no iron in it, thus the names "air-core" or "ironless". This type of linear motors are more suitable for precision applications due to the fact that they have no attraction forces and no cogging.

The next paragraphs show an example of an ultraprecision linear slide by using a linear motor (Fig. 11.10). It is composed of a six hydrostatic pad bearings slide using the slide's own weight as a preloading force, an air core linear motor in order to have no attraction forces, and a linear scale as measurement system.

The lineal slide is supported over hydrostatic bearings, using an integrated linear motor acting just at the centre of gravity of the moving slide, an optical linear scale Heidenhain® LIP 401, with a travel range of 100 mm with 0.1 μm of resolution with 4 μm of separation between lines, the linear speed is 5 mm/s, and there is 1 g of acceleration and stiffness values of 45 and 20.5 N/μm in vertical and lateral directions, respectively.

The stiffness of the slide is fully dependent of the bearing design. The slide is not flat and has two different slopes on its contact surface having six flat capillary hydrostatic bearings, four on the lowest slope side and two on the highest one. Figure 11.11 is a section of the 3D model in which the moving part (A) is shown in dark and the fixed part (B) is represented darker.

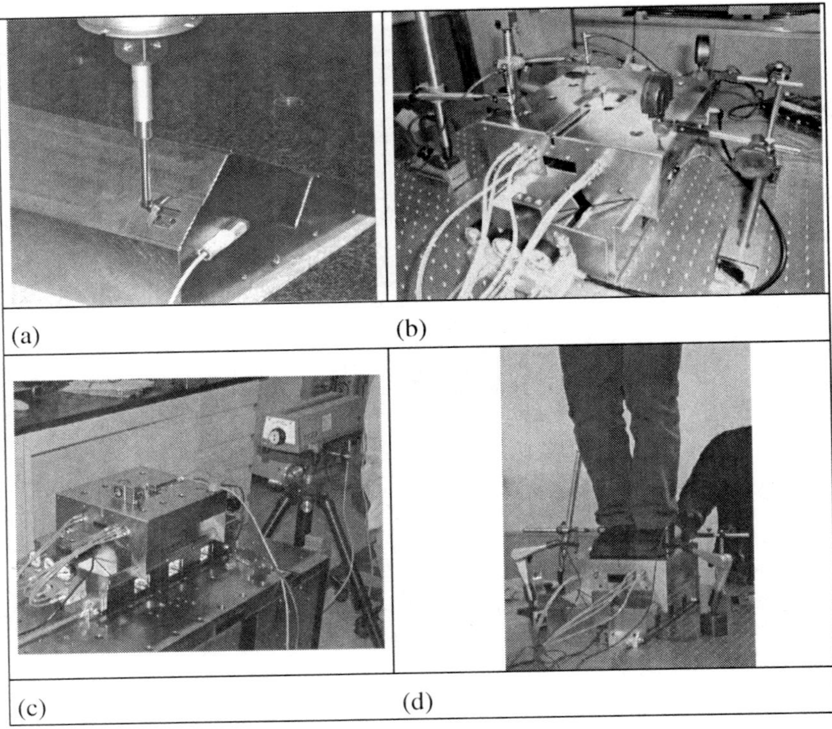

Fig. 11.11 Precision drive by linear motor. **a** Geometry verification. **b** Bearing calibration. **c** Slide error measurement. **d** Loading test by a "human weight"

11.4.4 New Tendencies: Hydrostatic Screws

An alternative existing in the market with a very good performance is the use of hydrostatic leadscrews (Fig. 11.12), which can be complemented also with hydrostatic support bearings. They have better performance than ball screws and clearly can compete with linear motors. The only drawback is the need of a hydraulic system for the oil supply. Their main advantages in comparison with ball screws and linear motors are as follows:

- Wear-free as no contact when in use, so no loss in accuracy, even when used for long periods under full load at maximum speed.
- Friction-free and no "stick slip" effect even at low speeds.
- No backlash when the direction changes, only the change of inertial force shall be considered.
- Translates even the smallest turning movement.
- No oscillation due to ball circulation which produces no fluctuations in frictional torque.

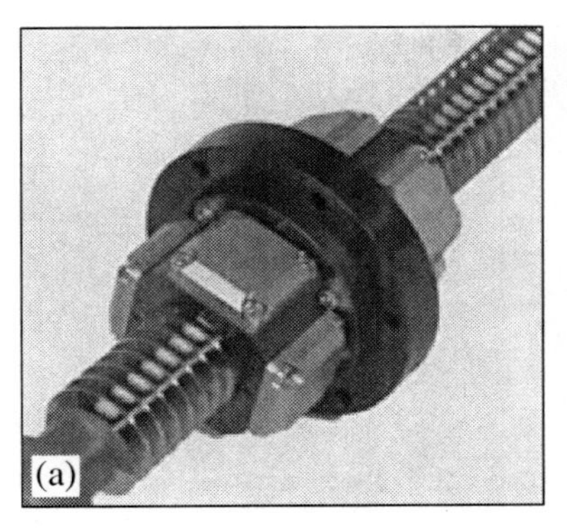

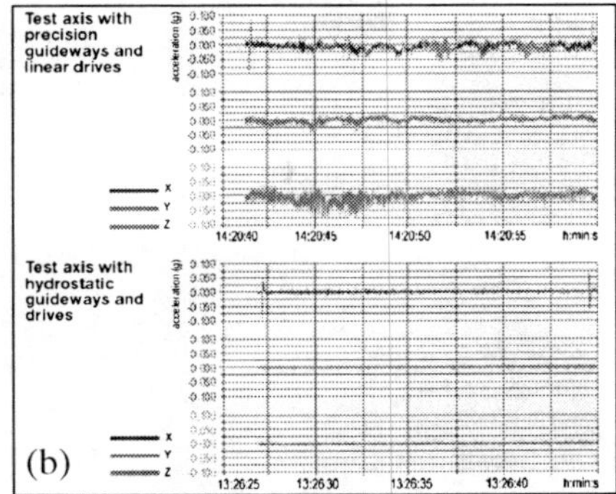

Fig. 11.12 Hydrostatic lead screw by Hyprostatik®. **a** Screw detail. **b** Comparison of performance in relation with linear motor slides; the good damping performance can be observed

- Higher axial stiffness similar to ball screws.
- Great improvement in damping, resulting in better surfaces and tool life.
- Also suitable for high speeds and acceleration.
- A cost-effective alternative to the linear motor, with high accuracy and much less heat created.
- Much lower losses and the absence of permanent magnets means no problems with metal chips.
- Much lower cooling power than linear motors.

11.5 Guideways

Although drives and guideways are presented in separate paragraphs, it is unthinkable to afford each of them in a separate way from the design point of view. Kinematic designs, no friction, no wear and no backlash are the key issues to achieve the required repeatability.

11.5.1 Special Rolling Guides Configurations

The use of cylindrical roller bearings without recirculation is moderately common in the motion of the linear carriages of micromilling machines. The absence of recirculation systems eliminates the related perturbations, at the expense of higher size and space demands. If submicrometric accuracies are to be achieved, solutions with recirculation become unthinkable and highly unlikely.

The embodiment of choice is 2 V-guides in opposition, in a semi-kinematic configuration with the rollers contacting two faces thanks to a constant preload. The preload force shall be calculated to fit the application and the related loads

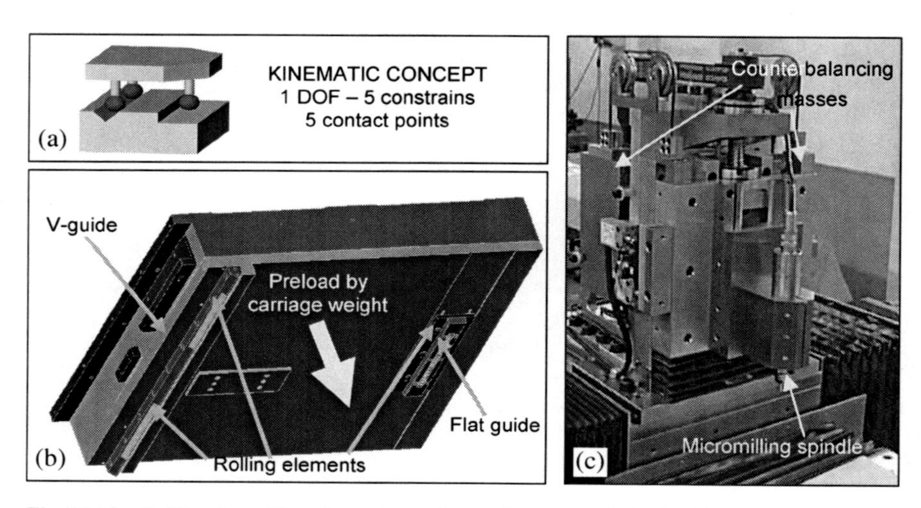

Fig. 11.13 Guiding by rolling elements. **a** Kinematic concept. **b** Preload by the carriage weight in a horizontal slide. **c** Preload by counterbalancing masses in a vertical slide

that the carriage is suffering during operation. Anyway, when high precision is the primary objective, a "constant preload" solution must be chosen above "constant displacement" solutions. Constant displacement solutions are easier to apply, but, on the contrary, create higher dispersions in repeatability due to the irregularities of the rolling surfaces. Constant force preloads can be achieved in different ways, e.g., created by the weight of the system itself in the case of horizontal carriages, by means of counterbalancing masses in vertical carriages (Fig. 11.13) or by spring-based locking systems that show well-known behaviours. With this configuration the following stiffness values have been achieved: 450 N/μm in the perpendicular and lateral directions of the carriage. The achieved straightness of the carriage is ± 0.2 μm in 150 mm for a vertical slide.

11.5.2 Aerostatic and Hydrostatic Guides

Hydrostatic and aerostatic guides are widely applied to precision feed tables, since their motion is accurate due to the lack of friction and wearing. Besides, the so- called averaging effects of the fluid (oil or air) average the motion errors. In the case of aerostatic guides, where load is supported over a thin film of air, the low viscosity and compressibility of air obliges the use of very high manufacturing and assembly tolerances. Typical gap values can be in the order of 10 μm for a standard air pressure supply of 6 bar flowing in a continuous way to the atmosphere. Their main advantage is the lack of friction, which assures a fine and precise movement even at low or high speeds. Their drawbacks are related with the low load capacity and the complexity of the design and manufacturing. These

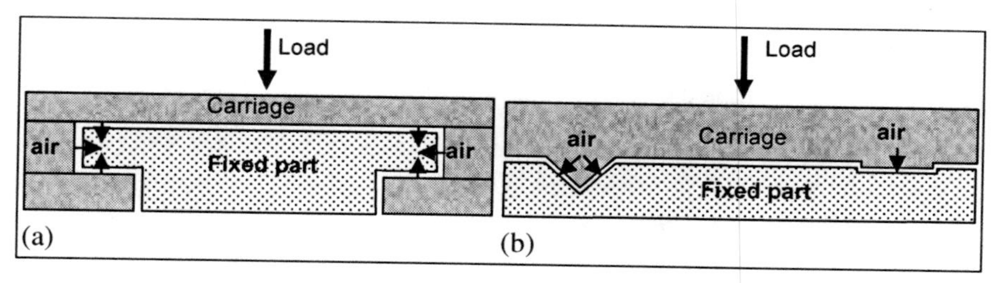

Fig. 11.14 Aerostatic guides configurations. **a** Bearing pads in opposition. **b** Preload by the carriage weight

limitations are solved by using hydrostatic guides as shown in the example in Sect. 11.4.3. Here possible configurations of aerostatic guides depending on the way the preload is done are described (Fig. 11.14).

11.5.2.1 Bearing Pads in Opposition

This configuration consists on a double set of bearing pads located in opposition, both in vertical and lateral directions as shown in Fig. 11.14a. An application case for micromilling machines is shown in Fig. 11.15. Air supply is done by an orifice in the centre of a compensation cavity. The optimal pressure ratio between cavity and supply is approximately ½. The accuracy of aerostatic guides is intrinsically high, but it depends a lot on the machining accuracy of the components. As a general rule, the maximum peak-to-valley value of roughness should be lower than ¼ of the air gap. Repeatability also depends strongly in the stability of air supply and in the accuracy of the air restrictors.

With such a configuration, the carriage of Fig. 11.15 has achieved a static stiffness value of 250 N/μm in the vertical direction and 150 N/μm in the lateral direction. The air consumption is 40 l/min and the straightness achieved in the movement is ±0.1 μm in the 150 mm of travel range.

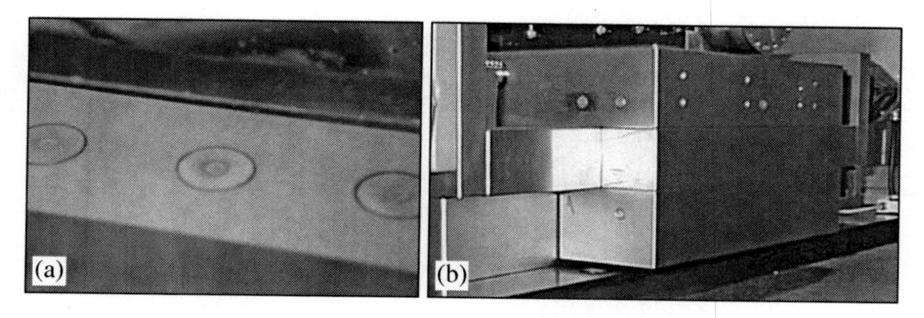

Fig. 11.15 Aerostatic guides in opposition. **a** Bearing pads detail. **b** General view

11.5.2.2 Preload by Carriage Weight

This configuration gives good accuracy due to the constant force of the preload, however the load capacity and stiffness is achieved only in one sense of the vertical direction, and lateral forces should be also very low. If preload is not constant the accuracy of the movement starts to get worse, but still is possible to design a relatively good accuracy slide in a cheap way (Fig. 11.16).

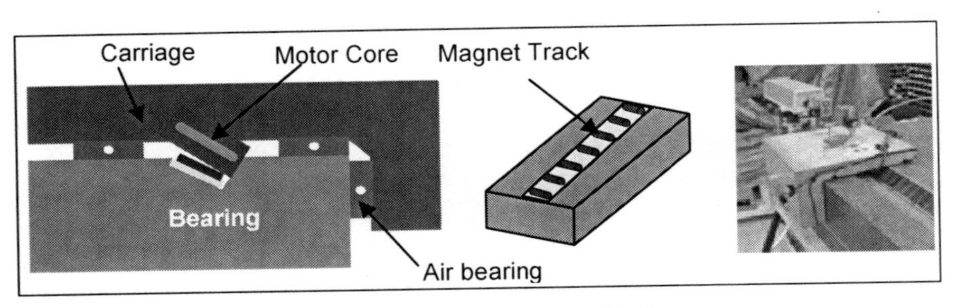

Fig. 11.16 "Cheap" aerostatic carriage with constant preload [16]

The goal is to use preload air bearings for minimal cost, using the magnetic attractive force of the motor, so air bearings need only ride on two surfaces instead of having to wrap around a beam; thus many precision tolerances to establish bearing gap can be eliminated. The magnet attraction force is $5\times$ greater than the motor force, so it can be positioned at an angle such that even preload is applied to all the bearings. As long as the magnet attraction net vertical and horizontal force are proportional to the bearing areas and is applied through the effective centres of the bearings, they will be evenly loaded without any applied moments. Its drawback is that the magnet pitch may cause the carriage to pitch as the motor's iron core windings pass over the magnets.

11.5.3 New Tendencies: Magnetic and Flexure Guidance Systems

There are other guiding systems which currently are appearing in the market, not only for micromilling machines, but in general for any kind of ultraprecision machine. Among them we would like to point out the magnetic guides and the guiding systems based on compliant mechanisms.

11.5.3.1 Magnetic Guides

Magnetic levitation guides have the advantage of no friction with its associated beneficial effects: no wear and no need of lubrication. In combination with linear

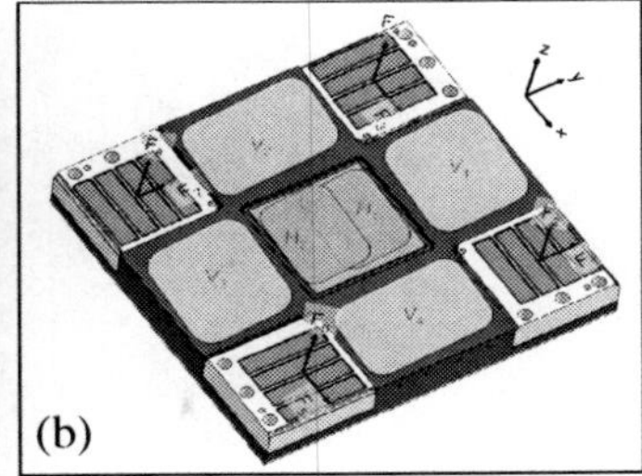

Fig. 11.17 Magnetically levitated 2D table. **a** Final assembly. **b** Magnets and forces location

motors, they allow avoiding any mechanical transmission between the moving slide and the fixed part. Their main drawbacks are the lack of damping and the complexity of the control system. However it is possible to found complete magnetically levitated and actuated tables, with moving capacity in two directions.

Figure 11.17 presents a two dimensional magnetic drive [5], where the moving part of the system is free of any contact and wires. The lightness of the moving part along with the big forces the 2D drive is capable of providing, makes feasible the obtaining of great accelerations. It has a compact sensoring system, and protected in a way that any collisions will not damage its important parts. The table has a range of movement of 160×125 mm. A resolution of 0.02 µm is obtained in each of the moving axes.

Force actuation in 6 degrees of freedom, or DOF. is needed to obtain the levitation of the system in addition to the propulsion in two of them. Required forces are generated by means of four 2D actuators, so eight independent forces can be obtained, creating an over constrained system. Each of the actuators is made of a kit of permanent magnets in the moving part, and by means of a winding in the static part of the system and facing the permanent magnets. The permanent magnet set is designed in such a way that the magnetic flux is projected in its maximum value towards the windings. This configuration is known as the "Halbach array".

It has been demonstrated that 2D drives can help to simplify the design of mechanical systems that require at least movement in two dimensions. This kind of solution, when mechanically well designed, can also increase the dynamic behaviour and the overall precision as they allow the integration of the measuring system in the centre of movement location.

11.5.3.2 Flexure Guidance Systems (Compliant Mechanisms)

Compliant mechanisms base their performance in the elastic properties of materials. Their main advantages as guiding solutions are no friction, no backlash and no stick-slip effects. Their main limitations are the low movement range and load capacity. But, with an adequate design they are able to achieve some tens of millimetres (Fig. 11.18).

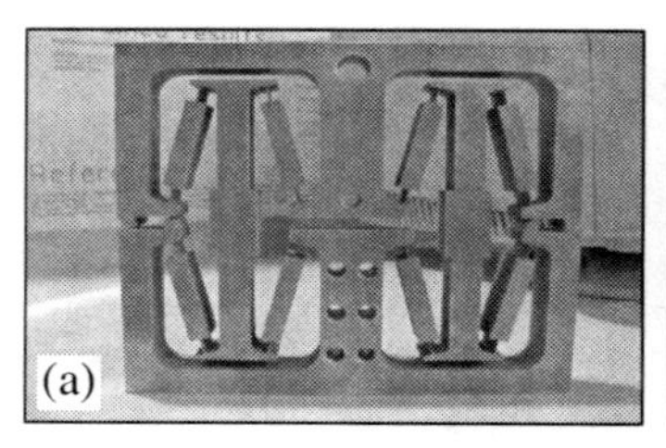
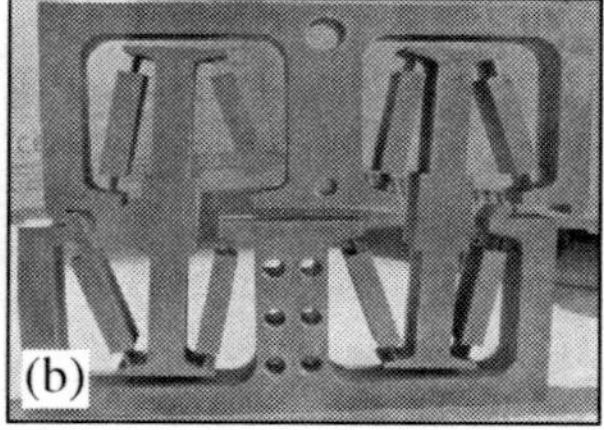

Fig. 11.18 Linear motion of 20 mm by a compliant mechanism. **a** Static position. **b** Deformed

Figure 11.18 shows a compliant mechanism for a lineal movement of 20 mm long, with a general size of 160×180 mm, and a parasitic movement perpendicular to the main direction of 0.1 μm, but highly repeatable, so it is easy to compensate.

Although basic principles of compliant mechanisms are well-known since years, design methods and applications have remained fragmented without a detailed methodology. The advance in precision machining by wire-EDM allows currently the production of complex monolithic structures with good accuracy and surface roughness. Recently, the growth of applications by compliant mechanisms (space, scientific instruments and ultraprecision machines) has produced a systematic approach to their design.

Figure 11.19 shows a 3D positioning device for a microEDM machine, with a travel range of $6 \times 6 \times 6$ mm, and a positioning accuracy of 0.1 by means of a parallel Delta-type kinematic robot.

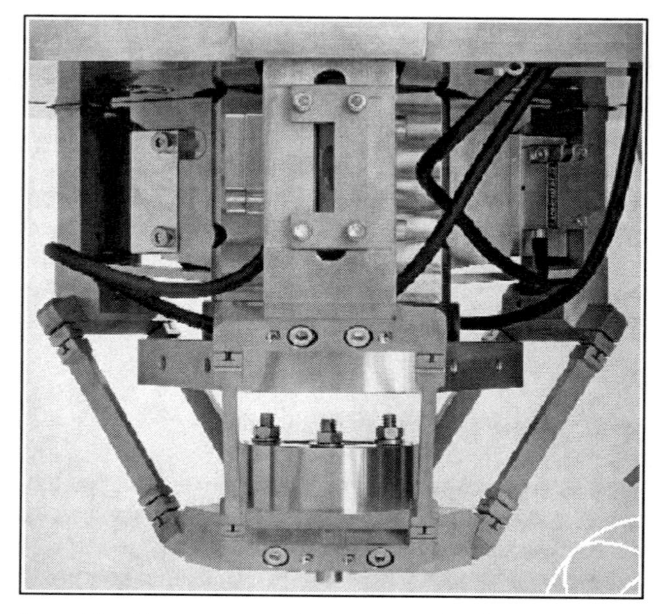

Fig. 11.19 3D compliant mechanism for a microEDM machine (courtesy of Agie®)

11.6 The High Speed Spindle and Collet

Standard solutions are achieved by high speed ball bearing spindles. Several builders are developing spindle designs that provide high rotational speeds and extreme stiffness so that very low runout is experienced at the tool tip.

Using small tools, the spindle must rotate at high revolutions (120,000 ~160,000 rev/min) to achieve the adequate cutting speed for most materials. Apart from the speed, the spindle must be stiff (> 25 N/μm) and must present a small runout (< 1 μm) in order to ensure high precision in the cutting process. To reach such high speeds, the spindles have ceramic ball bearings that are continuously refrigerated and lubricated. They are usually low power electro-spindles (200 ~500 W) or aerostatic spindles.

Usually the tool is clamped manually using special collets to reduce the runout. The most used collets are the precision "ER type" collets (clamp a small range of diameters close to a nominal value) and the super-precision ER type collets (only clamp the nominal diameter). The precision collets can present big runout errors that depend on the clamped diameter, and the super-precision collets present runout errors smaller than 2 μm. Tool wear during micromilling is quite high and that is the reason why it is usual to use 2 or more mills per operation (one for rough machining, other for finishing). Tool change is a critical operation because the tool runout, tool height and collet runout are modified; thus it must be performed carefully cleaning all the shank, collets, tools and nuts and applying controlled torques.

Figure 11.20 shows the case of a ceramic angular ball bearing spindle from Ibag™, with a rotational speed range from 14,000 to 140,000 rpm and a maximum

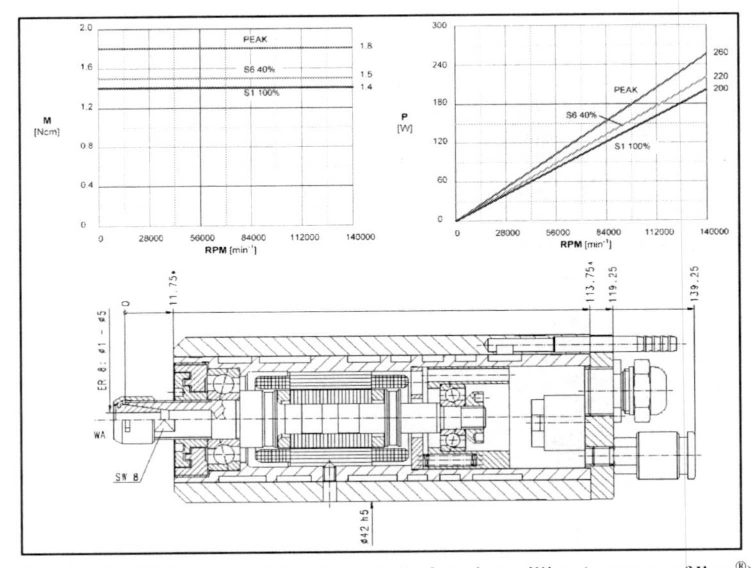

Fig. 11.20 High speed ball bearing spindle for micromilling (courtesy of Ibag®)

power of 200 W. Nominal torque is only 1.4 N·cm, but high enough for micro-milling applications. The bearings require continuous oil-air nebulised lubrication, and the stator is cooled with water.

11.6.1 Alternatives: Hydrostatic and Magnetic Spindles

Although current available rotational speeds are quite high, there is a need for higher speeds. Taking into account the small tool diameter, 0.1 mm or smaller, the achieved cutting speeds are already far away from high speed cutting condition. Vibration and runout are other limitations of existing spindles.

11.6.1.1 Hydrostatic Spindles

They are able to provide a soft and accurate rotation, with good vibration performance due to their high damping. However power losses due to the heat generated in the bearings limit the maximum achievable rotational speeds. Special spindles are available with a good compromise between speed and stiffness.

Figure 11.21 shows a high speed hydrostatic spindle with a maximum rotational speed of 30,000 rpm. The diameter of the spindle should be size limited in order to minimise peripheral speed in the bearings, and consequently the heat generated, which compromises the spindle stiffness. So, the main difficulties to achieve the expected requirements of precision and stiffness are the geometrical restrictions. Other limitations comes from the strict thermal requirements that force the use of low thermal expansion materials such as Invar that is less stiff than for example steel but has a ten times lower expansion coefficient in comparison with steel.

Fig. 11.21 Hydrostatic high speed spindle for micromilling

11.6.1.2 Magnetic Spindles

Since there is a growing interest in higher rotational speeds for micromilling, the conventional solution of ceramic bearings should be overcome. This challenge has been mainly polarised towards the development of motors and rotating devices without mechanical contact between the rotating and static parts, controlling the rotor's position by means of magnetic forces. Magnetic bearings [6] introduce some advantages to this kind of spindle, where they provide superior value compared to other types of bearings. Value is a function of the following: longer life, clean environment, extreme conditions, quieter operation, impact detection, and physical signal estimation (such as cutting forces and linear accelerations).

They have some drawbacks mainly that the spindle is bigger, since magnetic bearings need bigger volume for the same cutting force, and that the needed electric installation is also bigger. Product price could be another disadvantage for this kind of spindle, but as they are providing high reliability and long service intervals, and lower power consumption (due to frictionless bearings), the total operation cost should show better figures than ball bearing high speed spindles.

Apparently, micromilling should be an ideal application for magnetic spindles. They are widely used in other fields of application such as semiconductor manufacturing; vacuum pumps, natural gas pipeline compression equipment and energy storage flywheels. However they are not in widespread use in micromilling. The two main reasons are the low damping capacity, which determines its poor chatter-

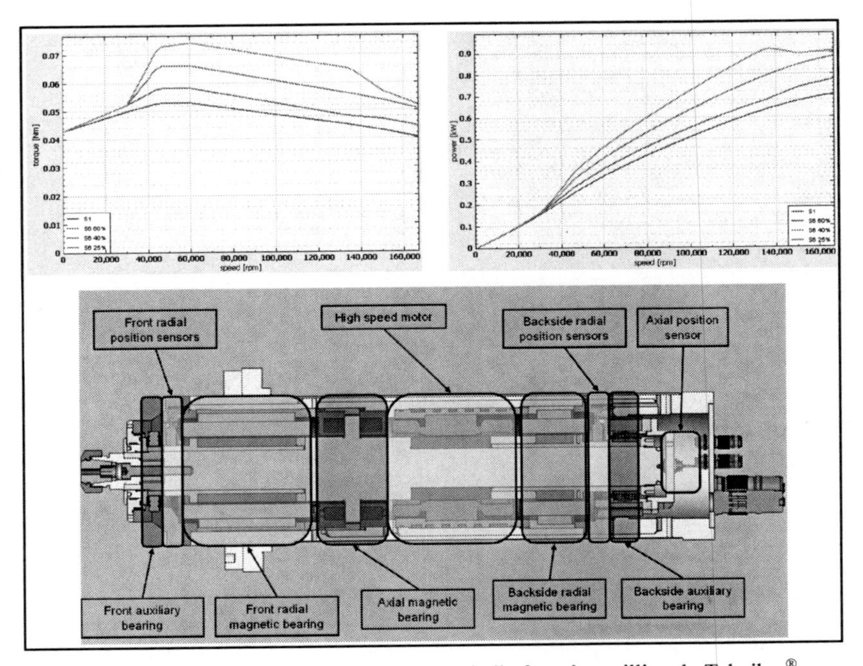

Fig. 11.22 High speed magnetically levitated spindle for micromilling, byTekniker®

free performance even more drastically than its stiffness, and the need of rotational speeds higher than natural frequencies of the spindle, which makes complex the control strategy. The literature shows cases of active magnetic bearings used for active damping or chatter suppression by means of appropriate control algorithms. However bandwidth of the magnetic bearings prevents active control of vibration in micromilling applications. Figure 11.22 shows the case of a magnetic levitated spindle from Tekniker®, with a maximum rotational speed range of 165,000 rpm and a maximum power of 700 W, and a nominal torque of 4.5 N·cm.

11.7 Measuring Systems

The first available micromilling machines used a laser interferometer for position control. For example, Fig. 11.23 shows the case of a three axes travelling column micromilling machine with a laser interferometer of the He-Ne Michelson type (wavelength 632 nm, red colour), heterodyne with planar optics and a 9 mm beam. That system allowed a resolution of 2.5 nm, with a final position uncertainty of 0.5 μm/m. However this system required a specific acquisition card, incompatible with commercial NC.

For such systems, the error of atmospheric compensation is the main source of uncertainty. The value of this error depends on the compensation method, on the atmosphere in which the laser is operating and on the atmospheric changes during laser operation.

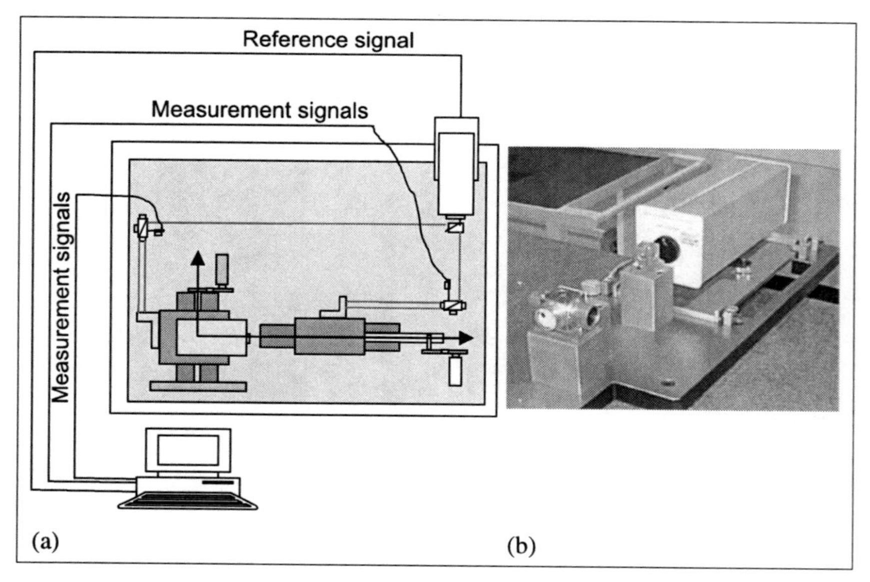

Fig. 11.23 Laser interferometer for micromilling. **a** The configuration. **b** The application in the machine

However, the latest developments in optical linear scales make them much more suitable for industrial application. They assure a bigger robustness of the system, no beam loss as in the laser; they are compatible with commercial NCs, which allow three and five axes linear and circular interpolation. Besides, the final accuracy of the machine is at the same level as using a laser interferometer. Commercial scales achieve resolution in the order of 0.01 μm and accuracy of 0.1 μm. Their main requirement is a careful mounting in the machine, and to assure a good level of cleanliness.

11.8 Examples

One of the first machines that we can find in the modern literature is the Nanocentre® from Cranfield University, with three lineal axes and position measurement by a laser interferometer system (resolution 1.25 nm), plus two rotational axes (resolution 0.35 arcsec). Each of the axes has an independent temperature control system with 0.001°C of resolution. It is ready to work only with monocrystalline diamond tools. Final result was a laboratory machine with a positioning accuracy of 10 nm in X and Y axes, and 100 nm in vertical Z axis. It has been used to machine mirrors up to Ø100 mm, with a final roughness of 2.4 nm *Ra* in aluminium and 0.8 nm *Ra* in germanium. But it should be considered as an excellent example of a lab prototype.

Nowadays there are a full range of manufacturers (Kern®, Fanuc®, Kugler®, Makino®, etc.) which offer also a wide range of models for micromilling. Hereinafter we show two examples of micromilling machines available in the market, and their main characteristics.

11.8.1 The Kern® Pyramid Nano

Among other models, the Pyramid Nano™ (Fig. 11.24) is a CNC machining centre for large chip production. The two main specific characteristics of this model are:

- All the possible rolling elements have been substituted by hydrostatic elements: guides, feed screw, and thrust axial and radial support bearing.
- Both the combined structural and thermal design assure an ultraprecision performance even in a conventional workshop.

It has been developed for applications which require maximum accuracy (±3 μm) and surface finish (0.05 μm *Ra*) on larger workpieces. An integrated automatic workpiece changer allows unmanned operations even for 5-axis simultaneous jobs.

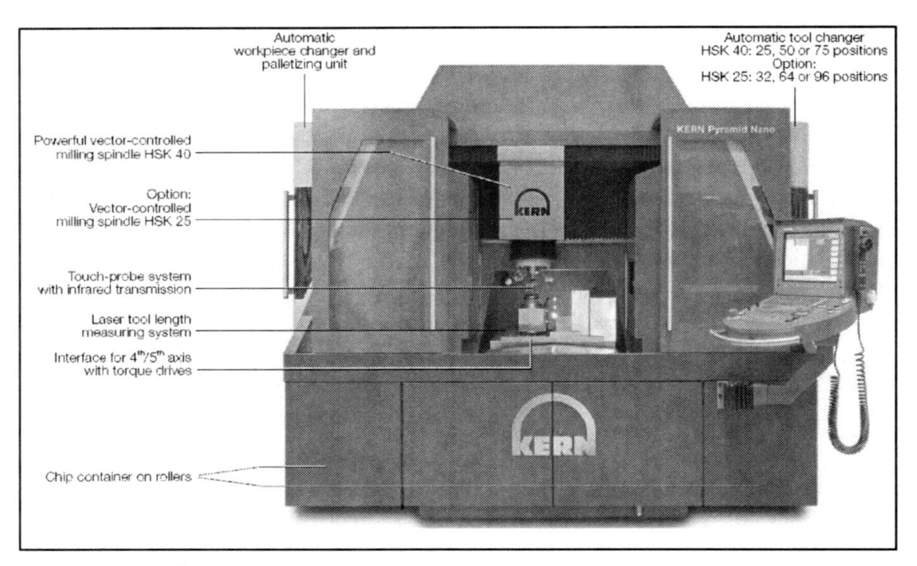

Fig. 11.24 Kern® Pyramid Nano showing the main components

The portal machine frame of the Kern Pyramid Nano is symmetric. This design concept allows for maximum rigidity and thermo-symmetric stability. High shear and tensile strength combined with an extraordinary vibration dampening are the main features of the new material Kern Armorith®. The low thermal conductivity of the material in combination with the integrated temperature management system of the machine desensitises the entire system against variations in temperatures. The intelligent temperature management (Fig. 11.25) system is composed of the following three separate circuits:

- A separate water cooling system for the milling spindle, hydraulic unit and the electrical cabinet.
- Permanently cooled circulating hydraulic oil flows through the machine base, guideways and drives of the axes.
- Temperature management of the central cooling tower within ± 0.25 K.

Hydrostatic drives and guideways of the X-, Y- and Z-axes allow maximum surface finishes in nano precision. The hydrostatic drives provide fast acceleration with maximum vibration dampening. The advantages of the hydrostatic system are as follows:

- The finest surface finishes with hydrostatic damping.
- The smallest movements of the axes at 0.1 μm.
- Wear-free guideways and drives.
- Almost frictionless movements – no "slip stick" effect, even when moving in small increments.
- Low sensitivity at high machining forces.

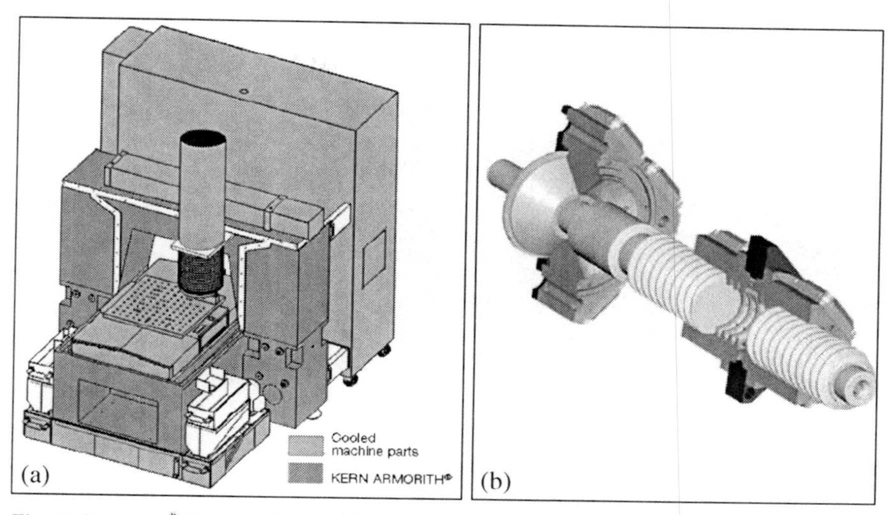

Fig. 11.25 Kern® Pyramid Nano. **a** Structure and thermal design. **b** Hydrostatic drive

- High dynamic stiffness.
- Permanently cooled circulating hydraulic oil flows through axes, drives and machine frame.
- The highly dynamic third axis construction does not require counter-balance weights.
- All motors are mounted outside of the axes – thus minimising possible temperature influence.
- Approximately 50% lower energy costs in comparison to linear drives.
- No unacceptable cogging effects, as seen with linear drives.

11.8.2 The Kugler® Microgantry nano 3/5X

The Microgantry nano 3/5X made by Kugler® (Fig. 11.26) is an ultraprecision machine, with 3 or 5 CNC axes which is able to combine microcutting (milling, grinding or drilling) with laser micromachining. This is one of the trends, the integration of different processes in the same machine, to produce microparts in a more effective and accurate way. That process integration also allows a higher versatility in the range of materials to machine.

The machine base is made of solid, fine-grained granite, which guarantees the highest thermal and mechanical long-term stability and, in combination with passive air damping elements, insulates vibration frequencies with high efficiency. The granite base serves as a mounting surface for the laser source and as a substructure for the beam guidance elements and the travel axes within the machining area.

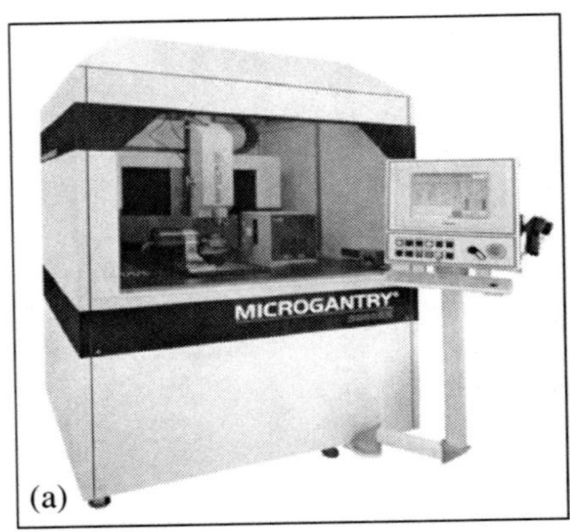

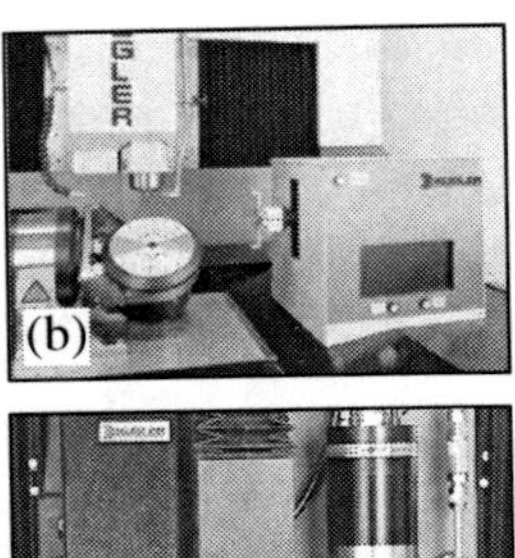

Fig. 11.26 Kugler® Microgantry nano 3/5X. **a** A general view. **b** Rotational CNC axes. **c** Laser and milling heads

All axes are designed as modular components and can be replaced quickly. The two horizontal air-bearing linear axes are composed of granite sleeves with slides in the shape of the surrounding cage. This design and the ironless linear motor drives gives them a high carrying capacity and stiffness as well as high motion dynamics. High resolution incremental scales are used in all the linear axes.

The machine enclosure has two functionalities: laser protection class 1 and heat isolation. All these construction details guarantees an absolute positioning accuracy in the whole X-Y area of $\pm 0.7\,\mu$m, and a travel accuracy of the main linear X axis less than $\pm 0.15\,\mu$m along the 300 mm of the travel range.

Acknowledgements Our thanks to all the companies cited by their pictures and information. Special thanks to Prof. L.N. López de Lacalle and Prof. A. Lamikiz for the time dedicated to discussing several aspects of this chapter.

References

[1] Alting L et al. (2003) Micro Engineering. Annals of the CIRP, 52/2:635–657

[2] Basuray PK et al. (1977) Transition from Ploughing to Cutting during Machining with Blunt Tools. Wear, 43(3):341–349

[3] Chuzhoy L et al. (2003) Machining Simulation of Ductile Iron and Its Constituents. J of Manufacturing Science and Engineering, 125(2):192–201

[4] Dow T.A. et al. (2004) Tool force and deflection compensation for small milling tools. Precision Engineering, 28:31–45

[5] Etxaniz I et al. (2005) Magnetic levitated 2D fast drive. LDIA Conf. 1:245–249

[6] Etxaniz I et al. (2006) Design and manufacture of a spindle with magnetic bearings for high speed machining. CIRP 5th Int. Conf. on HSM, 1:355–360

[7] Friedrich C (1998) Direct Fabrication of Deep X-Ray Lithography Masks by Micromechanical Milling. Precision Engineering, 22:164–173

[8] Fujimasa I (1996) Micromachines – A new era in mechanical engineering. Oxford Science Publications
[9] Herrero A et al. (2001) Development of a Three Axes Travelling Column Ultraprecision Milling Machine, Proc. of the 10th ICPE. Yokohama. Japan. 18–20 July
[10] Kim CJ (2004). Mechanisms of Chip Formation and Cutting Dynamics in the Micro-Scale Milling Process. PhD Thesis, University of Michigan
[11] Kussul EM et al. (1996) Micromechanical Engineering, A Basis for the Low-Cost Manufacturing of Mechanical Microdevices Using Microequipment. J of Micromechanics and Microengineering, 6(4):410–425
[12] Lucca DA et al. (1993) Effect of tool edge geometry on energy dissipation in ultra-precision machining. Annals of the CIRP, 42:83–86
[13] Masuzawa T (2000) State of the Art of Micromachining. Annals of the CIRP, 49/2
[14] Schaller T (1999) Microstructure Grooves with a Width of Less than 50 Micrometer Cut with Ground Hard Metal Micro End Mills. Precision Engineering, 23 229–235
[15] Schubert A et al. (2007) Improvement of micro-cutting accuracy by in-process application of 3D-sensors. EUSPEN Proc., Bremen, 368–371.
[16] Slocum A (2003) Linear motion carriage with aerostatic bearings preloaded by inclined iron core linear electric motor. Precision Engineering, 27(1):382–394
[17] Uriarte L et al. (2007) Error budget and stiffness chain assesment in a micromilling machine equipped with tools less than 0.3 mm in diameter. Precision Engineering, 31(1):1–12
[18] Uriarte L et al. (2006) Comparison between microfabrication technologies for metal tooling. Proc. IMechE Vol. 220 Part C: J. Mechanical Engineering Science: 1665–1676
[19] Vogler MP et al. (2003) Microstructure-Level Force Prediction Model for Micro-Milling of Multi-Phase Materials. J. of Manufacturing Science and Engineering, 125(2)
[20] Vogler MP (2003) On the Modelling and Analysis of Machining Performance in Micro-EndMilling. PhD Thesis, University of Illinois at Urbana-Champaign
[21] Weck M et al. (1997) Fabrication of Microcomponents Using Ultraprecision Machine Tools. Nanotechnology, 8:145–148
[22] Weule H et al. (2001) Micro-Cutting of Steel to Meet New Requirements in Miniaturization. Annals of the CIRP, 50(1):61–64
[23] Yuan ZJ et al. (1996) Effect of Diamond Tool Sharpness on Minimum Cutting Thickness and Cutting Surface Integrity in Ultraprecision Machining. J of Materials Processing Technology, 62(4) 327–330

Chapter 12
Machines for the Aeronautical Industry

J. Fernández and M. Arizmendi

Abstract This chapter highlights the role that machine tools play in the aerospace industry and is organised as follows. Firstly, a general description is made about how the aerospace industry is performing today and what the future looks like with respect to aerospace parts manufacturing. Secondly, the types of aerospace components in terms of geometry and function are described. Thirdly, the materials employed for the manufacturing of aerospace components are discussed. In the fourth section, the drives for machine tools currently operating in the manufacturing of aerospace parts are presented and finally, some trends in the machine tool construction are highlighted.

12.1 Aeronautical Business

This chapter deals with the role that manufacturing plays in the aerospace industrial sectors, both civil and military. These sectors have been strongly increasing throughout the world since the 1990s and it seems that they will continue to do so despite the recent economic setback of 2007.

One of the main driving forces of the aerospace industry is the increasing number of people travelling from one country to another, which increases the demand for aircraft.

To illustrate the good economic health of this sector, the turnover figures of the two principal world associations in aerospace, namely the Aerospace and Defence

J. Fernández and M. Arizmendi
Department of Mechanical Engineering, University of Navarra
TECNUN-School of Engineering, Paseo Manuel de Lardizábal 13, 20018 San Sebastián, Spain
jfdiaz@tecnun.es

L. N. López de Lacalle, A. Lamikiz, *Machine Tools for High Performance Machining*,

Industries Association of Europe (ASD) and the Aerospace Industries Association of USA (AIA) will be studied.

The European ASD represents over 2,000 aerospace and defence construction contractors and 80,000 suppliers companies which employ around 640,000 employees [1]. In contrast the AIA only represents 100 major aerospace and defence construction contractors and 175 leading aerospace and defence suppliers in the US which employ around 600,000 employees [2]. The US companies are much larger in size than the EU ones, because in the EU aerospace manufacturing is arranged in multi-national cooperative efforts and therefore the distribution of production is particularly broad.

In the year 2007 the European turnover reported by ASD was €120 billion while the turnover reported by the American AIA was greater than $200 billion.

The net profit in both Europe and the US aerospace companies has been growing every year by around 10% since the 1990s. Reliable sources forecast that by 2017 the number of running civil aircrafts will be between 25,000 and 30,000. This implies duplicating the actual number in 10 years and is congruent with an Airbus forecast that estimates a steady growth of 5% per year in the number of passengers from today to 2017 [3].

As far as spacecraft is concerned a growth rate of 50% is foreseen also for the next 10 years, as a result of the significant investment from the public sector.

Data from the European Union and the US indicate that aeronautics is a key sector for manufacturing companies and the manufacturing structure of both communities and the governments are actually stimulating aeronautics strongly through funded R&D programmes. Therefore this sector is in a good position and it seems that it will continue to be so for many years.

This fact brings about a beneficial effect for the machine tool companies. Presently, sales of machine tools for the aerospace sector is steadily increasing and there is a growing tendency, among the main machine tool companies, towards designing and building machines oriented towards machine specific aerospace parts instead of trying to sell general purpose machines adapted to aerospace.

As far as the type of company involved in aircraft manufacturing, 50% of the added value of a civil aircraft comes from the prime contractor or systems integrator (Boeing, Airbus/EADS, ...) that manufactures some structural parts and assembles the aircraft, and the other 50% comes from subcontractors that provide the equipment (30%) and that manufacture the engines (20%).

12.2 Aerospace Components

Manufacturing in the aerospace industry needs machine tools for mechanical components, clean rooms for electronic parts, and final-assembly facilities to integrate all subassemblies and systems. As was previously said, this final task is always performed by the prime contractor incorporating a complete range of hardware and software from suppliers that operate as subcontractors.

Here only the components manufactured by machine tools will be considered. Those components are usually classified in three groups, namely structures, engines and accessories and they will now be described in some detail.

12.2.1 *Aerospace Structures*

These components provide the aircraft with capacity to withstand mechanical stresses as a result of the weight, acceleration and wind. In their design, issues such as safety, reliability and, of course, cost, play an important role. Structures are geometrically complex and are usually manufactured out of "monolithic" aluminium alloy plates by removing a huge amount of material, sometimes up to 90% of the plate.

There are many structural components in an aircraft but here only the main ones will be considered, namely: wing ribs, stringers and longerons, spars, heavy frames, bulkheads and skins (Fig. 12.1).

Wing ribs (Fig. 12.1b) can be easily described through an analogy with the human body since they are related to the wing itself in the same way as human ribs are related to the spine. They are placed at equal intervals and give strength to the wing and the airfoil form to the metal sheet (*skin*) that is stretched on them.

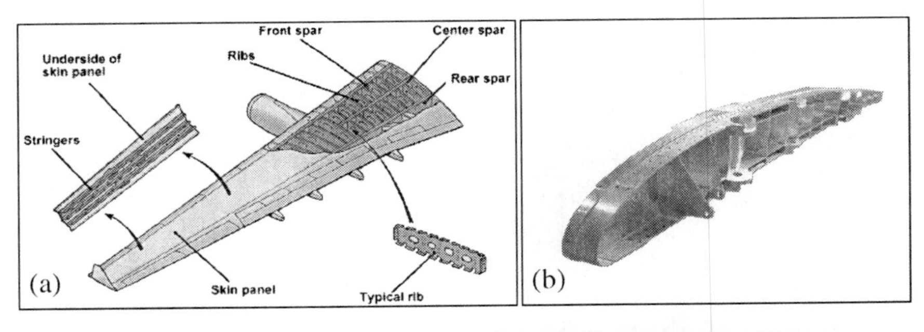

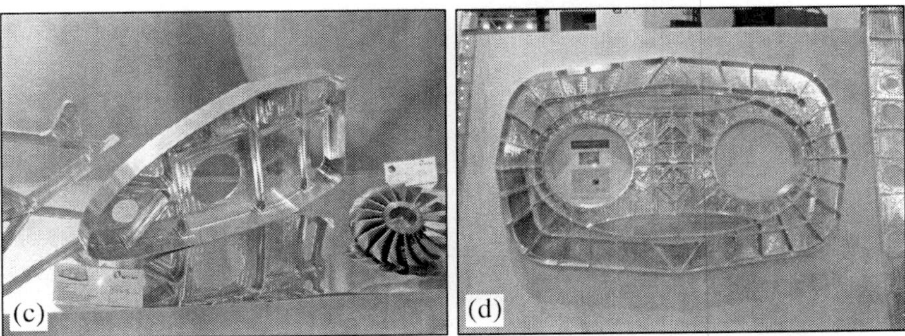

Fig. 12.1 **a** Airbus® wing structure from *www.sae.org*. **b** Wing rib made from aluminium for the Airbus® A380 machined on the HBZ Aerocell from Handtmann®. **c** Leading edge rib. **d** Bulkhead

Stringers are strips, placed perpendicular to the wing ribs and attached to them. For low-cost aircraft the stringers are made of formed sheet metal, either by extrusion or by bending. But when a higher quality is required, the stringer is either machined or cast.

The spar is the principal component of the wing structure. It is placed perpendicular to the fuselage and supports the bending of the wing both in air (lift force) and on the ground (wing weight). It is made out of aluminium or carbon fibre sheets.

Longerons are similar to stringers but they are placed in the fuselage instead of in the wings. They are also known as longitudinal stringers.

Heavy frames, also called formers, are laid perpendicular to the aircraft roll axis and give shape to the fuselage. They are attached to the longerons.

12.2.2 Aerospace Engines

In this section, the jet engine components will be considered. These are mainly a) housings, b) blades, c) blisks and d) rotors or discs (Fig. 12.2). Blisk stands for "bladed disc" and is composed of a rotor disc and blades. It can be made either by machining a solid cylinder or by assembling manufactured blades to manufactured discs.

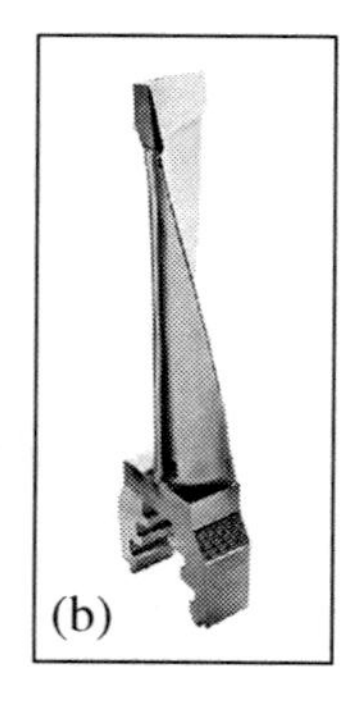

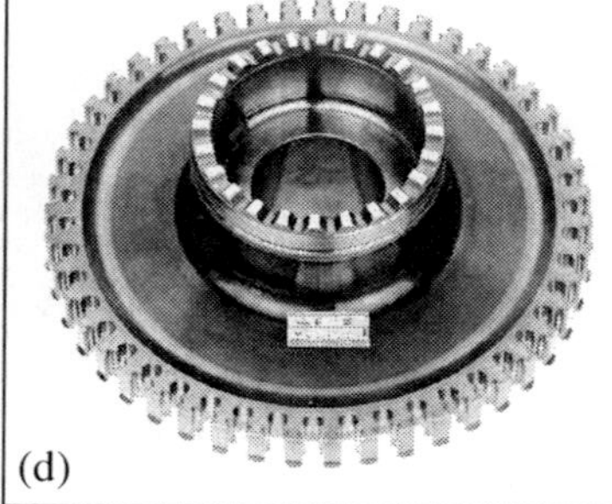

Fig. 12.2 Jet engine components. **a** Housing. **b** Turbine blade. **c** Blisks machined on an HX-243 from Starragheckert®. **d** Rotor disc

12.2.3 Accessories

Under this category all the metallic components that are neither engines nor structures are included. These are hydraulic and pneumatic systems, fuel control systems, mechanical systems such as landing gear, plane hinges, door frames, flap and slat tracks, brackets, seat rails, *etc.* (Fig. 12.3). About 40% of these components are cylindrical in shape and the remaining 60% are cubic forms manufactured out of either forgings or laminated products being the removed chip volume, not very large. The size of the accessories is usually smaller than 1 cubic metre, with the exception of some of the landing gear accessories that can be as long as 2.5 metres.

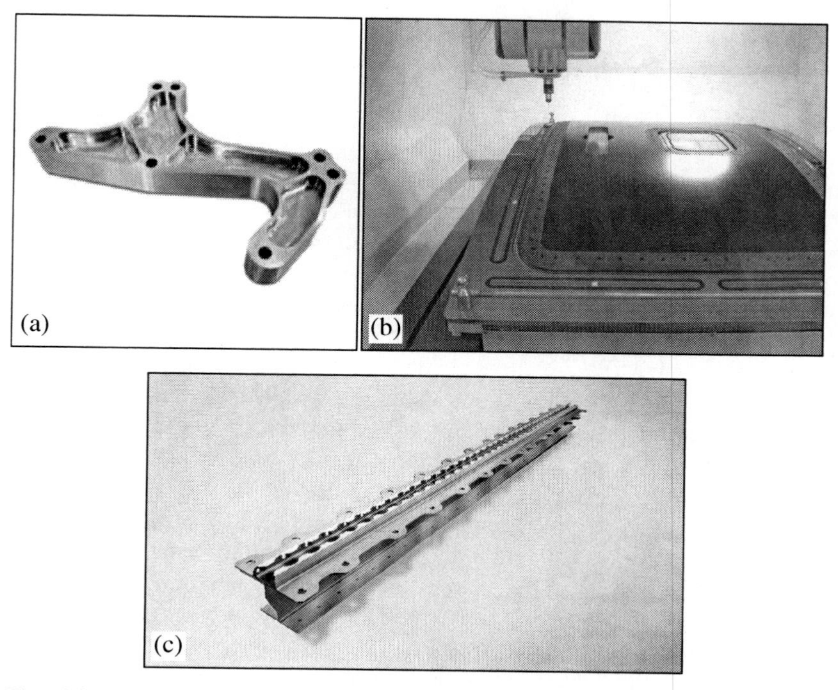

Fig. 12.3 Accessories. **a** Titanium landing gear bracket machined on the a81M machining centre from Makino®. **b** Door frame. **c** Seat rail machined on a MILL Series from Chiron®

12.3 Aerospace Materials

Titanium and aluminium alloys, heat resistant super alloys (HRSA) and composites are the main aerospace materials [6].

Wrought aluminium alloys are widely used for the airframe of most commercial aircraft [9, 13] because their density is much lower than that of steel and their strength, throughout the years, has improved and today some aluminium alloys are as strong as steel. Furthermore they present a better atmospheric corrosion resistance than that of steel. The most popular alloy series in aerospace is the AA7000 (the nomenclature of the American Aluminium Association) which is alloyed with zinc, magnesium and cooper and can be precipitation hardened to reach around 450 N/mm^2, the highest strength encountered in any aluminium alloy. Of this series the 7475, 7050 and 7075 are the most frequently used in the aerospace industry. These alloys "as received" present the following disadvantages: a) high residual stresses, b) high anisotropy, c) high spring-back recovery after machining and d) weld easily to the clearance and to the rake faces of the cutting tool.

Although new materials, such as polymer matrix composites, are also being increasingly used in modern commercial aircraft structures, experts think that aluminium alloys will continue to be the main material for aeronautical structures for a long time.

Titanium alloys are widely used in heavily loaded aeronautical structures such as bulkheads, wings and landing gear beams [13], due to their excellent combination of a high specific strength and a good corrosion resistance, which are both maintained at high temperatures. For these types of components titanium performs much better than steel and aluminium alloys, but on the other hand it is more difficult to machine.

Titanium is also used in the engine compressor stage (with a temperature below 500°) representing around 40% of the volume of an aeronautical engine.

Titanium alloys have a) a low thermal conductivity, b) a high chemical reactivity to tool materials and c) a low elastic modulus compared to that of steel. As a consequence of a) the heat tends to stay confined in the cutting area, raising local temperatures and facilitating welding to the tool rake face which damages the cutting edge geometry and leads to rapid deterioration of the tool. Although a generous application of cutting fluid improves cutting behaviour, titanium alloys machining is still very demanding both for the tool and for the workpiece.

Superalloys are nickel or cobalt alloys, employed for aerospace engines because of their a) high plastic deformation energy at high temperatures, b) chemical degradation resistance and c) wear resistance. However these materials have the lowest machinability index due to the following factors: the cutting forces and the temperature at the cutting zone are extremely high due to their high plastic deformation energy and low thermal conductivity, they are very ductile and the workpieces tend to move away from the tool creating geometric errors and they have a high strain hardening coefficient that makes hardness increase due to the plastic deformation inherent in the cutting process [5, 10].

Previous cold working should be taken into account because it increases hardness and makes machining more difficult, causing premature tool notch wear.

In order to reduce the strain hardening, large feeds, a low cutting speed and sharp tools are necessary.

Roughing operations of Ni-base alloys are done before hardening (25 HRC) and semifinishing and finishing are done after hardening (35–45 HRC). The most common tool material is a "hard metal" (i.e., tungsten carbide sintered with cobalt) (grades ISO K5–K10) of a very fine grain ($< 1\ \mu m$), and with a coating of PVD or CVD of the type TiN or TiAlN.

However more expensive tools such as PCBN (Polycrystalline cubic boron nitride) or whisker reinforced ceramics ($Al_2O_3 + CSi_W$), are also used.

In recent years, there is a growing interest in intermetallic materials such as alloys γ-TiAl [4, 12], which have recently emerged as an alternative to nickel based alloys in engine parts for working conditions below 800°C. It presents interesting properties in hardness, toughness and low density in comparison with conventional materials. However nowadays the use of these materials is still limited because of their very low machinability.

Besides aluminium and titanium alloys and superalloys, polymer-matrix composites are also being increasingly used in aerospace due to their stiffness, light weight, and heat resistance. Composites are made of a carbon (or hydrocarbon) fibre matrix, sometimes reinforced by metallic filaments that are bonded together by polymer resins. Some new application examples are: composites in laminated forms for wing skin, fuselage bulkheads, and solar array supports in satellites.

Fibre-wound forms, tubular or spherical shapes are used for rocket motor casings and for spherical containers of fuel, lubricants, and gases. These forms are manufactured by winding a continuous fibre on a spinning mandrel at a high speed while a liquid resin is injected as the part is formed. Later the resin is cured.

The increase in composite applications opens new roads for research [11] in topics such as static and dynamic analysis methods, manufacturing methods with enhanced reliability, methods and equipment for on-line inspection, and applications of embedded sensors to provide some intelligence to the composite, to mention only a few.

Composite consumption in aeronautics is going up at around 20% per year linked to the strong Airbus and Boeing programmes and could reach 22% of the total material used in aircrafts by 2010. For example 50% of the new Airbus 380 and Boeing "787 Dreamliner" are manufactured with this material. One example that highlights the increasing acceptance of composites to the detriment of aluminium alloys is the "LH-10 Ellipse" aircraft, a two-seat sport plane entirely designed in carbon fibre.

12.4 Costs, Weight and Precision in Machine Tools for Aerospace Machining

In this section the main topics that have driven or that are driving changes in the machine tools to make them more appealing to aerospace components manufacturers are examined.

Competition is strong in the aerospace industry, as in any other industry, and the main goals are to drive down costs and/or reduce aircraft weight [7, 8]. These two goals deserve to be considered in some detail.

12.4.1 The Drive to Reduce Aircraft Costs

The prevalent strategy for "small to middle range" aircraft is to cut costs. The reason is that all aircraft companies build good and reliable aircraft and therefore low price is a priority when it comes to buying a new aircraft.

The most straightforward strategy to cut costs consists of reducing process times and the simplest solution for this is to increase feed speeds. However this has two negative effects. On the one hand, a) it makes the cutting force to increase (roughly in the same proportion as the feed) and b) it increases surface roughness. However increasing cutting force also means increasing geometric error (a representative example is the thin wall machining of aircraft structures). And increasing roughness means poor surface quality. So increasing feed is not a good cost cutting strategy.

Another strategy to cut costs consists of increasing both feed speed and spindle speed, in the same proportion, so that feed per tooth (considering, for example, a milling process) and therefore cutting force and roughness, do not increase.

The later strategy is known as high speed milling (HSM) and is being increasingly applied in aerospace manufacturing.

Nevertheless increasing cutting speed presents the shortcoming of reducing tool life, and therefore increasing associated tool costs. To solve this problem tool coatings and more effective lubricant dispensation systems are being developed. Although this also increases the tool related cost, on the whole this increase in cost is compensated by the reduction in cutting times in many aerospace manufacturing operations and this is the reason why HSM is gaining momentum in aerospace manufacturing.

However the introduction of HSM implies running the spindle and the feed axis of the machine at higher speeds than the ones used in ordinary speed machining and this gives rise to some problems that imply the introduction of new or adapted systems in the machine tools:

a) Dissipation systems for the heat produced in the bearings, the rotor and the stator of the electro-spindles.
b) Automatic evacuation systems with high removal rates of chips.
c) Cutting fluid filters.
d) Mist handling, so that the high pressure cutting fluid does not impede the operator's ability to see the cutting zone.
e) Machine shielding to protect the operator from being hurt by chips at a high speed.

f) Feeding of the cutting fluid through the tool (otherwise tool tip cooling is not efficient in HSM).
g) Number of tools in the machine magazine to cope with the reduced tool life in HSM.

All these solutions have allowed HSM to be implemented by many component suppliers.

12.4.2 *The Drive to Reduce Aircraft Weight*

The prevalent strategy for “long range–big capacity” aircrafts is not to cut costs, as in small to middle range aircrafts (see Sect. 12.4.1). In the former, the number of constructors is smaller and competition is not so strong, so price is not as important when it comes to buying a new aircraft. Here the main issue is to minimize fuel consumption, and thereby reduce aircraft weight. There are two ways to reduce aircraft weight:

a) Improving performance of the aircraft materials or introducing new material components (the same mechanical properties with less weight) and
b) Reducing the mass of the aircraft components.

The role of the components supplier is apparently a secondary one, since the first approach concerns the “materials department” and the second one, the “design and engineering department” of the main contractor. Nevertheless, the component supplier actually plays an important role by working very closely with both departments in researching new solutions. Thanks to main contractors and component suppliers today’s aircrafts are lighter and relatively cheaper.

But besides aircraft cost and weight reduction, there are other topics that mark the way machine tools are designed and built. One is component precision that will be considered in the next section.

12.4.3 *The Drive for Aircraft Component Precision*

Precision is another “hot topic” in aerospace. It can be tackled mainly by:

a) Reducing the number of component set-ups on the machine, and
b) reducing component and machine distortions during machining.

12.4.3.1 Reducing the Number of Set-ups to Improve Precision

Component set-up on the machine table not only takes time but it is also a main source of component errors (parallelism, perpendicularity, position …). Therefore, to improve precision, a good strategy is to reduce the number of set-ups to a minimum.

For machining a particular component, the number of set-ups, increases with its geometrical complexity, but decreases with the number of machine tool CNC axes. Since a large amount of aerospace components are very complex in geometry and need more than one set-up, one strategy for improving precision consists of changing from a 3-axis machine to a 4 or 5-axis machine. Actually, this is the trend observed in many manufacturing subcontractors. 4/5-axis machines are more expensive and more difficult to program and control than the 3-axis machines. Nevertheless this extra cost is worthwhile in many components due to the improvement in precision.

Another advantage of the 4/5-axis machine is that machining can be done in many more possible ways and therefore better machining strategies can be selected. For example a 5-axis machine can perform an operation with a shorter overhang length tool than a 3-axis machine, and therefore with greater rigidity and precision.

12.4.3.2 Diminishing Distortion to Improve Precision

Apart from minimising the number of set-ups, which has just been considered, diminishing component and machine distortions during machining is another issue that allows improving precision and this will be now considered.

As was already mentioned (Sect. 12.2.1), structure components are usually machined out of large aluminium plates and, since the component geometry is usually strongly non-symmetrical, the natural equilibrium of the aluminium plate internal stresses is strongly broken and this generates component distortion. One way to minimize this danger is by checking the "as received" plates for internal stresses.

Another issue is the differences found in the shapes of components of the same batch caused by the changing thermal status of the machine tool over time as a result of different sources such as the warming up period of the machine or even difference between day and night temperatures in the shop. These thermal gradients give rise to distortions in the machine and generate differences in the component shape. There are two strategies to solve this problem: a) keeping the machine thermal status constant and b) estimating and subsequently correcting the state of distortions in the machine. Strategy a) demands good design and sometimes the installation of cooling devices close to the machine's thermal sources and represents the simplest solution. Some machine builders address also strategy b) by providing intelligence to the machine tool so that distortions can be estimated and corrected automatically. This topic and the improvement of the machine dynamics are probably the two most crucial research points for machine tool builders today.

12.5 Machine Tools for Aeronautical Components

Having considered how machine tools and machining methods have evolved to improve the price, weight and precision of aeronautical components, we may now address the evolution of machine tools and the machining methods related to the three main classes of components, structures, engines and accessories.

12.5.1 Machine Tools for Machining Aeronautical Structures

As was already mentioned (Sect. 12.2.1) aeronautical structures are geometrically complex and are manufactured mainly out of aluminium plates by removing a large amount of material (sometimes up to 90%). This large ratio aluminium plate mass/components mass, the so-called "buy to fly" ratio, implies converting a huge amount of raw material in chips and this implies considerable costs.

That geometrical complexity came into play little by little in aerospace gaining ground to components that initially were made out of many simpler parts joined by screws, rivets or welds. The disadvantages of the joining process were lower component strength and the amount of time spent in the assembly of the component. The complexity of the parts geometry demands a 4/5-axis machining centre. However tolerances are not an issue (around 0.2 mm).

As is mentioned in Chap. 1, three configurations are common in 5-axis machining centres, namely:

a) *LLLRR*, where the spindle head has two rotary axis,
b) *RLLLR*, with a rotary table and a swivelling spindle head and
c) *RRLLL*, with a double rotary table.

Besides the rotary and/or swivelling movements, all configurations also have 3 linear axes (X, Y, Z) between the head and the table.

The best configuration for a particular component will depend on its size which, for these components, varies considerably (between 1 to 12 metres).

12.5.1.1 The LLLRR Configuration

For machining large structural parts such as wing and fuselage skins, frames and stringers, LLLRR machining centres with three Cartesian axes (X, Y, Z) and two rotary axes (A, C) at the spindle head are used. The C-axis rotates the head while the A-axis tilts it.

This axes configuration is typical of large "gantry" machining centres, such as, for example, the dual Spindle gantry mill from Henri Liné® (Fig. 12.4a) and the Gantry TS machining centre from Handtmann® (Fig. 12.4b). Both have two moving vertical spindles so that either of the two components can be machined simultaneously or one component can be cut by two tools at the same time. In both cases an increase in the material removal rate, and hence in productivity, can be achieved.

Mtorres® has developed the 5-axis gantry milling machine TorresMill (Fig. 12.5) with a flexible tooling for the accurate positioning of parts for the routing and drilling of skin panels, stringers and frames.

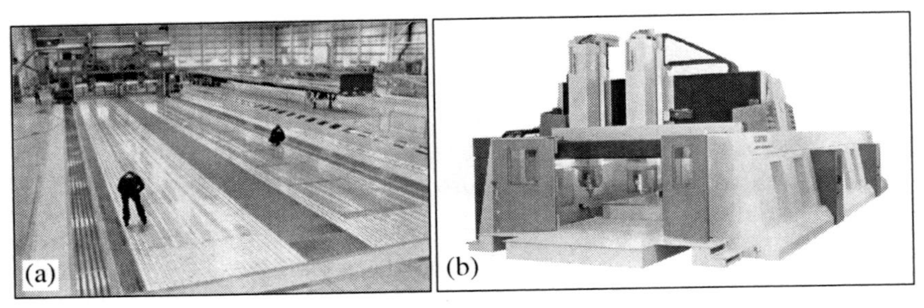

Fig. 12.4 **a** Machining of large wing skin on a dual Spindle Gantry mill from Henri Liné®. **b** Handtmann® Gantry TS with a twin spindle

A more specific machine has been developed by Fatronik® (Fig. 12.6), which is a crawling portable robot for drilling large frame panels and structures for fuselages and wings. The robot has the capacity of fixing itself to the frame panel by vacuum cups so that drilling can be performed. Compared to the previous machines, it provides flexibility and portability.

As was said before, machining of structural components generates huge amounts of chips that have to be evacuated from the working area. To solve this

Fig. 12.5 TorresMill from MTorres® for the routing and drilling of aerospace structural components

Fig. 12.6 Crawling portable robot from Fatronik® for the drilling of large panels and structures of fuselage and wings

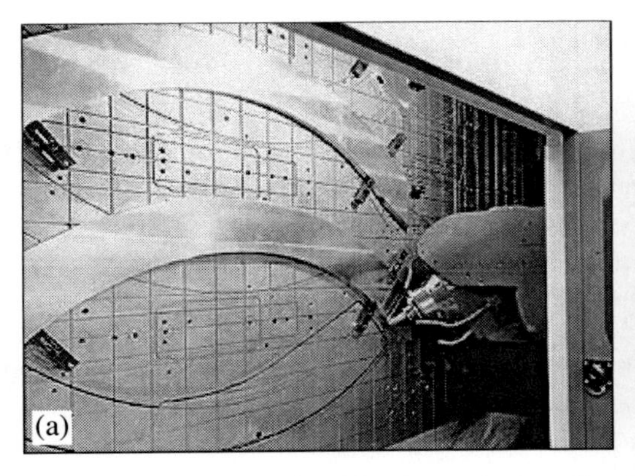

Fig. 12.7 **a** Horizontal machining centre Aerostar from Forest Liné®. **b** 5-axis hybrid machining centre Hera from Fatronik®

problem, horizontal machining centres with a travelling column, as the one shown in Fig. 12.7a, are commonly used. As in previous machines, the tool has two rotary axes (A, C). This configuration allows high chip volumes to be evacuated through a chip conveyor placed between the column and the vertical frontal panel where the part is placed.

Parallel kinematics has a niche for aeronautical structures and some 5-axis hybrid machining centres with a travelling column have been developed. As an example, the Hera from Fatronik® (Fig. 12.7b) has a 3 axes parallel kinematics module implemented in 2 serial axes for the machining of structural monolithic parts.

For medium size structures, 5-axis horizontal machining centres, with an LLLRR configuration, are common. As an example, Fig. 12.8 shows the MAG4 5-axis machining centre from Makino®. It has two A/C rotary axes at the spindle head where the A-axis rotates by +/–107.5° and C, by 360° allowing 5 side machining.

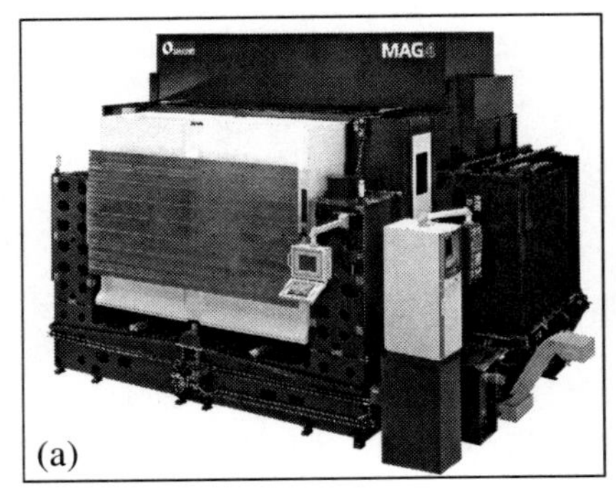

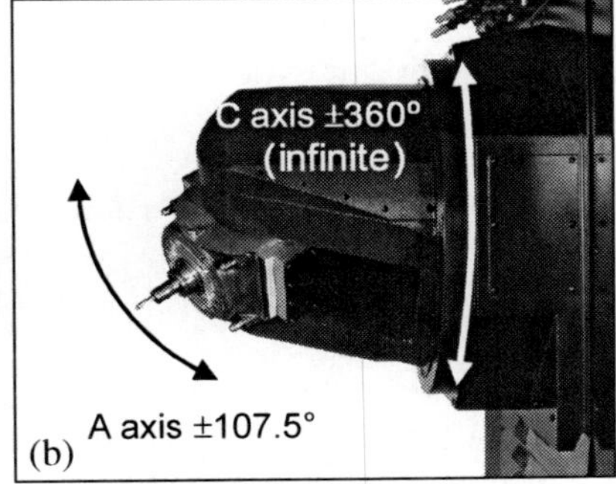

Fig. 12.8 **a** Five–axis horizontal machining centre MAG4 from Makino®. **b** A detail of the two-axis tilt/rotary spindle head

12.5.1.2 The RLLLR Configuration

So far, only the LLLRR configuration has been considered. This machine type is the most adequate for large components that do not require high precision but it cannot be employed to machine precise parts because the two rotary axes cannot provide enough stiffness to the tool. In this case, one of the rotary axes is transferred to the table and the tool stiffness, now with only one axis, can provide enough stiffness to fulfil the demanded precision in the component. This is the RLLLR configuration.

Medium-size components (such as wing ribs) usually demand more precision, and for this type of part machines with a RLLLR configuration are the right choice. Among the different possibilities of this configuration the most common one is a machine with a 360° rotary A-axis at the table and with a swivelling B-axis at the spindle head.

One example of this type of machines is the Mill Series from Chiron®. It is a 5-axis machining centre that can be seen in Fig. 12.9a. The combination of a 360° horizontal rotary (A) table and the (+/–100° range) swivelling (B) of the head allows complex parts, such as the one in the figure, to be machined in only one set-up.

(a)

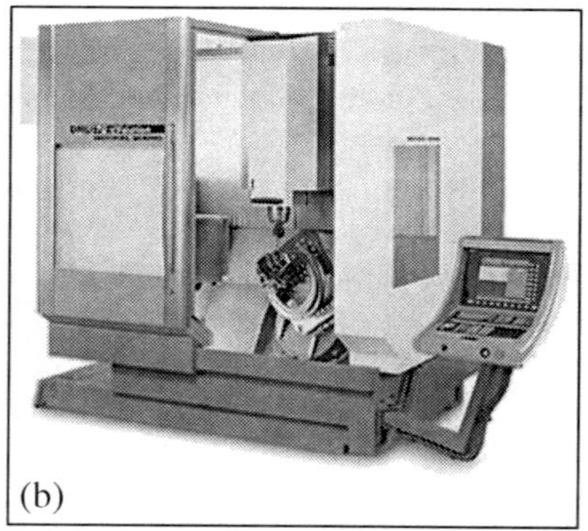
(b)

Fig. 12.9 **a** Mill Series from Chiron® with a B swivel head and A rotary table. **b** DMU 70 eVolution with a swivel rotary table from DMG®

12.5.1.3 The RRLLL Configuration

For small structural components, the configuration RRLLL with a swivel/rotary table is commonly used.

As an example of this configuration, Fig. 12.9b shows the vertical 5-axis machining centre DMU 70 eVolution. It has a table with a 360-degree rotary C-axis and a swivelling B-axis. It is a very flexible machine but it has the drawback of a small working area that limits the component size that can be machined.

As far as structures are concerned there is an emerging process of joining structural assemblies to make larger components: the so-called "friction stir welding" (FSW). FSW is easier and faster to implement in the shop than the other structural joining processes such as riveting or welding. It strongly reduces assembly time and eliminates rivets, thereby, reducing assembly costs and providing stronger and lighter joints. It is believed that, in the future, FSW will replace most of the rivets in assemblies of cabins, fuselages, wings and engines.

12.5.2 Machine Tools for Machining Engine Components

These components are machined out of castings or forgings of titanium alloys, HRSA and steel and undergo long machining operations to reduce weight. However, in contrast to structures, tolerances are demanding.

The machine tools employed to manufacture different engine components are: a) vertical lathes, machining centres of b) RRLLL and c) RLLLR configurations and d) the newly introduced multi-function or multi-tasking machines. All are described below.

12.5.2.1 Vertical Lathes

One of the engine components is the ring case (or housing), that can be as big as 2500 mm in diameter and needs external and internal turning operations plus milling operations to reduce weight. Commonly, the turning operations are performed in lathes and then the component goes to a machining centre for the thinning operations.

For shaping the internal and the external diameters of the ring type engine cases vertical lathes are employed (Fig. 12.10).

Fig. 12.10 Vertical turning lathe Contumat VCE from Dörries Scharmann Technologie® for turning and boring jet engine parts

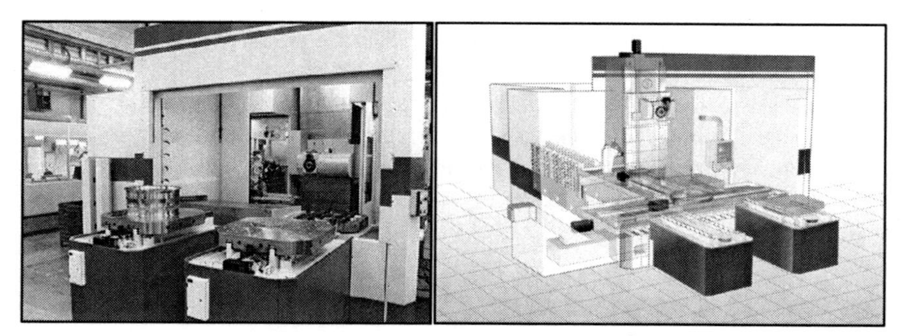

Fig. 12.11 Five-axis horizontal machining centre Starragheckert® ZT-800

For the subsequent weight reducing operations in engine cases, RRLLL and RLLLR configurations are commonly used. Apparently the machining centres are similar to the ones used for small and medium sized structural components but there are some differences because, due to the difficult-to-cut materials used in engine parts, much lower cutting speeds are employed and, as a consequence, main spindles with high torque at low speeds are used instead of high speed spindles that are commonly used for structural component machining.

As an example, Fig. 12.11 shows the 5-axis horizontal machining centre Starragheckert® ZT-800, that has an RLLLR configuration. The head spindle can move in the Y, Z and A axes and the table in the X and B axes. The machine has two loading/unloading stations to save cycle times.

12.5.2.2 Machining Centres with the RRLLL Configuration

For the machining of smaller engine components such as blisks and impellers, machining centres with swivelling and rotary table (RRLLL configuration) are used. The three linear axes (X, Y, Z) can be placed either at the tool side or distributed between the tool and the table.

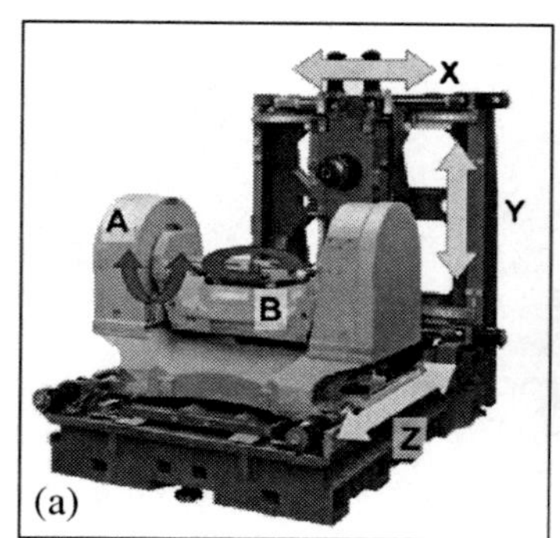

Fig. 12.12 a 5-axis horizontal machining centre NMH10000 DCG from Mori Seiki®. b Machining of a blisk

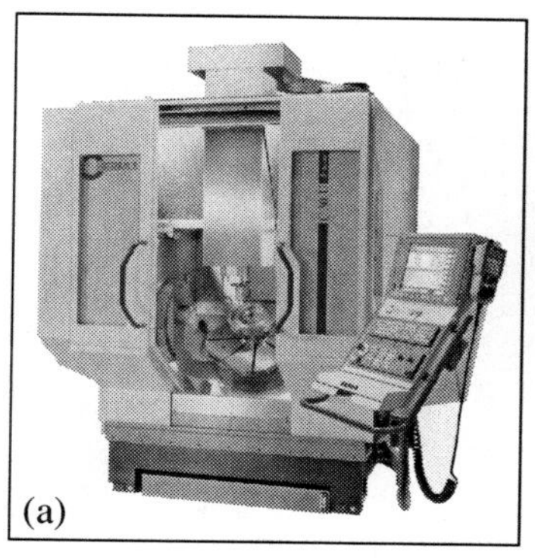

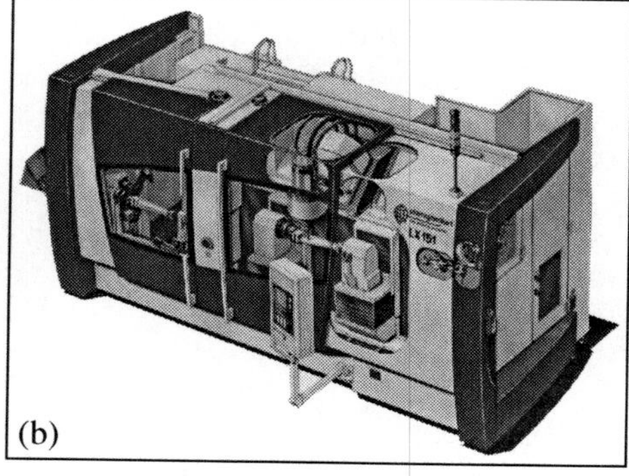

Fig. 12.13 **a** C30 U from Hermle®. **b** LX 051 from Starragheckert® for turbine blades machining

Figure 12.12a shows the NMH10000 DCG, a 5-axis horizontal machining centre from Mori Seiki that has a box-in-box frame architecture and a table with a linear Z axis, a B rotary axis and an A swivelling axis. Figure 12.12b illustrates how the machining of a blisk is performed.

Figure 12.13a shows another RRLLL configuration, a 5-axis vertical machining centre C30 U from Hermle®, with a gantry design instead a box-in-box frame. The tool travels in the three (X, Y and Z) axes and the component is placed on a swivelling (A) and rotary (C) table. The A-axis swivelling range is of +/–115°.

12.5.2.3 Machining Centres with the RLLLR Configuration

For the machining of turbine blades, as well as for case thinning, machining centres with an RLLLR configuration are the best choice. In Fig. 12.13b, a 5-axis vertical machining centre from Starragheckert® can be observed where the blade that is held between centres can rotate in the A-axis while the spindle head swivels on the B-axis (+/–50°).

12.5.2.4 Multi-function or Multi-tasking Machines

As seen above, some components such as the engine housings need to go through a vertical lathe, for external and internal turning, and later through a machining centre for weight reducing operations. This procedure has a time penalty since the component has to be loaded and unloaded in both machines.

To solve this problem new hybrid machines known as multi-function or multi-tasking machines have been shown in recent machine tools fairs. One representative example is a new DMG hybrid machine (100% lathe and 100% machining centre according to the builder) shown in Fig. 12.14a which is designed to machine engine components such as compressor housings, blisks and turbine blades in only one set-up.

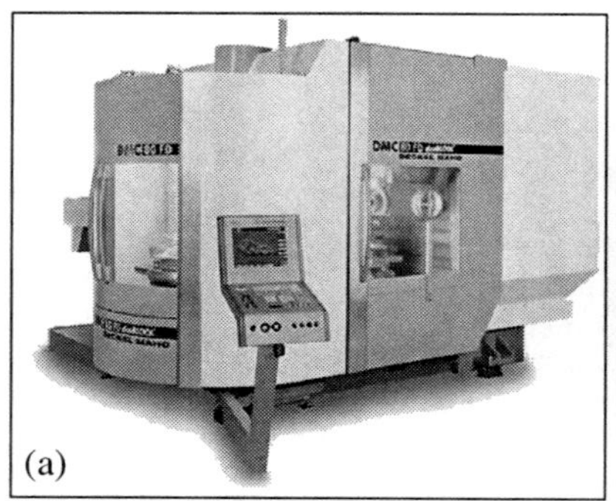

Fig. 12.14 **a** DMC 80 FD from DMG®. **b** Machining example

Figure 12.14b shows a boring operation on an engine housing performed in this "turning and milling" machine.

Besides the "turning and milling at the same machine" strategy other machine strategies are appearing for reducing set-up times. One of them is the "milling and grinding at the same time" strategy that is being applied to blades, turbine discs and also to housings. These components have milling and grinding features and therefore they have to visit both a milling and a grinding machine. This is the classic approach. But it implies an increasing number of set-ups, fixtures and cost. Some machine builders trying to reduce the machining cost of these components have incorporated grinding capabilities in their milling machines creating the so called "milling and grinding at the same time" strategy. One of them is Makino®. In Fig. 12.15 the "G5 iGrinder" horizontal machining centre is shown. This is another hybrid 5-axis machining centre suitable for machining special alloys for aerospace and engine components such as blades, turbine discs and housings.

Reviewing the manufacturing of aerospace engine components it can be concluded that milling is the principal operation followed by turning and grinding. But there are also other non-traditional processes such as laser, ultrasounds and

Fig. 12.15 Horizontal machining centre G5 iGrinder from Makino®

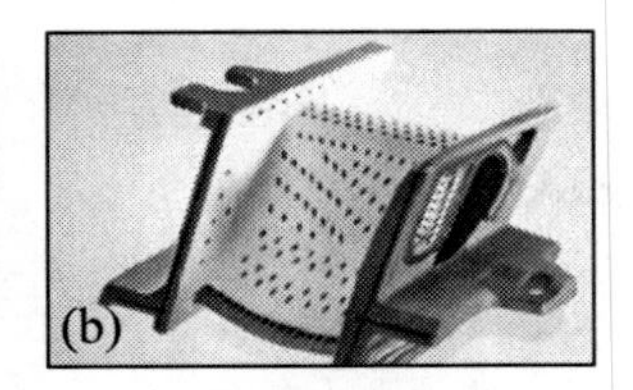

Fig. 12.16 **a** 5-axis laser precision drilling of holes in engine components with Lasertec 80 Powerdrill from DMG®. **b** Turbine vane made of Inconel. **c** Compressor ring made of stainless steel

electrodischarge machining (EDM), that have their small and very specific niches of applications in the manufacturing of engine components. In some cases these processes are the only way to perform certain operations.

One example of this is the Lasertec 80 Powerdrill from DMG® that can drill holes in difficult-to-cut materials (Fig.12.16) with better tolerances and times than other more traditional processes.

12.5.3 Machine Tools for Machining Accessories and Structure Fittings

Accessories and structure fittings are aerospace components, which are smaller than the aluminium structures. They are never larger than 1000 mm in length.

They are complex in geometry and also require complex fixtures to be fixed to the machine table. Their tolerances are usually rather demanding, ranging between 0.005 and 0.01 mm, and therefore grinding operations are not unusual. Complex geometry accessories require a large number of machining operations and to reduce cycle times: high-speed tool positioning, small tool changing times and magazines with a large number of tools are needed.

Because of the expensive fixtures and the demanding precision, it is also very important to minimise the number of set-ups and hybrid machines are a good solution because they have been designed to manufacture with fewer and simpler tools, simple and cheaper fixtures and fewer set-ups. And fewer set-ups means also more precision.

One representative example of these multi-task machines is the new 5 axis Integrex Mark IV Series machine from Mazak® (Fig. 12.17). This type of machine is a mixture of a turning and milling centre designed to produce components in one

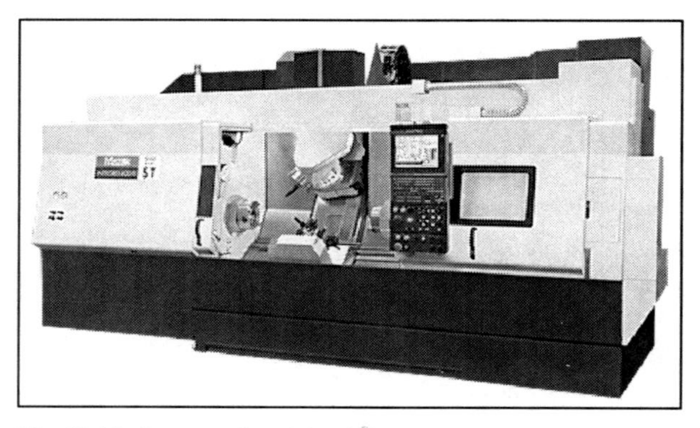

Fig. 12.17 Integrex from Mazak®

single set-up. It has a turning spindle (single or dual opposed) and also a milling spindle head with a linear Y-axis and a rotating B-axis. To shorten cycle times it has an optional additional tool turret for simultaneous cutting. The machine is very flexible and, according to the builder, can be competitive with different components such as: a) round parts with secondary operations, b) fully prismatic parts from a solid or a casting, c) or sculptured parts such as aerospace components and moulds.

Another example is the Millturn machine from WFL® shown in Fig. 12.18a. This is a turning-milling machine where the tool can be oriented through a B-axis to machine the workpiece at any angle. Figure 12.18b and c show two fully machined components from this turning-milling machine.

Typical batch sizes of accessories are less than 100 units/year for between 8 and 20 accessory types. Airbus is an exception with batches of 2,000 per year since the components are common for many aircraft models. To cope with small batches with many variants machine tool flexibility is a necessity. About 70% of those components (mainly valves and pistons) are made of difficult-to-machine superalloys and the remaining 30% are made of easy-to-machine materials, mainly

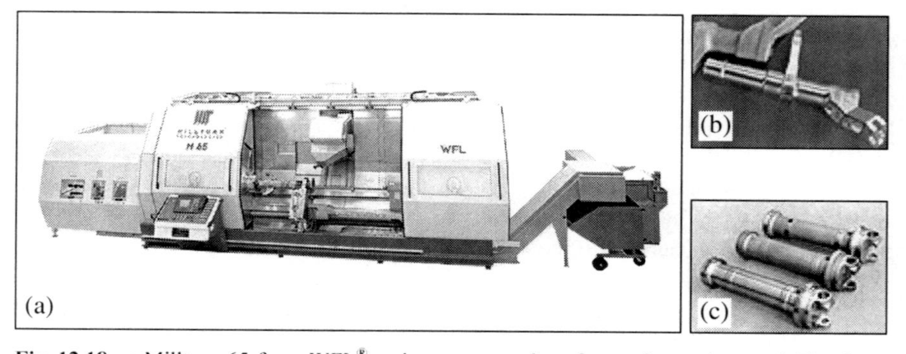

Fig. 12.18 **a** Millturn 65 from WFL® and some examples of manufactured parts. **b** Flap lever. **c** Landing gear

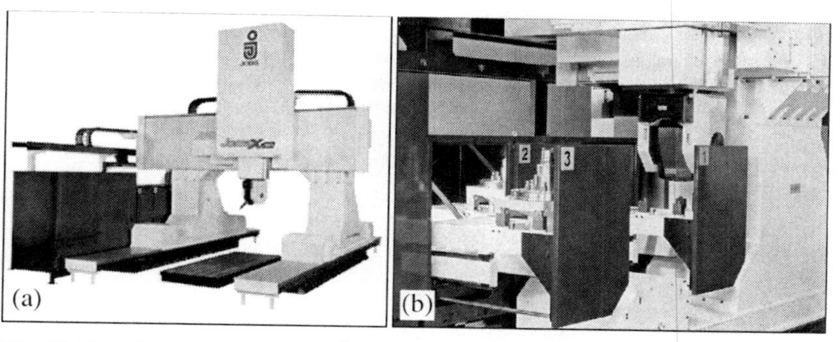

Fig. 12.19 **a** JomaX 265 from Jobs®. **b** Automatically interchangeable spindles

aluminium alloys. For light materials, a very large cutting speed can be used and a 40,000 rpm spindle would be a good choice, but this same spindle will be too expensive for machining superalloys since the cutting speed of this material group is low and a less expensive 6,000 rpm spindle will be sufficient. This problem has recently been addressed by some builders and they are offering machine tools equipped with two automatically interchangeable spindles such as the gantry type machining centre FRF (Q) from the Czech company Tos Kuřim® and the JomaX 265, a mobile gantry milling centre for 3/3 + 2/4/5 axis machining from the Italian company Jobs® (Fig. 12.19).

Acknowledgements Our thanks to Fatronik and Invema (Spanish Foundation for Machine Tool Research) for the information provided, and to all mentioned companies for the pictures.

References

[1] Aerospace and Defence Industries Association of Europe (ASD) http://www.asd-europe.org/. Accessed 28 March 2008
[2] Aerospace Industries Association. http://www.aia-aerospace.org/. Accessed 28 March 2008
[3] Airbus – Press Release – 7 February 2008 http://www.Airbus.com/en/presscentre/pressreleases/pressreleases_items/08_02_07_Airbus _forecast_2008.html, Accessed 28 March 2008
[4] Aspinwal DK, Dewes RC, Mantle AL (2005) The Machining of γ-TiAl Intermetallic Alloys. CIRP Ann- Manuf Techn, 54/1:99–104
[5] Ezugwu EO (2005) Key improvements in the machining of difficult-to-cut aerospace superalloys. Int J Mach Tools Manuf, 45/12–13:1353–1367
[6] Ezugwu EO, Bonney J, Yamane Y (2003) An overview of the machinability of aeroengine alloys. J Mater Process Technol 134:233–253
[7] Fatronik (2002) Internal Report: Tendair (Trends in Aircraft Industries)
[8] Invema (2007) Internal Report: Tendencias Tecnológicas y Oportunidades de Negocio (Business and Technological Opportunities, in Spanish)
[9] Lequeu P, Lassince P, Warner T, Raynaud GM (2001) Engineering for the future: weight saving and cost reduction initiatives. Aircraft Engineering and Aerospace Technology 73/2:147–159

[10] Smith RJ, Lewis GJ and Yates DH (2001) Development and application of nickel alloys in aerospace engineering. Aircraft Engineering and Aerospace Technology 73/2:138–146
[11] Teti R (2002) Machining of Composite Materials. CIRP Ann-Manuf Techn 51/2:611–634
[12] Weinert K, Biermann D, Bergmann S (2007) Machining of High Strength Light Weight Alloys for Engine Applications. CIRP Ann-Manuf Techn 56/1:105–108
[13] Williams JC, Starke EA (2003) Progress in structural materials for aerospace systems. Acta Materiala 51:5775–5799

Chapter 13
Machine Tools for the Automotive Industry

Ciro A. Rodríguez and Horacio Ahuett

Abstract This chapter describes the machine tools used in the automotive industry. The world trends in automotive production are discussed to stress the significance of this industrial sector in the global economy. A description of the typical automotive components with machining operations is also discussed. The use of flexible machines vs dedicated machines is analysed in the context of current trends in the automotive markets. Finally, new machine tool and process technologies of particular relevance to the automotive industry are described.

13.1 World Trends in Automotive Production

The automotive production could be considered the motor of the industrial development of the 20th century. Current trends indicate that such a significant influence will continue in the near future. This section presents the economic impact of the automotive industry and the corresponding automotive components that utilise significant amounts of machining operations.

13.1.1 The Economic Impact of the Automotive Industry

The automotive industry is one of the main drivers of the world's economy. Nearly 70 million cars, vans, trucks and buses were produced in 2006 alone. This level of output represents a global turnover of €1,900 billion, a figure that exceeds the GDP

Ciro A. Rodríguez and Horacio Ahuett
Centre for Innovation in Design and Technology, Tecnológico de Monterrey,
Av. Eugenio Garza Sada #2501 Sur, Monterrey, N.L. 64849 MÉXICO
{ciro.rodriguez, horacio.ahuett}@itesm.mx

L. N. López de Lacalle, A. Lamikiz, *Machine Tools for High Performance Machining*,

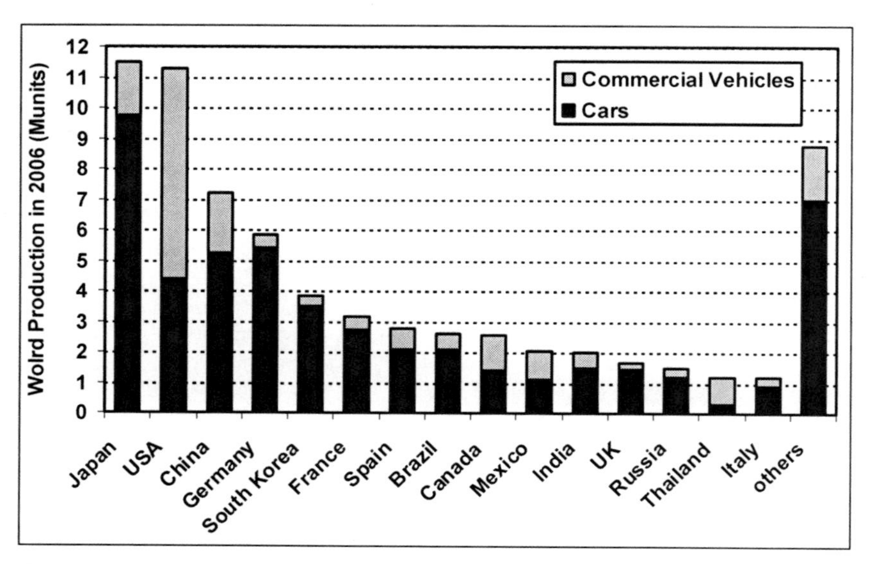

Fig. 13.1 World production of cars and commercial vehicles (adapted from [14])

(gross domestic product) of all but the five richest countries. Furthermore, this industry provides more than eight million direct jobs worldwide, or about 5% of the total manufacturing jobs, and approximately five times more are employed indirectly in related manufacturing and service supplies. In all, an estimated 50 million people earn their living from the manufacturing of cars, trucks, buses and coaches [14].

Worldwide, the automotive industry has grown at yearly rate of approximately 4% over the last few years. Furthermore, the production of cars and commercial vehicles is distributed all over the world. Figure 13.1 shows the top 15 countries in this industry during 2006. In addition to the top 15 countries, more than 25 countries contribute to the production of nearly 9 million vehicles [14].

Finally, the influence of the auto industry transcends the mere production of goods. The auto industry is also a leader in innovation, playing a key role in the technology level of other industries and of society as a whole. By current estimations, the industry invests about €85 billion in research and development per year, with several manufacturers found among the leading top 10 corporations in the world [14].

13.1.2 Machining Processes in Automotive Production

In the automotive industry, the general categories of products with significant machining operations include: a) engine components, b) transmission and drive components, c) axles and differential components and d) brakes and wheels.

In all cases, there is a significant amount of hole-making operations such as drilling, tapping, reaming and boring. There is also an extensive use of milling operations, such as face milling and pocket milling, and various turning opera-

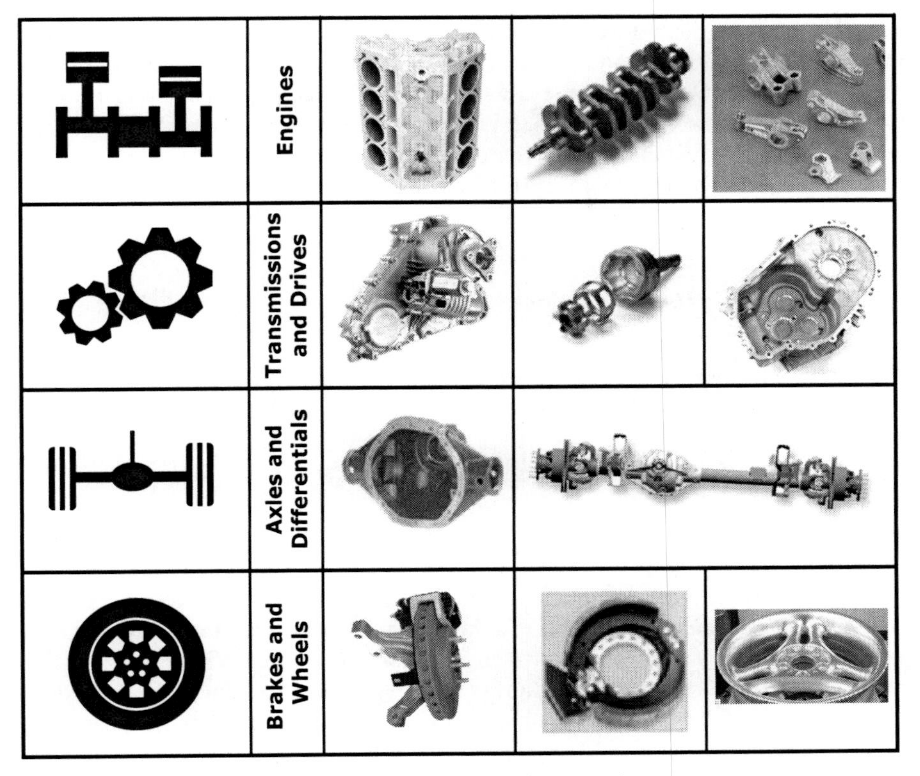

Fig. 13.2 Automotive components that require machining operations

tions. In some components such as crankshafts and camshafts, grinding operations are needed. Special metal cutting operations such as turn broaching and broaching are also applied for some components. The typical workpiece materials are forged steels, cast irons and cast aluminium silicon alloys.

Figure 13.2 presents typical examples of automotive components produced by first-tier suppliers such as Nemak® [8], Magna Powertrain®, Sisamex®, Delphi® and American Axle and Manufacturing®, among others. The following sections of this chapter concentrate on machine tools for milling and hole-making operations. In separate chapters of this book, technology trends associated with turning centres and grinding machines are discussed in depth.

13.2 Manufacturing System Architecture: High Volume Production Versus Flexibility

In the automotive industry, there is a constant push to maintain the low unit cost associated with high volume production, but at the same time to maintain flexibility. This need for flexibility comes from both a) the market demand shifts typical of the automotive sector and b) manufacturing operation changes associated with new product models. Clearly, the architecture of the machining system is designed to suit the production demands and ranges from transfer lines to individual machines in job shop arrangements. The choice of architecture in turn determines the structure of the machines that constitute the system.

13.2.1 Dedicated Machines

Dedicated machines are divided into linear transfer machines and rotary transfer machines. The following sections explain and illustrate both types of dedicated machines.

13.2.1.1 Linear Transfer Machines

The linear transfer machine or transfer line is an automated and interlocked group of stations that includes machining, deburring, cleaning and inspection operations. The whole transfer line must be designed and built as a unit. The various stations in the transfer machine perform manufacturing operations on the product simultaneously. By interlocking the machine stations and the material handing system, it is possible to

Operations Order In Foundry		Operations Order in Engine Plant	
OP 10C	Pre-machine oil pan face and crakcase features, finish-milling manufacturing lugs	OP 10	Mill bank faces, mill oil pan face and crankcase features, drill manufacturing holes
OP 20C	Pre-machine bank & end faces, core drill cup plug holes, drill oil gallery holes, de-flash cylinder bores	OP 20	Rough and finish-bore parent metal cylinder bores
OP 30C	Washer	OP 30	Preliminary washer
OP 40C	Leak test	OP 40	Cylinder block heater
		OP 50	Cylinder liner assembly
		OP 60	Cooling & storage system
		OP 70	Finish-mill, drill and tap bearing cap seats
		OP 80	Intermediate washer
		OP 90	Bearing cap assembly
		OP 100	Semi-finish mill, drill and tap oil pan face, finish bore manufacturing holes
		OP 110	Semi-finish mill, drill and tap end faces
		OP 120	Drill and tap deck faces, drill and ream oil feed and dip stick holes
		OP 130	Finish-bore cup plug holes, mill, drill & tap mounting
		OP 140	Rough, semi-finish and finish-bore crank bore, finish-bore water pump and dowel holes
		OP 150	Rough, semi-finish and finish-bore cylinder bores and finish-mill bank face
		OP 160	Finish-mill front, rear & pan faces. Finish-bore bank face dowels.
		OP 170	Hone
		OP 180	Final washer
		OP 190	Cup plug assembly and leak test

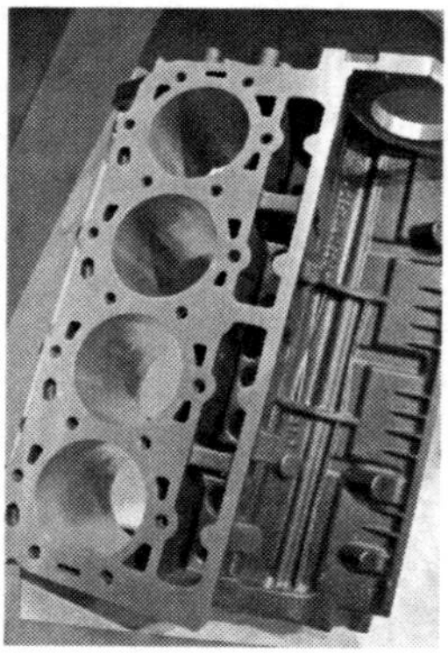

Fig. 13.3 Operations sequence for engine block machining in transfer line [1]

move a set of in-process products from station to station. As each product reaches the end of the transfer line, all the manufacturing operations have been performed [11].

The first transfer machines used in the automotive industry date back to the 1920s, both in the US and the United Kingdom. However, it wasn't until the 1960s that automobile manufactures started to develop some degree of flexibility through modularity in the transfer machines [11]. Through the years, transfer machine developers have increased the modularity of these systems to accommodate the shortening of product life cycle and market demand changes.

An example is presented in Fig. 13.3, which shows the typical operation sequence for engine block machining in a dedicated transfer line (the engine block picture is for illustration purposes only). In this case, the transfer line is designed to produce 650,000 engine blocks per year with a cycle time of 27 s. In general, for the production of large automotive components such as engine blocks and engine cylinder heads, a dedicated transfer line is recommended for volumes beyond 250,000 parts/year.

13.2.1.2 Rotary Transfer Machines

The rotary transfer machine, also known as the dial or carousel transfer machine, is better suited for small automotive parts. These machines typically handle working volume envelopes up to 0.027 m^3 *(300×300×300 mm)*. Rotary transfer machines can produce a large number of small automotive components, ranging from 100 to 550 parts/h. Considering a two-shift operation with a 75% uptime, these machines can produce between 300,000 and 1,650,000 parts/year (according to data from Kingsbury and [19]).

A typical design for a rotary transfer machine is presented by Gnutti®, which produces a rotary transfer machine with a structure that allows 25 stations with up to 33 high-speed spindles (see Fig. 13.4). This particular machine features 15 m/s^2 of axis acceleration.

An alternative design for rotary transfer machine is shown in Fig. 13.5. This machine concept allows for 3 to 6 stations, with various configurations for each station, such as horizontal, vertical, single spindle, double spindle and multi-head

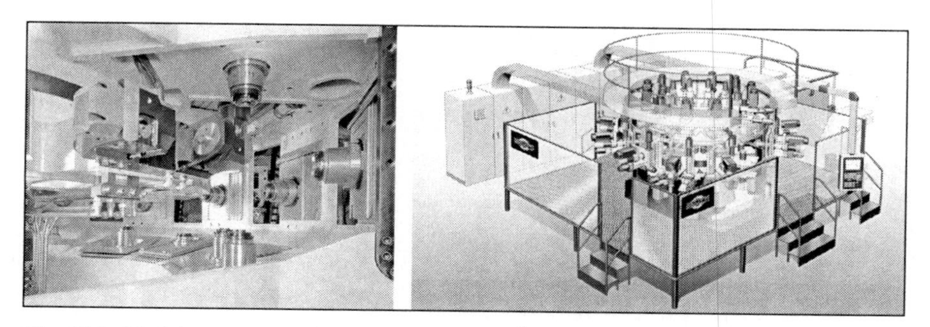

Fig. 13.4 Modular rotary transfer machine, by Gnutti®

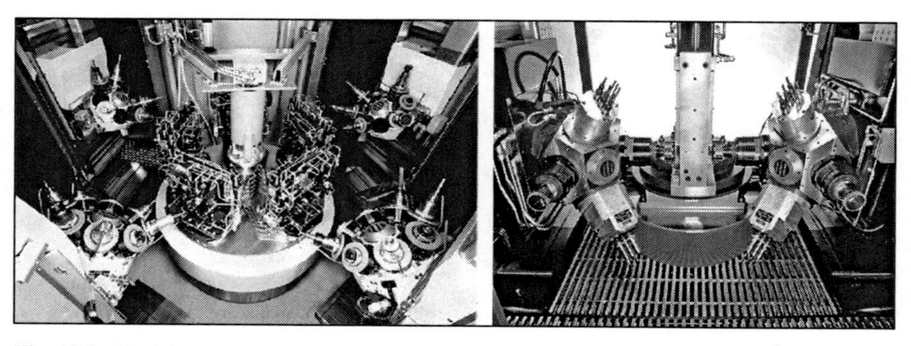

Fig. 13.5 Modular rotary transfer machines with a multi-head turret, by Kaufmann®

turret, which are shown in Fig. 13.6. In this type of machine, the use of multi-head turrets allows multiple operations in a single station. In addition, a single head position in the turret can be custom-built for the simultaneous drilling of multiple holes [16]. Alternatively, a single station can be fitted with an automatic tool changer, as shown in Fig. 13.7.

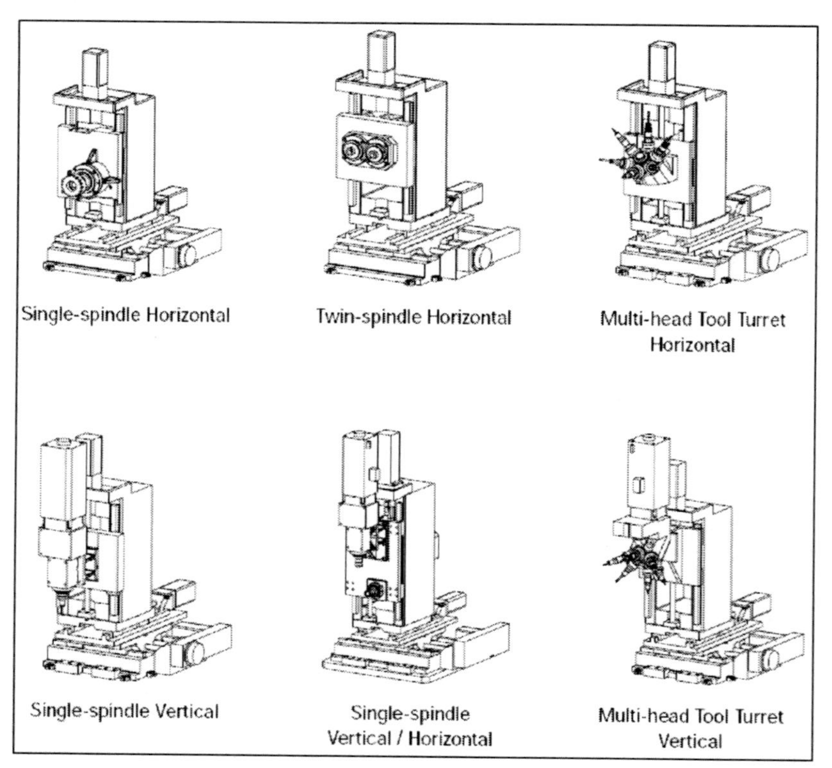

Fig. 13.6 Station configuration for modular rotary transfer machine, by Automation Tooling Systems®

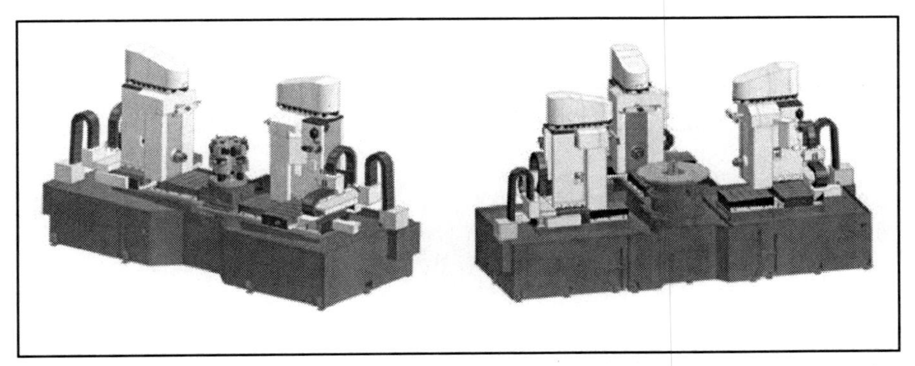

Fig. 13.7 Modular rotary transfer machine with automatic tool changer per station, by Ultra Tech Machinery®

13.2.2 Flexible Cells

For flexible cells, the most common type of machine configuration is a 4-axis machining centre with a horizontal spindle and pallet changer (a typical structure for this type of machine is shown in Fig. 13.8). Due to the need of high productivity,

Fig. 13.8 Structure of a 4-axis horizontal machining centre with an interchangeable pallet, by DMG® [3]

Table 13.1 Typical specifications of 4-axis high performance machining centres with horizontal spindle and pallet changer (spindle size CAT 50 o HSK 100) [10]

Specifications	Makino A100E	Heller MCH250	Okuma MA-500HB
X/Y/Z travel	1,700/1,350/1,400 mm	800/800/800 mm	700/900/780
Feed rate (program/rapid)	50 m/min 50 m/min	60 m/min 60 m/min	60 m/min 60 m/min
Acceleration	4 m/s^2	7 m/s^2	7 m/s^2
Spindle	12 and 18 k min^{-1} (HSK 100)	6 and 12 k min^{-1} (HSK 100)	6 and 12 k min^{-1} (Taper No. 50)
Chip to chip	5.5 s	4.1 s	4.4 s
ATC capacity	40/90/132/188/244/302	50/100/106/234/405	40/60/100/200/400

some of the key specifications for these machining centres are the rapid feed rate, the maximum programmable feed rate, the axis acceleration, and chip-to-chip time at the tool changer.

Representative specifications of high performance machining centres for automotive applications are shown in Table 13.1. An example of this type of machine is shown in Fig. 13.9. Due to the high frequency of spindle travel in air, especially

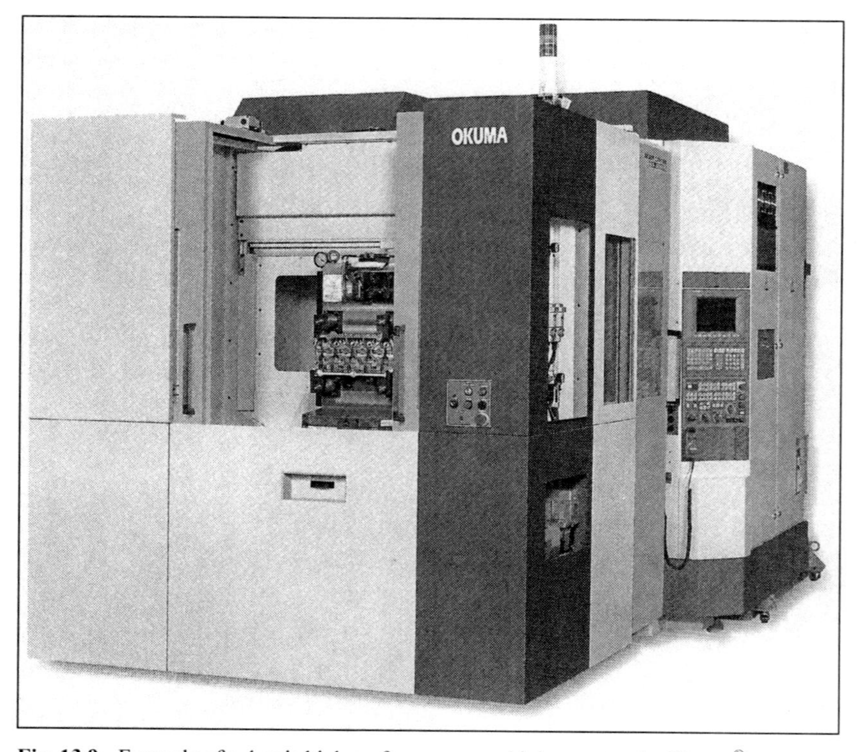

Fig. 13.9 Example of a 4-axis high performance machining centre, by Okuma®

during hole-making operations, the axis acceleration is a critical specification with a significant impact on cycle time [7]. As expected, larger machines have a lower acceleration capability. All machine models provide a basic spindle and a high-speed spindle, more appropriate for machining cast aluminium parts. The automatic tool changer (ATC) options in these machines can handle a large number of tools, with an extremely fast tool change time. The chip-to-chip time is measured as the total time needed to a) retract the spindle to the tool change position, b) switch tools and c) return the spindle to the workpiece.

The flexible cells are usually integrated through the use of pallet handling systems. Due to the generic nature of the machining centres used in flexible cells, this approach provides an excellent capability to incorporate product changes, within a given product family. The effective application of flexible cells, in machining operations of large automotive components, is recommended for production volumes up to 100,000 parts/year [17]. When selecting this option, the user must take into account the need for special work holding fixtures, which in principle should also be flexible (this facilitates product turnover when compared to dedicated fixtures).

13.2.3 Hybrid Systems

For large automotive components, when production volume falls between 200,000 and 400,000 parts/year, neither flexible cells nor dedicated transfer lines are the optimum choice. In this case, the most effective approach might be a hybrid system than integrates flexible cells into a dedicated transfer line. Hybrid machining systems are also sometimes referred to as *reconfigurable machining systems* [17]. Figure 13.10 illustrates the transition between the application of flexible cells and dedicated systems, in terms of production volume, based on case studies from the automotive industry (mainly engine blocks, cylinder heads and transmission cases).

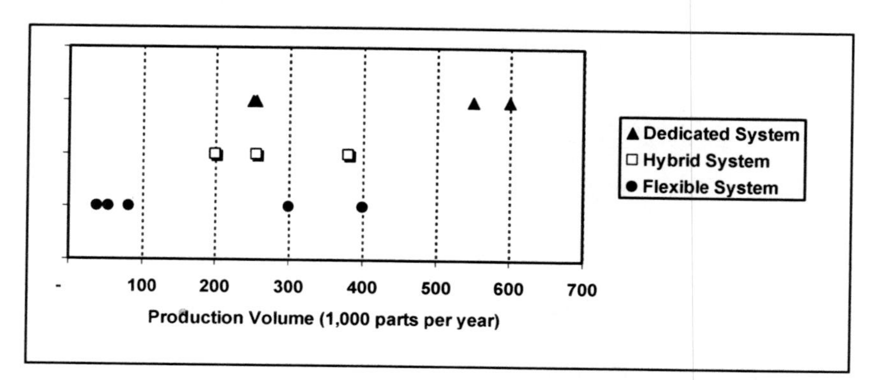

Fig. 13.10 Production volumes for flexible cells, hybrid systems and dedicated systems (adapted from case studies in [4, 20–23, 27], assuming 70% machine uptime)

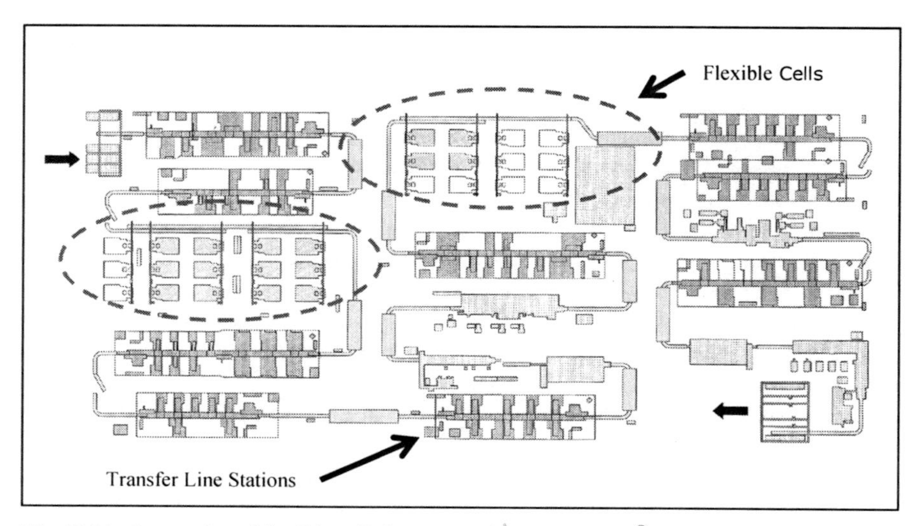

Fig. 13.11 Integration of flexible cells into a transfer line, by Grob® [9]

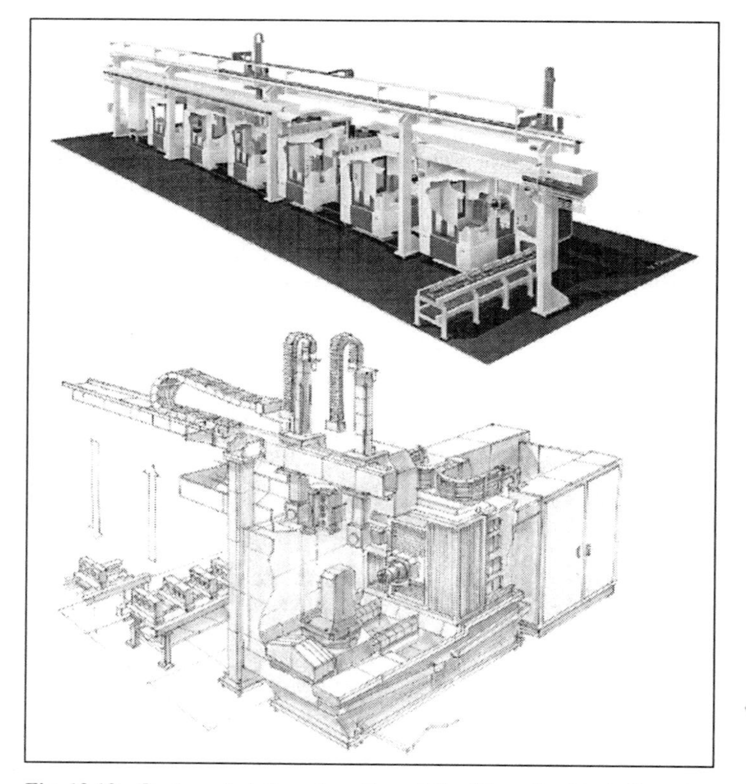

Fig. 13.12 Gantry robot for integration of flexible cell with dedicated transfer line, by Cross Hüller® [2]

Figure 13.11 presents a case in which some operations in the transfer line are replaced by flexible cells. Similar operations are simultaneously performed by several machines in parallel processing. Parts are handled by a Gantry robot placed between the machines in the flexible cells and the transfer line (see Fig. 13.12).

If one machine shuts down, due to malfunction or product changes, other machines in the cell continue the operation. In contrast, when a single station in a dedicated transfer machine shuts down, the entire production stops. Another advantage of the hybrid transfer line is the scalability, as cells can easily be expanded with additional machines. This allows the introduction of new machining operations that might be demanded by modifications to the product design. Again, to maintain flexibility, special fixturing systems might be required.

Each machine in the flexible cell is sometimes referred to as an *agile module*. The typical structure of the agile modules is shown in Fig. 13.13. These machines have a slender design that allows them to be stacked together as part of a flexible cell combined with a dedicated transfer line. Electrical cabinets and servo motors are all located behind the machine.

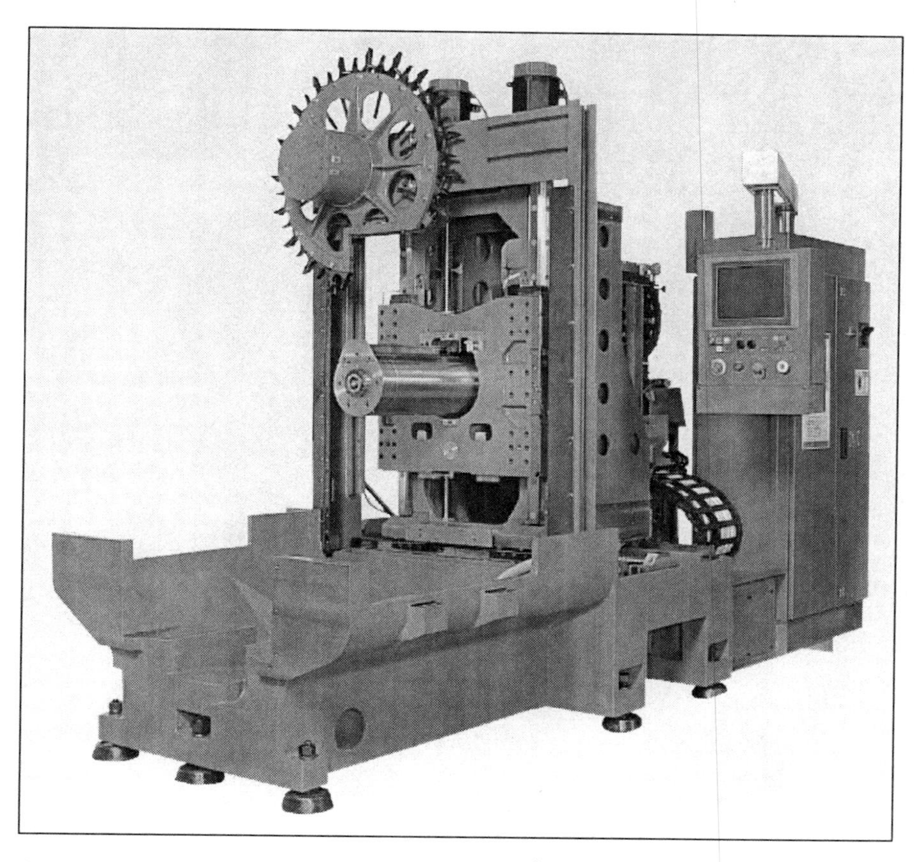

Fig. 13.13 Structure of 4-axis agile module, by Toyoda® [25]

Table 13.2 Typical specifications of 4-axis agile modules with horizontal spindle (spindle size CAT 40 o HSK 63)

Specifications	Makino J4M	Okuma MILLAC-44H
X/Y/Z travel	500/560/500 mm	410/460/470 mm
Feed rate (programmable/rapid)	50 m/min 60 m/min	20 m/min 50 m/min
Acceleration	12 m/s^2	10 m/s^2
Spindle	16 k min^{-1} (HSK 63)	12 and 20 k min^{-1} (HSK 63)
Chip to chip	3.0 s	3.0 s
ATC capacity	15/30	10/24
Dimensions (width × length)	1,500 × 4,075 mm	1,350 × 3,450 mm

Representative specifications of agile modules for automotive applications are shown in Table 13.2. As in the case of the high performance machining centres, the key requirements for these agile modules are the rapid feed rate, the maximum programmable feed rate, the axis acceleration, and chip-to-chip time at the tool changer. Additionally, the floor space is an important factor in the structure of agile modules. Because of their reduced flexibility, agile modules use tool chang-

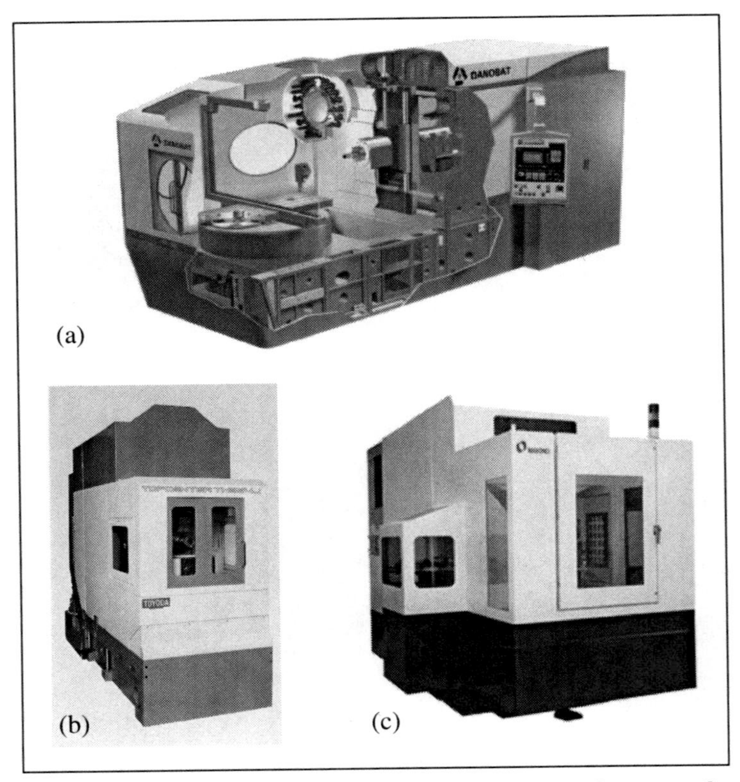

Fig. 13.14 Examples of agile modules. **a** Danobat®. **b** Toyoda®. **c** Makino®

ers with much less capacity than those offered in stand alone machining centres used in flexible cells. Examples of agile modules are shown in Fig. 13.14.

13.3 Technology Trends

The need for weight reduction, associated with fuel efficiency in automobiles, leads to the ever-increasing use of aluminium alloys in engines [24]. This trend in turn requires the use of machine components and cutting tools with a higher performance [18].

In terms of machine tool components, the need for a higher cutting speed for machining aluminium alloys requires high-speed spindles and more extensive use of linear drive motors. For example, the agile modules such as the Okuma® MILLAC-44H, the Makino® J55-ISO, the Toyoda® TOP J Transfer Line, and the Mazak® Ultra Narrow UN-600H include optional spindles with 20,000 rpm (in CAT 40/HSK 63 size).

Additionally, there is a continuous development of new and more effective materials and tool handing systems in machine tools [6]. The most common types of tool storage are drum and chain-based, with capacities up to 60 tool locations. For additional capacity, the tower and matrix systems are used, allowing up to 400 tool locations. Colombo Filippetti Torino®, Miksch® and Bertsche Engineering® are among the suppliers of tool changing systems.

The need to maintain high productivity has lead to the development of dual spindle machines, as shown in Fig. 13.15 (MAG® Powertrain SPECHT 550 DUO). This kind of machine has chip-to-chip times below 3 s. The dual spindle machine concept by Mikron® (Multistep XT-200, in Chap. 1, Fig. 1.15) can reach 1 s of chip-to-chip time.

Fig. 13.15 High performance machining centre with double spindle and 4-axis, by MAG® Powertrain

Fig. 13.16 Hole-making operations in cast aluminium, by Heller®

In automotive component machining, coolants represent up to 15% of the total operation cost [23]. Together with cost reduction objectives, the use of coolants represents an environmental and operator safety concern. Therefore, dry and minimum quantity lubrication machining are active research topics. Minimum quantity lubrication (MQL) is a near-dry machining method that may be applied to reduce coolant usage in operations such as drilling, which cannot be performed completely dry [5], as shown in Fig. 13.16. MQL has been incorporated into some commercial machine tool solutions for the automotive industry.

Finally, the use of polycrystalline diamond (PCD) cutting tools and various coatings for tungsten carbide tools has received significant attention. Some of the coatings under research for machining cast aluminium silicon alloys (A319, A356, A390) include carbon-based coatings or diamond like carbon (DLC) (commercially known as Dymon-iC™ and Graphit-iC™), MoS_2-based coatings (MoSTi), TiB_2, TiAlN/VN, TiAlCrYN, and nano-crystalline diamond [12, 13, 15, 26]. Under specific conditions, some of these coatings exhibit tool life performance comparable to that of the PCD tools.

Acknowledgements The authors acknowledge the support by all the companies cited in this chapter as well as the Association for Manufacturing Technology. Their valuable information and suggestions are greatly appreciated.

References

[1] Cross Hüller (2008). Global Transfer Lines for Cylinder Blocks. Brochure No. 6101e 2000 0704
[2] Cross Hüller (2008). Agile Cylinder Head Manufacturing System. Brochure No. 5101e 500 0903
[3] DMG (2002), Deckel Maho Gildemeister Bulletin 1
[4] Elkins DA, Huang NJ, Alden JM (2004) Agile manufacturing systems in the automotive industry. Int J Prod Econ, 91:201–214
[5] Filipovic A, Stephenson D (2006) Minimum Quantity Lubrication (MQL) applications in automotive power-train machining. Mach Sci Technol, 10:3–22
[6] Fleischer J, Denkena B, Winfough B, Mori M (2006) Workpiece and tool handling in metal cutting machines. Cirp Ann-Manuf Techn, 55:817–839
[7] Flores V, Ortega C, Alberti M, Rodriguez CA, de Ciurana J, Elias A (2007) Evaluation and modeling of productivity and dynamic capability in high-speed machining centers. Int J Adv Manuf Tech, 33:403–411
[8] Gibbs S (2007) Nemak Accelerates on the Global Track. Modern Casting, 33–37
[9] Grob Report 2002/03, Grobe-Werke
[10] Heller (2007), Brochure MCH-GB-10/03
[11] Hounshell DA (2000) Automation, transfer machinery, and mass production in the US automobile industry in the post-World War II era. Enterprise & Society, 1:100–138
[12] Hovsepian PE, Luo Q, Robinson G, Pittman M, Howarth M, Doerwald D, Tietema R, Sim WM, Deeming A, Zeus T (2006) TiAlN/VN superlattice structured PVD coatings: a new alternative in machining of aluminium alloys for aerospace and automotive components. Surf Coat Tech, 201:265–272
[13] Hu J, Chou YK, Thompson RG, Burgess J, Street S (2007) Characterizations of nanocrystalline diamond coating cutting tools. Surf Coat Tech, 202:1113–1117
[14] International Organization of Motor Vehicle Manufacturers, www.oica.net (accessed 2007)
[15] Kishawy HA, Dumitrescu M, Ng EG, Elbestawi MA (2005) Effect of coolant strategy on tool performance, chip morphology and surface quality during high-speed machining of A356 aluminum alloy. Int J Mach Tool Manufact, 45:219–227
[16] Lang JJ, Farkas EK (2002) Transfer machining with a twist. Manuf Eng, 128:74–87
[17] Mehrabi MG, Ulsoy AG, Koren Y, Heytler P (2002) Trends and perspectives in flexible and reconfigurable manufacturing systems. J Intell Manuf, 13:135–146
[18] Ng EG, Szablewski D, Dumitrescu M, Elbestawi MA, Sokolowski JH (2004) High speed face milling of a aluminium silicon alloy casting. Cirp Ann-Manuf Techn, 53:69–72
[19] Olexa R (2001) Where is the transfer line? Manuf Eng, 126:48–60
[20] Olexa R (2002) When cells make sense. Manuf Eng, 128:45–51
[21] Omar M, Hussain K, Wright S (1999) Simulation study of an agile high-speed machining system for automotive cylinder heads. P I Mech Eng B-J Eng, 213:491–499
[22] Owen JV (1999) Transfer lines get flexible. Manuf Eng, 122:42–50
[23] Quaile R (2007) What does automation cost? Manuf Eng, 138:175–181
[24] Shen CH (1996) The importance of diamond coated tools for agile manufacturing and dry machining. Surf Coat Tech, 86–7:672–677
[25] Toyoda (2007) Machine Tools and Mechatronics Line Up. Brochure Cat. No. M2001E
[26] Wain N, Thomas NR, Hickman S, Wallbank J, Teer DG (2005) Performance of low-friction coatings in the dry drilling of automotive Al-Si alloys. Surf Coat Tech, 200:1885–1892
[27] Waurzyniak P (2002) Building better diesels. Manuf Eng, 129:87–98

Index

N

O

P

R

S

Breinigsville, PA USA
20 December 2010

251735BV00005B/16/P

9 781848 003798